Spon's External Works and Landscape Price Book

Edited by
Davis Langdon, An AECOM Company

in association with
LandPro Ltd
Landscape Surveyors

2014

Thirty-third edition

CRC Press
Taylor & Francis Group

Free Updates

with three easy steps…

1. Register today on www.pricebooks.co.uk/updates

2. We'll alert you by email when new updates are posted on our website

3. Then go to www.pricebooks.co.uk/updates
    and download the update.

All four Spon Price Books – *Architects' and Builders'*, *Civil Engineering and Highway Works*, *External Works and Landscape* and *Mechanical and Electrical Services* – are supported by an updating service. Two or three updates are loaded on our website during the year, typically in November, February and May. Each gives details of changes in prices of materials, wage rates and other significant items, with regional price level adjustments for Northern Ireland, Scotland and Wales and regions of England. The updates terminate with the publication of the next annual edition.

As a purchaser of a Spon Price Book you are entitled to this updating service for this 2014 edition – free of charge. Simply register via the website www.pricebooks.co.uk/updates and we will send you an email when each update becomes available.

If you haven't got internet access or if you've some questions about the updates please write to us at Spon Price Book Updates, Spon Press Marketing Department, 3 Park Square, Milton Park, Abingdon, Oxfordshire, OX14 4RN.

Find out more about Spon books
Visit www.pricebooks.co.uk for more details.

Spon's External Works and Landscape Price Book 2014

www.pricebooks.co.uk/downloads

Scratch off access code:

YDMJ4APHMWEKWGQDGD23

DS010382

CORNWALL COLLEGE

Spon's
External Works
and Landscape
Price Book

2014

First edition 1978
Thirty-Second edition published 2014
by CRC Press
2 Park Square, Milton Park, Abingdon, Oxon, OX14 4RN

and by CRC Press
Taylor & Francis, Broken Sound Parkway, NW, Suite 300, Boca Raton, FL 33487

CRC Press is an imprint of the Taylor & Francis Group, an informa business

The publisher makes no representation, express or implied, with regard to the accuracy of the information contained in this book and cannot accept legal responsibility or liability for any errors or omissions that may be made.

British Library Cataloguing in Publication Data
A catalogue record for this book is available from the British Library

ISBN13: 978-1-4822-0412-4
Ebook: 978-1-4822-0413-1

712.3 LAN

ISSN: 0267-4181

Typeset in Arial by Taylor & Francis Books

Printed and bound in Great Britain by
TJ International Ltd, Padstow, Cornwall

Contents

Preface to the Thirty-Third Edition

Market conditions

'Flat' is probably the best term to describe the current market. Whilst some of the larger firms are enjoying good order books, there is generally strong competition for tendered work. Contractors have reduced margins in a rising labour cost market to secure work and protect their labour force. The landscape maintenance market is even more competitive at present with rumours of major contractors submitting term tenders at zero % to maintain their workforce.

Key influences

- Manufacturers and suppliers have advised us to keep our database prices for their products the same in many cases for this publication.
- The price of steel has fallen on last year's figures.
- The fuel price is lower than last year's price.
- The labour market has however risen slightly as landscape labour would otherwise migrate to other works areas. Employers have been forced to increase wages in line with indices in order to maintain their trained and valuable labour force.

Adjustment formulae for construction contracts

Price Adjustment Formulae indices, compiled by the Building Cost Information Service (previously by the Department for Business Innovation & Skills), are designed for the calculation of increased costs on fluctuating or Variation of Price contracts. They provide useful guidance on cost changes in various trades and industry sectors and on the differential movement of work sections in Spon's Price Books. Over the last 12 months between April 2012 and April 2013, the 60 Building Work categories recorded an average rise of just 1.4%. Those works categories relating to this publication showing the highest increases over the last 12 months have been:

- Excavation and Disposal 3.2%

Those categories showing the smallest increases or in some cases decreases are:

- Concrete: Reinforcement −3.8%
- Filling: Imported, Hardcore and Granular 0.8%
- Concrete: In Situ 1.8%
- Concrete: Formwork 2.0%

Changes to this edition

This year's publication omits the Minor works sections. Readers will find some of these rates incorporated to the previously named Major works section.

- Measurement Changes – New Rules of Measurement (NRM)
- The NRM is a suite of documents to provide a standard set of measurement rules that are understandable by all
- Those involved in a construction project.
- This year's edition has been re-formatted to move towards complying with NRM1 and NRM2. There will be further refinement over the next two or three editions as NRM becomes more embedded in general practice and workflows. There are three parts to the NRM suite. We have used towards implementing NRM 1 in this year's publication. Further information can be found on the RICS website

Supplier prices

Similarly to last year, there have been many suppliers who have only reflected minimal increases or even no increase in their prices especially for material supplies. The larger suppliers have increased their list costs by standard 5%–7.5% but we perceive that there is room to negotiate on material cost rates.

Labour rates

This year we have increased the price of the base cost of labour by £0.50p (Approximately £900.00 per annum).

This is due to our perception of a rising labour market during the coming 12 months and to ensure validity of the rates in this book. Readers may make adjustments to these rates in accordance with the instructions and the indices tables published in the general section following this preface.

The standard rate for this year's calculation's for Major works is £19.75 per hour an increase of 2.6 %.

Plant rates

The hired in heavy plant suppliers referenced in this publication have advised us during our update to keep their prices the same. The price of red diesel has decreased slightly and this would have some effect on the hourly rates of the plant used. Our supplier of prices to this publication comments that the large national cross-hiring companies are driving hire prices down.

Profit and overhead

Spon's External Works and Landscape prices do not allow for profit or site overhead. Readers should evaluate the market conditions in the sector or environment in which they operate and apply a percentage to these rates. Company overhead is allowed for within the labour rates used. Please refer to 'Part 1 General' for an explanation of the build-up of this year's labour rates.

Prices for suppliers and services

We acknowledge the support afforded to us by the suppliers of products and services who issue us with the base information used. Their contact details are published in the directory section at the front of this book. We advise that readers wishing to evaluate the cost of a product or service should approach these suppliers directly should they wish to confirm prices prior to submission of tenders or quotations.

Whilst every effort is made to ensure the accuracy of the information given in this publication, neither the editors nor the publishers in any way accept liability for loss of any kind resulting from the use made by any person of such information.

We remind readers that we would be grateful to hear from them during the course of the year with suggestions or comments on any aspect of the contents of this year's book and suggestions for improvements. Our contact details are shown below.

DAVIS LANGDON, An AECOM COMPANY
MidCity Place
71 High Holborn
London
WC1V 6QS

Tel: 02070 617000
e-mail: spons@davislangdon.com

SAM HASSALL
LANDPRO LTD
Landscape Surveyors
14 Upper Bourne Lane
Farnham
Surrey
GU10 4RQ

Tel: 01252 795030
e-mail: info@landpro.co.uk

Acknowledgements

This list has been compiled from the latest information available but as firms frequently change their names, addresses and telephone numbers due to reorganization, users are advised to check this information before placing orders.

ACO Technologies Plc
ACO Business Park
Hitchin Road
Shefford
Bedfordshire SG17 5TE
Tel: 01462 816666
Website: www.aco.co.uk
E-mail: technologies@aco.co.uk
Linear drainage

Addagrip Surface Treatments UK Ltd
Addagrip House
Bell Lane Industrial Estate
Uckfield
East Sussex TN22 1QL
Tel: 01825 761333
Website: www.addagrip.co.uk
E-mail: sales@addagrip.co.uk
Resin bound & bonded surfacing

Agripower Ltd
Broomfield Farm
Rignall Road
Great Missenden
Bucks HP16 9PE
Tel: 01494 866776
Website: www.agripower.co.uk
E-mail: info@agripower.co.uk
Land drainage and sports grounds

Amenity Land Solutions Ltd (ALS)
Allscott Park
Allscott
Telford
Shropshire TF6 5DY
Tel: 01952 641949
Website: www.amenity.co.uk
E-mail: sales@amenity.co.uk
Horticultural products supplier

Architectural Heritage Ltd
Taddington Manor
Cutsdean
Cheltenham
Gloucestershire GL54 5RY
Tel: 01386 584414
Website: www.architectural-heritage.co.uk
E-mail: puddy@architectural-heritage.co.uk
Ornamental stone buildings and water features

AVS Fencing Supplies Ltd
Unit 1 AVS Trading Park
Chapel Lane
Milford
Surrey GU8 5HU
Tel: 01483 410960
Website: www.avsfencing.co.uk
E-mail: sales@avsfencing.co.uk
Fencing

Bauder Ltd
70 Landseer Road
Ipswich
Suffolk IP3 0DH
Tel: 01473 257671
Website: www.bauder.co.uk
E-mail: info@bauder.co.uk
Green roof systems

BBS Green Roofing Ltd
Vitis House
50 Dickens Street
Battersea
London SW8 3EQ
Tel: 0207 622 6225
Website: http: //www.green-roofing.co.uk/
Email: bbs@green-roofing.co.uk
Specialist Green Roof Consultants/Installers

Bituchem Group
Forest Vale Industrial Estate
Cinderford
Gloucestershire GL14 2YH
Tel: 01594 826768
Website: www.bituchem.com
E-mail: info@bituchem.com
Coloured hard landscaping materials

Boddingtons Ltd
Fiberweb Geosynthetics Ltd
Blackwater Trading Estate
The Causeway
Maldon
Essex CM9 4GG
Tel: 01621 874200
Website: www.boddintons-ltd.com
E-mail: sales@boddingtons-ltd.com
Erosion control, soil stabilization, plant protection

Breedon Special Aggregates
(A division of Breedon Aggregates)
Breedon Quarry
Breedon-on-the-Hill
Derby
Derbyshire DE73 8 AP
Tel: 01332 694001
Website: www.breedon-special-aggregates.co.uk
E-mail: specialaggs@breedonaggregates.com
Specialist gravels

British Seed Houses Ltd
Camp Road
Witham St Hughs
Lincoln LN6 9QJ
Tel: 01522 868714
Website: www.bshamenity.com
E-mail: seeds@bshlincoln.co.uk
Grass and wildflower seed

British Sugar Topsoil
Sugar Way
Peterborough PE2 9 AY
Tel: 0870 240 2314
Website: www.bstopsoil.co.uk
E-mail: topsoil@britishsugar.com
Topsoil

Broxap Street Furniture
Rowhurst Industrial Estate
Chesterton
Newcastle-under-Lyme
Staffordshire ST5 6BD
Tel: 0844 800 4085
Website: www.broxap.com
E-mail: sales@broxap.com
Street furniture

Capital Garden Products Ltd
Gibbs Reed Barn
Pashley Road
Ticehurst
East Sussex TN5 7HE
Tel: 01580 201092
Website: www.capital-garden.com
E-mail: sales@capital-garden.com
Plant containers

CED Ltd
728 London Road
West Thurrock
Grays
Essex RM20 3LU
Tel: 01708 867237
Website: www.ced.ltd.uk
E-mail: sales@ced.ltd.uk
Natural stone

Cedar Nursery
Horsley Road
Cobham
Surrey KT11 3JX
Tel: 01932 862473
Website: www.landscaping.co.uk
E-mail: sales@landscaping.co.uk
Aggregate containment grids

Charcon Hard Landscaping
Hulland Ward
Ashbourne
Derbyshire DE6 3ET
Tel: 01335 372222
Website: www.aggregate.com
E-mail: ukenquiries@aggregate.com
Paving and street furniture

Cleartrack (EvL) Ltd
Ploughmans Barn
Station Farm
Denton Road
Horton
Northants NN7 2BG
Tel: 01604 871360
Website: www.cleartrack.co.uk
E-mail: info@cleartrack.co.uk
Ground clearance

Combined Harvesters Ltd
Britannia House
Dock Road
Birkenhead
Wirral CH41 1DF
Tel: 0151 639 0880
Website: www.combinedharvesters.co.uk
E-mail: info@combinedharvesters.co.uk
Rainwater harvesting

Cooper Clarke Civils and Lintels
Bloomfield Road
Farnworth
Bolton BL4 9LP
Tel: 01204 862222
Website: www.civilsandlintels.co.uk
E-mail: farnworthspecials@civilsandlintels.co.uk
Wall drainage and erosion control

Craft Pegg
34 Riverside Building
Trinity Buoy Wharf
London E14 0JY
Tel: 0207 538 9010
Website: www.craftpegg.com
E-mail: mail@craftpegg.com
Landscape architects

CU Phosco Ltd
Charles House
Lower Road
Great Amwell
Ware
Hertfordshire SG12 9TA
Tel: 01920 860600
Website: www.cuphosco.co.uk
E-mail: sales@cuphosco.co.uk
Lighting

Deepdale Trees Ltd
Tithe Farm
Hatley Road
Sandy
Bedfordshire SG19 2DX
Tel: 01767 262636
Website: www.deepdale-trees.co.uk
E-mail: mail@deepdale-trees.co.uk
Trees, multi-stem and hedging

DLF Trifolium Ltd
Thorn Farm
Evesham Road
Inkberrow
Worcestershire WR7 4LJ
Tel: 01386 791102
Website: www.dlf.co.uk
E-mail: amenity@dlf.co.uk
Grass and wildflower seed

Earth Anchors Ltd
15 Campbell Road
Croydon
Surrey CR0 2SQ
Tel: 020 8684 9601
Website: www.earth-anchors.com
E-mail: sales@earth-anchors.com
Anchors for site furniture

Elliott
Manor Drive
Peterborough PE4 7 AP
Tel: 01733 298700
Website: www.elliottuk.com
E-mail: hirediv@elliottuk.com
Site offices

English Woodlands
Burrow Nursery
Cross in Hand
Heathfield
East Sussex TN21 0UG
Tel: 01435 862992
Website: www.ewburrownursery.co.uk
E-mail: sales@ewburrownursery.co.uk
Plant protection

Everris Limited
Epsilon House
West Road
Ipswich
Suffolk IP3 9FJ
Tel: 01473 237100
Website: www.everris.com
E-mail: prof.sales@everris.com
Landscape chemicals

Eve Trakway
Bramley Vale
Chesterfield
Derbyshire S44 5GA
Tel: 08700 767676
Website: www.evetrakway.co.uk
E-mail: marketing@evetrakway.co.uk
Portable roads

Exterior Decking
Exterior Solutions Ltd
Unit 5b
Reed Industrial Estate
28 Plantation Road
Amersham
Buckinghamshire HP6 6HJ
Tel: 01494 722204
Website: www.exteriordecking.co.uk
E-mail: office@exteriordecking.co.uk
Decking

Fairwater Ltd
Lodge Farm
Malthouse Lane
Ashington
West Sussex RH20 3BU
Tel: 01903 892228
Website: www.fairwater.co.uk
E-mail: info@fairwater.co.uk
Water garden specialists

Farmura Environmental Products Ltd
Stone Hill
Egerton
Ashford
Kent TN27 9DU
Tel: 01233 756241
Website: www.farmura.com
E-mail: info@farmura.com
Organic fertilizer suppliers

Forticrete Ltd
Anstone Works
Kiverton Park Station
Kiverton Park
Sheffield S26 6NP
Tel: 0870 9034015
Website: www.forticrete.co.uk
E-mail: info@forticrete.com
Concrete retaining wall systems

FP McCann Ltd
Whitehill Road
Coalville
Leicester LE67 1ET
Tel: 01530 240000
Website: www.fpmccann.co.uk
E-mail: info@fpmccann.co.uk
Precast concrete pipes

Furnitubes Street Furniture
Meridian House
Royal Hill
Greenwich
London SE10 8RT
Tel: 020 8378 3200
Website: www.furnitubes.com
E-mail: spons@furnitubes.com
Street furniture

Gardenlink Ltd
27 Lesbourne Road
Reigate
Surrey
RH2 7JS
Tel: 01737 243 224
Website: www.gardenlink.co.uk
E-mail: enquiries@gardenlink.co.uk
Steel edgings

The Garden Trellis Company
Unit 1 Brunel Road
Gorse Lane Industrial Estate
Clacton-On-Sea
Essex CO15 4LU
Tel: 01255 688361
Website: www.gardentrellis.co.uk
E-mail: info@gardentrellis.co.uk
Garden joinery specialists

Geosynthetics Ltd
Fleming Road
Harrowbrook Ind Est
Hinckley
Leicestershire LE10 3DU
Tel: 01455 617139
Website: www.geosyn.co.uk
E-mail: sales@geosyn.co.uk
Cellular containment systems

Grass Concrete Ltd
Duncan House
142 Thornes Lane
Thornes
West Yorkshire WF2 7RE
Tel: 01924 379443
Website: www.grasscrete.com
E-mail: info@grasscrete.com
Grass block paving

Greenfix Soil Stabilization and Erosion Control Ltd
Allens West
Durham Lane
Eaglescliffe
Stockton on Tees TS16 ORW
Tel: 01642 888693
Website: www.greenfix.co.uk
E-mail: stockton@greenfix.co.uk
Seeded erosion control mats

Greenleaf Horticulture
Ivyhouse Industrial Estate
Haywood Way
Hastings TN35 4PL
Tel: 01424 717797
E-mail: enquiries@greenleaftrees.co.uk
Root directors

Griffin Nurseries
New Barn Farm
Rake road
Milland
Liphook
Hants GU30 7JU
Tel: 01428 741655
Website: www.griffinnurseries.co.uk
E-mail: enquiries@griffinnurseries.co.uk
Hedging

Grundon Waste Management Ltd
Goulds Grove
Ewelme
Wallingford OX10 6PJ
Tel: 0870 4438278
Website: www.grundon.com
E-mail: sales@grundon.com
Gravel

Guncast Swimming Pools Ltd
Unit L
The Old Bakery
Golden Square
Petworth
West Sussex GU28 0 AP
Tel: 01798343725
Website: www.guncast.com
E-mail: info@guncast.com
Swimming pools

Haddonstone Ltd
The Forge House
Church Lane
East Haddon
Northampton NN6 8DB
Tel: 01604 770711
Website: www.haddonstone.com
E-mail: info@haddonstone.co.uk
Landscape ornaments and architectural cast stone

Harrison External Display Systems
Borough Road
Darlington
Co Durham DL1 1SW
Tel: 01325 355433
Website: www.harrisoneds.com
E-mail: sales@harrisoneds.com
Flagpoles

Havells Sylvania Fixtures UK Ltd
Avis Way
Newhaven
East Sussex BN9 0ED
Tel: 0870 6062030
Website: www.havells-sylvania.com
E-mail: info.concord@havells-sylvania.com
Lighting

Headland Amenity Ltd
1010 Cambourne Business Park
Cambourne
Cambridgeshire CB23 6DP
Tel: 01223 597834
Website: www.headlandamenity.com
E-mail: info@headlandamenity.com
Chemicals

Heicom UK
4 Frog Lane
Tunbridge Wells
Kent TN1 1YT
Tel: 01892 522360
Website: www.treesand.co.uk
E-mail: mike.q@treesand.co.uk
Tree sand

Hepworth Wavin Plc
Hazelhead
Crow Edge
Sheffield S36 4HG
Tel: 0870 4436000
Website: www.hepworthdrainage.co.uk
E-mail: info@hepworthdrainage.co.uk
Drainage

HSS Hire
Unit 3
14 Wates Way
Mitcham
Surrey CR4 4HR
Tel: 020 8685 9500
Website: www.hss.com
E-mail: hire@hss.com
Tool and plant hire

Inturf
The Chestnuts
Wilberfoss
York YO41 5NT
Tel: 01759 321000
Website: www.inturf.com
E-mail: info@inturf.co.uk
Turf

J Toms Ltd
7 Marley Farm
Headcorn Road
Smarden
Kent TN27 8PJ
Tel: 01233 770066
Website: www.jtoms.co.uk
E-mail: jtoms@btopenworld.com
Plant protection

Jacksons Fencing
Stowting Common
Ashford
Kent TN25 6BN
Tel: 01233 750393
Website: www.jacksons-fencing.co.uk
E-mail: sales@jacksons-fencing.co.uk
Fencing

James Coles & Sons (Nurseries) Ltd
The Nurseries
Uppingham Road
Thurnby
Leicester LE7 9QB
Tel: 01162 412115
Website: www.colesnurseries.co.uk
E-mail: sales@colesnurseries.co.uk
Nurseries

Johnsons Wellfield Quarries Ltd
Crosland Hill
Huddersfield
West Yorkshire HD4 7 AB
Tel: 01484 652311
Website: www.johnsons-wellfield.co.uk
E-mail: sales@johnsons-wellfield.co.uk
Natural Yorkstone pavings

Jones of Oswestry
Whittington Road
Oswestry
Shropshire SY11 1HZ
Tel: 01691 653251
Website: www.jonesofoswestry.com
E-mail: sales@jonesofoswestry.com
Channels, gulleys, manhole covers

Kinley Systems
Haywood Way
Hastings
East Sussex TN35 4PL
Tel: 01424 201111
Website: www.kinleysystems.com
E-mail: sales@exceledge.co.uk
Aluminium edgings

Kompan Ltd
20 Denbigh Hall
Bletchley
Milton Keynes
Bucks MK3 7QT
Tel: 01908 642466
Website: www.kompan.com
E-mail: kompan.uk@kompan.com
Play equipment

Landline Ltd
1 Bluebridge Industrial Estate
Halstead
Essex CO9 2EX
Tel: 01787 476699
Website: www.landline.co.uk
E-mail: sales@landline.co.uk
Pond and lake installation

Lappset UK Ltd
Lappset House
Henson Way
Kettering
Northants NN16 8PX
Tel: 01536 412612
Website: www.lappset.com
E-mail: customerserviceuk@lappset.com
Play equipment

LDC Limited
The Charcoal House
Blacksmith Lane
Guildford
Surrey GU4 8NQ
Tel: 01483 573817
Website: www.ldclandscape.co.uk
E-mail: info@ldc.co.uk
Willow walling

Leaky Pipe Systems Ltd
Frith Farm
Dean Street
East Farleigh
Maidstone
Kent ME15 0PR
Tel: 01622 746495
Website: www.leakypipe.co.uk
E-mail: sales@leakypipe.co.uk
Irrigation systems

Lorenz von Ehren
Maldfeldstrasse 4
D−21077 Hamburg
Germany
Tel: 0049 40 761080
Website: www.lve.de
E-mail: sales@lve.de
Topiary

Maccaferri Ltd
7400 The Quorum
Oxford Business Park North
Garsington Road
Oxford OX4 2JZ
Tel: 01865 770555
Website: www.maccaferri.co.uk
E-mail: oxford@maccaferri.co.uk
Gabions

Marshalls Plc
Landscape House
Premier Way
Lowfields Business Park
Elland HX5 9HT
Tel: 01422 312000
Website: www.marshalls.co.uk
Hard landscape materials, street furniture, linear drainage

Matta Products
19 Triumph Way
Woburn Road Industrial Estate
Kempston
Bedford MK42 7QB
Tel: 01234 848484
Website: www.matta-products.com
E-mail: mattasales@phs.co.uk
Safety surfacing

Maxit Building Products Ltd
Heath Business Park
Runcorn
Cheshire WA7 4QX
Tel: 01928 565656
Website: www.maxit-uk.co.uk/2367
E-mail: sales@maxit-uk.co.uk
Expanded clay aggregate

McArthur Group Ltd
Foundry Lane
Bristol BS5 7UE
Tel: 0117 943 0500
Website: www.mcarthur-group.com
E-mail: marketing@mcarthur-group.com
Security fencing

Melcourt Industries Ltd
Boldridge Brake
Long Newnton
Tetbury
Gloucestershire GL8 8RT
Tel: 01666 502711
Website: www.melcourt.co.uk
E-mail: mail@melcourt.co.uk
Mulch and compost

Milton Precast
Cooks Lane
Milton Regis
Sittingbourne
Kent ME10 2QF
Tel: 01795 425191
Website: www.miltonprecast.com
E-mail: sales@miltonprecast.com
Soakaway rings

Mobilane UK
PO Box 449
Stoke-on-Trent
Staffordshire ST6 9AE
Tel: 07711 895261
Website: www.mobilane.co.uk
E-mail: sales@mobilane.co.uk
Green screens

Neptune Outdoor Furniture Ltd
Thompsons Lane
Marwell
Winchester
Hampshire SO21 1JH
Tel: 01962 777799
Website: www.nofl.co.uk
E-mail: info@nofl.co.uk
Street furniture

Nomix Enviro
A Division of Frontier Agriculture Ltd
The Grain Silos
Weyhill Road
Andover
Hants SP10 3NT
Tel: 01264 388050
Website: www.nomixenviro.co.uk
E-mail: nomixenviro@frontierag.co.uk
Herbicides

Norris and Gardiner Ltd
Lime Croft Road
Knaphill
Woking
Surrey GU21 2TH
Tel: 01483 289111
Website: www.norrisandgardiner.co.uk
E-mail: rich@norg.co.uk
Grounds maintenance

Oakover Nurseries Ltd
Maidstone Road
Hothfield
Ashford
Kent TN26 1AR
Tel: 01233 713016
Website: www.oakovernurseries.co.uk
E-mail: enquiries@oakovernurseries.co.uk
Native plant nurseries

Orchard Street Furniture Ltd
Whistler House
51 The Green North
Warborough
Oxfordshire OX10 7DW
Tel: 01491 642123
Website: www.orchardstreet.co.uk
E-mail: sales@orchardstreet.co.uk
Street furniture

PBA Solutions
Bryn
Rake Road
Liss
Hants GU33 7HB
Tel: 01730 893460
Website: www.pba-solutions.com
E-mail: info@pba-solutions.com
Japanese Knotweed consultants

PC Landscapes Ltd
Abbott House
Hale Road
Farnham
Surrey GU9 9QH
Tel: 01252 891150
Website: www.pclandscapes.co.uk
E-mail: info@pclandscapes.co.uk
Landscaping services

Phi Group
Harcourt House
13 Royal Crescent
Cheltenham
Gloucestershire GL50 3DA
Tel: 01242 707600
Website: www.phigroup.co.uk
E-mail: southern@phigroup.co.uk
Retaining wall specialists

Plastech Southern
Home Park
Lyon Way
Frimley
Surrey GU16 7ER
Tel: 01276 24765
Website: www.plastechtitan.com
E-mail: sales@plastechsouthern.co.uk
Ducting

Platipus Anchors Ltd
Kingsfield Business Centre
Philanthropic Road
Redhill
Surrey RH1 4DP
Tel: 01737 762300
Website: www.platipus-anchors.com
E-mail: info@platipus-anchors.com
Tree anchors

Practicality Brown Ltd
Iver Stud Nursery
Iver
Bucks SLO 9LA
Tel: 01753 652022
Website: www.pracbrown.co.uk
E-mail: sales@pracbrown.co.uk
Hedging

Rigby Taylor Ltd
The Riverway Estate
Portsmouth Road
Peasmarsh
Guildford
Surrey GU3 1LZ
Tel: 01483 446900
Website: www.rigbytaylor.com
E-mail: sales@rigbytaylor.com
Horticultural supply

RIW Ltd
Arc House
Terrace Road South
Binfield
Bracknell
Berkshire RG42 4PZ
Tel: 01344 397777
Website: www.riw.co.uk
E-mail: enquiries@riw.co.uk
Waterproofing products

Road Equipment Ltd
28–34 Feltham Road
Ashford
Middlesex TW15 1DL
Tel: 01784 256565
E-mail: roadequipment@aol.com
Plant hire

Rolawn Ltd
York Road
Elvington
York YO41 4XR
Tel: 01904 608661
Website: www.rolawn.co.uk
E-mail: info@rolawn.co.uk
Industrial turf

RTS Ltd
UK Sales
Daisy Dene
Inglewhite Road
Goosnargh
Preston PR3 2EB
Tel: 01772 780234
Website: www.rtslimited.uk.com
E-mail: rtslimited@hotmail.co.uk
Permaloc edging

Sawhorse Ltd
Unit 4 Willows Gate
Stratton Audley
Bicester
Oxon OX27 9 AU
Tel: 07989 578661
Website: www.3rdspace.co.uk
E-mail: info@sawhorse-ltd.co.uk
Garden buildings

Scotscape Ltd
Unit 1
Wandle Technology Park
Mill Green Road
Mitcham Junction
Surrey CR4 4HZ
Tel: 02082 545000
Website: www.scotscape.net
E-mail: sales@scotscape.net
Living walls

SMP Playgrounds Ltd
Ten Acre Lane
Thorpe
Egham
Surrey TW20 8RJ
Tel: 01784 489100
Website: www.smp.co.uk
E-mail: sales@smp.co.uk
Playground equipment

Southern Conveyors
Unit 2, Denton Slipways Site
Wharf Road
Gravesend DAI2 2RU
Tel: 01474 564145
Website: www.southernconveyors.co.uk
E-mail: sales@southernconveyors.co.uk
Conveyors

Spadeoak (Duracourt) Ltd
Town Lane
Wooburn Green
High Wycombe
Buckinghamshire HP10 0PD
Tel: 01628 529421
Website: www.duracourt.co.uk
E-mail: info@duracourt.co.uk
Tennis courts

Spadeoak Construction Co Ltd
Town Lane
Wooburn Green
High Wycombe
Bucks HP10 0PD
Tel: 01628 529421
Website: www.spadeoak.co.uk
E-mail: email@spadeoak.co.uk
Macadam contractors

Steelway Brickhouse
Brickhouse Lane
West Bromwich
West Midlands B70 0DY
Tel: 01215 214500
Website: www.steelway.co.uk
E-mail: sales@steelwaybrickhouse.co.uk
Access covers

Steelway Fensecure
Queensgate Works
Bilston Road
Wolverhampton
West Midlands WV2 2NJ
Tel: 01902 490919
Website: www.steelway.co.uk
E-mail: sales@steelway.co.uk
Fencing

SteinTec
728 London Road
West Thurrock
Essex RM20 3LU
Tel: 01708 860049
Website: www.steintec.co.uk
E-mail: info@steintec.co.uk
Specialized mortars

Stonebank Ironcraft Ltd
Parkway Farm
Fossbridg
Northleach
Cheltenham
Gloucestershire GL54 3JL
Tel: 01285 720 737
Website: www.stonebank-ironcraft.co.uk
E-mail: info@stonebank-ironcraft.co.uk
Solid steel products

SureSet UK Limited
32 Deverill Road
Trading Estate
Sutton Veny
Warminster BA12 7BZ
Tel: 01985 841873
Website: www.sureset.co.uk
E-mail: mail@sureset.co.uk
Solid steel products

Tensar International
Cunningham Court
Shadsworth Business Park
Blackburn
Lancashire BB2 4PJ
Tel: 01254 262431
Website: www.tensar-international.com
E-mail: sales@tensar.co.uk
Erosion control, soil stabilization

Terranova Lifting Ltd
Terranova House
Bennett Road
Reading
Berks RG2 OQX
Tel: 0118 931 2345
Website: www.terranova-lifting.co.uk
E-mail: cranes@terranovagroup.co.uk
Crane hire

Townscape Products Ltd
Fulwood Road South
Sutton-in-Ashfield
Nottinghamshire NG17 2JZ
Tel: 01623 513355
Website: www.townscape-products.co.uk
E-mail: sales@townscape-products.co.uk
Hard landscaping

Trulawn Ltd
Unit 7E Vulcan Way
Sandhurst
Berkshire GU47 9DB
Tel: 01252 819695
Website: www.trulawn.co.uk
E-mail: sales@trulawn.co.uk
Artificial grass

Tubex Ltd
Aberaman Park
Aberaman
South Wales CF44 6DA
Tel: 01685 888000
Website: www.tubex.com
E-mail: plantcare@tubex.com
Plant protection, tree guards

Turf Management Systems
Seven Hills Road
Iver
Buckinghamshire SL0 0PA
Tel: 01895 834411
Liquid sod

Wade International Ltd
Third Avenue
Halstead
Essex CO9 2SX
Tel: 01787 475151
Website: www.wadedrainage.co.uk
E-mail: sales@wade.eu
Stainless steel drainage

Wavin Plastics Ltd
Parsonage Way
Chippenham
Wiltshire SN15 5PN
Tel: 01249 766600
Website: www.wavin.co.uk
E-mail: info@wavin.co.uk
Drainage products

Wicksteed Leisure Ltd
Digby Street
Kettering
Northamptonshire NN16 8YJ
Tel: 01536 517028
Website: www.wicksteed.co.uk
E-mail: sales@wicksteed.co.uk
Play equipment

Willowbank Services
Curload
Stoke St Gregory
Taunton Somerset TA3 6JD
Tel: 01823 690113
Website: www.willowbankservices.co.uk
E-mail: info@willowbankservices.co.uk
Waterside bank stabilization

Woodscape Ltd
Church Works
Church Street
Church
Lancashire BB5 4JT
Tel: 01254 383322
Website: www.woodscape.co.uk
E-mail: sales@woodscape.co.uk
Street furniture

Wybone Ltd
Mason Way
Platts Common Industrial Estate
Barnsley
South Yorkshire S74 9TF
Tel: 01226 744010
Website: www.wybone.co.uk
E-mail: sales@wybone.co.uk
Street furniture

Yeoman Aggregates Ltd
Stone Terminal
Horn Lane
Acton
London W3 9EH
Tel: 020 8896 6820
Website: www.foster-yeoman.co.uk
E-mail: sales@yeoman-aggregates.co.uk
Aggregates

Item	Unit	Total rate £
Excluding site overheads and profit		

HOW TO USE THIS BOOK

How this book is presented

All prices in this book exclude overhead and profit.

There is one pricing section to this book which is then split into two sub-sections as follows:

Approximate Estimates – This is presented in a 3 column format. This section combines prices from the Measured Works items to form a composite rate for the work described.

Measured Works – This is presented in an 8 column format and is a detailed analysis of single tasks commonly employed throughout the external works and landscape sector.

Column formats

Description of the column formats used in this book.

Approximate Estimates

Only the 3 column format is used; users of the book cannot see the components which make up the prices in the book format. An electronic version is supplied free with this book. Users of the electronic system can see more detail in the electronic format.

Measured Works

Where an 8 column analysis is shown these are tasks carried out directly by the external works or landscape contractor and are shown at cost.

Where prices are shown in the 8 column section without analysis, prices have been supplied by specialist subcontractors and are assumed to have the specialist subcontractor's selling price of the item included.

Explanation of the columns used in the Measured Works analysis

Prime Cost

Commonly known as the 'PC'. Prime Cost is the actual price of the material item being addressed such as paving, shrubs, bollards or turves, as sold by the supplier. Prime Cost is given 'per square metre', 'per 100 bags' or 'each' according to the way the supplier sells his product. In researching the material prices for the book we requested that the suppliers price for full loads of their product delivered to a site close to the M25 in London. Spons rates do not include VAT. Some companies may be able to obtain greater discounts on the list prices than those shown in the book. Prime Cost prices for those products and plants which have a wide cost range will be found under the heading of Market Prices in the main sections of this book, so that the user may select the product most closely related to their specification.

Item	Unit	Total rate £
Excluding site overheads and profit		

Column formats – cont

Explanation of the columns used in the Measured Works analysis – cont

Labour Hours

This is the total amount of man hours required to carry out the quantity and unit of the measured works item which is being analysed.

Labour £

This is the value of the labour used. Please refer to the labour rates used in this edition.

Plant

This covers the cost of machinery used to work out the cost of the item addressed. Plant can range from heavy excavators to light electrical breakers. All plant is assumed as hired in except for concrete mixers and normal site tools which a contractor would be expected to carry in his or her company vehicle (hand drills, grinders etc.). Plant includes for fuel and down time. The rates used are shown in the section of this book 'Computation of Mechanical Plant Costs' which follows this section.

Delivery and removal to site costs are not included in any rates.

Materials

The PC material plus the additional materials required to fix the PC material. Every job needs materials for its completion besides the product bought from the supplier. Paving needs sand for bedding, expansion joint strips and cement pointing; fencing needs concrete for post setting and nails or bolts; tree planting needs manure or fertilizer in the pit, tree stakes, guards and ties. If these items were to be priced out separately, *Spon's External Works and Landscape Price Book* (and the Bill of Quantities) would be impossibly unwieldy, so they are put together under the heading of Materials.

Delivery Prices

These are not generally included. Exceptions would be aggregates or loose materials which are delivered by the load.

Unit

The unit being used to measure the item.

Total Price £

The total price 'At Cost' – without profit of the item being addressed.

Where only a figure is shown in the columns

This is either a subcontract rate which has been received from a specialist subcontractor.

This rate would generally include the subcontractor's overhead and profit but not the main contractors oncosts

Where a figure is shown in the Material column only

This is generally the market price of a single material only.

Item Excluding site overheads and profit	Unit	Total rate £
Worked examples using data in this edition The following operations describe the trench excavation, pipe laying and backfilling operations as contained in section R12 of this edition.		
Example 1		
Pipe laying 100 m		
Remove topsoil 150 × 300 mm wide	100 m	48.00
Excavate for drain 150 × 450 mm; excavated material to spoil heaps on site by machine	100 m	740.00
Lay flexible plastic pipe 100 mm dia. wide	100 m	160.00
Backfilling with gravel rejects, blinding with sand and topping with 150 mm topsoil	100 m	340.00
TOTAL FOR ALL OF THE ABOVE COMBINED	100 m	1275.00
This total above item is a composite of all of the above 'Measured Works' items combined and is typical of an entry in the Approximate Estimates section of this book.		
Example 2		
The following individual items are shown as components of a tree planting specification for one tree. These individual items are all extracted from the paving section of this book.		
The PC of 900 × 600 × 50 mm precast concrete flags is shown in the PC (Prime cost) column as:	nr	4.45
This equates to the rate for the PC material of:	m²	8.20
Costs of bedding and pointing are added to give a total Materials cost of:	m²	4.90
Further costs for labour and mechanical plant give a resultant price of:	m²	21.00
If a quotation of PC £8.00/m² was received from a supplier of the flags the resultant price would be calculated as:		
Original price of the flags:	m²	8.20
Gives an extra over of:	m²	−0.24
Less the original cost PC (£7.48) plus the revised cost (£8.00) to give:	m²	21.00
Example 2		
The following samples are extracted from the tree planting section of this edition.		
The prime cost of an Acer platanoides 8 −10 cm bare root tree is given as:	nr	12.00
To which the cost of handling and planting only to treepits excavated separately is added:	nr	7.90
Giving an overall cost of:	nr	19.90
The following items are also shown as component prices in the Measured Works section		
labour and mechanical plant for mechanical excavation of a 600 × 600 × 600 mm pit	nr	3.60
a single tree stake	nr	5.45
importing Topgrow compost in 75 litre bags	nr	2.15
backfilling with imported topsoil	nr	8.70
transporting the excavated material to spoil heaps on site 25 m distant	nr	1.00
and disposing of the excavated material	nr	4.90

Item Excluding site overheads and profit	Unit	Total rate £
Worked examples using data in this edition – cont		
The following samples are extracted from the tree planting section of this edition. – cont		
The following price which is a composite of all the above might be shown in the Approximate Estimates section:		
Tree planting; supply and plant Acer platanoides 8–10 cm in 600 × 600 mm tree pit excavated mechanically; allow for a single tree stake and Topgrow compost and backfilling with imported topsoil; all excavated material removed from site	nr	47.00
ADJUSTMENT AND VARIATION OF THE RATES IN THIS BOOK		
It will be appreciated that a variation in any one item in any group will affect the final Measured Works price. Any cost variation must be weighed against the total cost of the contract. A small variation in Prime Cost where the items are ordered in thousands may have more effect on the total cost than a large variation on a few items. A change in design that necessitates the use of earth moving equipment which must be brought to the site for that one job will cause a dramatic rise in the contract cost. Similarly, a small saving on multiple items will provide a useful reserve to cover unforeseen extras. Using the tree planting example above, if the tree size was to be increased from a 8–10 cm tree to a 10–12 cm tree of the same variety, this might necessitate an increase in tree pit size, more compost, a larger tree stake and more excavated material being disposed of off site.		
The resultant variation would again be referenced from the tree planting section as follows:		
The revised PC (prime cost) of an Acer platanoides 10–12 cm bare root tree is given as:	nr	19.50
To which the cost of handling and planting only to treepits excavated separately is added:	nr	11.50
Giving an overall cost of:	nr	31.00
Add to this the following variations as described:		
Tree pit increased in size from 600 × 600 × 600 mm to 900 × 900 × 600 mm	nr	8.00
Additional compost	nr	7.15
Increased staking requirements to two stakes	nr	7.55
Increased volume of imported topsoil	nr	19.60
increased disposal volume	nr	12.00
Giving a resultant price for the increased tree size as	nr	78.00
The variation therefore from the original specification for the 8–10 cm tree to the revised 10–12 cm tree is calculated as an additional sum per tree of	nr	31.00
Labour Rates Used in this Edition		
Major Works generally		
General labour rate for all works	hr	19.80
Major Works maintenance		
General labour rate for all works	hr	17.50

Item Excluding site overheads and profit	Unit	Total rate £
Minor Works generally		
General labour rate for all works	hr	21.00
Computation of Mechanical Plant Costs		
Hired Plant; Road Equipment Ltd		
Figures in Brackets reflect the actual working hours per week of the hired machinery		
Dumper 3 tonne and operator composite	hr	27.00
Dumper 3 tonne Thwaites self drive (32 hours/week)	hr	45.00
Dumper 5 tonne Thwaites self drive (32 hours/week)	hr	29.00
Dumper 6 tonne Thwaites self drive (32 hours/week)	hr	4.40
Excavator 360 Tracked 21 ton operated (32 hours/week)	hr	83.00
Excavator 360 Tracked 7 ton self drive (32 hours/week)	hr	66.00
Excavator Tracked 5 ton self drive (32 hours/week)	hr	54.00
Fork lift telehandler self drive (24 hours/week)	hr	32.00
JCB 3CX 4 × 4 Sitemaster + breaker operated (32 hours/week)	hr	39.00
JCB 3CX 4 × 4 Sitemaster operated (32 hours/week)	hr	37.50
Mini Excavator 1.5 tonne self drive (32 hours/week)	hr	4.05
Mini Excavator JCB 803 Rubber tracks self drive (32 hours/week)	hr	7.00
Mini Excavator JCB 803 Steel tracks self drive (32 hours/week)	hr	7.50
Skip loader 1 tonne (32 hours/week)	hr	3.45
Fuel charge Red diesel (32 hours/week)	l	1.05
HSS Ltd		
Access tower; alloy; 5.2 m	day	37.00
Cultivator; 110 kg	hr	9.40
Diamond blade consumable; 450 mm; concrete	mm	21.00
Heavy-duty breaker; 110 v; 2200 w	hr	5.80
Heavy-duty breaker; petrol; 5 hrs/day; single tool	hr	7.05
Mesh fence; temporary security fencing; 2.85 × 2.0 m high	1 m/ wk	2.50
Oxy-acetylene cutting kit	hr	5.00
Petrol masonry saw bench; 350 mm	hr	6.40
Petrol poker vibrator + 50 mm head	hr	3.10
Post hole borer; 1 man; weekly rate	hr	4.70
Vibrating plate compactor	hr	3.00
Vibrating roller; 136 kg; 10.1kN; 4 hrs/day	hr	6.60
Vibration damped breaker (light); 1600 w; 110v	hr	3.30

How this Book is Compiled

INTRODUCTION

First-time users of *Spon's External Works and Landscape Price Book* and others who may not be familiar with the way in which prices are compiled may find it helpful to read this section before starting to calculate the costs of landscape works.

Please also refer to the 'How to use this book' notes on page xxi.

The cost of an item of landscape construction or planting is made up of many components:

- the cost of the product
- the labour and additional materials needed to carry out the job
- the cost of running the contractor's business

These are described more fully below.

IMPORTANT NOTES ON THE PROFIT ELEMENT OF RATES IN THIS BOOK

The rates shown in the Approximate Estimates and Measured Works sections of this book do not contain a profit element unless the rate has been provided by a subcontractor. Prices are shown at cost. This is the cost including the costs of company overhead required to perform this task or project. For reference please see the tables on pages 00–00.

Analysed Rates Versus Subcontractor Rates

As a general rule if a rate is shown as an analysed rate in the Measured Works section, i.e. it has figures shown in the columns other than the 'Unit' and 'Total Rate' column, it can be assumed that this has no profit or overhead element and that this calculation is shown as a direct labour/material supply/plant task being performed by a contractor.

On the other hand, if a rate is shown as a Total Rate only, this would normally be a subcontractor's rate and would contain the profit and overhead for the subcontractor.

The foregoing applies for the most part to the Approximate Estimates section, however in some items there may be an element of subcontractor rates within direct works build-ups.

As an example of this, to excavate, lay a base and place a macadam surface, a general landscape contractor would normally perform the earthworks and base installation. The macadam surfacing would be performed by a specialist subcontractor to the landscape contractor.

The Approximate Estimate for this item uses rates from the Measured Works section to combine the excavation, disposal, base material supply and installation with all associated plant. There is no profit included on these elements. The cost of the surfacing, however, as supplied to us in the course of our annual price enquiry from the macadam subcontractor, would include the subcontractor's profit element.

The landscape contractor would add this to his section of the work but would normally apply a mark up at a lower percentage to the subcontract element of the composite task. Users of this book should therefore allow for the profit element at the prevailing rates. Please see the worked examples below and the notes on overheads for further clarification.

INTRODUCTORY NOTES ON PRICING CALCULATIONS USED IN THIS BOOK

There are two pricing sections to this book:

Approximate Estimating Rates – Average contract value £100,000.00 – £300,000.00

Measured Works – Average contract value £100,000.00 – £300,000.00

Typical Project Profile Used in this Book

Contract value	£100,000.00 – £300,000.00
Labour rate (see page 3)	£19.75 per hour
Labour rate for maintenance contracts	£17.25 per hour
Number of site staff	35
Project area	6000 m²
Project location	Outer London
Project components	50% hard landscape 50% soft landscape and planting
Access to works areas	Very good
Contract	Main contract
Delivery of materials	Full loads

As explained in more detail later the prices are generally based on wage rates and material costs current at June 2013. They do not allow for preliminary items, which are dealt with in the Preliminaries section of this book, or for any Value Added Tax (VAT) which may be payable.

Adjustments should be made to standard rates for time, location, local conditions, site constraints and any other factors likely to affect the costs of a specific scheme.

Term contracts for general maintenance of large areas should be executed at rates somewhat lower than those given in this section.

There is now a facility available to readers that enables a comparison to be made between the level of prices given and those for projects carried out in regions other than outer London; this is dealt with on page 23.

Units of Measurement

The units of measurement have been varied to suit the type of work and care should be taken when using any prices to ascertain the basis of measurement adopted.

The prices per unit of area for executing various mechanical operations are for work in the following areas and under the following conditions:

Prices per m^2 relate to areas not exceeding $100\,m^2$
     (any plan configuration)

Prices per $100\,m^2$ relate to areas exceeding $100\,m^2$ but not exceeding 1/4 ha
     (generally clear areas but with some subdivision)

Prices per ha relate to areas over 1/4 ha
     (clear areas suitable for the use of tractors and tractor-operated equipment)

The prices per unit area for executing various operations by hand generally vary in direct proportion to the change in unit area.

Approximate Estimating

These are combined Measured Work prices which give an approximate cost for a complete section of landscape work. For example, the construction of a car park comprises excavation, levelling, roadbase, surfacing and marking parking bays. Each of these jobs is priced separately in the Measured Works section, but a comprehensive price is given in the Approximate Estimates section, which is intended to provide a quick guide to the cost of the job. It will be seen that the more items that go to make up an approximate estimate, the more possibilities there are for variations in the PC prices and the user should ensure that any PC price included in the estimate corresponds to his specification.

The figures given in *Spon's External Works and Landscape Price Book* are intended for general guidance, and if a significant variation appears in any one of the cost groups, the Price for Measured Work should be recalculated.

Measured Works

Prime Cost: Commonly known as the 'PC'. Prime Cost is the actual price of the material item being addressed such as paving, shrubs, bollards or turves, as sold by the supplier. Prime Cost is given 'per square metre', 'per 100 bags' or 'each' according to the way the supplier sells his product. In researching the material prices for the book we requested that the suppliers price for full loads of their product delivered to a site close to the M25 in London. Spon's rates do not include VAT. Some companies may be able to obtain greater discounts on the list prices than those shown in the book. Prime Cost prices for those products and plants which have a wide cost range will be found under the heading of Market Prices in the main sections of this book, so that the user may select the product most closely related to his specification.

Materials: The PC material plus the additional materials required to fix the PC material. Every job needs materials for its completion besides the product bought from the supplier. Paving needs sand for bedding, expansion joint strips and cement pointing; fencing needs concrete for post setting and nails or bolts; tree planting needs manure or fertilizer in the pit, tree stakes, guards and ties. If these items were to be priced out separately, *Spon's External Works and Landscape Price Book* (and the Bill of Quantities) would be impossibly unwieldy, so they are put together under the heading of Materials.

Labour: This figure covers the cost of planting shrubs or trees, laying paving, erecting fencing etc. and is calculated on the wage rate (skilled or unskilled) and the time needed for the job. Extras such as highly skilled craft work, difficult access, intermittent working and the need for labourers to back up the craftsman all add to the cost. Large regular areas of planting or paving are cheaper to install than smaller intricate areas, since less labour time is wasted moving from one area to another.

Plant, consumable stores and services: This rather impressive heading covers all the work required to carry out the job which cannot be attributed exactly to any one item. It covers the use of machinery ranging from JCBs to shovels and compactors, fuel, static plant, water supply (which is metered on a construction site), electricity and rubbish disposal. The cost of transport to site is deemed to be included elsewhere and should be allowed for elsewhere or as a preliminary item. Hired plant is calculated on an average of 36 hours working time per week.

Subcontract rates: Where there is no analysis against an item, this is deemed to be a rate supplied by a subcontractor. In most cases these are specialist items where most external works contractors would not have the expertise or the equipment to carry out the task described. It should be assumed that subcontractor rates include for the subcontractor's profit. An example of this may be found for the tennis court rates in section Q26.

Labour Rates Used in this Edition

The rates for labour used in this edition have been based on surveys carried out on a cross section of external works contractors. These rates include for company overheads such as employee administration, transport insurance, and oncosts such as National Insurance. Please see the calculation of these rates shown on page 0. The rates do not include for profit.

Overheads: An allowance for this is included in the labour rates which are described above. The general overheads of the contract such as insurance, site huts, security, temporary roads and the statutory health and welfare of the labour force are not directly assignable to each item, so they are distributed as a percentage on each, or as a separate preliminary cost item. The contractor's and subcontractor's profits are not included in this group of costs. Site overheads, which will vary from contract to contract according to the difficulties of the site, labour shortages, inclement weather or involvement with other contractors, have not been taken into account in the build up of these rates, while overhead (or profit) may have to take into account losses on other jobs and the cost to the contractor of unsuccessful tendering.

ADJUSTMENT AND VARIATION OF THE RATES IN THIS BOOK

It will be appreciated that a variation in any one item in any group will affect the final Measured Works price. Any cost variation must be weighed against the total cost of the contract. A small variation in Prime Cost where the items are ordered in thousands may have more effect on the total cost than a large variation on a few items. A change in design that necessitates the use of earth moving equipment which must be brought to the site for that one job will cause a dramatic rise in the contract cost. Similarly, a small saving on multiple items will provide a useful reserve to cover unforeseen extras.

Worked examples

These are shown on page xxiii

HOW THIS BOOK IS UPDATED EACH YEAR

The basis for this book is a database of Material, Labour, Plant and Subcontractor resources each with its own area of the database.

Material, Plant and Subcontractor Resources

Each year the suppliers of each material, plant or subcontract item are approached and asked to update their prices to those that will prevail in September of that year.

These resource prices are individually updated in the database. Each resource is then linked to one or many tasks. The tasks in the task library section of the database is automatically updated by changes in the resource library. A quantity of the resource is calculated against the task. The calculation is generally performed once and the links remain in the database.

On occasions where new information or method or technology are discovered or suggested, these calculations would be revisited. A further source of information is simple time and production observations made during the course of the last year.

Labour Resource Update

Most tasks, except those shown as subcontractor rates (see above), employ an element of labour. The Data Department at Davis Langdon conducts ongoing research into the costs of labour in various parts of the country.

Tasks or entire sections are then re-examined and recalculated. Comments on the rates published in this book are welcomed and may be submitted to the contact addresses shown in the Preface.

Davis Langdon

An AECOM Company

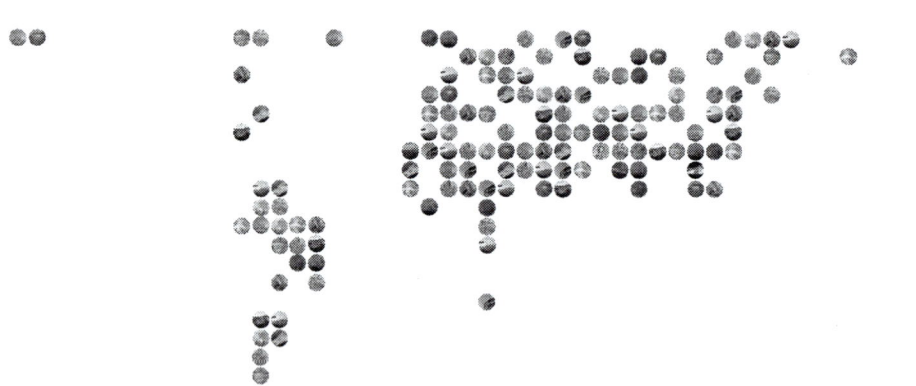

We at Davis Langdon have decided to change our name. From October 2013 we will be known as AECOM.

The time is right.

AECOM

At the heart of AECOM, we join
the dots that create, enhance and
sustain built, natural and social
environments.

SPON'S PRICEBOOKS 2014

Spon's Architects' and Builders' Price Book 2014

DAVIS LANGDON

The most detailed, professionally relevant source of UK construction price information currently available anywhere.

What's new in this year's Spon's A&B Price Book? There are new cost models for out of town retail developments, office to residential conversion, and a museum fit out; there's a re-formatted Preliminaries example, and typical hire rates for common preliminaries items. We're also giving overhauled coverage of automated car parking systems (from simple stack systems to fully automatic ones), Plasmor concrete blocks, diamond drilling, ETFE roofing, and flowing screeds. And an extended range of underground drainage

Hbk & electronic package 824pp approx.
978-1-4822-0406-3 £155.00
ebook & electronic package
978-1-4822-0407-0 £155 .00
(inc. sales tax where appropriate)

Spon's External Works and Landscape Price Book 2014

DAVIS LANGDON

Now in its 33rd edition, the EW&L Price Book has a heavily enhanced approximate estimates section, with most measured works items being also presented as composite items. It covers wildflower meadows – reflecting current interest – and urban and street tree planting systems, Sudscape porous paving systems, and maintenance operations for long term cyclic maintenance contractors.

Hbk & electronic package 720pp approx.
978-1-4822-0412-4 £135.00
ebook & electronic package
978-1-4822-0413-1 £135 .00
(inc. sales tax where appropriate)

Spon's Civil Engineering and Highway Works Price Book 2014

DAVIS LANGDON

With output in civils starting to rise, and buoyancy in highways, rail and electricity a number of prices are increasing. Labour rates have increased at the beginning of the year and further increases are expected in 2014. Pressure is continuing on material costs with a slight fall in materials prices as some of the rail projects come on stream.

Hbk & electronic package 680pp approx.
978-1-4822-0410-0 £165.00
ebook & electronic package
978-1-4822-0411-7 £165.00
(inc. sales tax where appropriate)

Spon's Mechanical and Electrical Services Price Book 2014

DAVIS LANGDON

Our M&E price book continues to be the most comprehensive and best annual services engineering price book currently available. This year the methodology for estimating year Building Management Systems (BMS) has been examined. BIM guidance, Feed-In Tariffs and Carbon Trading sections have been brought up to date with the current rates and processes. The book also gives the usual market update of labour rates and daywork rates, material costs and prices for measured works, and all-in-rates and elemental rates in the Approximate Estimating section. Prices for measured works are brought into line with RICS New Rules of Measurement NRM2.

Hbk & electronic package 912pp approx.
978-1-4822-0414-8 £155.00
ebook & electronic package
978-1-4822-0415-5 £155.00
(inc. sales tax where appropriate)

Receive our online data viewing package free when you order any hard copy or ebook Spon 2014 Price Book

Visit www.pricebooks.co.uk

To order:
Tel: 01235 400524 Fax: 01235 400525
Post: Taylor & Francis Customer Services, Bookpoint Ltd, 200 Milton Park, Abingdon, Oxon, OX14 4SB, UK
Email: book.orders@tandf.co.uk
A complete listing of all our books is on www.sponpress.com

CRC CRC Press
Taylor & Francis Group

Estimator's Pocket Book

Duncan Cartlidge

The Estimator's Pocket Book is a concise and practical reference covering the main pricing approaches, as well as useful information such as how to process sub-contractor quotations, tender settlement and adjudication. It is fully up-to-date with NRM2 throughout, features a look ahead to NRM3 and describes the implications of BIM for estimators.

It includes instructions on how to handle:

- the NRM order of cost estimate;
- unit-rate pricing for different trades;
- pro rata pricing and dayworks
- builders' quantities;
- approximate quantities.

Worked examples show how each of these techniques should be carried out in clear, easy-to-follow steps. This is the indispensible estimating reference for all quantity surveyors, cost managers, project managers and anybody else with estimating responsibilities. Particular attention is given to NRM2, but the overall focus is on the core estimating skills needed in practice.

May 2013 186x123: 310pp
Pb: 978-0-415-52711-8: £19.99

To Order: Tel: +44 (0) 1235 400524 Fax: +44 (0) 1235 400525
or Post: Taylor and Francis Customer Services,
Bookpoint Ltd, Unit T1, 200 Milton Park, Abingdon, Oxon, OX14 4TA UK
Email: book.orders@tandf.co.uk

For a complete listing of all our titles visit:
www.tandf.co.uk

Understanding JCT Standard Building Contracts
Ninth Edition

David Chappell

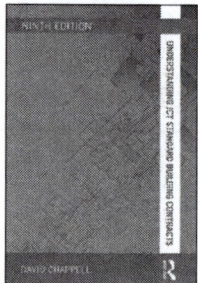

This ninth edition of David Chappell's bestselling guide has been revised to take into account changes made in 2011 to payment provisions, and elsewhere. This remains the most concise guide available to the most commonly used JCT building contracts: Standard Building Contract with quantities, 2011 (SBC11), Intermediate Building Contract 2011 (IC11), Intermediate Building Contract with contractor's design 2011 (ICD11), Minor Works Building Contract 2011 (MW11), Minor Works Building Contract with contractor's design 2011 (MWD11) and Design and Build Contract 2011 (DB11).

Chappell avoids legal jargon but writes with authority and precision. Architects, quantity surveyors, contractors and students of these professions will find this a practical and affordable reference tool arranged by topic.

April 2012: 234 x 156: 160 pp
Pb: 978-0-415-50890-2: £24.99

To Order: Tel: +44 (0) 1235 400524 Fax: +44 (0) 1235 400525
or Post: Taylor and Francis Customer Services,
Bookpoint Ltd, Unit T1, 200 Milton Park, Abingdon, Oxon, OX14 4TA UK
Email: book.orders@tandf.co.uk

For a complete listing of all our titles visit:
www.tandf.co.uk

PART 1

General

This part of the book contains the following sections:

Landscape Architect's Pocket Book
Second Edition

Siobhan Vernon, Rachel Tennant & Nicola Garmory

An indispensable tool for all landscape architects, this time-saving guide answers the most frequently asked questions in one pocket-sized volume. It is a concise, easy-to-read reference that gives instant access to a wide range of information needed on a daily basis, both out on site and in the office.

Covering all the major topics, including hard landscaping, soft landscaping as well as planning and legislation, the pocket book also includes a handy glossary of important terms, useful calculations and helpful contacts. Not only an essential tool for everyday queries on British standards and procedures, this is a first point of reference for those seeking more extensive, supplementary sources of information, including websites and further publications.

This new edition incorporates updates and revisions from key planning and environmental legislation, guidelines and national standards.

April 2013 186x123: 340pp
Pb: 978-0-415-63084-9: £19.99

To Order: Tel: +44 (0) 1235 400524 Fax: +44 (0) 1235 400525
or Post: Taylor and Francis Customer Services,
Bookpoint Ltd, Unit T1, 200 Milton Park, Abingdon, Oxon, OX14 4TA UK
Email: book.orders@tandf.co.uk

For a complete listing of all our titles visit:
www.tandf.co.uk

Taylor & Francis
Taylor & Francis Group

Labour Rates Used in this Edition

Based on surveys carried out on a cross section of external works contractors, the following rates for labour have been used in this edition.

These rates include for company overheads such as employee administration, transport insurance, and oncosts such as National Insurance.

The rates used do not include for profit.

The rates used may include a subcontractor's profit where subcontractor's rates are used to build up the costs of an item in the book.

The rates for all labour used in this edition are as follows:

General contracting	£19.75/hour
Maintenance contracting	£17.50/hour

Guidelines for Landscape and Visual Impact Assessment
Third Edition

Landscape Institute & I.E.M.A.

Landscape and Visual Impact Assessment (LVIA) can be key to planning decisions by identifying the effects of new developments on views and on the landscape itself.

This fully revised edition of the industry standard work on LVIA presents an authoritative statement of the principles of assessment. Offering detailed advice on the process of assessing the landscape and visual effects of developments and their significance, it also includes a new expanded chapter on cumulative effects and updated guidance on presentation.

Written by professionals for professionals, the third edition of this widely respected text provides an essential tool for landscape practitioners, developers, legal advisors and decision-makers.

March 2010: 276x219: 170pp
Hb: 978-0-415-68004-2: £49.99

To Order: Tel: +44 (0) 1235 400524 Fax: +44 (0) 1235 400525
or Post: Taylor and Francis Customer Services,
Bookpoint Ltd, Unit T1, 200 Milton Park, Abingdon, Oxon, OX14 4TA UK
Email: book.orders@tandf.co.uk

For a complete listing of all our titles visit:
www.tandf.co.uk

Taylor & Francis
Taylor & Francis Group

Computation of Labour Rates Used in this Edition

Different organizations will have varying views on rates and costs, which will in any event be affected by the type of job, availability of labour and the extent to which mechanical plant can be used. However this information should assist the reader to:

(1)   compare the prices to those used in his own organization
(2)   calculate the effect of changes in wage rates or prices of materials
(3)   calculate prices for work similar to the examples given

From September 2013 – assumed basic weekly rates of pay for craft and general operatives are £416.13 and £313.17 respectively; to these rates have been added allowances for the items below in accordance with the recommended procedure of the Chartered Institute of Building in its 'Code of Estimating Practice'. The resultant hourly rates are £14.06 and £10.46 for craft operatives and general operatives respectively.

The items for which allowances have been made are:

- Lost time
- Construction Industry Training Board Levy
- Holidays with pay
- Accidental injury, retirement and death benefits scheme
- Sick pay
- National Insurance
- Severance pay and sundry costs
- Employer's liability and third party insurance

The tables that follow illustrate how the above hourly rates have been calculated. Productive time has been based on a total of 1801.8 hours worked per year for daywork calculations above.

How the Labour Rate has been calculated

A survey of typical landscape/external works companies indicates that they are currently paying above average wages for multi-skilled operatives regardless of specialist or supervisory capability. In our current overhaul of labour constants and costs we have departed from our previous policy of labour rates based on national awards and used instead a gross rate per hour for all rate calculations. Estimators can readily adjust the rate if they feel it is inappropriate for their work.

The productive labour of any organization must return their salary plus the cost of the administration which supports the labour force.

The labour rate used in this edition is calculated as follows:

Team size

A three man team with annual salaries as shown and allowances for National Insurance, uniforms, site tools and a company vehicle, returns a basic labour rate of £16.24 per hour to which company overhead is added as per the following tables.

Basic labour rate calculation for Spon's 2014

Standard Working Hours per year (2013/4)
    =   1802

Less allowance for sick leave and down time
    =   −39

Actual Working Hours per year (2013/4)
    =   1763

Working Hours per year 2013/2014		1763 *Allows sick and down time of 39 hours*					
	Number of	**Labour team**	**NI**	**Net Cost**	**Clothing**	**Site Tools**	**TOTAL 3 Man Team**
Foreman	1	25750.00	3296.00	29046.00	200.00	150.00	
Craftsman	1	21200.00	2713.60	23913.60	200.00	150.00	
Labourer	1	16450.00	2105.60	18555.60	200.00	150.00	
	3.00	63400.00	8115.20	71515.20	600.00	450.00	£ 72,565.20
Overtime Hours							
nominal hours	40.00	21.91	2.80	988.52			
/annum	40.00	18.04	2.31	813.85			
at 1.5 times normal	40.00	14.00	1.79	631.50			
standard rate							
Total staff in team	3.00			73949.07	600.00	450.00	£ 74,999.07
Vehicle Costs Inclusive of Fuel Insurances etc.	**Working Days**	**£ /Day**					
	240.00	45.50 >>>>>>>>>>>>>>>>>>>>>>>>>>>>>>>>					£ 10,920.00
						TOTAL	£ 85,919.07
Net Average Cost per man Hour			16.24				

The basic average labour rate excluding overhead costs (below) is

$$\frac{£85919.07}{\text{Total staff in team (3)} \times \text{Working hours per year (1763)}} = £16.24$$

Add to the above the overhead costs as per the table on the next page

Overhead costs

Add to the above basic rate the company overhead costs for a small company as per the table below. These costs are absorbed by the number of working men multiplied by the number of working hours supplied in the table below. This then generates an hourly overhead rate which is added to the net cost rate above.

Illustrative overhead for company employing 12–50 landscape operatives

COST CENTRE	Number of	Cost	Total
MD: Salary only: excludes profits + NI + Vehicle	1	54000.00	54000.00
Senior contracts managers + NI + Vehicle	1	40500.00	40500.00
Other contracts managers + NI + Vehicle	1	30500.00	30500.00
Business administrator/ secretary	1	20500.00	20500.00
Book Keeper	1	19000.00	19000.00
Rental	12	1000.00	12000.00
Insurances	1	5000.00	5000.00
Telephone and Mobiles	12	250.00	3000.00
Office Equipment	1	1000.00	1000.00
Stationery	12	50.00	600.00
Advertising	12	200.00	2400.00
Other vehicles not allocated to contract teams	1	9000.00	9000.00
Other consultants	1	6000.00	6000.00
Accountancy	1	2500.00	2500.00
Lights heating water	12	200.00	2400.00
Other expenses	1	8000.00	10000.00
TOTAL OFFICE OVERHEAD			218400.00

The final labour rate used is then generated as follows:

$$\text{Nett labour rate} \quad + \quad \frac{\text{Total Office Overhead (£218400.00)}}{\text{Working hours per year (1763)} \times \text{Labour resources employed}}$$

Total nr of Site Staff	Admin Cost per hour	Total Rate per Man Hour
12	10.32	26.57
15	8.26	24.50
18	6.88	23.13
20	6.19	22.44
25	4.96	21.20
35	3.54	19.78
40	3.10	19.34
45	2.75	19.00
50	2.48	18.72

Hourly labour rates used for Spon's External Works and Landscape 2014

This year's rates are calculated on the rounded value of a 35 man working team rounded to the nearest £0.25

General contracting	£19.75
Maintenance contracting	£17.25

Smaller organizations

Smaller organizations should adjust the rates shown in this book to suit a labour rate which is generally slightly higher due to the balance of company overhead and number of employees. The costs for smaller organizations are no longer shown in theis edition but users may adjust the rates using the resultant labour rates shown below. The calculation of this labour rate is as follows

- A typical smaller works labour calculation is based on the above but the administration cost of the organization is £107,440.00 per annum
- The owner of the business does not carry out site works and is assumed to be paid a salary of £40,000.00 plus National insurance plus a motor vehicle per annum. The balance of his income is taken in dividend and profit share figure constitutes part of the £107440.00 shown
- There are between 6 and 15 site operatives

Illustrative overhead for small company employing 6–15 landscape operatives

Cost Centre	Number of	Cost	Total
MD: Salary only: excludes profits + NI + Vehicle	1	53100.00	53100.00
Secretary	1	17000.00	17000.00
Book Keeper	1	8000.00	8000.00
Rental	12	300.00	3600.00
Insurances	1	4000.00	4000.00
Telephone and Mobiles	12	200.00	2400.00
Office Equipment	1	1000.00	1000.00
Stationery	12	20.00	240.00
Advertising	12	200.00	2400.00
Other vehicles not allocated to contract teams	1	4000.00	4000.00
Other consultants	1	1000.00	1000.00
Accountancy	1	1500.00	1500.00
Lights heating water	12	100.00	1200.00
Other expenses	1	8000.00	8000.00
Total office overhead			£107,440.00

The final labour rate used is then generated as follows:

$$\text{Nett labour rate} \quad + \quad \frac{\text{Total Office Overhead (£107440.00)}}{\text{Working hours per year (1763)} \times \text{Labour resources employed}}$$

Total nr of Site Staff	Admin Cost per hour	Total Rate per Man Hour
3	20.41	36.21
6	10.20	26.45
9	6.80	23.05
12	5.10	21.35
15	4.08	20.33

Hourly labour rates for a typical small company working in London for 2014 could be calculated on the rounded value of a 9 man working team (£23.05). We would suggest using a rate of 22.50 due to indications of labour rate pressures which we predict for the following 12 month period.

General and Maintenance Contracting (Smaller organizations)	£22.50

Calculation of Annual Hours Worked

Normal hours worked		
Normal hours worked per week		39 hours
Normal hours worked per year	39 hrs × 52 wks	2028 hours
Less holidays		
Annual holidays	39 hrs × 4.2 wks	−163.8 hours
Public holidays (includes diamond jubilee day)	8 days @7.8 hours	−62.4 hours
Less sick and downtime	39 hrs × 1 wk	−39 hours
Actual hours worked per annum		1763 hours

Computation of the Cost of Materials

Percentages of default waste are placed against material resources within the supplier database. These range from 2.5% (for bricks) to 20% (for topsoil). An allowance for the cost of unloading, stacking etc. should be added to the cost of materials.

The following are typical hours of labour for unloading and stacking some of the more common building materials.

Material	Unit	Labourer hours
Cement	tonne	0.67
Lime	tonne	0.67
Common bricks	1000	1.70
Light engineering bricks	1000	2.00
Heavy engineering bricks	1000	2.40

Spon's Asia-Pacific Construction Costs Handbook

Fourth Edition

Davis Langdon

Spon's Asia Pacific Construction Costs Handbook includes construction cost data for twenty countries. This new edition has been extended to include Pakistan and Cambodia. Australia, UK and America are also included, to facilitate comparison with construction costs elsewhere.

Information is presented for each country in the same way, as follows:

- key data on the main economic and construction indicators.
- an outline of the national construction industry, covering structure, tendering and contract procedures, materials cost data, regulations and standards
- labour and materials cost data
- measured rates for a range of standard construction work items
- approximate estimating costs per unit area for a range of building types
- price index data and exchange rate movements against £ sterling, $US and Japanese Yen.

The book also includes a Comparative Data section to facilitate country-to-country comparisons. Figures from the national sections are grouped in tables according to national indicators, construction output, input costs and costs per square metre for factories, offices, warehouses, hospitals, schools, theatres, sports halls, hotels and housing.

This unique handbook will be an essential reference for all construction professionals involved in work outside their own country and for all developers or multinational companies assessing comparative development costs.

April 2010: 234x156: 480pp
Hb: 978-0-415-46565-6: £140.00

To Order: Tel: +44 (0) 1235 400524 Fax: +44 (0) 1235 400525
or Post: Taylor and Francis Customer Services,
Bookpoint Ltd, Unit T1, 200 Milton Park, Abingdon, Oxon, OX14 4TA UK
Email: book.orders@tandf.co.uk

For a complete listing of all our titles visit:
www.tandf.co.uk

Landfill Tax

The Tax

The Landfill Tax came into operation on 1 October 1996. It is levied on operators of licensed landfill sites at the following rates with effect from 1 April 2013:

- Inactive or inert wastes   £2.50 per tonne   Included are soil, stones, brick, plain and reinforced concrete, plaster and glass – lower rate
- All other taxable wastes   £72 per tonne   Included are timber, paint and other organic wastes generally found in demolition work and builders skips – standard rate

The standard (higher) rate for 'all other taxable wastes' will be increased by £8 per tonne each year at least until 2014 when the rate will be £80 per tonne. There will also be a floor under the standard rate, so that the rate will not fall below £80 per tonne from 2014–15 to 2019–20. The lower rate for 'inactive or inert wastes' will be frozen at £2.50 per tonne to 31 March 2014.

The Landfill Tax (Qualifying Material) Order 2011 came into force on 1 April 2011. This has amended the qualifying criteria that are eligible for the lower rate of landfill tax. The revisions introduced arose primarily from the need to reflect changes in wider environmental policy and legislation since 1996, such as the implementation of the European Landfill Directive.

A waste will be lower rated for landfill tax from 1 April 2011 only if it is listed as a qualifying material in the Landfill Tax (Qualifying Material) Order 2011.

The principal changes to qualifying material are:

- Rocks and sub-soils that are currently lower rated will remain so.
- Topsoil and peat will be removed from the lower rate, as these are natural resources that can always
- be recycled/re-used.
- Used foundry sand, which has in practice been lower rated by extra-statutory concession since the tax's introduction in 1996, will now be included in the lower rate Order.
- Definitions of qualifying ash arising from the burning of coal and petroleum coke (including when burnt with biomass) will be clarified.
- The residue from titanium dioxide manufacture will qualify, rather than titanium dioxide itself reflecting industry views.
- Minor changes will be made to the wording of the calcium sulphate group of wastes to reflect the implementation of the Landfill Directive since 2001.
- Water will be removed from the lower rate – water is now banned from landfill so its inclusion in the list of lower rated wastes is unnecessary; where water is used as a waste carrier the water is not waste and therefore not taxable.

Calculating the Weight of Waste

There are two options:

- If licensed sites have a weighbridge, tax will be levied on the actual weight of waste.
- If licensed sites do not have a weighbridge, tax will be levied on the permitted weight of the lorry based on an alternative method of calculation based on volume to weight factors for various categories of waste.

Effect on Prices

The tax is paid by Landfill site operators only. Tipping charges reflect this additional cost.

Exemptions

The following disposals are exempt from Landfill Tax subject to meeting certain conditions:

- dredgings which arise from the maintenance of inland waterways and harbours
- naturally occurring materials arising from mining or quarrying operations
- pet cemeteries
- material from the reclamation of contaminated land (see below)
- inert waste used to restore landfill sites and to fill working and old quarries where a planning condition or obligation is in existence
- waste from visiting NATO forces

The exemption for waste from contaminated land has been phased out completely from 1 April 2012 and no new applications for landfill tax exemption are now accepted.

For further information contact the National Advisory Service, Telephone: 0845 010 9000.

The Aggregates Levy

The Aggregates Levy came into operation on 1 April 2002 in the UK, except for Northern Ireland where it has been phased in over 5 years from 2003.

It was introduced to ensure that the external costs associated with the exploitation of aggregates are reflected in the price of aggregate, and to encourage the use of recycled aggregate. There continues to be strong evidence that the levy is achieving its environmental objectives, with sales of primary aggregate down and production of recycled aggregate up. The Government expects that the rates of the levy will at least keep pace with inflation over time, although it accepts that the levy is still bedding in.

The rate of the levy increased to £2.10 per tonne from 1 April 2012 and is levied on anyone considered to be responsible for commercially exploiting virgin aggregates in the UK and should naturally be passed by price increase to the ultimate user.

All materials falling within the definition of Aggregates are subject to the levy unless specifically exempted.

It does not apply to clay, soil, vegetable or other organic matter.

The intention is that it will:

- Encourage the use of alternative materials that would otherwise be disposed of to landfill sites.
- Promote development of new recycling processes, such as using waste tyres and glass
- Promote greater efficiency in the use of virgin aggregates
- Reduce noise and vibration, dust and other emissions to air, visual intrusion, loss of amenity and damage to wildlife habitats

Definitions

'Aggregates' means any rock, gravel or sand which is extracted or dredged in the UK for aggregates use. It includes whatever substances are for the time being incorporated in it or naturally occur mixed with it.

'Exploitation' is defined as involving any one or a combination of any of the following:

- Being removed from its original site
- Becoming subject to a contract or other agreement to supply to any person
- Being used for construction purposes
- Being mixed with any material or substance other than water, except in permitted circumstances

Incidence

It is a tax on primary aggregates production – i.e. virgin aggregates won from a source and used in a location within the UK territorial boundaries (land or sea). The tax is not levied on aggregates which are exported or on aggregates imported from outside the UK territorial boundaries.

It is levied at the point of sale.

Exemption from tax

An aggregate is exempt from the levy if it is:

- Material which has previously been used for construction purposes
- Aggregate that has already been subject to a charge to the Aggregates Levy
- Aggregate which was previously removed from its originating site before the start date of the levy
- Aggregate which is being returned to the land from which it was won
- Aggregate won from a farm land or forest where used on that farm or forest
- Rock which has not been subjected to an industrial crushing process
- Aggregate won by being removed from the ground on the site of any building or proposed building in the course of excavations carried out in connection with the modification or erection of the building and exclusively for the purpose of laying foundations or of laying any pipe or cable
- Aggregate won by being removed from the bed of any river, canal or watercourse or channel in or approach to any port or harbour (natural or artificial), in the course of carrying out any dredging exclusively for the purpose of creating, restoring, improving or maintaining that body of water
- Aggregate won by being removed from the ground along the line of any highway or proposed highway in the course of excavations for improving, maintaining or constructing the highway otherwise than purely to extract the aggregate
- Drill cuttings from petroleum operations on land and on the seabed
- Aggregate resulting from works carried out in exercise of powers under the New Road and Street Works Act 1991, the Roads (Northern Ireland) Order 1993 or the Street Works (Northern Ireland) Order 1995
- Aggregate removed for the purpose of cutting of rock to produce dimension stone, or the production of lime or cement from limestone
- Aggregate arising as a waste material during the processing of the following industrial minerals:
    - ball clay
    - barytes
    - calcite
    - china clay
    - coal, lignite, slate or shale
    - feldspar
    - flint
    - fluorspar
    - fuller's earth
    - gems and semi-precious stones
    - gypsum
    - any metal or the ore of any metal
    - muscovite
    - perlite
    - potash
    - pumice
    - rock phosphates
    - sodium chloride
    - talc
    - vermiculite

However, the levy is still chargeable on any aggregates arising as the spoil or waste from or the by-products of the above exempt processes. This includes quarry overburden.

Anything that consists 'wholly or mainly' of the following is exempt from the levy (note that 'wholly' is defined as 100% but 'mainly' as more than 50%, thus exempting any contained aggregates amounting to less than 50% of the original volumes:

- clay, soil, vegetable or other organic matter
- coal, slate or shale
- china clay waste and ball clay waste

Relief from the levy either in the form of credit or repayment is obtainable where:

- it is subsequently exported from the UK in the form of aggregate
- it is used in an exempt process
- where it is used in a prescribed industrial or agricultural process
- it is waste aggregate disposed of by dumping or otherwise, e.g. sent to landfill or returned to the originating site

The Aggregates Levy Credit Scheme (ALCS) for Northern Ireland was suspended with effect from 1 December 2010 following a ruling by the European General Court.

A new exemption for aggregate obtained as a by-product of railway, tramway and monorail improvement, maintenance and construction was introduced in 2007.

Discounts

From 1 July 2005 the standard added water percentage discounts listed below can be used. Alternatively a more exact percentage can be agreed and this must be done for dust dampening of aggregates.

- washed sand = 7%
- washed gravel = 3.5%
- washed rock/aggregate = 4%

Impact

The British Aggregates Association suggests that the additional cost imposed by quarries is more likely to be in the order of £3.40 per tonne on mainstream products, applying an above average rate on these in order that by-products and low grade waste products can be held at competitive rates, as well as making some allowance for administration and increased finance charges.

With many gravel aggregates costing in the region of £16.00 to £18.00 per tonne, there is a significant impact on constructioncosts.

Avoidance

An alternative to using new aggregates in filling operations is to crush and screen rubble which may become available during the process of demolition and site clearance as well as removal of obstacles during the excavation processes.

Example:

Assuming that the material would be suitable for fill material under buildings or roads, a simple cost comparison would be as follows (note that for the purpose of the exercise, the material is taken to be 1.80 tonne per m³ and the total quantity involved less than 1,000 m³):

	£/m³	£/tonne
Importing fill material:		
Cost of 'new' aggregates delivered to site	31.23	17.35
Addition for Aggregates Tax	3.78	2.10
Total cost of importing fill materials	35.01	19.45
Disposing of site material:		
Cost of removing materials from site materials	21.52	11.95
Crushing site materials:		
Transportation of material from excavations or demolition to stockpiles	3.00	1.67
Transportation of material from temporary stockpiles to the crushing plant	4.00	2.22
Establishing plant and equipment on site; removing on completion	2.00	1.11
Maintain and operate plant	9.00	5.00
Crushing hard materials on site	13.00	7.22
Screening material on site	2.00	1.11
Total cost of crushing site materials	33.00	18.33

From the above it can be seen that potentially there is a great benefit in crushing site materials for filling rather than importing fill materials.

Setting the cost of crushing against the import price would produce a saving of £2.10 per m³. If the site materials were otherwise intended to be removed from the site, then the cost benefit increases by the saved disposal cost to £23.53 per m³.

Even if there is no call for any or all of the crushed material on site, it ought to be regarded as a useful asset and either sold on in crushed form or else sold with the prospects of crushing elsewhere.

Specimen Unit rates	unit³	£
Establishing plant and equipment on site; removing on completion		
Crushing plant	trip	1,200.00
Screening plant	trip	600.00
Maintain and operate plant		
Crushing plant	week	7,200.00
Screening plant	week	1,800.00
Transportation of material from excavations or demolition places to temporary stockpiles	m³	3.00
Transportation of material from temporary stockpiles to the crushing plant	m³	2.40
Breaking up material on site using impact breakers		
mass concrete	m³	14.00
reinforced concrete	m³	16.00
brickwork	m³	6.00
Crushing material on site		
mass concrete not exceeding 1000 m³	m³	13.00
mass concrete 1000–5000 m³	m³	12.00
mass concrete over 5000 m³	m³	11.00
reinforced concrete not exceeding 1000 m³	m³	15.00
reinforced concrete 1000–5000 m³	m³	14.00
reinforced concrete over 5000 m³	m³	13.00
brickwork not exceeding 1000 m³	m³	12.00
brickwork 1000–5000 m³	m³	11.00
brickwork over 5000 m³	m³	10.00
Screening material on site	m³	2.00

More detailed information can be found on the HMRC website (www.hmrc.gov.uk) in Notice AGL 1 Aggregates Levy published in May 2009.

Spon's First Stage Estimating Handbook

Third Edition

Bryan Spain

Have you ever had to provide accurate costs for a new supermarket or a pub "just an idea...a ballpark figure..." ?

The earlier a pricing decision has to be made, the more difficult it is to estimate the cost and the more likely the design and the specs are to change. And yet a rough-and-ready estimate is more likely to get set in stone.

Spon's First Stage Estimating Handbook is the only comprehensive and reliable source of first stage estimating costs. Covering the whole spectrum of building costs and a wide range of related M&E work and landscaping work, vital cost data is presented as:

- costs per square metre
- elemental cost analyses
- principal rates
- composite rates.

Compact and clear, Spon's First Stage Estimating Handbook is ideal for those key early meetings with clients. And with additional sections on whole life costing and general information, this is an essential reference for all construction professionals and clients making early judgements on the viability of new projects.

January 2010: 216x138: 244 pp
Pb: 978-0-415-54715-4: £46.99

To Order: Tel: +44 (0) 1235 400524 Fax: +44 (0) 1235 400525
or Post: Taylor and Francis Customer Services,
Bookpoint Ltd, Unit T1, 200 Milton Park, Abingdon, Oxon, OX14 4TA UK
Email: book.orders@tandf.co.uk

For a complete listing of all our titles visit:
www.tandf.co.uk

Taylor & Francis
Taylor & Francis Group

Cost Indices

The purpose of this section is to show changes in the cost of carrying out landscape work (hard surfacing and planting) since 1990. It is important to distinguish between costs and tender prices: the following table reflects the change in cost to contractors but does not necessarily reflect changes in tender prices. In addition to changes in labour and material costs, which are reflected in the indices given below, tender prices are also affected by factors such as the degree of competition at the time of tender and in the particular area where the work is to be carried out, the availability of labour and materials, and the general economic situation. This can mean that in a period when work is scarce, tender prices may fall despite the fact that costs are rising, and when there is plenty of work available, tender prices may increase at a faster rate than costs.

The Constructed Cost Index

A Constructed Cost Index based on PSA Price Adjustment Formulae for Construction Contracts (Series 2). Cost indices for the various trades employed in a building contract are published monthly by HMSO and are reproduced in the technical press.

The indices comprise 49 Building Work indices plus seven 'Appendices' and other specialist indices. The Building Work indices are compiled by monitoring the cost of labour and materials for each category and applying a weighting to these to calculate a single index.

Although the PSA indices are prepared for use with price-adjustment formulae for calculating reimbursement of increased costs during the course of a contract, they also present a time series of cost indices for the main components of landscaping projects. They can therefore be used as the basis of an index for landscaping costs.

The method used here is to construct a composite index by allocating weightings to the indices representing the usual work categories found in a landscaping contract, the weightings being established from an analysis of actual projects. These weightings totalled 100 in 1976 and the composite index is calculated by applying the appropriate weightings to the appropriate PSA indices on a monthly basis, which is then compiled into a quarterly index and rebased to 1976 = 100.

Constructed Landscaping Cost Index 1976 = 100

Year	First Quarter	Second Quarter	Third Quarter	Fourth Quarter	Annual Average
2000	493	496	513	514	504
2001	512	514	530	530	522
2002	531	540	573	574	555
2003	577	582	603	602	591
2004	602	610	639	639	623
2005	641	648	684	683	664
2006	688	692	710	710	710
2007	714	718	737	742	728
2008	752	771	803	794	780
2009	789	792	796	802	795
2010	804	810	814	817	811
2011	825	835	845	852	839
2012	858	858	860	862	860
2013	875*	875*			

* *Provisional*

This index is updated every quarter in Spon's Price Book Update. The updating service is available, free of charge, to all purchasers of Spon's Price Books.

Regional Variations

Prices in *Spon's External Works and Landscape Price Book* are based upon conditions prevailing for a competitive tender in the Outer London area. For the benefit of readers, this edition includes regional variation adjustment factors which can be used for an assessment of price levels in other regions.

Special further adjustment may be necessary when considering city centre or very isolated locations.

Region	Adjustment Factor
Outer London	1.00
Inner London	1.09
South East	1.03
South West	0.96
East Midlands	0.92
West Midlands	0.92
East Anglia	1.00
Yorkshire & Humberside	0.92
Northern	0.86
North West	0.86
Scotland	0.91
Wales	0.90
Northern Ireland	0.56
Channel Islands	1.61

The following example illustrates the adjustment of prices for regions other than Outer London, by use of regional variation adjustment factors.

A. Value of items priced using *Spon's External Works and Landscape Price Book* for Outer London   £100,000

B. Adjustment to value of A. to reflect Northern Region price level 100,000 × 0.86   £ 86,000

Quantity Surveyor's Pocket Book

2nd Edition

D Cartlidge

This second edition of the Quantity Surveyor's Pocket Book is fully updated in line with NRM1, NRM2 and JCT(11), and remains a must-have guide for students and qualified practitioners. Its focussed coverage of the data, techniques, and skills essential to the quantity surveying role make it an invaluable companion for everything from initial cost advice to the final account stage.

Key features include:

- the structure of the construction industry

- cost forecasting and feasibility studies

- measurement and quantification, with NRM2 and SMM7 examples

- estimating and bidding

- whole life costs

- contract selection

- final account procedure.

June 2012: 186x123: 440pp
Pb: 978-0-415-50110-1: £18.99

To Order: Tel: +44 (0) 1235 400524 Fax: +44 (0) 1235 400525
or Post: Taylor and Francis Customer Services,
Bookpoint Ltd, Unit T1, 200 Milton Park, Abingdon, Oxon, OX14 4TA UK
Email: book.orders@tandf.co.uk

For a complete listing of all our titles visit:
www.tandf.co.uk

PART 2

Rates of Wages – Building Industry

Building Regulations in Brief
Seventh Edition

Ray Tricker & Sam Alford

The most popular and trusted guide to the building regulations, Building Regulations in Brief is updated regularly to reflect constant changes. Now in its seventh edition, it has sold over 28,000 copies since its first publication in 2003.

This new edition includes the latest on all the significant amendments to Building Regulations, Planning Permission and the Approved Documents that occurred in October 2010 and includes changes to Parts F and L, as well as Approved Documents A, C, and J. There are also changes reflecting the consolidation of the building regulations included.

The no-nonsense approach has made it a firm favourite with all involved in the building industry including designers, building surveyors and inspectors, students and architects. A ready reference giving practical information, it enables compliance in the simplest and most cost-effective manner possible. Building Regulations in Brief cuts through the confusion to explain the meaning of the regulations, their history, current status, requirements, associated documentation and how local authorities view their importance, as well as emphasizing the benefits and requirements of each regulation. It's an essential purchase for anyone needing to comply with the building regulations.

February 2012: 234 x156: 1,034 pp
Pb: 978-0-415-80969-6: £28.99

To Order: Tel: +44 (0) 1235 400524 Fax: +44 (0) 1235 400525
or Post: Taylor and Francis Customer Services,
Bookpoint Ltd, Unit T1, 200 Milton Park, Abingdon, Oxon, OX14 4TA UK
Email: book.orders@tandf.co.uk

For a complete listing of all our titles visit:
www.crcpress.com

BUILDING INDUSTRY WAGES

Note

The following data is published for information only and is not used for within the calculation of rates for this publication. The exception is the calculation shown for the working hours per year.

The Working Rule Agreement includes a pay structure with general operative and additional skilled rates of pay as well as craft rate. Plus rates and additional payments will be consolidated into basic pay to provide the following rates (for a normal 39 hour week) which will come into effect from the following dates:

Effective from 7 January 2013

The following basic rates of pay are:

	Rate per 39-hour week (£)	Rate per hour (£)
Craft Rate	416.13	10.67
Skill Rate 1	396.63	10.17
Skill Rate 2	381.81	9.79
Skill Rate 3	357.24	9.16
Skill Rate 4	337.35	8.65
General operative	313.17	8.03

BUILDING CRAFT AND GENERAL OPERATIVES

Effective from 7 January 2013

			Craft operatives		General operatives	
			£	£	£	£
Wages at standard basic rate						
Productive time	44.30	wks	416.13	18,434.56	313.17	13,873.43
Lost time allowance	0.90	wks	416.13	367.15	313.17	281.85
Overtime	0.00	wks	0.00	0.00	0.00	0.00
				18,809.08		14,155.28
Extra payments under National Working Rules	45.00	wks		–	–	–
Sick pay	1.00	wk		–	–	–
CITB Allowance (0.50% of payroll)	1.00	year		106.11		79.86
Holiday pay	4.20	wks	416.13	1,747.75	313.17	1,315.31
Public Holiday pay	1.60	wks	416.13	665.81	313.17	501.07
Employer's contributions to:						
EasyBuild Stakeholder Pension						
(Death and accident cover is provided free)	52.00	wks	5.00	260.00	5.00	260.00
National Insurance (average weekly payment)	52.00	wks	35.90	1,866.80	21.96	1,141.92
				23,455.55		17,453.44
Severance pay and sundry costs	Plus		1.50%	351.83	0.02	261.80
				23,087.38		17,715.24
Employer's Liability and Third Party Insurance	Plus		2.00%	476.15	0.02	354.30
Total cost per annum				**24,283.53**		**18,069.54**
Total cost per hour				**14.06**		**10.46**

DAYWORK RATES – BUILDING OPERATIVES

Example Calculations of Prime Cost of Labour in Daywork

Effective from 7 January 2013

EXAMPLE ONE

		Craft operative		General operative	
		Rate £	Annual cost £	Rate £	Annual cost £
Basic Wages	46.2 wks	416.13	19,225.21	313.17	14,468.45
Extra Payments: Where applicable			0.00	0.00	0.00
Sub Total:			19,225.21		14,468.45
Employer's National Insurance contribution					
(13.8% after the first £148 per week)			2,046.64		1,057.14
			21,271.85		15,525.59
Employer's Contributions to:					
CITB levy (0.5% of payroll)			108.19		81.42
Holiday Pay	226.2 hrs	10.67	2,413.55	8.03	1,816.39
Welfare Benefit – stamps	52 wks	11.39	592.28	11.39	592.28
Annual labour cost:			**24,385.87**		**18,015.68**
Hourly base rates as defined			**13.53**		**10.00**

EXAMPLE TWO

		Craft operative		General operative	
		Rate £	Annual cost £	Rate £	Annual cost £
Basic Wages	46.2 wks	416.13	19,225.21	313.17	14,468.45
Extra Payments: Where applicable			0.00	0.00	0.00
Sub Total:			19,225.21		14,468.45
Employer's National Insurance contribution					
(13.8% after the first £148 per week)			1,713.57		1,057.14
			20,938.78		15,525.59
Employer's Contributions to:					
CITB Levy (0.5% of payroll)			96.13		72.34
Annual labour cost:			**21,034.91**		**15,597.93**
Hourly base Rates			**11.67**		**8.66**

NOTES:

1. Calculated following Definition of Prime Cost of Daywork carried out under a Building Contract, published by the Royal Institution of Chartered Surveyors and the Construction Confederation.

2. Standard basic rates effective from 7 January 2013.
3. Standard working hours per annum calculated as follows:

52 weeks @ 39 hours	=	2028
Less		
4.2 weeks holiday @ 39 hours	=	163.8
8 days public holidays @ 7.8 hours	=	62.4
		226.2
		1801.8

4. All labour costs incurred by the contractor in his capacity as an employer, other than those contained in the hourly base rate, are to be taken into account under Section 6.
5. The above example is for guidance only and does not form part of the Definition; all the basic costs are subject to re-examination according to the time when and in the area where the daywork is executed.
6. N.I. payments are at not-contracted out rates applicable from April 2013.
7. Basic rate and GMB number of weeks = 52 weeks – 4.2 weeks annual holiday – 1.6 weeks public holiday = 46.2 weeks.

PART 3

Approximate Estimating Rates

Prices in this section are based upon the Prices for Measured Works, but allow for incidentals which would normally be measured separately in a Bill of Quantities. They do not include for Preliminaries which are priced elsewhere in this book.

Items shown as subcontract or specialist rates would normally include the specialist's overhead and profit. All other items which could fall within the scope of works of general landscape and external works contractors would not include profit.

Based on current commercial rates, profits of 10% to 35% may be added to these rates to indicate the likely 'with profit' values of the tasks below. The variation quoted above is dependent on the sector in which the works are taking place – domestic, public or commercial.

European Gardens
History, Philosophy and Design

Tom Turner

Garden design and usage has been a feature of human civilisation as far back as Neolithic times, when the first gardens began to be used for residential, horticultural and sacred tasks. Tom Turner follows the entire history of the European garden from its prehistoric rootsright up to the present day in this beautifully illustrated book.

European Gardens is divided into ten periods of history and garden development, detailing the advancement of land usage for over 10,000 years. Some of the topics covered in this comprehensive book include the Egyptian gardens of the Pharaohs, the castle gardens of medieval times, eclectic gardens of the nineteenth century and abstract gardens of the last 100 years. The geographical scope of this book covers the whole of the European continent, and touches the garden designs of North Africa and the Middle East.

Tom Turner is a skilled landscape architect and garden historian, who supports his engaging writing with his own detailed plans and diagrams. European Gardens also features almost 1,000 colour photographs from across the continent allowing the reader to see for themselves how the design and structure of gardens has developed over time.

A companion to the Asian Gardens book, published by Routledge in 2010, European Gardens is a development of the original Garden History book from 2004

December 2010: 250x250: 424pp
Hb: 978-0-415-49684-1: £35.00

To Order: Tel: +44 (0) 1235 400524 Fax: +44 (0) 1235 400525
or Post: Taylor and Francis Customer Services,
Bookpoint Ltd, Unit T1, 200 Milton Park, Abingdon, Oxon, OX14 4TA UK
Email: book.orders@tandf.co.uk

For a complete listing of all our titles visit:
www.tandf.co.uk

Taylor & Francis
Taylor & Francis Group

PRELIMINARIES

Item Excluding site overheads and profit	Unit	Total rate £
Contract Administration and Management		
Prepare tender bid for external works or landscape project; measured works contract; inclusive of bid preparation, provisional program and method statements		
Measured works bid; measured bills provided by the employer; project value		
£30,000.00	nr	315.00
£50,000.00	nr	520.00
£100,000.00	nr	1150.00
£200,000.00	nr	1800.00
£500,000.00	nr	2325.00
£1,000,000.00	nr	3350.00
Lump sum bid; scope document, drawings and specification provided by the employer; project value		
£30,000.00	nr	670.00
£50,000.00	nr	880.00
£100,000.00	nr	1725.00
£200,000.00	nr	2950.00
£500,000.00	nr	4050.00
£1,000,000.00	nr	6200.00
Prepare and maintain Health and Safety file; prepare risk and Coshh assessments, method statements, works programmes before the start of works on site; project value		
£35,000.00	nr	160.00
£75,000.00	nr	240.00
£100,000.00	nr	315.00
£200,000.00 – £500,000.00	nr	790.00
Contract management		
Site management by site surveyor carrying out supervision and administration duties only; inclusive of oncosts		
full time management	week	1200.00
managing one other contract	week	600.00
Parking		
Parking expenses where vehicles do not park on the site area; per vehicle		
metropolitan area; city centre	week	200.00
Metropolitan area; outer areas	week	160.00
suburban restricted parking areas	week	40.00
Congestion charging		
London only	week	40.00

Approximate Estimates

PRELIMINARIES

Item Excluding site overheads and profit	Unit	Total rate £
Site Setup and Site Establishment		
Site fencing; supply and erect temporary protective fencing and remove at completion of works		
Cleft chestnut paling		
1.20 m high 75 mm larch posts at 3 m centres	100 m	930.00
1.50 m high 75 mm larch posts at 3 m centres	100 m	1075.00
Heras fencing		
2 week hire period	100 m	970.00
4 week hire period	100 m	1475.00
8 week hire period	100 m	2450.00
12 week hire period	100 m	3450.00
16 week hire period	100 m	4450.00
24 week hire period	100 m	6400.00
Site compounds		
Establish site administration area on reduced compacted Type 1 base; 150 mm thick; allow for initial setup of site office and welfare facilities for site manager and maximum 8 site operatives		
temporary base or hard standing; 20 × 10 m	200 m²	2450.00
To hardstanding area above; erect office and welfare accommodation; allow for furniture, telephone connection, and power; restrict access using temporary safety fencing to the perimeter of the site administration area		
Establishment; delivery and collectioncosts only		
site office and chemical toilet; 20 m of site fencing	nr	850.00
site office and chemical toilet; 40 m of site fencing	nr	920.00
site office and chemical toilet; 60 m of site fencing	nr	980.00
Weekly hire rates of site compound equipment		
site office 2.4 × 3.6 m; chemical toilet; 20 m of site fencing	week	115.00
site office 4.8 × 2.4 m; chemical toilet; 20 m of site fencing; armoured store 2.4 × 3.6 m	week	130.00
site office 2.4 × 3.6 m; chemical toilet; 40 m of site fencing	week	165.00
site office 2.4 × 3.6 m; chemical toilet; 60 m of site fencing	week	215.00
Removal of site compound		
Remove base; fill with topsoil and reinstate to turf on completion of site works	200 m²	1675.00
Surveying and setting out		
Note: In all instances a quotation should be obtained from a surveyor. The following cost projections are based on the costs of a surveyor using EDM (electronic distance measuring equipment) where an average of 400 different points on site are recorded in a day. The surveyed points are drawn in a CAD application and the assumption used that there is one day of drawing up for each day of on-site surveying.		
Survey existing site using laser survey equipment and produce plans of existing site conditions		
site 1,000 to 10,000 m² with sparse detail and features; boundary and level information only; average five surveying stations	nr	750.00
site up to 1,000 m² with dense detail such as buildings, paths, walls and existing vegetation; average five survey stations	nr	1500.00
site up to 4000 m² with dense detail such as buildings, paths, walls and existing vegetation; average 10 survey stations	nr	2250.00

8.1.1 SITE CLEARANCE

Item Excluding site overheads and profit	Unit	Total rate £
CLEARING VEGETATION		
Clear existing scrub vegetation including shrubs and hedges and stack on site		
By machine		
light scrub and grasses	100 m²	21.00
light scrub and grasses but remove vegetation to licensed tip	100 m²	42.00
undergrowth brambles and heavy weed growth	100 m²	51.00
small shrubs	100 m²	51.00
By hand		
light scrub and grasses to spoil heaps 25 m distance	100 m²	63.00
light scrub and grasses to spoil heap; subsequently disposed off site	100 m²	77.00
light scrub and grasses to tip	100	60.00
Fell trees on site; grub up roots; all by machine; remove debris to licensed tip		
trees 600 mm girth	each	290.00
trees 1.5–3.00 m girth	each	890.00
trees over 3.00 m girth	each	1425.00
Chip existing vegetation on site; remove to stockpile 100 m		
light scrub and bramble; average height 500 mm	100 m²	23.00
dense scrub and bramble; average height 1.00 m	100 m²	46.50
mixed scrub and shrubs average 1.50 m high	100 m²	75.00
dense overgrown shrubs and bramble; average 2.00 m high	100 m²	100.00
Turf stripping for preservation		
Cut and strip existing turf area by machine; turves 50 mm thick; turf to be lifted for preservation		
load to dumper or palette by hand; transport on site not exceeding 100 m travel to new location	100 m²	80.00
all by hand; including barrowing to stacking locally 10 m distance	100 m²	460.00
Turf stripping for disposal; spray turf with glyphosate and return to site after 3 weeks; strip dead grass; load to stockpile on site; rotavate to prepare area for new surface treatments to within 50 mm of final levels		
By knapsack sprayer; excavator and dumper; pedestrian rotavator		
5 tonne excavator and 6 tonne dumper; stock pile distance 25 m	100 m²	63.00
5 tonne excavator and 3 tonne dumper; stock pile distance 50 m	100 m²	65.00
5 tonne excavator and 3 tonne dumper; stock pile distance 100 m	100 m²	69.00
By tractor drawn equipment		
spraying with glyphosate; 2 passes	100 m²	510.00

8.1.1 SITE CLEARANCE

Item Excluding site overheads and profit	Unit	Total rate £
CLEARING FENCES GATES AND LIGHT STRUCTURES		
CLEARING FENCING AND GATES		
Clear site; demolish existing closeboard fence; clear shrubs 50% in shrub beds and 50% free-standing; spray turf and return to clear ground of sprayed vegetation; chip vegetation on site and transport to spoil heaps 50 m; works by mechanical excavator		
Rectangular site with 70% turf and 30% shrub area		
1000 m^2	nr	1850.00
2000 m^2	nr	3350.00
4000 m^2	nr	6300.00
Cleartrack (Evl) Ltd; as above but clearance by mechanical spraying and in situ mulcher; vegetation only to spoil heap; turf mulched in		
5000 m^2	nr	2175.00
Clear site; demolish existing chain-link fence; clear shrubs 50% in shrub beds and 50% free-standing; spray turf and return to clear ground of sprayed vegetation; chip vegetation on site and transport to spoil heaps 50 m; works by high capacity mechanical mulcher		
Cleartrack (Evl) Ltd		
overgrown areas of weeds and brambles with maximum stem diameter of 100 mm	ha	3600.00
overgrown areas of weeds and brambles with maximum stem diameter of 150 mm	ha	7000.00
Rectangular site with 60% turf grass/scrub area and 40% shrubs and overgrowth		
site area; 1000 m^2	nr	2075.00
site area; 2000 m^2	nr	3900.00
site area; 4000 m^2	nr	7200.00
Demolish fencing and gates		
Demolish fencing; break out posts, remove fencing stack and load mechanically for removal to tip		
Chain link fencing; vegetation not present on angle iron stakes driven in	100 m	220.00
Chain link fencing; overgrown with vegetation on angle iron stakes driven in	100 m	610.00
DEMOLITIONS OF EXTERNAL STRUCTURES AND SURFACES		
CLEARING STRUCTURES		
Demolish existing free-standing walls; grub out foundations; remove arisings to tip		
Brick wall; 112 mm thick		
300 mm high	m	12.30
500 mm high	m	18.40
Brick wall; 225 mm thick		
300 mm high	m	15.40
500 mm high	m	19.90
1.00 m high	m	32.00
1.20 m high	m	34.00
1.50 m high	m	41.50
1.80 m high	m	47.00

8.1.1 SITE CLEARANCE

Item Excluding site overheads and profit	Unit	Total rate £
Demolish existing structures		
Timber sheds on concrete base 150 mm thick inclusive of disposal and backfilling with excavated material; plan area of building		
10 m²	nr	210.00
15 m²	nr	300.00
Building of cavity wall construction; insulated with tiled roof; inclusive of foundations; allow for disconnection of electrical water and other services; grub out and remove 10 m of electrical cable, water and waste pipes		
10 m²	nr	1600.00
20 m²	nr	2150.00
CLEARING SURFACES		
Demolish existing surfaces		
Break up plain concrete slab; remove to licensed tip		
150 mm thick	m²	9.35
200 mm thick	m²	12.50
300 mm thick	m²	18.70
Break up plain concrete slab and preserve arisings for hardcore; break down to maximum size of 200 × 200 mm and transport to location; maximum distance 50 m		
150 mm thick	m²	6.10
200 mm thick	m²	8.10
300 mm thick	m²	12.10
Break up reinforced concrete slab and remove to licensed tip		
150 mm thick	m²	13.80
200 mm thick	m²	18.40
300 mm thick	m²	27.50
Break out existing surface and associated 150 mm thick granular base load to 20 tonne muck away vehicle for removal off site		
macadam surface; 70 mm thick	m²	8.40
macadam surface; 150 mm thick	m²	12.10
block paving; 50 mm thick	m²	8.30
block paving; 80 mm thick	m²	9.00

8.1.2 PREPARATORY GROUNDWORKS

Item Excluding site overheads and profit	Unit	Total rate £
EXCAVATION OF BANKS		
Excavate existing slope to vertical to form flat area; grade bank behind to 45°; no allowance for removal or backfilling of excavated materials;		
Existing slope angle 10°; excavation by 7 tonne machine; excavated material to location on-site 50 m distant; linear measurement of flat area at the base of the slope;		
2.00 m wide; height 350 mm	m	3.70
4.00 m wide; height 860 mm	m	14.60
6.00 m wide; height 1.28 m	m	33.00
8.00 m wide; height 1.71 m	m	59.00
10.00 m wide; height 2.14 m	m	92.00
Existing slope angle 20°; excavation by 7 tonne machine; excavated material to loctaion on-site 50 m distant; linear measurement of flat area at the base of the slope;		
2.00 m wide; height 1.14 m	m	9.80
4.00 m wide; height 2.29 m	m	39.00
6.00 m wide; height 3.43 m	m	88.00
8.00 m wide; height 4.58 m	m	155.00
10.00 m wide; height 5.72 m	m	240.00
Excavate and grade banks to grade to receive stabilization treatments below; removal of excavated material not included		
By 21 tonne excavator	m³	0.35
By 7 tonne excavator	m³	2.40
By 5 tonne excavator	m³	2.80
By 3 tonne excavator	m³	4.20
Excavate sloped bank to vertical to receive retaining or stabilization treatment (priced separately); allow 1.00 m working space at top of bank; backfill working space and remove balance of arisings off site; measured as length along the top of the bank		
Bank height 1.00 m; excavation by 5 tonne excavator		
10 degree	m	71.00
30 degree	m	22.00
45 degree	m	12.70
60 degree	m	7.35
Bank height 1.00 m; excavation by 5 tonne excavator; arisings moved to stockpile on site		
10 degree	m	12.30
30 degree	m	6.25
45 degree	m	3.60
60 degree	m	2.10
Bank height 2.00 m; excavation by 5 tonne excavator		
10 degree	m	290.00
30 degree	m	86.00
45 degree	m	51.00
60 degree	m	29.00
Bank height 2.00 m; excavation by 5 tonne excavator but arisings moved to stockpile on site		
10 degree	m	81.00
30 degree	m	24.50
45 degree	m	14.40
60 degree	m	8.30

8.1.2 PREPARATORY GROUNDWORKS

Item	Unit	Total rate £
Excluding site overheads and profit		
Bank height 3.00 m; excavation by 21 tonne excavator		
20 degree	m	265.00
30 degree	m	170.00
45 degree	m	92.00
60 degree	m	50.00
As above but arisings moved to stockpile on site		
20 degree	m	41.00
30 degree	m	26.00
45 degree	m	10.40
60 degree	m	5.70
Excavating generally		
Prices for excavating and reducing to levels are for work in light or medium soils; multiplying factors for other soils are as follows		
Clay	1.5	–
Compact gravel	1.2	–
Soft chalk	2.0	–
Hard rock	3.0	–
Excavating topsoil for preservation		
Remove topsoil average depth 300 mm; deposit in spoil heaps not exceeding 100 m travel to heap; turn heap once during storage		
by machine	m³	13.40
by hand	m³	120.00
Excavate to reduce levels; remove spoil to stockpile not exceeding 100 m travel; all by machine		
Average 0.25 m deep	m²	2.05
Average 0.25 m deep	m³	8.30
Average 0.30 m deep	m²	2.50
Average 0.30 m deep	m³	7.70
Average 1.0 m deep; using 21 tonne 360 tracked excavator	m³	4.80
Average 1.0 m deep; using JCB	m³	7.70
Excavate to reduce levels; remove spoil to stockpile not exceeding 100 m travel; all by hand		
Average 0.10 m deep	m²	10.60
Average 0.20 m deep	m²	21.00
Average 0.30 m deep	m²	32.00
Average 100–300 mm deep	m³	105.00
Average 1.0 m deep	m³	120.00
Extra for carting spoil to licensed tip off site (20 tonnes)	m³	23.00
Extra for carting spoil to licensed tip off site (by skip; machine loaded)	m³	49.00
Extra for carting spoil to licensed tip off site (by skip; hand loaded)	m³	105.00
Spread excavated material to levels in layers not exceeding 150 mm; using scraper blade or bucket		
Average thickness 100 mm	m²	0.60
Average thickness 100 mm but with imported topsoil	m²	4.30
Average thickness 200 mm	m²	1.25

8.1.2 PREPARATORY GROUNDWORKS

Item Excluding site overheads and profit	Unit	Total rate £
Excavating generally – cont		
Spread excavated material to levels in layers not exceeding 150 mm – cont		
Average thickness 200 mm but with imported topsoil	m²	8.55
Average thickness 250 mm	m²	1.60
Average thickness 200–250 mm	m³	6.30
Average thickness 250 mm but with imported topsoil	m²	10.70
Average thickness 250 mm but with imported topsoil	m³	43.00
Extra for work to banks exceeding 30 degree slope	30%	-
Rip subsoil using approved sub-soiling machine to a depth of 600 mm below topsoil at 600 mm centres; all tree roots and debris over 150 × 150 mm to be removed; cultivate ground where shown on drawings to depths as shown		
100 mm	100 m²	9.55
200 mm	100 m²	9.90
300 mm	100 m²	10.50
400 mm	100 m²	15.10
Form landform from imported material; mounds to falls and grades to receive turf or seeding treatments; material dumped by 20 tonne vehicle to location; volume 100 m³ spread over area; thickness varies from 150–500 mm; prices to not include for surface preparations		
Imported standard grade topsoil		
by large excavator	m³	34.00
by 5 tonne excavator	m³	36.50
by 3 tonne excavator	m³	38.00
Form landform from excavated material; mounds to falls and grades to receive turf or seeding treatments; material moved by dumper to location maximum 50 m; volume 100 m³ spread over a 200 m² area; thickness varies from 150–500 mm; prices to not include for surface preparations		
Imported recycled topsoil		
by large excavator	m³	3.60
by 5 tonne excavator	m³	6.05
by 3 tonne excavator	m³	7.45
EXCAVATION OF TRENCHES		
Excavate trenches for foundations; trenches 225 mm deeper than specified foundation thickness; pour plain concrete foundations GEN 1 10 N/mm and to thickness as described; disposal of excavated material off site; dimensions of concrete foundations		
250 mm wide × 250 mm deep	m	11.30
300 mm wide × 300 mm deep	m	15.10
400 mm wide × 600 mm deep	m	37.00
600 mm wide × 400 mm deep	m	38.50

8.1.2 PREPARATORY GROUNDWORKS

Item Excluding site overheads and profit	Unit	Total rate £
Excavate trenches for foundations; trenches 225 mm deeper than specified foundation thickness; pour plain concrete foundations GEN 1 10 N/mm and to thickness as described; disposal of excavated material off site; dimensions of concrete foundations		
250 mm wide × 250 mm deep	m³	210.00
300 mm wide × 300 mm deep	m³	165.00
600 mm wide × 400 mm deep	m³	160.00
400 mm wide × 600 mm deep	m³	140.00
Excavate trenches for services; grade bottoms of excavations to required falls; remove 20% of excavated material off site		
Trenches 300 mm wide		
600 mm deep	m	6.20
750 mm deep	m	8.00
900 mm deep	m	9.60
1.00 m deep	m	10.60
Ha-ha		
Excavate ditch 1200 mm deep × 900 mm wide at bottom battered one side 45 degree slope; excavate for foundation to wall and place GEN 1 concrete foundation 150 × 500 mm; construct one and a half brick wall (brick PC £300.00/1000) battered 10° from vertical; laid in cement: lime: sand (1:1:6) mortar in English garden wall bond; precast concrete coping weathered and throated set 150 mm above ground on high side; rake bottom and sides of ditch and seed with low maintenance grass at 35 g/m²		
wall 600 mm high	m	260.00
wall 900 mm high	m	325.00
wall 1200 mm high	m	405.00
Excavate ditch 1200 mm deep × 900 mm wide at bottom battered one side 45 degree slope; place deer or cattle fence 1.20 m high at centre of excavated trench; rake bottom and sides of ditch and seed with low maintenance grass at 35 g/m²		
fence 1200 mm high	m	20.00
Soil stabilization		
Please refer to slope excavation above for preparation of banks		
REVETMENTS		
Gabion mattress revetments		
Construct revetment of Maccaferri Ltd Reno mattress gabions laid on firm level ground; tightly packed with broken stone or concrete and securely wired; all in accordance with manufacturer's instructions		
gabions 6 × 2 × 0.17 m; one course	m²	36.00
gabions 6 × 2 × 0.17 m; two courses	m²	60.00

8.1.2 PREPARATORY GROUNDWORKS

Item Excluding site overheads and profit	Unit	Total rate £
Soil stabilization – cont		
Grass concrete; bank revetments; excluding bulk earthworks		
Bring bank to final grade by machine; lay regulating layer of Type 1; lay filter membrane; lay 100 mm drainage layer of broken stone or approved hardcore 28–10 mm size; blind with 25 mm sharp sand; lay grass concrete surface (price does not include edgings or toe beams); fill with approved topsoil and fertilizer at 35 g/m²; seed with dwarf rye based grass seed mix		
Grasscrete in situ reinforced concrete surfacing GC 1; 100 mm thick	m²	58.00
Grasscrete in situ reinforced concrete surfacing GC 2; 150 mm thick	m²	70.00
Grassblock 103 mm open matrix blocks; 406 × 406 × 103 mm	m²	54.00
Grade slope to even grade; Stabilize river bank edge with woven willow walling; lay bank stabilization mat pinned to graded embankment 5 m long(measured per m of river bank edge		
Slope edge to river bank 5 m long seeded with Greenfix stabilization matting		
Live willow walling 1.20 m high along river bank	m	155.00
Brushwood faggott and pre-planted coir revetment edge	m	170.00
Brushwood faggott revetment edge	m	150.00
Neoweb: cellular slope retention; clear existing bank of light scrub, shrubs and grasses; spray with glyphosate; grade surface only to even grade by machine and final hand grade; lay soil retaining system		
Backfilled with excavated material		
100 mm deep	m²	14.40
200 mm deep	m²	24.00
Backfilled with ballast		
100 mm deep	m²	16.60
200 mm deep	m²	28.00
Backfilled with concrete 10Nmm		
100 mm deep	m²	20.50
200 mm deep	m²	37.00
Backfilled with imported topsoil and seeded		
100 mm deep	m²	19.50
SOIL REINFORCEMENT		
Geogrid soil reinforcement (Note: measured per metre run at the top of the bank)		
Backfill excavated bank; lay Tensar geogrid; cover with excavated material as work proceeds		
2.00 m high bank; geogrid at 1.00 m vertical lifts		
10 degree slope	m	600.00
20 degree slope	m	335.00
30 degree slope	m	240.00
2.00 m high bank; geogrid at 0.50 m vertical lifts		
45 degree slope	m	61.00
60 degree slope	m	48.00

8.1.2 PREPARATORY GROUNDWORKS

Item Excluding site overheads and profit	Unit	Total rate £
3.00 m high bank		
10 degree slope	m	600.00
20 degree slope	m	335.00
30 degree slope	m	240.00
45 degree slope	m	520.00
EROSION CONTROL		
Bank stabilization; erosion control mats Excavate fixing trench for erosion control mat 300 × 300 mm; backfill after placing mat selected below	m	6.85
To anchor trench above; place mat to slope to be retained; allow extra over to the slope for the area of matting required to the anchor trench By machine; final grade only of excavated bank by excavator; lay erosion control solution; sow with low maintenance grass at 35 g/m²; spread imported topsoil 25 mm thick incorporating medium grade sedge peat at 3 kg/m² and fertilizer at 30 g/m²; water lightly		
unseeded Eromat Light	m²	2.80
seeded Covamat Standard	m²	3.40
lay Greenfix biodegradable pre-seeded erosion control mat; fix with 6 × 300 mm steel pegs at 1.0 m centres	m²	3.90
Tensar open textured erosion mat	m²	5.45
By hand; final grade only of excavated bank by hand; lay erosion control solution; sow with low maintenance grass at 35 g/m²; spread imported topsoil 25 mm thick incorporating medium grade sedge peat at 3 kg/m² and fertilizer at 30 g/m²; water lightly		
unseeded Eromat Light	m²	3.25
seeded Covamat Standard	m²	3.90
lay Greenfix biodegradable pre-seeded erosion control mat; fix with 6 × 300 mm steel pegs at 1.0 m centres	m²	4.40
Tensar open textured erosion mat	m²	5.90
Cellular retaining systems; final grade to excavated bank by machine and by hand; lay bank retaining system Cellweb; filling with angular aggregate;	m	24.00
Haul road for site enablement Reduce existing soil to stockpile on site; Lay Tensar Biaxial geogrid on 200 mm compacted hardcore base; backfill with 200 mm Type 1 granular fill and compact	m²	19.60
Reduce existing soil to stockpile on site; Lay 200 mm compacted hardcore base blinded with 50 mm sharp sand; Lay Cellweb load support	m²	34.50

8.2.1 ROADS PATHS PAVINGS AND SURFACINGS

Item Excluding site overheads and profit	Unit	Total rate £
Excavation kerbs and edgings to paved surfaces		
Excavate for pathways or roadbed to receive surface 100 mm thick priced separately; grade; 100 mm hardcore; kerbs PC 125 × 255 mm on both sides; including foundations haunched		
Excavation 350 deep; Type 1 base 150 thick		
1.50 m wide	m	89.00
2.00 m wide	m	100.00
3.00 m wide	m	120.00
4.00 m wide	m	145.00
5.00 m wide	m	165.00
6.00 m wide	m	185.00
7.00 m wide	m	210.00
Excavation 450 mm deep; Type 1 base 250 mm thick		
1.50 m wide	m	170.00
2.00 m wide	m	205.00
3.00 m wide	m	280.00
4.00 m wide	m	350.00
5.00 m wide	m	425.00
6.00 m wide	m	500.00
7.00 m wide	m	570.00
EDGINGS TO PATHS AND PAVINGS		
Excavation for pathway; Concrete base to support edging 150 thick; Lay edgings to pathway as below; lay base to patheway of 150 mm type 1;		
xxxPathways 1.00 m wide		
PRECAST CONCRETE KERBS		
Note: excavation is by machine unless otherwise mentioned		
Excavate trench and construct concrete foundation 300 mm wide × 300 mm deep; lay precast concrete kerb units bedded in semi-dry concrete; slump 35 mm maximum; haunching one side; disposal of arisings off site		
Kerbs laid straight; site mixed concrete bases		
125 mm high × 125 mm thick; bullnosed; type BN	m	26.50
125 × 255 mm; ref HB2; SP	m	29.00
150 × 305 mm; ref HB1	m	34.50
Kerbs laid straight; ready mixed concrete bases		
125 mm high × 125 mm thick; bullnosed; type BN	m	25.00
125 × 255 mm; ref HB2; SP	m	27.00
150 × 305 mm; ref HB1	m	33.00
Straight kerbs 125 mm × 255 mm laid to radius ready mixed concrete bases		
0.9 m radius	m	38.00
1.8 m radius	m	37.00
3.0 m radius	m	35.50
6.1 m radius	m	34.00
9.15 m radius	m	33.50
12.20 m radius	m	33.00

8.2.1 ROADS PATHS PAVINGS AND SURFACINGS

Item Excluding site overheads and profit	Unit	Total rate £
Dropper kerbs		
125 mm × 255 −150 mm; left or right handed	m	30.50
Quadrants		
305 mm radius	nr	34.00
455 mm radius	nr	36.00
Excavate trench and construct concrete foundation 300 mm wide × 150 mm deep; lay precast concrete kerb units bedded in semi-dry concrete; slump 35 mm maximum; haunching one side; disposal of arisings off site		
Conservation kerbs laid straight		
225 mm wide × 150 mm high;	m	50.00
155 mm wide × 255 mm high	m	49.00
Conservation kerbs laid to radius; 145 mm × 255 mm		
radius 3.25 m	m	61.00
radius 6.50 m	m	55.00
radius 9.80 m	m	55.00
solid quadrant; 305 × 305 mm	nr	58.00
STONE KERBS		
Excavate and construct concrete foundation 450 mm wide × 150 mm deep; lay kerb units bedded in semi-dry concrete; slump 35 mm maximum; haunching one side;		
Granite kerbs		
straight; 125 × 250 mm	m	57.00
curved; 125 × 250 mm	m	69.00
Precast concrete edging on concrete foundation 100 × 150 mm deep and haunching one side 1:2:4 including all necessary excavation disposal and formwork		
Rectangular, chamfered or bullnosed; to one side of straight path		
50 × 150 mm	m	24.00
50 × 200 mm	m	30.00
50 × 250 mm	m	30.00
Rectangular, chamfered or bullnosed as above but to both sides of straight paths		
50 × 150 mm	m	52.00
50 × 200 mm	m	60.00
50 × 250 mm	m	61.00
Timber edgings; softwood		
Straight		
150 × 38 mm	m	6.80
150 × 50 mm	m	8.60
Curved; over 5 m radius		
150 × 38 mm	m	8.50
150 × 50 mm	m	11.00
Curved; 4–5 m radius		
150 × 38 mm	m	9.20
150 × 50 mm	m	11.70

8.2.1 ROADS PATHS PAVINGS AND SURFACINGS

Item Excluding site overheads and profit	Unit	Total rate £
Excavation kerbs and edgings to paved surfaces – cont		
Timber edgings – cont		
Curved; 3–4 m radius		
150 × 38 mm	m	10.20
150 × 50 mm	m	13.10
Curved; 1–3 m radius		
150 × 38 mm	m	12.00
150 × 50 mm	m	14.80
Brick or concrete block edge restraint; excavate for foundation; lay concrete 1:2:4 250 mm wide × 150 mm deep; on 50 mm thick sharp sand bed; lay blocks or bricks		
Blocks 200 × 100 × 60 mm; PC £8.17/m^2; butt jointed		
header course	m	28.00
stretcher course	m	25.00
Bricks 215 × 112.5 × 50 mm; PC £300.00/1000; with mortar joints		
header course	m	29.50
header course; brick on edge	m	38.00
stretcher course; flat or brick on edge	m	29.00
stretcher course; 2 rows brick on edge	m	40.00
Bricks 215 × 112.5 × 50 mm; PC £600.00/1000; with mortar joints		
header course	m	32.50
header course; brick on edge	m	41.00
stretcher course; flat or on edge	m	23.00
stretcher course; 2 rows brick on edge	m	59.00
STONE EDGINGS		
Sawn yorkstone edgings; excavate for foundations; lay concrete 1:2:4 150 mm deep × 33.3% wider than the edging; on 35 mm thick mortar bed		
Yorkstone 50 mm thick		
100 mm wide × random lengths	m	42.00
100 × 100 mm	m	40.00
100 mm wide × 200 mm long	m	48.00
250 mm wide × random lengths	m	47.50
500 mm wide × random lengths	m	74.00
Granite edgings; excavate for footings; lay concrete 1:2:4 150 mm deep × 33.3% wider than the edging; on 35 mm thick mortar bed;		
Granite 50 mm thick		
100 mm wide × 200 mm long	m	43.50
100 mm wide × random lengths	m	45.50
250 mm wide × random lengths	m	52.00
setts; 100 × 100 mm	m	56.00
300 mm wide × random lengths	m	57.00

8.2.1 ROADS PATHS PAVINGS AND SURFACINGS

Item Excluding site overheads and profit	Unit	Total rate £
Bases for paving		
Excavate ground and reduce levels to receive 38 mm thick slab and 25 mm mortar bed; dispose of excavated material off site		
Lay granular fill Type 1; limestone aggregate 150 mm thick laid to falls and compacted		
all by machine	m²	13.90
all by hand except disposal by grab	m²	47.00
Lay granular fill Type 1; recycled material; 150 mm thick laid to falls and compacted		
all by machine	m²	11.80
all by hand except disposal by grab	m²	45.00
Lay 1:2:4 concrete base 150 mm thick laid to falls		
all by machine	m²	21.00
all by hand except disposal by grab	m²	58.00
150 mm hardcore base with concrete base 150 mm deep		
concrete 1:2:4 site mixed	m²	32.00
concrete PAV 1 35 ready mixed	m²	33.00
Hardcore base 150 mm deep; concrete reinforced with A142 mesh		
site mixed concrete 1:2:4; 150 mm deep	m²	49.00
site mixed concrete 1:2:4; 250 mm deep	m²	67.00
concrete PAV 1 35 N/mm² ready mixed; 150 mm deep	m²	39.00
concrete PAV 1 35 N/mm² ready mixed; 250 mm deep	m²	56.00
Channels		
Excavate and construct concrete foundation 450 mm wide × 150 mm deep; lay precast concrete kerb units bedded in semi-dry concrete; slump 35 mm maximum; haunching one side; lay channel units bedded in 1:3 sand mortar; jointed in cement: sand mortar 1:3		
Kerbs; 125 mm high × 125 mm thick; bullnosed; type BN		
dished channel; 125 × 225 mm; Ref CS	m	49.50
square channel; 125 × 150 mm; Ref CS2	m	47.00
bullnosed channel; 305 × 150 mm; Ref CBN	m	58.00
Brick channel; class B engineering bricks		
Excavate and construct concrete foundation 600 mm wide × 200 mm deep; lay channel to depths and falls bricks to be laid as headers along the channel; bedded in 1:3 sand mortar bricks close jointed in cement: sand mortar 1:3		
3 courses wide	m	64.00
3 coursed granite setts; dished		
Excavate and construct concrete foundation 400 mm wide × 200 mm deep; lay channel to depths and falls ; bedded in 1:3: sand mortar		
3 courses wide	m	67.00

8.2.1 ROADS PATHS PAVINGS AND SURFACINGS

Item Excluding site overheads and profit	Unit	Total rate £
CONCRETE ROADS		
Excavate weak points of excavation by hand and fill with well rammed Type 1 granular material		
100 mm thick	m²	11.00
200 mm thick	m²	13.90
Reinforced concrete roadbed; excavation and disposal off site; reinforcing mesh A142; expansion joints at 50 m centres; on 150 mm hardcore blinded sand; kerbs 155 × 255 mm to both sides; haunched one side.		
Concrete 150 mm thick		
4.00 m wide	m	195.00
5.00 m wide	m	220.00
6.00 m wide	m	300.00
7.00 m wide	m	315.00
Concrete 250 mm thick		
4.00 m wide	m	255.00
5.00 m wide	m	295.00
6.00 m wide	m	350.00
7.00 m wide	m	400.00
FOOTPATHS		
Note: For pathway surfaces; select appropriate finish from the Measured Works section		
Excavate 250 mm; disposal off site; grade to levels and falls; lay 150 mm Type 1 granular material; all disposal off site; Lay or fix edge as described		
Lay Permaloc Asphalt edge 51 mm metal edgings; path width		
1.50 m wide	m	44.00
3.00 m wide	m	63.00
Lay timber edging 150 × 38 mm; path width		
1.50 m wide	m	28.00
3.00 m wide	m	46.50
Lay precast edgings kerbs 50 × 150 mm on both sides; including foundations haunched one side in 11.5 N/mm² concrete with all necessary formwork; path width		
1.50 m wide	m	69.00
3.00 m wide	m	87.00
Lay concrete edging Keyblock header edging ; path width		
1.50 m wide	m	43.50
3.00 m wide	m	62.00
Lay concrete edging Keyblock stretcher edging ; path width		
1.50 m wide	m	46.00
3.00 m wide	m	79.00
Lay brick on flat header edging; brick £300.00/1000; path width		
1.50 m wide	m	70.00
3.00 m wide	m	88.00

8.2.1 ROADS PATHS PAVINGS AND SURFACINGS

Item Excluding site overheads and profit	Unit	Total rate £
Lay brick on flat header edging; brick £600.00/1000; path width		
1.50 m wide	m	76.00
3.00 m wide	m	94.00
Lay brick on edge header edging; brick £300.00/1000; path width		
1.50 m wide	m	83.00
3.00 m wide	m	100.00
Lay brick on edge header edging; brick £600.00/1000; path width		
1.50 m wide	m	91.00
3.00 m wide	m	110.00
Lay brick on flat or brick on edge stretcher edging; brick £300.00/1000; path width		
1.50 m wide	m	60.00
3.00 m wide	m	78.00
Lay brick on edge strecher edging; brick £600.00/1000; path width		
1.50 m wide	m	63.00
3.00 m wide	m	81.00
Lay new granite setts 100 × 100 × 100 mm single row; path width		
1.50 m wide	m	110.00
3.00 m wide	m	125.00
Lay new granite setts 200 × 100 × 100 mm single row; path width		
1.50 m wide	m	86.00
3.00 m wide	m	100.00
BRICK PAVING		
WORKS BY MACHINE		
Excavate and lay base Type 1 150 mm thick remove arisings; all by machine; lay clay brick paving		
200 × 100 × 50 mm thick; butt jointed on 50 mm sharp sand bed		
PC £300.00/1000	m²	65.00
PC £600.00/1000	m²	80.00
200 × 100 × 50 mm thick; 10 mm mortar joints on 35 mm mortar bed		
PC £300.00/1000	m²	81.00
PC £600.00/1000	m²	94.00
Excavate and lay base Type 1 250 mm thick all by machine; remove arisings; lay clay brick paving		
200 × 100 × 50 mm thick; 10 mm mortar joints on 35 mm mortar bed		
PC £300.00/1000	m²	89.00
Excavate and lay 150 mm readymix concrete base reinforced with A393 mesh; all by machine; remove arisings; lay clay brick paving		
200 × 100 × 50 mm thick; 10 mm mortar joints on 35 mm mortar bed; running or stretcher bond		
PC £300.00/1000	m²	98.00
PC £600.00/1000	m²	110.00
200 × 100 × 50 mm thick; 10 mm mortar joints on 35 mm mortar bed; butt jointed; herringbone bond		
PC £300.00/1000	m²	82.00
PC £600.00/1000	m²	97.00

8.2.1 ROADS PATHS PAVINGS AND SURFACINGS

Item Excluding site overheads and profit	Unit	Total rate £
BRICK PAVING – cont		
Excavate and lay base readymix concrete base 150 mm thick reinforced with A393 mesh; all by machine; remove arisings; lay clay brick paving 215 × 102.5 × 50 mm thick; 10 mm mortar joints on 35 mm mortar bed		
PC £300.00/1000; herringbone	m²	100.00
PC £600.00/1000; herringbone	m²	115.00
PATHS		
Brick paved paths; bricks 215 × 112.5 × 65 mm with mortar joints; prices are inclusive of excavation, disposal off site, 100 mm hardcore, 100 mm 1:2:4 concrete bed; edgings of brick 215 mm wide, haunched; all jointed in cement: lime: sand mortar (1:1:6)		
Path 1015 mm wide laid stretcher bond; edging course of headers		
rough stocks PC £450.00/1000	m	90.00
engineering brick	m	85.00
Path 1015 mm wide laid stack bond		
rough stocks PC £450.00/1000	m	95.00
engineering brick	m	90.00
Path 1115 mm wide laid header bond		
rough stocks PC £450.00/1000	m	99.00
engineering brick	m	94.00
Path 1330 mm wide laid basketweave bond		
rough stocks PC £450.00/1000	m	115.00
engineering brick	m	110.00
Path 1790 mm wide laid basketweave bond		
rough stocks PC £450.00/1000	m	150.00
engineering brick	m	110.00
Brick paved paths; brick paviors 200 × 100 × 50 mm chamfered edge with butt joints; prices are inclusive of excavation, 100 mm Type 1 and 50 mm sharp sand; jointing in kiln dried sand brushed in; exclusive of edge restraints		
Path 1000 mm wide laid stretcher bond; edging course of headers		
rough stocks PC £450.00/1000	m	80.00
engineering brick	m	86.00
Path 1330 mm wide laid basketweave bond		
rough stocks PC £450.00/1000	m	115.00
engineering brick	m	110.00
Path 1790 mm wide laid basketweave bond		
rough stocks PC £450.00/1000	m	150.00
engineering brick	m	110.00

8.2.1 ROADS PATHS PAVINGS AND SURFACINGS

Item Excluding site overheads and profit	Unit	Total rate £
WORKS BY HAND		
Excavate and lay base Type 1 150 mm thick by hand; arisings barrowed to spoil heap maximum distance 25 m and removal off site by grab; lay clay brick paving 200 × 100 × 50 mm thick; butt jointed on 50 mm sharp sand bed		
PC £300.00/1000	m²	99.00
PC £600.00/1000	m²	130.00
Excavate and lay 150 mm concrete base; 1:3:6: site mixed concrete reinforced with A393 mesh; remove arisings to stockpile and then off site by grab; lay clay brick paving 215 × 102.5 × 50 mm thick; 10 mm mortar joints on 35 mm mortar bed		
PC £300.00/1000	m²	98.00
PC £600.00/1000	m²	130.00
BLOCK PAVING		
CONCRETE SETT PAVING		
Excavate ground and bring to levels; supply and lay 150 mm granular fill Type 1 laid to falls and compacted; supply and lay setts bedded in 50 mm sand and vibrated; joints filled with dry sand and vibrated; all excluding edgings or kerbs measured separately Marshalls Mono; Tegula precast concrete setts		
random sizes 60 mm thick	m²	51.00
single size 60 mm thick	m²	48.50
random size 80 mm thick	m²	55.00
single size 80 mm thick	m²	68.00
CAR PARKS		
Excavate 450 mm and prepare base of 100 hardcore and 250 mm Type 1; Work to falls, crossfalls and cambers not exceeding 15° mark out car parking bays 5.0 × 2.4 m with thermoplastic road paint; surfaces all mechanically laid; edgings not included Marshalls Keyblock concrete blocks 200 × 100 × 80 mm; grey		
per bay; 5.0 × 2.4 m	each	660.00
gangway	m²	57.00
Marshalls Keyblock concrete blocks 200 × 100 × 80 mm; colours		
per bay; 5.0 × 2.4 m	each	680.00
gangway	m²	59.00
Marshalls Keyblock concrete blocks 200 × 100 × 60 mm; grey		
per bay; 5.0 × 2.4 m	each	650.00
gangway	m²	55.00
Marshalls Keyblock concrete blocks 200 × 100 × × 60 mm; colours		
per bay; 5.0 × 2.4 m	each	640.00
gangway	m²	56.00

8.2.1 ROADS PATHS PAVINGS AND SURFACINGS

Item Excluding site overheads and profit	Unit	Total rate £
BLOCK PAVING – cont		
Excavate 450 mm and prepare base of 100 hardcore and 250 mm Type 1 – cont		
Marshalls Tegula concrete blocks 200 × 100 × 80 mm; random sizes		
per bay; 5.0 × 2.4 m	each	890.00
gangway	m²	76.00
Marshalls Tegula concrete blocks 200 × 100 × 80 mm; single size		
per bay; 5.0 × 2.4 m	each	860.00
gangway	m²	74.00
BLOCK PAVING EDGINGS		
Edge restraint to block paving; excavate for groundbeam; lay concrete 1:2:4 200 mm wide × 150 mm deep; on 50 mm thick sharp sand bed; inclusive of haunching one side		
Blocks 200 × 100 × 60 mm; PC £8.17/m²; butt jointed		
header course	m	15.60
stretcher course	m	12.20
Bricks 215 × 112.5 × 50 mm; PC £300.00/1000; with mortar joints		
header course	m	17.00
stretcher course	m	16.10
Excavate ground; supply and lay granular fill Type 1 150 mm thick laid to falls and compacted; supply and lay block pavers; laid on 50 mm compacted sharp sand; vibrated; joints filled with loose sand excluding edgings or kerbs measured separately		
Concrete blocks		
200 × 100 × 60 mm	m²	34.00
200 × 100 × 80 mm	m²	37.00
Reduce levels; lay 150 mm granular material Type 1; lay vehicular block paving to 90 degree herringbone pattern; on 50 mm compacted sand bed; vibrated; jointed in sand and vibrated; excavate and lay precast concrete edging 50 × 150 mm; on concrete foundation 1:2:4		
Blocks 200 × 100 × 60 mm		
1.0 m wide clear width between edgings	m	83.00
1.5 m wide clear width between edgings	m	100.00
2.0 m wide clear width between edgings	m	120.00
Blocks 200 × 100 × 60 mm but blocks laid 45 degree herringbone pattern including cutting edging blocks		
1.0 m wide clear width between edgings	m	89.00
1.5 m wide clear width between edgings	m	105.00
2.0 m wide clear width between edgings	m	120.00
Blocks 200 × 100 × 60 mm but all excavation by hand disposal off site by grab		
1.0 m wide clear width between edgings	m	195.00
1.5 m wide clear width between edgings	m	220.00
2.0 m wide clear width between edgings	m	250.00

8.2.1 ROADS PATHS PAVINGS AND SURFACINGS

Item Excluding site overheads and profit	Unit	Total rate £
MACADAM SURFACES		
Macadam roadway over 1000 m²		
Excavate 350 mm for pathways or roadbed; bring to grade; lay 100 mm well-rolled hardcore; lay 150 mm Type 1 granular material; lay precast edgings kerbs 50 × 150 mm on both sides; including foundations haunched one side in 11.5 N/mm² concrete with all necessary formwork; machine lay surface of 90 mm macadam 60 mm base course and 30 mm wearing course; all disposal off site		
1.50 m wide	m	115.00
2.00 m wide	m	120.00
3.00 m wide	m	180.00
4.00 m wide	m	220.00
5.00 m wide	m	260.00
6.00 m wide	m	305.00
7.00 m wide	m	350.00
Macadam roadway 400–1000 m²		
Excavate 350 mm for pathways or roadbed; bring to grade; lay 100 mm well-rolled hardcore; lay 150 mm Type 1 granular material; lay precast edgings kerbs 50 × 150 mm on both sides; including foundations haunched one side in 11.5 N/mm² concrete with all necessary formwork; machine lay surface of 90 mm macadam 60 mm base course and 30 mm wearing course; all disposal off site		
1.50 m wide	m	120.00
2.00 m wide	m	145.00
3.00 m wide	m	190.00
4.00 m wide	m	240.00
5.00 m wide	m	285.00
6.00 m wide	m	330.00
7.00 m wide	m	380.00
MACADAM FOOTPATH		
Excavate footpath to reduce level; remove arisings to tip on site maximum distance 25 m; lay Type 1 granular fill 100 mm thick; lay base course of 28 mm size dense bitumen macadam 50 mm thick; wearing course of 10 mm size dense bitumen macadam 20 mm thick; timber edge 150 × 38 mm		
Areas over 1000 m²		
1.0 m wide	m	39.00
1.5 m wide	m	53.00
2.0 m wide	m	68.00
Areas 400–1000 m²		
1.0 m wide	m	43.00
1.5 m wide	m	60.00
2.0 m wide	m	95.00

8.2.1 ROADS PATHS PAVINGS AND SURFACINGS

Item Excluding site overheads and profit	Unit	Total rate £
MACADAM SURFACES – cont		
CAR PARKS		
Excavate 450 mm and prepare base of 100 hardcore and 250 mm Type 1; Work to falls, crossfalls and cambers not exceeding 15° mark out car parking bays 5.0 × 2.4 m with thermoplastic road paint; surfaces all mechanically laid; edgings not included		
Macadam base course 60 mm thick and wearing course 40 mm thick		
per bay; 5.0 × 2.4 m	each	690.00
Gangway	m²	58.00
Bay for disabled people; Macadam base course 60 mm thick and wearing course 40 mm thick		
per bay; 5.80 × 3.25 ambulant	each	1050.00
per bay; 6.55 × 3.80 wheelchair	each	1375.00
IN SITU CONCRETE PAVINGS AND SURFACES		
CONCRETE PAVING		
Pedestrian areas		
Excavate to reduce levels; lay 100 mm Type 1; lay PAV 1 air entrained concrete; joints at maximum width of 6.0 m cut out and sealed with sealant; inclusive of all formwork and stripping		
100 mm thick	m²	48.00
Trafficked areas		
Excavate to reduce levels; lay 150 mm Type 1 lay PAV 2 40 N/mm² air entrained concrete; reinforced with steel mesh 200 × 200 mm square at 2.22 kg/m²; joints at max width of 6.0 m cut out and sealed with sealant		
150 mm thick	m²	66.00
Reinforced concrete paths; excavate path and dispose of excavated material off site; bring to grade; lay reinforcing mesh A142 lapped and joined; lay in situ reinforced concrete PAV1 35 N/mm²; with 15 impregnated fibreboard expansion joints sealant at 50 m centres; on 100 mm hardcore blinded sand; edgings 150 × 50 mm to both sides; including foundations haunched one side in 11.5 N/mm² concrete with all necessary formwork		
Concrete 150 mm thick		
1.00 m wide	m	61.00
2.00 m wide	m	97.00
3.00 m wide	m	130.00

8.2.1 ROADS PATHS PAVINGS AND SURFACINGS

Item	Unit	Total rate £
Excluding site overheads and profit		
WORKS BY HAND; Ready mixed concrete – mixed on site		
Reinforced concrete paths; excavate path by hand barrow to spoil heap 25 m and dispose by grab offsite; bring to grade; lay reinforcing mesh A142 lapped and joined; lay in situ reinforced ready mixed concrete mixed on site (1:2:4:); with 15 impregnated fibreboard expansion joints sealant at 50 m centres; on 100 mm hardcore blinded sand; edgings 150 × 50 mm to both sides; including foundations haunched one side in 11.5 N/mm² concrete with all necessary formwork		
Concrete 150 mm thick		
1.00 m wide	m	87.00
2.00 m wide	m	160.00
3.00 m wide	m	225.00
Grass concrete		
IN SITU GRASS CONCRETE TO CAR PARKS		
Grass Concrete Grasscrete in situ continuously reinforced cellular surfacing; fill with topsoil and peat (5:1) and fertilizer at 35 g/m²; seed with dwarf rye grass at 35 g/m²; Excludes marking		
GC1; 100 mm thick for cars and light traffic; per bay; 5.0 × 2.4 m	each	840.00
GC1; 100 mm thick for cars and light traffic; gangway	m²	69.00
GC2; 150 mm thick for HGV traffic including dust carts		
per bay; 5.0 × 2.4 m	each	990.00
gangway	m²	82.00
Car park as above but with interlocking concrete blocks 200 × 100 × 60 mm; grey; but mark out bays		
per bay; 5.80 × 3.25 ambulant	each	360.00
per bay; 6.55 × 3.80 wheelchair	each	480.00
gangway	m²	19.20
MODULAR GRASS CONCRETE PAVING		
Reduce levels; lay 150 mm Type 1 granular fill compacted; supply and lay precast grass concrete blocks on 20 sand and level by hand; fill blocks with 3 mm sifted topsoil and pre-seeding fertilizer at 50 g/m²; sow with perennial ryegrass/ chewings fescue seed at 35 g/m²		
Grass concrete		
GB103 406 × 406 × 103 mm	m²	44.50
GB83 406 × 406 × 83 mm	m²	43.00
extra for geotextile fabric underlayer	m²	0.75
Firepaths		
Excavate to reduce levels; lay 300 mm well rammed hardcore; blinded with 100 mm type 1; supply and lay Marshalls Mono Grassguard 180 precast grass concrete blocks on 50 mm sand and level by hand; fill blocks with sifted topsoil and pre-seeding fertilizer at 50 g/m²		
firepath 3.8 m wide	m	275.00
firepath 4.4 m wide	m	320.00
firepath 5.0 m wide	m	365.00
turning areas	m²	73.00
Charcon Hard Landscaping Grassgrid; 366 × 274 × 100 mm	m²	37.00

8.2.1 ROADS PATHS PAVINGS AND SURFACINGS

Item Excluding site overheads and profit	Unit	Total rate £
Resin bonded and Resin bound surfaces		
Resin bonded surfaces; excavate 230 mm; bring to grade; lay 150 mm Type 1 granular material; hand lay base course of 60 mm macadam and resin bonded wearing course; all disposal off site; edgings excluded		
Natratex resin bonded macadam wearing course		
general paved area	m²	51.00
pathway 1.50 m wide	m	80.00
pathway 3.00 m wide	m	160.00
Addagrip; resin bonded aggregate; 1–3 mm golden pea gravel with buff adhesive		
general paved area	m²	58.00
pathway 1.50 m wide	m	91.00
pathway 3.00 m wide	m	180.00
Addagrip; resin bonded aggregate; 6 mm chocolate gravel 18 mm depth		
general paved area	m²	90.00
pathway 1.50 m wide	m	140.00
pathway 3.00 m wide	m	280.00
Addagrip; resin bonded aggregate; Cobalt blue recycled glass 15 mm depth		
general paved area	m²	110.00
pathway 1.50 m wide	m	170.00
pathway 3.00 m wide	m	340.00
Resin bound surfaces; excavate 230 mm; bring to grade; lay 150 mm Type 1 granular material;hand lay base course of 60 mm macadam and resin bound wearing course; all disposal off site; edgings excluded		
Sureset; resin bound permeable surfacing; natural gravel 6 mm aggregate 18 mm thick		
paved area	m²	79.00
pathway 1.50 m wide	m	120.00
pathway 3.00 m wide	m	240.00
Sureset; resin bound permeable surfacing; crushed rock 6 mm aggregate 18 mm thick		
paved area	m²	99.00
pathway 1.50 m wide	m	125.00
pathway 3.00 m wide	m	410.00
Sureset; resin bound permeable surfacing; marble 6 mm aggregate 18 mm thick		
paved area	m²	84.00
pathway 1.50 m wide	m	130.00
pathway 3.00 m wide	m	260.00
Sureset; resin bound permeable surfacing; recycled glass 6 mm aggregate 18 mm thick		
paved area	m	96.00
pathway 1.50 m wide	m	150.00
pathway 3.00 m wide	m	295.00

8.2.1 ROADS PATHS PAVINGS AND SURFACINGS

Item Excluding site overheads and profit	Unit	Total rate £
GRAVEL PATHS		
Reduce levels and remove spoil to dump on site; lay 150 mm hardcore well rolled; lay geofabric; fix timber edge 150 × 38 mm to both sides of straight paths		
Lay Cedec gravel 50 mm thick; watered and rolled		
1.0 m wide	m	32.50
1.5 m wide	m	43.00
2.0 m wide	m	59.00
Lay Breedon gravel 50 mm thick; watered and rolled		
1.0 m wide	m	30.50
1.5 m wide	m	41.00
2.0 m wide	m	52.00
Reduce levels and remove spoil to dump on site; lay 150 mm Type 1 well rolled; lay geofabric; fix timber edge 150 × 38 mm to both sides of straight paths		
Lay Cedec gravel 50 mm thick; watered and rolled		
1.0 m wide	m	36.00
1.5 m wide	m	48.00
2.0 m wide	m	66.00
Lay Breedon gravel 50 mm thick; watered and rolled		
1.0 m wide	m	34.00
1.5 m wide	m	46.00
2.0 m wide	m	58.00
Lay 20 mm shingle; 30 mm thick		
1.0 m wide	m	23.00
1.5 m wide	m	29.50
2.0 m wide	m	41.00
Cellular confined surfaces		
Excavate ground and reduce levels to receive cellular confined paving; dispose of excavated material off site		
Lay granular fill Type 1; limestone aggregate 150 mm thick laid to falls and compacted; lay confined cellular surface;		
Cellweb 100 thick	m²	37.50
Nidagravel honecomb cellular gravel grid; 25 mm deep for foot traffic and bicycles filled with shingle	m²	34.00
Nidagravel honecomb cellular gravel grid; 25 mm deep for foot traffic and bicycles filled with decorative aggregate	m²	35.00
Nidagravel honecomb cellular gravel grid 40 mm deep for vehicular traffic; filled with shingle	m²	38.00
Nidagravel honecomb cellular gravel grid 40 mm deep for vehicular traffic; filled with decorative aggregate	m²	39.00

8.2.1 ROADS PATHS PAVINGS AND SURFACINGS

Item Excluding site overheads and profit	Unit	Total rate £
SLAB PAVINGS		
SLAB/BRICK PATHS		
Stepping stone path inclusive of hand excavation and mechanical disposal, 100 mm Type 1 and sand blinding; slabs laid 100 mm apart to existing turf		
600 mm wide; 600 × 600 × 50 mm slabs		
natural finish	m	51.00
coloured	m	51.00
exposed aggregate	m	62.00
900 mm wide; 600 × 900 × 50 mm slabs		
natural finish	m	56.00
coloured	m	58.00
exposed aggregate	m	74.00
Pathway inclusive of mechanical excavation and disposal; 100 mm Type 1 and sand blinding; slabs close butted		
Straight butted path 900 mm wide; 600 × 900 × 50 mm slabs		
natural finish	m	23.00
coloured	m	24.00
exposed aggregate	m	35.00
Straight butted path 1200 mm wide; double row; 600 × 900 × 50 mm slabs; laid stretcher bond		
natural finish	m	40.50
coloured	m	43.00
exposed aggregate	m	67.00
Straight butted path 1200 mm wide; one row of 600 × 600 × 50 mm; one row 600 × 900 × 50 mm slabs		
natural finish	m	39.00
coloured	m	41.50
exposed aggregate	m	74.00
Straight butted path 1500 mm wide; slabs of 600 × 900 × 50 mm and 600 × 600 × 50 mm; laid to bond		
natural finish	m	48.50
coloured	m	51.00
exposed aggregate	m	74.00
Straight butted path 1800 mm wide; two rows of 600 × 900 × 50 mm slabs; laid bonded		
natural finish	m	54.00
coloured	m	64.00
exposed aggregate	m	100.00
Straight butted path 1800 mm wide; three rows of 600 × 900 × 50 mm slabs; laid stretcher bond		
natural finish	m	59.00
coloured	m	69.00
exposed aggregate	m	105.00

8.2.1 ROADS PATHS PAVINGS AND SURFACINGS

Item Excluding site overheads and profit	Unit	Total rate £
FLAG PAVING TO PEDESTRIAN AREAS		
Prices are inclusive of all mechanical excavation and disposal.		
Supply and lay precast concrete flags; excavate ground and reduce levels; treat substrate with total herbicide; supply and lay granular fill Type 1 150 mm thick laid to falls and compacted		
Standard precast concrete flags bedded and jointed in lime: sand mortar (1:3)		
450 × 450 × 70 mm; chamfered	m²	46.50
450 × 450 × 50 mm; chamfered	m²	42.00
600 × 300 × 50 mm	m²	40.50
600 × 450 × 50 mm	m²	40.00
600 × 600 × 50 mm	m²	37.00
750 × 600 × 50 mm	m²	36.00
900 × 600 × 50 mm	m²	35.00
Coloured flags bedded and jointed in lime: sand mortar (1:3)		
600 × 600 × 50 mm	m²	39.00
450 × 450 × 70 mm; chamfered	m²	53.00
400 × 400 × 65 mm	m²	53.00
750 × 600 × 50 mm	m²	38.50
900 × 600 × 50 mm	m²	37.00
Marshalls Saxon; textured concrete flags; reconstituted yorkstone in colours; butt jointed bedded in lime: sand mortar (1:3)		
300 × 300 × 35 mm	m²	73.00
450 × 450 × 50 mm	m²	62.00
600 × 300 × 35 mm	m²	57.00
600 × 600 × 50 mm	m²	54.00
Tactile flags; Marshalls blister tactile pavings; red or buff; for blind pedestrian guidance laid to designed pattern		
450 × 450 mm	m²	53.00
400 × 400 mm	m²	60.00
PEDESTRIAN DETERRENT PAVING		
Excavate ground and bring to levels; treat substrate with total herbicide; supply and lay granular fill Type 1 150 mm thick laid to falls and compacted; supply and lay precast deterrent paving units bedded in lime: sand mortar (1:3) and jointed in lime: sand mortar (1:3)		
Marshalls Mono		
Lambeth pyramidal paving; 600 × 600 × 75 mm	m²	44.00
Townscape		
Abbey square cobble pattern pavings; reinforced; 600 × 600 × 60 mm	m²	48.50
Geoset chamfered studs; 600 × 600 × 60 mm	m²	43.50

8.2.1 ROADS PATHS PAVINGS AND SURFACINGS

Item Excluding site overheads and profit	Unit	Total rate £
SLAB PAVINGS – cont		
IMITATION YORKSTONE PAVING		
Excavate ground and bring to levels; treat substrate with total herbicide; supply and lay granular fill Type 1 150 mm thick laid to falls and compacted; supply and lay imitation yorkstone paving laid to coursed patterns bedded in lime: sand mortar (1:3) and jointed in lime: sand mortar (1:3)		
Marshalls Heritage; square or rectangular		
300 × 300 × 38 mm	m²	76.00
600 × 300 × 30 mm	m²	62.00
600 × 450 × 38 mm	m²	59.00
450 × 450 × 38 mm	m²	56.00
600 × 600 × 38 mm	m²	51.00
As above but laid to random rectangular patterns		
various sizes selected from the above	m²	65.00
Imitation yorkstone laid random rectangular as above but on concrete base 150 mm thick		
by machine	m²	73.00
by hand	m²	100.00
BARK PAVINGS		
Excavate to reduce levels; remove all topsoil to dump on site; treat area with herbicide; lay 150 mm clean hardcore; blind with sand; lay 0.7 mm geotextile filter fabric water flow 50 l/m²/sec; supply and fix treated softwood edging boards 50 × 150 mm to bark area on hardcore base extended 150 mm beyond the bark area; boards fixed with galvanized nails to treated softwood posts 750 × 50 × 75 mm driven into firm ground at 1.0 m centres; edging boards to finish 25 mm above finished bark surface; tops of posts to be flush with edging boards and once weathered		
Supply and lay 100 mm Melcourt conifer walk chips 10–40 mm size		
1.00 m wide	m	28.00
2.00 m wide	m	43.00
3.00 m wide	m	58.00
4.00 m wide	m	70.00
Extra over for wood fibre	m²	−1.55
Extra over for hardwood chips	m²	−1.36
COBBLE PAVINGS		
BEACH COBBLE PAVING		
Excavate ground and bring to levels and fill with compacted Type 1 fill 100 mm thick; lay GEN 1 concrete base 100 mm thick; supply and lay cobbles individually laid by hand bedded in cement: sand mortar (1:3) 25 mm thick minimum; dry grout with 1:3 cement: sand grout; brush off surplus grout and water in; sponge off cobbles as work proceeds; all excluding formwork edgings or kerbs measured separately		
Scottish beach cobbles 50–75 mm	m²	120.00
Scottish beach cobbles 100–200 mm	m²	145.00
Kidney flint cobbles 75–100 mm	m²	110.00

8.2.1 ROADS PATHS PAVINGS AND SURFACINGS

Item Excluding site overheads and profit	Unit	Total rate £
NATURAL STONE PAVING		
GRANITE SETT PAVING – PEDESTRIAN		
Excavate and grade; 100 mm hardcore ; Type 1 50 mm thick; 100 mm concrete 1:2:4; supply and lay granite setts 100 × 100 × 100 mm bedded in cement: sand mortar (1:3) 25 mm thick minimum; jointed in mortar 1:3		
Setts laid to bonded pattern and jointed; bedded in cement: sand mortar 1:3 25 mm thick		
new setts 100 × 100 × 100 mm	m²	110.00
second-hand cleaned 100 × 100 × 100 mm	m²	115.00
Setts laid to bonded pattern and jointed; bedded in SteinTec mortar 25 mm thick		
new setts 100 × 100 × 100 mm	m²	110.00
second-hand cleaned 100 × 100 × 100 mm	m²	115.00
Setts laid in curved pattern		
new setts 100 × 100 × 100 mm	m²	120.00
second-hand cleaned 100 × 100 × 100 mm	m²	120.00
GRANITE SETT PAVING – TRAFFICKED AREAS		
Excavate to reduce levels levels; lay 100 mm hardcore; compact; Type 150 mm thick; lay 150 mm site mixed concrete 1:2:4 reinforced with steel fabric; ; granite setts 100 × 100 × 100 mm bedded in cement: sand mortar (1:3) 25 mm thick minimum;jointed in mortar 1:3; all excluding edgings or kerbs measured separately		
Site mixed concrete		
new setts 100 × 100 × 100 mm	m²	120.00
second-hand cleaned 100 × 100 × 100 mm	m²	120.00
Ready mixed concrete		
new setts 100 × 100 × 100 mm	m²	120.00
second-hand cleaned 100 × 100 × 100 mm	m²	125.00
NATURAL STONE SLAB PAVING		
WORKS BY MACHINE		
Excavate ground by machine and reduce levels; to receive 65 mm thick slab and 35 mm mortar bed; dispose of excavated material off site; treat substrate with total herbicide; lay granular fill Type 1 150 mm thick laid to falls and compacted; lay to random rectangular pattern on 35 mm mortar bed		
New riven slabs		
laid random rectangular on 35 mm 1:3 mortar bed	m²	135.00
laid random rectangular on 30 mm thick Steintec mortar bed	m²	140.00
New riven slabs; but to 150 mm plain concrete base		
laid random rectangular on 35 mm 1:3 mortar bed	m²	140.00
laid random rectangular on 30 mm thick Steintec mortar bed	m²	150.00
New riven slabs; disposal by grab		
laid random rectangular	m²	140.00

8.2.1 ROADS PATHS PAVINGS AND SURFACINGS

Item Excluding site overheads and profit	Unit	Total rate £
NATURAL STONE PAVING – cont		
Excavate ground by machine and reduce levels – cont		
Reclaimed Cathedral grade riven slabs		
laid random rectangular	m²	180.00
Reclaimed Cathedral grade riven slabs; but to 150 mm plain concrete base		
laid random rectangular	m²	180.00
Reclaimed Cathedral grade riven slabs; disposal by grab		
laid random rectangular	m²	180.00
New slabs sawn 6 sides		
laid random rectangular	m²	110.00
three sizes; laid to coursed pattern	m²	115.00
New slabs sawn 6 sides; but to 150 mm plain concrete base		
laid random rectangular	m²	120.00
three sizes; laid to coursed pattern	m²	120.00
New slabs sawn 6 sides; disposal by grab		
laid random rectangular	m²	110.00
3 sizes, laid to coursed pattern	m²	120.00
WORKS BY HAND		
Excavate ground by hand and reduce levels, to receive 65 mm thick slab and 35 mm mortar bed; barrow all materials and arisings 25 m; dispose of excavated material off site by grab; treat substrate with total herbicide; lay granular fill Type 1 150 thick laid to falls and compacted; lay to random rectangular pattern on 35 mm mortar bed		
New riven slabs		
laid random rectangular	m²	170.00
New riven slabs laid random rectangular; but to 150 mm plain concrete base		
laid random rectangular	m²	180.00
New riven slabs; but disposal to skip		
laid random rectangular	m²	190.00
Reclaimed Cathedral grade riven slabs		
laid random rectangular	m²	210.00
Reclaimed Cathedral grade riven slabs; but to 150 mm plain concrete base		
laid random rectangular	m²	225.00
Reclaimed Cathedral grade riven slabs; but disposal to skip		
laid random rectangular	m²	230.00
New slabs sawn 6 sides		
laid random rectangular	m²	145.00
three sizes; sawn 6 sides laid to coursed pattern	m²	150.00
New slabs sawn 6 sides; but to 150 mm plain concrete base		
laid random rectangular	m²	160.00
three sizes; sawn 6 sides laid to coursed pattern	m²	160.00

8.2.1 ROADS PATHS PAVINGS AND SURFACINGS

Item	Unit	Total rate £
Excluding site overheads and profit		
TIMBER DECKING		
HARDWOOD DECKS		
Timber decking; Exterior Decking Limited; timber decking to roof gardens; cranage or movement of materials to roof top not included; coated one coat with Seasonite		
Joist size 47 × 47 mm		
Ipe; pre-drilled and countersunk; screw fixed	m^2	155.00
Ipe; Exterpark; 21 mm invisibly fixed	m^2	180.00
Ipe; Exterpark; 28 mm invisibly fixed	m^2	240.00
Ipe; Exterpark; 35 mm invisibly fixed	m^2	305.00
Rustic Teak; 21 mm invisibly fixed	m^2	195.00
Merbau; 21 mm invisibly fixed	m^2	160.00
Merbau; 28 mm invisibly fixed	m^2	130.00
Timber decking to commercial applications; double beams at 300 mm centres; coated one coat with Seasonite		
Joist size 47 × 150 mm		
Ipe; pre-drilled and countersunk; screw fixed	m^2	210.00
Ipe; Exterpark; 21 mm invisibly fixed	m^2	235.00
Ipe; Exterpark; 28 mm invisibly fixed	m^2	295.00
Ipe; Exterpark; 35 mm invisibly fixed	m^2	360.00
Rustic Teak; 21 mm invisibly fixed	m^2	250.00
Merbau; 21 mm invisibly fixed	m^2	215.00
Merbau; 28 mm invisibly fixed	m^2	180.00
SOFTWOOD DECKS		
Timber decking; AVS Ltd support structure of timber joists for decking laid laid at 400 mm centres on blinded base (measured separately)		
Yellow balau		
joists 38 × 88 mm	10 m^2	120.00
joists 47 × 100 mm	10 m^2	115.00
joists 47 × 150 mm	10 m^2	120.00
As above but boards 141 × 28 mm thick		
joists 47 × 100 mm	10 m^2	92.00
joists 47 × 150 mm	10 m^2	97.00
As above but decking boards in red cedar 150 mm wide × 32 mm thick		
joists 38 × 88 mm	10 m^2	76.00
joists 47 × 100 mm	10 m^2	70.00
joists 47 × 150 mm	10 m^2	75.00
Add to all of the above for handrails fixed to posts 100 × 100 × 1370 mm high		
square balusters at 100 mm centres	m	75.00
square balusters at 300 mm centres	m	48.00
turned balusters at 100 mm centres	m	95.00
turned balusters at 300 mm centres	m	55.00

8.2.1 ROADS PATHS PAVINGS AND SURFACINGS

Item Excluding site overheads and profit	Unit	Total rate £
STEPS		
Excavate for new steps; grade to correct levels; remove excavated material from site and lay concrete base; Construct formwork; supply and lay site mixed concrete; Fix treads and risers to steps		
Stone clad steps; 1.20 m wide; tread depth 225 mm; Granite cut off- site		
1 risers	nr	245.00
2 risers	nr	415.00
3 risers	nr	590.00
4 risers	nr	810.00
6 risers	nr	1150.00
Stone clad steps; 1.20 m wide; tread depth 225 mm; Riven Yorkstone cut on site		
1 risers	nr	230.00
2 risers	nr	390.00
3 risers	nr	550.00
4 risers	nr	750.00
6 risers	nr	1075.00
Stone clad steps; 2.40 m wide; tread depth 225 mm; Granite cut off- site		
1 risers	nr	470.00
2 risers	nr	800.00
3 risers	nr	1150.00
4 risers	nr	1575.00
6 risers	nr	2275.00
Play area for Ball games		
Excavate for playground; bring to grade; lay 225 mm consolidated hardcore; blind with 100 mm type 1; lay macadam base of 40 mm size dense bitumen macadam to 75 mm thick; lay wearing course 30 mm thick		
Over 1000 m²	100 m²	4900.00
400–1000 m²	100 m²	5100.00
Excavate for playground; bring to grade; lay 100 mm consolidated hardcore; blind with 100 mm type 1; lay macadam base of dense bitumen macadam to 50 mm thick; lay wearing course 20 mm thick		
Over 1000 m²	100 m²	3700.00
400–1000 m²	100 m²	4600.00
Excavate for playground; bring to grade; lay 100 mm consolidated hardcore; blind with 100 mm type 1; lay macadam base of dense bitumen macadam to 50 mm thick; lay wearing course of Addagrip resin coated aggregate 3 mm dia. × 6 mm thick		
Over 1000 m²	100 m²	5800.00
400–1000 m²	100 m²	6500.00

8.2.1 ROADS PATHS PAVINGS AND SURFACINGS

Item	Unit	Total rate £
Excluding site overheads and profit		
SAFETY SURFACED PLAY AREAS		
SAFETY SURFACING		
Excavate ground and reduce levels to receive safety surface; dispose of excavated material off site; treat substrate with total herbicide; lay granular fill Type 1 150 mm thick laid to falls and compacted; lay macadam base 40 mm thick; supply and lay Rubaflex wet pour safety system to thicknesses as specified		
All by machine except macadam by hand		
black		
15 mm thick	m²	55.00
35 mm thick	m²	69.00
60 mm thick	m²	84.00
coloured		
15 mm thick	m²	93.00
35 mm thick	m²	90.00
60 mm thick	m²	110.00
All by hand except disposal by grab		
black		
15 mm thick	m²	80.00
35 mm thick	m²	95.00
60 mm thick	m²	110.00
coloured		
15 mm thick	m²	120.00
35 mm thick	m²	125.00
60 mm thick	m²	140.00
Excavate for playground; bring to grade; lay with 150 mm type 1; lay macadam base of dense bitumen macadam to 50 mm thick; lay safety surface		
Wetpour coloured; 0.50 m critical fall height	m	98.00
Wetpour black; 0.50 m critical fall height	m	55.00
Wetpour coloured; 1.50 m critical fall height	m	110.00
Wetpour black; 1.50 m critical fall height	m	83.00
Excavate for playground; bring to grade; lay with 150 mm type 1; lay safety surface tiles Play Matta		
natural colours; 0.50 m critical fall height	m	81.00
bright colours; 0.50 m critical fall height	m	93.00
natural colours; 1.70 m critical fall height	m	86.00
bright colours; 1.70 m critical fall height	m	96.00
natural colours; 3.20 m critical fall height	m	100.00
bright colours; 3.20 m critical fall height	m	110.00
natural colours; 3.20 m critical fall height	m	110.00
bright colours; 3.20 m critical fall height	m	120.00
BARK PLAY AREA		
Excavate playground area to 450 mm depth; lay 150 mm broken stone or clean hardcore; lay filter membrane; lay bark surface		
Melcourt Industries		
Playbark 10/50; 300 mm thick	m²	36.00
Playbark 8/25; 300 mm thick	m²	43.00

8.2.2 SPECIAL SURFACINGS AND PAVINGS

Item Excluding site overheads and profit	Unit	Total rate £
Sports Field construction		
Plain sports pitches; site clearance, grading and drainage not included		
Agripower Ltd; cultivate ground and grade to levels; apply pre-seeding fertilizer at 900 kg/ha; apply pre-seeding selective weedkiller; seed in two operations with sports pitch type grass seed at 350 kg/ha; harrow and roll lightly; including initial cut; size		
association football; senior 114 × 72 m	each	3300.00
association football; junior 106 × 58 m	each	2450.00
rugby union pitch; 156 × 81 m	each	5100.00
rugby league pitch; 134 × 60 m	each	3200.00
hockey pitch; 95 × 60 m	each	2275.00
shinty pitch; 186 × 96 m	each	7300.00
men's lacrosse pitch; 100 × 55 m	each	2500.00
women's lacrosse pitch; 110 × 73 m	each	3200.00
target archery ground; 150 × 50 m	each	2900.00
cricket outfield; 160 × 142 m	each	3100.00
cycle track outfield; 160 × 80 m	each	5100.00
polo ground; 330 × 220 m	each	29000.00
Agripower Ltd; cricket square; excavate to depth of 100 mm; lay imported marl or clay loam; bring to accurate levels; apply pre-seeding fertilizer at 50 g/m; apply selective weedkiller; seed with cricket square type g grass seed at 50 g/m²; rake in and roll lightly; erect and remove temporary protective chestnut fencing; allow for initial cut and watering three times		
22.8 × 22.8 m	each	14000.00
JOGGING TRACK		
Excavate track 250 mm deep; lay filter membrane; lay 100 mm depth gravel waste or similar; lay 100 mm compacted gravel	100 m²	7700.00
Extra for treated softwood edging 50 × 150 mm on 50 × 50 mm posts		
both sides	m	9.70
Bowling Greens		
Agripower Ltd; bowling green; complete		
Excavate 300 mm deep and grade to level; excavate and install 100 mm land drain to perimeter and backfill with shingle; install 60 mm land drain to surface at 4.5 m centres backfilled with shingle; install 50 m non perforated pipe 50 m long; install 100 mm compacted and levelled drainage stone, blind with grit and sand; spread 150 mm imported 70:30 sand: soil accurately compacted and levelled; lay bowling green turf and top dress twice luted into surface; exclusive of perimeter ditches and bowls protection		
6 rink green 38.4 × 38.4 m	each	54000.00
install Toro automatic bowling green irrigation system with pump, tank, controller, electrics, pipework and 8 nr Toro 780 sprinklers; excluding pumphouse (optional)	each	8600.00
Supply and install to perimeter of green; Sportsmark preformed bowling green ditch channels		
Ultimate Design 99 steel reinforced concrete channel; 600 mm long section	each	8000.00

8.2.2 SPECIAL SURFACINGS AND PAVINGS

Item Excluding site overheads and profit	Unit	Total rate £
Supply and fit bowls protection material to rear hitting face of Ultimate channels 1 and 2 above		
Curl Grass artificial grass; 0.45 m wide	each	2200.00
Astroturf artificial grass; 0.45 m wide	each	2050.00
bowls protection ditch liner laid loose; 300 mm wide	each	1250.00

8.3.1 SOFT LANDSCAPING SEEDING AND TURFING

Item Excluding site overheads and profit	Unit	Total rate £
SURFACE PREPARATION FOR SEEDING AND TURFING		
BY MACHINE		
Mechanized preparation; Stone burier and cultivator; Treat area with systemic herbicide in advance of operations; rip subsoil using subsoiling machine to a depth of 500 mm at 1.20 m centres in light to medium soils; rotavate to 200 mm deep in two passes;		
Larger areas		
light to medium soils	100 m²	24.00
heavier soils	100 m²	44.00
Rip area by tractor and subsoiler; soil preparation by pedestrian operated rotavator; clearance and raking by hand; herbicide application by knapsack sprayer		
light to medium soils	100 m²	55.00
heavier soils	100 m²	57.00
Spread and lightly consolidate topsoil brought from spoil heap not exceeding 100 m; in layers not exceeding 150 mm; grade to specified levels; remove stones over 25 mm; all by machine		
100 mm thick	100 m²	83.00
Extra to the above for imported topsoil		
topsoil PC £33.60 m³ allowing for 20% settlement		
100 mm thick	100 m²	335.00
topsoil to BS3882 PC £33.00 m³ allowing for 20% settlement		
100 mm thick	100 m²	620.00
Excavate existing topsoil and remove off site; fill with new good quality topsoil to BS3882; add 50 mm mushroom compost (delivered in 65 m³ loads) and cultivate in; grade to level removing stones		
For turfing		
200 mm deep	m²	21.00
300 mm deep	m²	30.00
SURFACE PREPARATION BY HAND		
To area preavious cleared and sprayed; cultivate existing topsoil to receive new ornamental planting; rotavate using pedestrian rotavator. dog over by hand 1 spit deep; rake and grade to level to receive new planting		
For areas to be seeded or turfed	m²	0.80
Spread only and lightly consolidate topsoil brought from spoil heap in layers not exceeding 150 mm; grade to specified levels; remove stones over 25 mm; all by hand		
100 mm thick	100 m²	395.00

8.3.1 SOFT LANDSCAPING SEEDING AND TURFING

Item	Unit	Total rate £
Excluding site overheads and profit		
As above but inclusive of loading to location by barrow maximum distance 25 m; finished topsoil depth		
100 mm thick	100 m²	870.00
As above but loading to location by barrow maximum distance 100 m; finished topsoil depth		
100 mm thick	100 m²	890.00
Extra to above for imported topsoil PC £33.60 m³ allowing for 20% settlement		
100 mm thick	100 m²	335.00
SEEDING		
Spray with systemic herbicide; Cultivate topsoil 200 mm a fine tilth by tractor; remove stones over 25 mm by mechanical stone rake and bring to final tilth by harrow; apply pre-seeding fertilizer at 50 g/m² and work into top 50 mm during final cultivation; seed with certified grass seed in two operations; roll seedbed lightly after sowing		
General amenity grass areas; 35 g/m²; BSH A3		
general grass seeded areas	100 m²	54.00
lawn areas for public or domestic general use	100 m²	73.00
high grade lawn areas	100 m²	89.00
General amenity grass at 35 g/m²; DLF Trifolium Promaster 10		
general turfed areas	100 m²	65.00
lawn areas for public or domestic general use	100 m²	81.00
high grade lawn areas	100 m²	97.00
Shaded areas; BSH A6 at 50 g/m²	100 m²	63.00
Motorway and road verges; DLF Trifolium Promaster 120 at 25–35 g/m²	100 m²	84.00
Spray with systemic herbicide; Cultivate topsoil 200 mm a fine tilth by tractor; remove stones over 25 mm by mechanical stone rake and bring to final tilth by harrow; Add 50 mm of imported topsoil over surface of area to be seeded; apply pre-seeding fertilizer at 50 g/m² and work into top 50 mm during final cultivation; seed with certified grass seed in two operations; roll seedbed lightly after sowing		
General amenity grass areas; 35 g/m²; BSH A3		
general turfed areas	100 m²	240.00
lawn areas for public or domestic general use	100 m²	260.00
high grade lawn areas	100 m²	280.00
Spray with systemic herbicide; Cultivate topsoil 200 mm a fine tilth by tractor; remove stones over 25 mm by mechanical stone rake and bring to final tilth by harrow; Add 50 mm of topsoil from site spoil heaps maximum distance 100 m over surface of area to be seeded; apply pre-seeding fertilizer at 50 g/m² and work into top 50 mm during final cultivation; seed with certified grass seed in two operations; roll seedbed lightly after sowing		
General amenity grass areas; 35 g/m²; BSH A3		
general turfed areas	100 m²	97.00
lawn areas for public or domestic general use	100 m²	115.00
high grade lawn areas	100 m²	130.00

8.3.1 SOFT LANDSCAPING SEEDING AND TURFING

Item Excluding site overheads and profit	Unit	Total rate £
SEEDING – cont		
Bring top 200 mm of topsoil to a fine tilth using pedestrian operated rotavator; remove stones over 25 mm; apply pre-seeding fertilizer at 50 mm g/m² and work into top 50 mm during final hand cultivation; seed with certified grass seed in two operations; rake and roll seedbed lightly after sowing		
General amenity grass areas; 35 g/m²; BSH A3		
general grass seeded areas	100 m²	120.00
lawn areas for public or domestic general use	100 m²	135.00
high grade lawn areas	100 m²	145.00
Extra to above for using imported topsoil spread by machine		
100 mm minimum depth	100 m²	370.00
150 mm minimum depth	100 m²	550.00
extra for slopes over 30°	50%	–
Extra to above for using imported topsoil spread by hand; maximum distance for transporting soil 100 m		
100 mm minimum depth	m²	730.00
150 mm minimum depth	m²	1075.00
extra for slopes over 30°	50%	–
Extra for mechanically screening top 25 mm of topsoil through 6 mm screen and spreading on seedbed; debris carted to dump on site not exceeding 100 m	m³	6.60
Bring top 200 mm of topsoil to a fine tilth; remove stones over 50 mm; apply pre-seeding fertilizer at 50 g/m² and work into top 50 mm of topsoil during final cultivation; seed with certified grass seed in two operations; harrow and roll seedbed lightly after sowing		
Areas inclusive of fine levelling by specialist machinery		
outfield grass at 350 kg/ha; DLF Trifolium Promaster 40	ha	8900.00
outfield grass at 350 kg/ha; DLF Trifolium Promaster 70	ha	8700.00
sportsfield grass at 300 kg/ha; DLF Trifolium J Pitch	ha	8800.00
Seeded areas prepared by chain harrow		
low maintenance grass at 350 kg/ha	ha	9700.00
verge mixture grass at 150 kg/ha	ha	9100.00
Extra for wild flora mixture at 30 kg/ha		
BSH WSF 75 kg/ha	ha	3300.00
extra for slopes over 30°	50%	–
Cut existing turf to 1.0 × 1.0 × 0.5 m turves; roll up and move to stack not exceeding 100 m		
by pedestrian operated machine; roll up and stack by hand	100 m²	74.00
all works by hand	100 m²	210.00
Extra for boxing and cutting turves	100 m²	3.85
TURFING		
Cultivate topsoil in two passes; remove stones over 25 mm and bring to final tilth; apply pre-seeding fertilizer at 50 g/m² and work into top 50 mm during final cultivation; roll turf bed lightly		
Using tractor drawn implements and mechanical stone rake	100 m²	69.00
Cultivation by pedestrian rotavator; all other operations by hand		
Light or medium soils	m²	94.00
heavy wet or clay soils	m²	110.00

8.3.1 SOFT LANDSCAPING SEEDING AND TURFING

Item Excluding site overheads and profit	Unit	Total rate £
Cultivate topsoil in two passes; remove stones over 25 mm and bring to final tilth; apply pre-seeding fertilizer at 50 g/m² and work into top 50 mm during final cultivation; lay imported turves maximum distance fromn turf stack 25 m		
using tractor drawn implements and mechanical stone rake	m²	395.00
cultivation by pedestrian rotavator; all other operations by hand	m²	2.20
Cultivate topsoil in two passes; remove stones over 25 mm and bring to final tilth; apply pre-seeding fertilizer at 50 g/m² and work into top 50 mm during final cultivation; roll turf bed supply and lay imported turf; Rolawn Medallion		
using tractor drawn implements and mechanical stone rake	m²	4.35
cultivation by pedestrian rotavator; all other operations by hand	m²	3.95
Turfed areas; watering on two occasions and carrying out initial cut		
by ride on triple mower	100 m²	1.20
by pedestrian mower	100 m²	5.10
by pedestrian mower; box cutting	100 m²	5.70
Extra for using imported topsoil spread and graded by machine		
25 mm minimum depth	m²	1.00
75 mm minimum depth	m²	2.80
100 mm minimum depth	m²	3.70
150 mm minimum depth	m²	5.55
Extra for using imported topsoil spread and graded by hand; distance of barrow run 25 m		
25 mm minimum depth	m²	1.90
75 mm minimum depth	m²	4.95
100 mm minimum depth	m²	6.30
150 mm minimum depth	m²	9.50
Extra for work on slopes over 30° including pegging with 200 galvanized wire pins	m²	1.85
Inturf Big Roll		
Supply, deliver in one consignment, fully prepare the area and install in Big Roll format Inturf 553, a turfgrass comprising dwarf perennial ryegrass, smooth stalked rneadowgrass and fescues; installation by tracked machine		
Preparation by tractor drawn rotavator	m²	3.40
Cultivation by pedestrian rotavator; all other operations by hand	m²	3.25
Erosion control		
On ground previously cultivated, bring area to level, treat with herbicide; lay 20 mm thick open texture erosion control mat with 100 mm laps, fixed with 8 × 400 mm steel pegs at 1.0 m centres; sow with low maintenance grass suitable for erosion control on slopes at 35 g/m²; spread imported topsoil 25 mm thick and fertilizer at 35 g/m²; water lightly using sprinklers on two occasions	m²	7.30
as above but hand-watering by hose pipe maximum distance from mains supply 50 m	m²	7.40

8.3.1 SOFT LANDSCAPING SEEDING AND TURFING

Item Excluding site overheads and profit	Unit	Total rate £
ARTIFICIAL TURF		
Excavate ground and reduce levels to receive 20 mm thick artifial grass surface on a compacted 150 thick Type 1 base; dispose of excavated material off site		
Lay artificial turf; by machine		
Trulawn Value; 16 mm	m²	49.00
Trulawn Play; 20 mm; entry level; lawn and play areas	m²	51.00
Trulawn Continental; 20 mm; realistic appearance; lawns and patios	m²	46.50
Trulawn Optimum; 26 mm; softest pile; lush green; lawns and patios	m²	58.00
Trulawn Luxury; 30 mm; deep pile; realistic lawn	m²	61.00
All by hand except disposal by grab		
Trulawn Value; 16 mm	m²	77.00
Trulawn Play; 20 mm; entry level; lawn and play areas	m²	79.00
Trulawn Continental; 20 mm; realistic appearance; lawns and patios	m²	75.00
Trulawn Optimum; 26 mm; softest pile; lush green; lawns and patios	m²	86.00
Trulawn Luxury; 30 mm; deep pile; realistic lawn	m²	89.00
Maintenance of turfed/seeded areas		
Maintenance executed as part of a landscape construction contract		
Grass cutting		
Grass cutting; fine turf; using pedestrian guided machinery; arisings boxed and disposed of off site		
per occasion	100 m²	5.15
per annum 26 cuts	100 m²	135.00
per annum 18 cuts	100 m²	94.00
Grass cutting; standard turf; using self propelled three gang machinery		
per occasion	100 m²	0.40
per annum 26 cuts	100 m²	10.90
per annum 18 cuts	100 m²	7.55
Grass cutting for one year; recreation areas, parks, amenity grass areas; using tractor drawn machinery		
per occasion	ha	31.00
per annum 26 cuts	ha	810.00
per annum 18 cuts	ha	560.00
Maintain grass area inclusive of grass cutting; fertilization twice per annum; leaf and litter clearance		
using pedestrian guided machinery; arisings boxed and disposed of off site		
26 cuts; public areas	100 m²	170.00
26 cuts; private areas	100 m²	155.00
18 cuts; public areas	100 m²	130.00
18 cuts; private areas	100 m²	115.00

8.3.1 SOFT LANDSCAPING SEEDING AND TURFING

Item Excluding site overheads and profit	Unit	Total rate £
Maintain grass area inclusive of grass cutting; fertilization twice per annum; leaf and litter clearance		
using ride-on three cylinder machinery; arisings left on site		
26 cuts; public areas	100 m²	42.50
26 cuts; private areas	100 m²	29.00
18 cuts; public areas	100 m²	40.00
18 cuts; private areas	100 m²	26.50
Aeration of turfed areas		
Aerate ground with spiked aerator; apply spring/summer fertilizer once; apply autumn/winter fertilizer once; cut grass, 16 cuts; sweep up leaves twice		
as part of a landscape contract, defects liability	ha	2200.00
as part of a long term maintenance contract	ha	1125.00

8.3.2 EXTERNAL PLANTING

Item Excluding site overheads and profit	Unit	Total rate £
SURFACE PREPARATIONS FOR PLANTING		
SURFACE PREPARATION BY MACHINE		
Treat area with systemic non selective herbicide 1 month before starting cultivation operations; rip up subsoil using subsoiling machine to a depth of 250 mm below topsoil at 1.20 m centres in light to medium soils; rotavate to 200 mm deep in two passes; cultivate with chain harrow; roll lightly; clear stones over 50 mm		
By tractor	100 m²	10.00
As above but carrying out operations in clay or compacted gravel	100 m²	11.40
As above but ripping by tractor rotavation by pedestrian operated rotavator; clearance and raking by hand; herbicide application by knapsack sprayer	100 m²	26.00
As above but carrying out operations in clay or compacted gravel	100 m²	27.50
Spread and lightly consolidate topsoil brought from spoil heap not exceeding 100 m; in layers not exceeding 150 mm; grade to specified levels; remove stones over 25 mm; all by machine		
100 mm thick	100 m²	0.80
150 mm thick	100 m²	1.20
300 mm thick	100 m²	9.30
450 mm thick	100 m²	3.50
Extra to the above for imported topsoil		
topsoil PC £33.60 m³ allowing for 20% settlement		
100 mm thick	100 m²	335.00
150 mm thick	100 m²	500.00
300 mm thick	100 m²	1000.00
450 mm thick	100 m²	1500.00
500 mm thick	100 m²	1675.00
600 mm thick	100 m²	2025.00
750 mm thick	100 m²	2500.00
1.00 m thick	100 m²	3350.00
topsoil to BS3882 PC £33.00 m³ allowing for 20% settlement		
100 mm thick	100 m²	620.00
150 mm thick	100 m²	920.00
300 mm thick	100 m²	1850.00
450 mm thick	100 m²	2800.00
500 mm thick	100 m²	3100.00
600 mm thick	100 m²	3700.00
750 mm thick	100 m²	4600.00
1.00 m thick	100 m²	6200.00
Excavate existing topsoil and remove off site; fill with new good quality topsoil to BS3882; add 50 mm mushroom compost (delivered in 65 m³ loads) and cultivate in; grade to level removing stones		
For planting		
300 mm deep	m²	29.50
400 mm deep	m²	40.00
500 mm deep	m²	49.00

8.3.2 EXTERNAL PLANTING

Item Excluding site overheads and profit	Unit	Total rate £
Extra for incorporating mushroom compost at 50 mm/m² into the top 150 mm of topsoil (compost delivered in 20 m³) loads		
manually spread; mechanically rotavated	100 m²	180.00
mechanically spread and rotavated	100 m²	130.00
Extra for incorporating manure at 50 mm/m² into the top 150 mm of topsoil loads		
manually spread; mechanically rotavated; 20 m³ loads	100 m²	280.00
mechanically spread and rotavated; 60 m³ loads	100 m²	170.00
SURFACE PREPARATION BY HAND		
To area preavious cleared and sprayed; cultivate existing topsoil to receive new ornamental planting; rotavate using pedestrian rotavator. dog over by hand 1 spit deep; rake and grade to level to receive new planting For ornamental planting beds	m²	0.55
Spread only and lightly consolidate topsoil brought from spoil heap in layers not exceeding 150 mm; grade to specified levels; remove stones over 25 mm; all by hand		
100 mm thick	100 m²	395.00
150 mm thick	100 m²	590.00
300 mm thick	100 m²	1175.00
450 mm thick	100 m²	1775.00
As above but inclusive of loading to location by barrow maximum distance 25 m; finished topsoil depth		
100 mm thick	100 m²	870.00
150 mm thick	100 m²	1300.00
300 mm thick	100 m²	2600.00
450 mm thick	100 m²	3900.00
500 mm thick	100 m²	4350.00
600 mm thick	100 m²	5200.00
As above but loading to location by barrow maximum distance 100 m; finished topsoil depth		
100 mm thick	100 m²	890.00
150 mm thick	100 m²	1325.00
300 mm thick	100 m²	2650.00
450 mm thick	100 m²	4000.00
500 mm thick	100 m²	4450.00
600 mm thick	100 m²	5300.00
Extra to the above for incorporating mushroom compost at 50 mm/m² into the top 150 mm of topsoil; by hand	100 m²	210.00
Extra to above for imported topsoil PC £33.60 m³ allowing for 20% settlement		
100 mm thick	100 m²	335.00
150 mm thick	100 m²	500.00
300 mm thick	100 m²	1000.00
450 mm thick	100 m²	1500.00
500 mm thick	100 m²	1675.00
600 mm thick	100 m²	2025.00
750 mm thick	100 m²	2500.00
1.00 m thick	100 m²	3350.00

8.3.2 EXTERNAL PLANTING

Item Excluding site overheads and profit	Unit	Total rate £
TREE PLANTING		
Excavate tree pit by hand; fork over bottom of pit; plant tree with roots well spread out; backfill with excavated material incorporating treeplanting compost at 1 m³ per 3 m³ of soil; one tree stake and two ties; tree pits square in sizes shown		
Light standard bare root tree in pit; PC £9.75		
600 × 600 mm deep	each	35.00
900 × 600 mm deep	each	51.00
Standard bare root tree in pit; PC £16.00		
600 × 600 mm deep	each	47.00
900 × 600 mm deep	each	56.00
Standard root balled tree in pit; PC £28.00		
600 × 600 mm deep	each	59.00
900 × 600 mm deep	each	68.00
1.00 × 1.00 deep	each	91.00
Selected standard bare root tree in pit; PC £22.50		
900 × 900 mm deep	each	61.00
1.00 × 1.00 deep	each	89.00
Selected standard root ball tree in pit; PC £40.00		
900 × 900 mm deep	each	88.00
1.00 × 1.00 deep	each	125.00
Heavy standard bare root tree in pit; PC £39.25		
900 × 900 mm deep	each	100.00
1.00 × 1.00 deep	each	110.00
1.50 m × 750 mm deep	each	150.00
Heavy standard root ball tree in pit; PC £60.26		
900 × 900 mm deep	each	110.00
1.00 × 1.00 deep	each	125.00
1.50 m × 750 mm deep	each	160.00
Extra heavy standard bare root tree in pit; PC £63.00		
1.00 × 1.00 deep	each	135.00
1.20 × 1.00 deep	each	135.00
1.50 m × 750 mm deep	each	180.00
Extra heavy standard root ball tree in pit; PC £72.00		
1.00 × 1.00 deep	each	150.00
1.50 m × 750 mm deep	each	180.00
1.50 × 1.00 deep	each	215.00
Excavate tree pit by machine; fork over bottom of pit; plant tree with roots well spread out; backfill with excavated material, incorporating organic manure at 1 m³ per 3 m³ of soil; one tree stake and two ties; tree pits square in sizes shown		
Light standard bare root tree in pit; PC £9.75		
600 × 600 mm deep	each	30.00
900 × 900 mm deep	each	43.00

8.3.2 EXTERNAL PLANTING

Item Excluding site overheads and profit	Unit	Total rate £
Standard bare root tree in pit; PC £16.00		
600 × 600 mm deep	each	42.00
900 × 600 mm deep	each	44.00
900 × 900 mm deep	each	51.00
Standard root balled tree in pit; PC £28.00		
600 × 600 mm deep	each	54.00
900 × 600 mm deep	each	56.00
900 × 900 mm deep	each	62.00
Selected standard bare root tree in pit; PC £22.50		
900 × 900 mm deep	each	61.00
1.00 × 1.00 deep	each	72.00
Selected standard root ball tree in pit; PC £40.00		
900 × 900 mm deep	each	88.00
1.00 × 1.00 deep	each	110.00
Heavy standard bare root tree in pit; PC £39.25		
900 × 900 mm deep	each	100.00
1.00 × 1.00 deep	each	92.00
1.20 × 1.00 deep	each	115.00
Heavy standard root ball tree in pit; PC £60.26		
900 × 900 mm deep	each	110.00
1.00 × 1.00 deep	each	110.00
1.20 × 1.00 deep	each	120.00
Extra heavy standard bare root tree in pit; PC £63.00		
1.00 × 1.00 deep	each	120.00
1.20 × 1.00 deep	each	135.00
1.50 × 1.00 deep	each	165.00
Extra heavy standard root ball tree in pit; PC £72.00		
1.00 × 1.00 deep	each	130.00
1.20 × 1.00 deep	each	145.00
1.50 × 1.00 deep	each	175.00
SEMI MATURE TREE PLANTING		
Trees in soft landscape environments		
Excavate tree pit deep by machine; fork over bottom of pit; plant rootballed tree Acer platanoides Emerald Queen using telehandler where necessary; backfill with excavated material, incorporating Melcourt Topgrow bark/manure mixture at 1 m³ per 3 m³ of soil; Platipus underground guying system; tree pits 1500 × 1500 × 1500 mm deep inclusive of Platimats; excavated material not backfilled to treepit spread to surrounding area		
16–18 cm girth; PC £85.00	each	260.00
18–20 cm girth; PC £100.00	each	290.00
20–25 cm girth; PC £125.00	each	380.00
25–30 cm girth; PC £165.00	each	480.00
30–35 cm girth; PC £300.00	each	730.00
As above but treepits 2.00 × 2.00 × 1.5 m deep		
40–45 cm girth; PC £500.00	each	1175.00
45–50 cm girth; PC £700.00	each	1450.00
55–60 cm girth; PC £1,200.00	each	1850.00
67–70 cm girth; PC £2,200.00	each	3050.00
75–80 cm girth; PC £4,500.00	each	5700.00

8.3.2 EXTERNAL PLANTING

Item Excluding site overheads and profit	Unit	Total rate £
TREE PLANTING – cont		
Excavate tree pit deep by machine – cont		
Extra to the above for imported topsoil moved 25 m from tipping area and disposal off site of excavated material		
Tree pits 1500 × 1500 × 1500 mm deep		
16–18 cm girth	each	220.00
18–20 cm girth	each	210.00
20–25 cm girth	each	195.00
25–30 cm girth	each	185.00
30–35 cm girth	each	180.00
Tree pit in Trafficked environments		
Excavate tree pit; remove excavated material; break up bottom of pit; line pit with root barrier; Lay drainage layer 200 mm thick; backfill treepit with cellular load supporting cells to 7.5 tonne traffic weight limit; Plant tree with underground guying system; inclusive of root director Add rootball irrigation system; backfill over rootball with topsoil enriched with composts and fertilizers		
Tree pit tree size		
2.00 × 2.00 × 1.00deep; 30–35 cm tree	nr	1575.00
2.50 × 2.50 × 1.00deep; 40–45 cm tree	nr	2250.00
Excavate tree pit; remove excavated material; break up bottom of pit; line pit with root barrier; Lay drainage layer 200 mm thick; backfill treepit with cellular load supporting cells to 40 tonne traffic weight limit; Plant tree with underground guying system; inclusive of root director Add rootball irrigation system; backfill over rootball with topsoil enriched with composts and fertilizers		
Tree pit tree size		
2.00 × 2.00 × 1.00deep; 30–35 cm tree	nr	2050.00
2.50 × 2.50 × 1.00deep; 40–45 cm tree	nr	2950.00
TREE PLANTING WITH MOBILE CRANES		
Excavate treepit 1.50 × 1.50 × 1.00deep; supply and plant semi mature trees delivered in full loads; trees lifted by crane; inclusive of backfilling tree pit with imported topsoil, compost, fertilizers and underground guying using Platipus anchors		
Self managed lift; local authority applications, health and safety, traffic management or road closures not included; tree size and distance of lift; 35 tonne crane		
25–30 cm; maximum 25 m distance	each	700.00
30–35 cm; maximum 25 m distance	each	790.00
35–40 cm; maximum 25 m distance	each	1150.00
55–60 cm; maximum 15 m distance	each	1850.00
80–90 cm; maximum 10 m distance	each	6900.00
Managed lift; inclusive of all local authority applications, health and safety, traffic management or road closures all by crane hire company; tree size and distance of lift; 35 tonne crane		
25–30 cm; maximum 25 m distance	each	700.00
30–35 cm; maximum 25 m distance	each	810.00
35–40 cm; maximum 25 m distance	each	1175.00
55–60 cm; maximum 15 m distance	each	1875.00
80–90 cm; maximum 10 m distance	each	7100.00

8.3.2 EXTERNAL PLANTING

Item	Unit	Total rate £
Excluding site overheads and profit		
Self managed lift; local authority applications, health and safety, traffic management or road closures not included; tree size and distance of lift; 80 tonne crane		
25–30 cm; maximum 40 m distance	each	700.00
30–35 cm; maximum 40 m distance	each	800.00
35–40 cm; maximum 40 m distance	each	1150.00
55–60 cm; maximum 33 m distance	each	1875.00
80–90 cm; maximum 23 m distance	each	7000.00
Managed lift; inclusive of all local authority applications, health and safety, traffic management or road closures all by crane hire company; tree size and distance of lift; 35 tonne crane		
25–30 cm; maximum 40 m distance	each	710.00
30–35 cm; maximum 40 m distance	each	810.00
35–40 cm; maximum 40 m distance	each	1175.00
55–60 cm; maximum 33 m distance	each	1925.00
80–90 cm; maximum 23 m distance	each	7100.00
PLANTING SHRUBS AND GROUNDCOVERS		
SHRUBS		
Excavate planting holes 250 × 250 × 300 mm deep to area previously ripped and rotavated; excavated material left alongside planting hole		
By mechanical auger		
250 mm centres (16 plants per m^2)	m^2	11.40
300 mm centres (11.11 plants per m^2)	m^2	7.90
400 mm centres (6.26 plants per m^2)	m^2	4.45
450 mm centres (4.93 plants per m^2)	m^2	3.50
500 mm centres (4 plants per m^2)	m^2	2.80
600 mm centres (2.77 plants per m^2)	m^2	2.00
750 mm centres (1.77 plants per m^2)	m^2	1.25
900 mm centres (1.23 plants per m^2)	m^2	0.85
1.00 m centres (1 plant per m^2)	m^2	0.70
1.50 m centres (0.44 plants per m^2)	m^2	0.30
As above but excavation by hand		
250 mm centres (16 plants per m^2)	m^2	11.40
300 mm centres (11.11 plants per m^2)	m^2	7.90
400 mm centres (6.26 plants per m^2)	m^2	4.45
450 mm centres (4.93 plants per m^2)	m^2	3.50
500 mm centres (4 plants per m^2)	m^2	2.80
600 mm centres (2.77 plants per m^2)	m^2	2.00
750 mm centres (1.77 plants per m^2)	m^2	1.25
900 mm centres (1.23 plants per m^2)	m^2	0.85
1.00 m centres (1 plant per m^2)	m^2	0.70
1.50 m centres (0.44 plants per m^2)	m^2	0.30

8.3.2 EXTERNAL PLANTING

Item Excluding site overheads and profit	Unit	Total rate £
PLANTING SHRUBS AND GROUNDCOVERS – cont		
Clear light vegetation from planting area and remove to dump on site; dig planting holes; plant whips with roots well spread out; backfill with excavated topsoil; including one 38 × 38 mm treated softwood stake, two tree ties and mesh guard 1.20 m high; planting matrix 1.5 × 1.5 m; allow for beating up once at 10% of original planting, cleaning and weeding round whips once, applying fertilizer once at 35 gm/m²; using the following mix of whips, bare rooted		
Plant bare root plants average PC £0.41 each to a required matrix		
plant mix as above	100 m²	260.00
Plant bare root plants average PC £0.75 each to a required matrix		
plant mix as above	100 m²	280.00
Cultivate and grade shrub bed; bring top 300 mm of topsoil to a fine tilth, incorporating mushroom compost at 50 mm and Enmag slow release fertilizer; rake and bring to given levels; remove all stones and debris over 50 mm; dig planting holes average 300 × 300 × 300 mm deep; supply and plant specified shrubs in quantities as shown below; backfill with excavated material as above; water to field capacity and mulch 50 mm bark chips 20–40 mm size; water and weed regularly for 12 months and replace failed plants		
Shrubs 3 l PC £3.40; ground covers 9 cm PC £1.50		
100% shrub area		
300 mm centres	100 m²	6700.00
400 mm centres	100 m²	4000.00
500 mm centres	100 m²	2750.00
600 mm centres	100 m²	2075.00
100% groundcovers		
200 mm centres	100 m²	7500.00
300 mm centres	100 m²	3650.00
400 mm centres	100 m²	2275.00
500 mm centres	100 m²	1675.00
groundcover 30%/shrubs 70% at the distances shown below		
200/300 mm	100 m²	6900.00
300/400 mm	100 m²	3900.00
300/500 mm	100 m²	3000.00
400/500 mm	100 m²	2600.00
groundcover 50%/shrubs 50% at the distances shown below		
200/300 mm	100 m²	7100.00
300/400 mm	100 m²	3800.00
300/500 mm	100 m²	3200.00
400/500 mm	100 m²	2500.00
Bulb planting		
Cultivate ground by machine and rake to level; plant bulbs as shown; bulbs PC £25.00/100		
15 bulbs per m²	100 m²	670.00
25 bulbs per m²	100 m²	1100.00
50 bulbs per m²	100 m²	2175.00

8.3.2 EXTERNAL PLANTING

Item Excluding site overheads and profit	Unit	Total rate £
Cultivate ground by machine and rake to level; plant bulbs as shown; bulbs PC £13.00/100		
15 bulbs per m²	100 m²	470.00
25 bulbs per m²	100 m²	770.00
50 bulbs per m²	100 m²	1500.00
Form holes in grass areas and plant bulbs using bulb planter, backfill with organic manure and turf plug; bulbs PC £13.00/100		
15 bulbs per m²	100 m²	690.00
25 bulbs per m²	100 m²	1150.00
50 bulbs per m²	100 m²	2275.00
Hedge planting		
Works by machine; excavate trench for hedge 300 mm wide × 300 mm deep; deposit spoil alongside and plant hedging plants in single row at 200 mm centres; backfill with excavated material		
Trench 300 mm × 300 mm; bare root hedging plants PC £0.48 per plant; single row		
200 mm centres	m	6.20
300 mm centres	m	5.25
400 mm centres	m	4.60
600 mm centres	m	3.90
800 mm centres	m	3.55
Trench 600 mm × 300 mm deep; bare root hedging plants PC £0.48 per plant; double staggered row		
300 mm centres	m	9.70
400 mm centres	m	3.85
500 mm centres	m	7.10
600 mm centres	m	6.75
800 mm centres	m	5.85
1.00 mm centres	m	5.45
Works by hand; excavate trench for hedge 300 mm wide × 450 mm deep; deposit spoil alongside and plant hedging plants in single row at 200 mm centres; backfill with excavated material incorporating organic manure at 1 m³ per 5 m³; carry out initial cut; including delivery of plants from nursery		
Trench 300 mm × 300 mm; bare root hedging plants PC £0.48 per plant; single row		
200 mm centres	m	8.55
300 mm centres	m	7.65
400 mm centres	m	6.95
600 mm centres	m	6.30
800 mm centres	m	5.95
Trench 600 mm × 300 mm deep; bare root hedging plants PC £0.48 per plant; double staggered row		
300 mm centres	m	15.70
400 mm centres	m	13.90
500 mm centres	m	13.10
600 mm centres	m	12.80
800 mm centres	m	11.90
1.00 mm centres	m	11.50

8.3.2 EXTERNAL PLANTING

Item Excluding site overheads and profit	Unit	Total rate £
Hedge planting – cont		
Practicality Brown Ltd; Fully preformed hedge planting excavate trench by machine 700 mm wide × 500 mm deep; add compost at 100 mm/m² mixed to excavated material; plant mature hedge plants; backfill with excavated material and compost; allow for disposal of 50% of excavated material to spoil heaps 50 m distant		
Beech or hornbeam hedging		
1.50 m × 500 mm wide at 500 mm centres	m	260.00
1.50 m × 500 mm wide at 750 mm centres	m	170.00
2.00 m × 500 mm wide at 500 mm centres	m	350.00
2.00 m × 500 mm wide at 750 mm centres	m	250.00
1.50 m × 500 mm wide at 500 mm centres	m	260.00
Yew (Taxus) hedging		
1.50 m × 500 mm wide at 500 mm centres	m	340.00
1.50 m × 500 mm wide at 750 mm centres	m	225.00
1.50 m × 500 mm wide at 1.00 m centres	m	175.00
1.75 m × 500 mm wide at 500 mm centres	m	390.00
1.75 m × 500 mm wide at 750 mm centres	m	250.00
1.75 m × 500 mm wide at 1.00 m centres	m	200.00
2.00 m × 500 mm wide at 500 mm centres	m	450.00
2.00 m × 500 mm wide at 750 mm centres	m	310.00
2.00 m × 500 mm wide at 1.00 mm centres	m	230.00
Laurel (Prunus) hedging		
2.00 m × 500 mm wide at 1.00 m centres	m	10.70
1.40/1.60 m high × 400 mm wide	m	170.00
1.60/1.80 m high × 500 mm wide	m	185.00
1.80/2.00 m high × 500 mm wide	m	200.00
Box (Buxus) hedging		
800 mm/1.00 m high × 300 mm wide	m	150.00
1.00 mm/1.20 m high × 300 mm wide	m	170.00
Practicality Brown Ltd; pre-clipped and preformed hedges planting; excavate trench by machine 500 mm wide × 500 mm deep; add compost at 100 mm/ m² mixed to excavated material; plant mature hedge plants; backfill with excavated material and compost; allow for disposal of 50% of excavated material to spoil heaps 50 m distant		
Beech or Hornbeam hedging		
1.75 m high × 300 mm wide at 400 mm centres	m	95.00
1.75 m high × 300 mm wide at 500 mm centres	m	77.00
1.75 m high × 300 mm wide at 750 mm centres	m	57.00
1.75 m high × 300 mm wide at 900 mm centres	m	46.50
2.00 m high × 300 mm wide at 400 mm centres	m	170.00
2.00 m high × 300 mm wide at 500 mm centres	m	135.00

8.3.2 EXTERNAL PLANTING

Item Excluding site overheads and profit	Unit	Total rate £
Elveden hedges; strip hedge planting; excavate trench 700 mm wide × 500 mm deep by machine; add compost at 100 mm/m^2 mixed to excavated material; plant instant strip hedging; backfill with excavated material and compost; allow for disposal of 50% of excavated material to spoil heaps 50 m distant		
Beech hedging		
1.40/1.60 m high × 400 mm wide	m	175.00
1.60/1.80 m high × 500 mm wide	m	190.00
Yew (Taxus) hedging		
1.40/1.60 m high × 400 mm wide	m	215.00
1.60/1.80 m high × 500 mm wide	m	240.00
1.80/2.00 m high × 500 mm wide	m	265.00
Laurel (Prunus) hedging		
1.40/1.60 m high × 400 mm wide	m	170.00
1.60/1.80 m high × 500 mm wide	m	185.00
1.80/2.00 m high × 500 mm wide	m	200.00
Box (Buxus) hedging		
800 mm/1.00 m high × 300 mm wide	m	150.00
1.00/1.20 m high × 300 mm wide	m	170.00
Griffin Nurseries; individual feathered hedge planting; excavate trench by machine 500 mm wide × 500 mm deep; add compost at 100 mm/m^2 mixed to excavated material; plant mature hedge plants; backfill with excavated material and compost; allow for disposal of 50% of excavated material to spoil heaps 50 m distant		
Yew hedging		
1.0–1.20 high × 300 mm wide at 2 / m	m	60.00
1.25 m–1.50 high × 300 mm wide at 3 / m	m	130.00
1.50–1.75 m high at 450 mm centres	m	165.00
1.50–1.75 m high at 600 mm centres	m	125.00
2.00–2.25 m high at 600 mm centres	m	205.00
2.00–2.25 m high at 900 mm centres	m	150.00
Griffin nurseries; Works by machine; excavate trench for hedge 300 mm wide × 300 mm deep; deposit spoil alongside and plant hedging plants in single row; backfill with excavated material		
Box hedging; rootballed		
400–500 mm high; 200 mm centres	m	37.50
400–500 mm high; 300 mm centres	m	26.00
400–500 mm high; 500 mm centres	m	16.80
500–600 mm mm high; 200 mm centres	m	57.00
500–600 mm mm high; 300 mm centres	m	39.00
500–600 mm mm high; 600 mm centres	m	25.00
600–800 mm high; 250 mm centres	m	58.00
600–800 mm high; 400 mm centres	m	37.50
800 mm–1.0 m high; 300 mm centres	m	76.00
800 mm–1.0 m high; 500 mm centres	m	47.50
1.0–1.25 m high; 600 mm centres	m	54.00
1.0–1.25 m high; 900 mm centres	m	37.50

8.3.2 EXTERNAL PLANTING

Item Excluding site overheads and profit	Unit	Total rate £
ANNUAL BEDDING		
BEDDING		
Spray surface with glyphosate; lift and dispose of turf when herbicide action is complete; cultivate new area for bedding plants to 400 mm deep; spread compost 100 mm deep and chemical fertilizer Enmag and rake to fine tilth to receive new bedding plants; remove all arisings		
Existing turf area		
disposal to skip	m²	5.35
disposal to compost area on site; distance 25 m	m²	5.75
Plant bedding to existing planting area; bedding planting PC £0.25 each		
Clear existing bedding; cultivate soil to 230 mm deep; incorporate compost 75 mm and rake to fine tilth; collect bedding from nursery and plant at 100 mm centres; irrigate on completion; maintain weekly for 12 weeks		
mass planted; 100 mm centres	m²	34.00
to patterns; 100 mm centres	m²	37.50
mass planted; 150 mm centres	m²	19.90
to patterns; 150 mm centres	m²	23.50
mass planted; 200 mm centres	m²	19.90
to patterns; 200 mm centres	m²	23.50
Watering of bedding by hand held hose pipe; The prices shown here are per occasion; Bedding may require up to 40 visits during a dry season		
Flow rate 25 litres/minute; per occasion		
10 litres/m²	100 m²	0.10
15 litres/m²	100 m²	0.20
20 litres/m²	100 m²	0.25
25 litres/m²	100 m²	0.30
Flow rate 40 litres/minute		
10 litres/m²	100 m²	0.10
15 litres/m²	100 m²	0.10
20 litres/m²	100 m²	0.15
25 litres/m²	100 m²	0.20
Living Wall		
Living wall; Scotscape Ltd; Design and installation of planted modules with automatic irrigation systems		
ANS; soil-based systems		
Walls over 35 m²	m²	680.00
Walls under 35 m²	m²	790.00
Biotecture hydroponic lightweight systems		
Walls over 35 m²	m²	680.00
Walls under 35 m²	m²	740.00
Preliminary costs for all living walls	item	700.00
Annual maintence costs for living walls		
Walls over 35 m²	item	3650.00
Walls under 35 m²	item	1875.00

8.3.2 EXTERNAL PLANTING

Item Excluding site overheads and profit	Unit	Total rate £
Planters		
To brick planter; coat insides with 2 coats RIW liquid asphaltic composition; fill with 50 mm shingle and cover with geofabric; fill with screened topsoil incorporating 25% Topgrow compost and Enmag		
Planters 1.00 m deep		
1.00 × 1.00	each	110.00
1.00 × 2.00	each	190.00
1.00 × 3.00	each	270.00
Planters 1.50 m deep		
1.00 × 1.00	each	170.00
1.00 × 2.00	each	250.00
1.00 × 3.00	each	405.00
Container planting; fill with 50 mm shingle and cover with geofabric; fill with screened topsoil incorporating 25% Topgrow compost and Enmag		
Planters 1.00 m deep		
400 × 400 × 400 mm deep	each	8.20
400 × 400 × 600 mm deep	each	10.20
1.00 m × 400 mm wide × 400 mm deep	each	18.00
1.00 m × 600 mm wide × 600 mm deep	each	27.00
1.00 m × 100 mm wide × 400 mm deep	each	28.00
1.00 m × 100 mm wide × 600 mm deep	each	40.50
1.00 m × 100 mm wide × 1.00 m deep	each	66.00
1.00 m dia. × 400 mm deep	each	22.00
1.00 m dia. × 1.00 m deep	each	53.00
2.00 m dia. × 1.00 m deep	each	200.00
Street planters		
Supply and locate in position precast concrete planters; fill with topsoil placed over 50 mm shingle and terram; plant with assorted 5 litre and 3 litre shrubs at to provide instant effect		
970 mm dia. × 470 mm high; white exposed aggregate finish	each	2.00
LANDSCAPE MAINTENANCE		
As part of a term maintenance contract		
MAINTENANCE OF ORNAMENTAL SHRUB BEDS		
Ornamental shrub beds; private gardens or business parks and resticted commercial environments		
Hand weed ornamental shrub bed during the growing season; clear leaves and litter; edge beds when necessary; prune shrubs or hedges and remove replace mulch to 50 mm at end of each year; (mulched areas only) planting less than 2 years old		

8.3.2 EXTERNAL PLANTING

Item Excluding site overheads and profit	Unit	Total rate £
LANDSCAPE MAINTENANCE – cont		
Mulched beds; weekly visits; 12 month period planting centres shown; public areas; shopping areas		
600 mm centres	100 m²	145.00
400 mm centres	100 m²	130.00
300 mm centres	100 m²	110.00
ground covers	100 m²	97.00
Mulched beds; weekly visits; 12 month period planting centres shown; private areas ; business or office parks or similar		
600 mm centres	100 m²	130.00
400 mm centres	100 m²	105.00
300 mm centres	100 m²	88.00
ground covers	100 m²	70.00
Non-mulched beds; weekly visits; planting centres		
600 mm centres	100 m²	120.00
400 mm centres	100 m²	110.00
300 mm centres	100 m²	110.00
ground covers	100 m²	100.00
Non-mulched beds; monthly visits; planting centres		
600 mm centres	100 m²	94.00
400 mm centres	100 m²	87.00
300 mm centres	100 m²	83.00
ground covers	100 m²	79.00
Maintain rose bed		
Weed around base of plants; dead head regularly; fertilize once per annum with Enmag; prune once per annum; edge with half moon edging tool; add compost to 50 mm once per annum; clear of litter and leaves; private or restricted access beds		
private or restricted access beds	100 m²	250.00
public areas	100 m²	275.00
Remulch planting bed at the start of the planting season; top up mulch 25 mm thick; Melcourt Ltd		
Larger areas maximum distance 25 m; 80 m³ loads		
Ornamental bark mulch	100 m²	150.00
Melcourt Bark Nuggets	100 m²	140.00
Amenity Bark	100 m²	105.00
Forest biomulch	100 m²	100.00
Smaller areas; maximum distance 25 m; 25 m³ loads		
Ornamental bark mulch	100 m²	185.00
Melcourt Bark Nuggets	100 m²	175.00
Amenity Bark	100 m²	140.00
Forest biomulch	100 m²	130.00
Shrub beds in ornamental gardens		
Maintain mature shrub bed; shrubs average 1.00 m–3.00 m high with occasional mature trees in bed or in proximity; prune one third of all planting by 30% every year in rotation; fertilse; add compost; annually; add fertilizers; clear leaves; edge bed with half moon tool; water with hand held hose on 6 occasions		
massed plants; 1.00 high	m²	385.00
plants up to 2 m high;	m²	435.00
plants 2 m–3 m high;	m²	560.00

8.3.2 EXTERNAL PLANTING

Item Excluding site overheads and profit	Unit	Total rate £
VEGETATION CONTROL		
Native planting; roadside railway or forestry planted areas.		
Post planting maintenance; control of weeds and grass; herbicide spray applications; maintain weed free circles 1.00 m dia. to planting less than 5 years old in roadside, rail or forestry planting environments and the like; strim grass to 50–75 mm; prices per occasion (three applications of each operation normally required)		
Knapsack spray application; glyphosate; planting at		
1.50 mm centres	ha	1025.00
1.75 mm centres	ha	630.00
2.00 mm centres	ha	680.00
Maintain planted areas; control of weeds and grass; maintain weed free circles 1.00 m dia. to planting less than 5 years old in roadside, rail or forestry planting environments and the like; strim surrounding grass to 50–75 mm; prices per occasion (three applications of each operation normally required)		
Herbicide spray applications; CDA (controlled droplet application) glyphosate and strimming; plants planted at the following centres		
1.50 mm centres	ha	550.00
1.75 mm centres	ha	630.00
2.00 mm centres	ha	680.00
Post planting maintenance; control of weeds and grass; herbicide spray applications; CDA (controlled droplet application); Xanadu glyphosate/diuron; maintain weed free circles 1.00 m dia. to planting less than 5 years old in roadside, rail or forestry planting environments and the like; strim grass to 50–75 mm; prices per occasion (1.5 applications of herbicide and three strim operations normally required)		
Plants planted at the following centres		
1.50 mm centres	ha	590.00
1.75 mm centres	ha	660.00
2.00 mm centres	ha	710.00
MAINTENANCE AND DEFECTS LIABILITY AS PART OF A LANDSCAPE CONSTRUCTION PROJECT		
Clean and weed plant beds; fork over as necessary; water with full liability for the duration of the first growing season using moveable sprinklers; clear leaf fall; top up mulch at the end of the 12 month period		
1 st year after planting		
weekly visits to high profile areas new planting areas	100 m²	330.00
monthly visits to new planting areas	100 m²	300.00

8.3.3 IRRIGATION SYSTEMS

Item Excluding site overheads and profit	Unit	Total rate £
IRRIGATION SYSTEMS		
LANDSCAPE IRRIGATION		
Automatic irrigation; KAR UK Ltd		
Large garden consisting of 24 stations; 7000 m² irrigated area		
turf only	nr	11000.00
70/30% turf/shrub beds	nr	14000.00
Medium sized garden consisting of 12 stations; 3500 m² irrigated area		
turf only	nr	7700.00
70/30% turf/shrub beds	nr	11000.00
Medium sized garden consisting of 6 stations of irrigated area; 1000 m²		
turf only	nr	5200.00
70/30% turf/shrub beds	nr	5800.00
50/50% turf/shrub beds	nr	6300.00
Leaky pipe irrigation; Leaky Pipe Ltd		
Works by machine; main supply and connection to laterals; excavate trench for main or ring main 450 mm deep; supply and lay pipe; backfill and lightly compact trench		
20 mm LDPE	100 m	275.00
16 mm LDPE	100 m	230.00
Works by hand; main supply and connection to laterals; excavate trench for main or ring main 450 mm deep; supply and lay pipe; backfill and lightly compact trench		
20 mm LDPE	100 m	1425.00
16 mm LDPE	100 m	1375.00
Turf area irrigation; laterals to mains; to cultivated soil; excavate trench 150 mm deep using hoe or mattock; lay moisture leaking pipe laid 150 mm below ground at centres of 350 mm		
low leak	100 m²	590.00
high leak	100 m²	530.00
Landscape area irrigation; laterals to mains; moisture leaking pipe laid to the surface of irrigated areas at 600 mm centres		
low leak	100 m²	300.00
high leak	100 m²	270.00
Landscape area irrigation; laterals to mains; moisture leaking pipe laid to the surface of irrigated areas at 900 mm centres		
low leak	100 m²	200.00
high leak	100 m²	180.00
multistation controller	each	415.00
solenoid valves connected to automatic controller	each	530.00

8.3.3 IRRIGATION SYSTEMS

Item	Unit	Total rate £
Excluding site overheads and profit		
RAINWATER HARVESTING		
Combined Harvesters Ltd; rainwater tanks; soft landscape areas installed below ground		
Excavate in soft landscape area by machine to install rainwater harvesting tank; grade base and lay compacted granular base to level; connect filters to rainwater down pipe of building 10 m distance and lay 110 mm uPVC pipe in trench. Install rainwater tank complete with filters for irrigation and submersible pump. cover with geofabric and backfill with excavated material compacting lightly. remove surplus material off -site		
Columbus tank 3700 litre	nr	2950.00
Columbus tank 4500 litre	nr	3100.00
Columbus tank 6500 litre	nr	3400.00
Columbus tank 9000 litre	nr	4200.00
Columbus tank 13000 litre; 2 nr 6500 tanks	nr	5300.00
Excavate in soft landscape area by hand to install rainwater harvesting tank; grade base and lay compacted granular base to level; connect filters to rainwater down pipe of building 10 m distance and lay 110 mm uPVC pipe in trench. Install low profile – rainwater tank complete with filters for irrigation and submersible pump. cover with geofabric and backfill with excavated material compacting lightly. remove surplus material off-site		
Flat tank 1500 litre	nr	3300.00
Flat tank 3000 litre	nr	4500.00

Approximate Estimates

8.4.1 FENCING AND RAILINGS

Item Excluding site overheads and profit	Unit	Total rate £
Temporary fencing		
Site fencing; supply and erect temporary protective fencing and remove at completion of works		
Cleft chestnut paling		
1.20 m high 75 mm larch posts at 3 m centres	100 m	930.00
1.50 m high 75 mm larch posts at 3 m centres	100 m	1075.00
Heras fencing		
2 week hire period	100 m	970.00
4 week hire period	100 m	1475.00
8 week hire period	100 m	2450.00
12 week hire period	100 m	3450.00
16 week hire period	100 m	4450.00
24 week hire period	100 m	6400.00
Chail link and Wire fencing		
Chain link fencing; supply and erect chain link fencing; form post holes and erect concrete posts and straining posts with struts at 50 m centres all set in 1:3:6 concrete; fix line wires		
3 mm galvanized wire 50 mm chainlink fencing		
900 mm high	m	22.00
1200 mm high	m	27.00
1800 mm high	m	36.00
Plastic coated 3.15 gauge galvanized wire mesh		
900 mm high	m	19.30
1200 mm high	m	24.00
1800 mm high	m	35.00
Extra for additional concrete straining posts with 1 strut set in concrete		
900 mm high	each	66.00
1200 mm high	each	70.00
1800 mm high	each	84.00
Extra for additional concrete straining posts with 2 struts set in concrete		
900 mm high	each	95.00
1200 mm high	each	98.00
1800 mm high	each	120.00
Extra for additional angle iron straining posts with 2 struts set in concrete		
900 mm high	each	91.00
1200 mm high	each	105.00
1400 mm high	each	125.00
1800 mm high	each	130.00
AGRICULTURAL FENCING		
Rabbit fencing		
Construct rabbit-stop fencing; erect galvanized wire netting; mesh 31 mm; 900 mm above ground, 150 mm below ground turned out and buried; on 75 mm dia. treated timber posts 1.8 m long driven 700 mm into firm ground at 4.0 m centres; netting clipped to top and bottom straining wires 2.63 mm dia.; straining post 150 mm dia. × 2.3 m long driven into firm ground at 50 m intervals		
turned in 150 mm	100 m	1125.00
buried 150 mm	100 m	1200.00

8.4.1 FENCING AND RAILINGS

Item Excluding site overheads and profit	Unit	Total rate £
Deer fence Construct deer-stop fencing; erect five 4 mm dia. plain galvanized wires and five 2 ply galvanized barbed wires at 150 mm spacing on 45 × 45 × 5 mm angle iron posts 2.4 m long driven into firm ground at 3.0 m centres, driven into firm ground at 3.0 m centres, with timber droppers 25 × 38 mm × 1.0 m long at 1.5 m centres	100 m	1575.00
Forestry fencing Supply and erect forestry fencing of three lines of 3 mm plain galvanized wire tied to 1700 × 65 mm dia. angle iron posts at 2750 m centres with 1850 × 100 mm dia. straining posts and 1600 × 80 mm dia. struts at 50.0 m centres driven into firm ground		
1800 mm high; three wires	100 m	1175.00
1800 mm high; three wires including cattle fencing	100 m	1475.00
Timber fencing		
Clear fence line of existing evergreen shrubs 3 m high average; grub out roots by machine chip on site and remove off site; erect closeboarded timber fence in treated softwood, pales 100 × 22 mm lapped, 150 × 22 mm gravel boards Concrete posts 100 × 100 mm at 3.0 m centres set into ground in 1:3:6 concrete		
900 mm high	m	67.00
1500 mm high	m	71.00
1800 mm high	m	77.00
As above but with softwood posts; three arris rails		
1350 mm high	m	68.00
1650 mm high	m	76.00
1800 mm high	m	78.00
Erect chestnut pale fencing; cleft chestnut pales; two lines galvanized wire, galvanized tying wire, treated softwood posts at 3.0 m centres and straining posts and struts at 50 m centres driven into firm ground 900 mm high; posts 75 mm dia. × 1200 mm long	m	6.90
1200 mm high; posts 75 mm dia. × 1500 mm long	m	9.10
Construct timber rail; horizontal hit and miss type; rails 150 × 25 mm; posts 100 × 100 mm at 1.8 m centres; twice stained with coloured wood preservative; including excavation for posts and concreting into ground (C7P) In treated softwood		
1800 mm high	m	54.00
Construct cleft oak rail fence with rails 300 mm minimum girth tennoned both ends; 125 × 100 mm treated softwood posts double mortised for rails; corner posts 125 × 125 mm, driven into firm ground at 2.5 m centres 3 rails	m	17.60
4 rails	m	21.50
Birdsmouth fencing; timber 600 mm high	m	22.00
900 mm high	m	23.00

Approximate Estimates

8.4.1 FENCING AND RAILINGS

Item Excluding site overheads and profit	Unit	Total rate £
Security fencing		
Supply and erect chainlink fence; 51 × 3 mm mesh; with line wires and stretcher bars bolted to concrete posts at 3.0 m centres and straining posts at 10 m centres; posts set in concrete 450 × 450 mm × 33% of height of post deep; fit straight extension arms of 45 × 45 × 5 mm steel angle with three lines of barbed wire and droppers; all metalwork to be factory hot-dip galvanized for painting on site		
Galvanized 3 mm mesh		
900 mm high	m	26.00
1200 mm high	m	32.50
1800 mm high	m	42.50
As above but with PVC coated 3.15 mm mesh (dia. of wire 2.5 mm)		
900 mm high	m	23.50
1200 mm high	m	30.00
1800 mm high	m	41.50
Add to fences above for base of fence to be fixed with hairpin staples cast into concrete ground beam 1:3:6 site mixed concrete; mechanical excavation disposal to on site spoil heaps		
125 × 225 mm deep	m	5.40
Add to fences above for straight extension arms of 45 × 45 × 5 mm steel angle with three lines of barbed wire and droppers	m	2.35
Supply and erect palisade security fence Jacksons Barbican 2500 mm high with rectangular hollow section steel pales at 150 mm centres on three 50 × 50 × 6 mm rails; rails bolted to 80 × 60 mm posts set in concrete 450 × 450 × 750 mm deep at 2750 mm centres; tops of pales to points and set at 45 degree angle; all metalwork to be hot-dip factory galvanized for painting on site	m	120.00
Supply and erect single gate to match above complete with welded hinges and lock		
1.0 m wide	each	1400.00
4.0 m wide	each	1600.00
8.0 m wide	pair	2250.00
Supply and erect Orsogril proprietary welded steel mesh panel fencing on steel posts set 750 mm deep in concrete foundations 600 × 600 mm; supply and erect proprietary single gate 2.0 m wide to match fencing		
930 mm high	100 m	6700.00
1326 mm high	100 m	8600.00
1722 mm high	100 m	10500.00
Ball stop fencing		
Supply and erect plastic coated 30 × 30 mm netting fixed to 60.3 mm dia. 12 mm solid bar lattice galvanized dual posts; top, middle and bottom rails with 3 horizontal rails on 60.3 mm dia. nylon coated tubular steel posts at 3.0 m centres and 60.3 mm dia. straining posts with struts at 50 m centres set 750 mm into FND2 concrete footings 300 × 300 × 600 mm deep; include framed chain link gate 900 × 1800 mm high to match complete with hinges and locking latch		
4500 mm high	100 m	11000.00
5000 mm high	100 m	14000.00
6000 mm high	100 m	15500.00

8.4.1 FENCING AND RAILINGS

Item Excluding site overheads and profit	Unit	Total rate £
Railings		
Conservation of historic railings; Eura Conservation Ltd		
Remove railings to workshop off site; shotblast and repair mildly damaged railings; remove rust and paint with three coats; transport back to site and re-erect		
railings with finials 1.80 m high	m	570.00
railings ornate cast or wrought iron	m	890.00
Supply and erect mild steel bar railing of 19 mm balusters at 115 mm centres welded to mild steel top and bottom rails 40 × 10 mm; bays 2.0 m long bolted to 51 × 51 mm ms hollow section posts set in C15P concrete; all metal work galvanized after fabrication		
900 mm high	m	82.00
1200 mm high	m	110.00
1500 mm high	m	120.00
Supply and erect mild steel pedestrian guard rail Class A; rails to be rectangular hollow sealed section 50 × 30 × 2.5 mm, vertical support 25 × 19 mm central between intermediate and top rail; posts to be set 300 mm into paving base; all components factory welded and factory primed for painting on site		
Panels 1000 mm high × 2000 mm wide with 150 mm toe space and 200 mm visibility gap at top	m	92.00
GATES		
STOCK GATE		
Erect stock gate, ms tubular field gate, diamond braced 1.8 m high hung on tubular steel posts set in concrete (C7P); complete with ironmongery; all galvanized		
Width 3.00 m	each	330.00
Width 4.20 m	each	360.00

8.4.2 WALLS AND SCREENS

Item Excluding site overheads and profit	Unit	Total rate £
FOUNDATIONS FOR WALLS		
IN SITU CONCRETE FOUNDATIONS		
Excavate foundation trench mechanically; remove spoil off site; lay 1:3:6 site mixed concrete foundations; distance from mixer 25 m; depth of trench to be 225 mm deeper than foundation to allow for three underground brick courses priced separately		
Concrete poured to blinded exposed ground		
200 mm deep × 400 mm wide	m	21.50
300 mm deep × 500 mm wide	m	35.00
400 mm deep × 400 mm wide	m	34.50
400 mm deep × 600 mm wide	m	52.00
600 mm deep × 600 mm wide	m	72.00
Concrete poured to formwork		
200 mm deep × 400 mm wide	m	45.50
300 mm deep × 500 mm wide	m	60.00
300 mm deep × 600 mm wide	m	65.00
300 mm deep × 800 mm wide	m	81.00
400 mm deep × 400 mm wide	m	58.00
400 mm deep × 600 mm wide	m	89.00
600 mm deep × 600 mm wide	m	110.00
Excavate foundation trench by hand; remove spoil off site; lay 1:3:6 site mixed concrete foundations; distance from mixer 25 m; depth of trench to be 225 mm deeper than foundation to allow for three underground brick courses priced separately		
Disposal to spoil heap 25 m by barrow and off site by grab vehicle		
200 mm deep × 400 mm wide	m	61.00
300 mm deep × 500 mm wide	m	87.00
400 mm deep × 400 mm wide	m	87.00
400 mm deep × 600 mm wide	m	130.00
600 mm deep × 600 mm wide	m	170.00
Excavate foundation trench mechanically; remove spoil off site; lay ready mixed concrete GEN1 discharged directly from delivery vehicle to location; depth of trench to be 225 mm deeper than foundation to allow for three underground brick courses priced separately; foundation size		
200 mm deep × 400 mm wide	m	17.40
300 mm deep × 500 mm wide	m	27.50
400 mm deep × 400 mm wide	m	27.00
400 mm deep × 600 mm wide	m	40.50
400 mm deep × 800 mm wide	m	51.00
400 mm deep × 1000 mm wide	m	64.00
400 mm deep × 1200 mm wide	m	77.00
600 mm deep × 600 mm wide	m	58.00
Reinforced concrete wall to foundations above (site mixed concrete)		
Up to 1.00 m high × 200 mm thick	m²	63.00
Up to 1.00 m high × 300 mm thick	m²	63.00

8.4.2 WALLS AND SCREENS

Item Excluding site overheads and profit	Unit	Total rate £
Reinforced concrete wall to foundations above (ready mix concrete RC35)		
1.00 m high × 200 mm thick	m²	93.00
1.00 m high × 300 mm thick	m²	105.00
Grading; excavate to reduce levels for concrete slab; lay waterproof membrane, 100 mm hardcore and form concrete slab reinforced with A142 mesh in 1:2:4 site mixed concrete to thickness; remove excavated material off site		
Concrete 1:2:4 site mixed		
100 mm thick	m²	45.50
150 mm thick	m²	53.00
250 mm thick	m²	67.00
300 mm thick	m²	75.00
Concrete ready mixed GEN 2		
100 mm thick	m²	44.00
150 mm thick	m²	51.00
250 mm thick	m²	64.00
300 mm thick	m²	71.00
BLOCK WALLING		
Notes: Measurements allow for works above ground only		
Concrete block walls; including excavation of foundation trench 450 mm deep; remove spoil off site; lay GEN 1 concrete foundations 600 × 300 mm thick; 1 block below ground		
Solid blocks 7 N/mm²; 100 mm thick		
500 mm high	m²	53.00
750 mm high	m²	62.00
1.00 m high	m²	70.00
1.25 m high	m²	78.00
1.50 m high	m²	87.00
1.80 m high	m²	97.00
140 mm thick		
140 mm thick	m²	70.00
100 mm blocks; laid on flat		
500 mm high	m²	79.00
750 mm high	m²	100.00
100 mm blocks laid flat; 2 courses underground		
1.00 m high	m²	130.00
1.20 m high	m²	145.00
1.50 m high	m²	170.00
1.80 m high	m²	195.00
Hollow blocks filled with concrete; 440 × 215 × 215 mm		
500 mm high	m²	80.00
750 mm high	m²	98.00
1.00 m high	m²	115.00
1.25 m high	m²	130.00
1.50 m high	m²	150.00
1.80 m high	m²	170.00

8.4.2 WALLS AND SCREENS

Item Excluding site overheads and profit	Unit	Total rate £
BLOCK WALLING – cont		
Concrete block retaining walls; 2 courses underground; including excavation of foundation trench 450 mm deep; remove spoil off site; lay GEN 1 concrete foundations 1200 × 400 mm thick		
Solid blocks 7 N/mm²; 100 mm thick		
1.00 m high	m	140.00
1.50 m high	m	165.00
1.50 m high	m	180.00
100 mm blocks; laid on flat		
1.00 m high	m²	205.00
1.50 m high	m²	245.00
1.80 m high	m²	270.00
Hollow blocks filled with concrete; 215 mm thick		
1.00 m high	m²	225.00
1.50 m high	m²	275.00
1.80 m high	m²	290.00
Hollow blocks with steel bar cast into the foundation		
1.00 m high	m²	250.00
1.50 m high	m²	305.00
1.80 m high	m²	325.00
Concrete block wall with brick face 112.5 mm thick to stretcher bond; including excavation of foundation trench 450 mm deep; remove spoil off site; lay GEN 1 concrete foundations 600 × 300 mm thick; place stainless steel ties at 4 nr/m² of wall face		
Solid blocks 7 N/mm² with stretcher bond brick face; reclaimed bricks PC £800.00/1000		
100 mm thick	m²	2050.00
140 mm thick	m²	2050.00
BRICK WALLING		
One brick thick wall (225 mm thick); excavation 400 mm deep; remove arisings off site; lay GEN 1 concrete foundations 450 mm wide × 250 mm thick; all in English Garden Wall bond; laid in cement: lime: sand (1:1:6) mortar with flush joints, fair face one side; DPC two courses engineering brick in cement: sand (1:3) mortar; precast concrete coping 152 × 75 mm		
Wall 900 mm high above DPC		
engineering brick (class B)	m	150.00
brick PC £300.00/1000	m	220.00
brick PC £800.00/1000	m	210.00
Wall 1200 mm high above DPC		
engineering brick (class B)	m	190.00
brick PC £300.00/1000	m	280.00
brick PC £800.00/1000	m	395.00

8.4.2 WALLS AND SCREENS

Item Excluding site overheads and profit	Unit	Total rate £
Wall 1800 mm high above DPC		
engineering brick (class B)	m	255.00
brick PC £300.00/1000	m	415.00
brick PC £800.00/1000	m	4700.00
One and a half brick wall; excavate 450 mm deep; remove arisings off site; lay GEN 1 concrete foundations 600 × 300 mm thick; two thick brick piers at 3.0 m centres; all in English Garden Wall bond; laid in cement: lime: sand (1:1:6) mortar with flush joints; fair face one side; DPC two courses engineering brick in cement: sand (1:3) mortar; coping of headers on edge		
Wall 900 mm high above DPC		
engineering brick (class B)	m	580.00
brick PC £300.00/1000	m	230.00
Wall 1200 mm high above DPC		
engineering brick (class B)	m	330.00
brick PC £300.00/1000	m	290.00
Wall 1800 mm high above DPC		
engineering brick (class B)	m	410.00
brick PC £300.00/1000	m	450.00
brick PC £800.00/1000	m	980.00
BRICK WALLING WITH PIERS		
Half brick thick wall; 102.5 mm thick; with one brick piers at 2.00 m centres; excavate foundation trench 500 mm deep; remove spoil off site; lay site mixed concrete foundations 1:3:6 350 × 150 mm thick; laid in cement: lime: sand (1:1:6) mortar with flush joints; fair face one side; DPC two courses underground; engineering brick in cement: sand (1:3) mortar; coping of headers on end		
Wall 900 mm high above DPC		
engineering brick (class B)	m	170.00
brick PC £300.00/1000	m	105.00
brick PC £800.00/1000	m	160.00
One brick thick wall; 225 mm thick; with one and a half brick piers at 3.0 m centres; excavation 400 mm deep; remove arisings off site; lay GEN 1 concrete foundations 450 mm wide × 250 mm thick; all in English Garden Wall bond; laid in cement: lime: sand (1:1:6) mortar with flush joints; fair face one side; DPC two courses engineering brick in cement: sand (1:3) mortar; precast concrete coping 152 × 75 mm		
Wall 900 mm high above DPC		
engineering brick (class B)	m	290.00
brick PC £300.00/1000	m	220.00
brick PC £800.00/1000	m	435.00
Wall 1200 mm high above DPC		
engineering brick (class B)	m	320.00
brick PC £300.00/1000	m	280.00
brick PC £800.00/1000	m	3250.00
Wall 1800 mm high above DPC		
engineering brick (class B)	m	470.00
brick PC £300.00/1000	m	415.00
brick PC £800.00/1000	m	4700.00

8.4.2 WALLS AND SCREENS

Item Excluding site overheads and profit	Unit	Total rate £
Faced block walling		
Concrete block wall with brick face 112.5 mm thick to bond pattern using snap headers; including excavation of foundation trench 450 mm deep; remove spoil off site; lay GEN 1 concrete foundations 600 × 300 mm thick; place stainless steel ties at 4 nr/m² of wall face; lay coping of brick on edge in engineering brick		
Solid blocks 7 N/mm² with snap header brick face; bricks PC £800.00/1000		
500 mm high	m²	120.00
750 mm high	m²	150.00
1.00 m high	m²	170.00
1.50 m high	m²	235.00
1.80 m high	m²	290.00
PIERS		
BLOCK PIERS		
Block piers; blocks 440 × 215 × 100 mm thick laid on flat piers on concrete footings 600 × 600 × 250 mm thick; inclusive of excavations and disposal off site; coping of engineering brick on edge; 2 courses underground		
Pier 450 × 450 mm; to receive cladding treatment priced separately		
500 mm high	nr	45.50
750 mm high	nr	58.00
1.00 m high	nr	77.00
1.25 m high	nr	90.00
1.50 m high	nr	100.00
1.80 m high	nr	125.00
BRICK PIERS		
Brick piers on concrete footings 500 × 500 × 250 mm thick; inclusive of excavations and disposal off site; coping of engineering brick on edge; 2 courses underground		
Pier 215 × 215 mm; one brick thick; brick PC £300/1000		
500 mm high	nr	45.00
750 mm high	nr	56.00
1.00 m high	nr	67.00
1.25 m high	nr	78.00
1.50 m high	nr	88.00
1.80 m high	nr	100.00
Pier 215 × 215 mm; one brick thick; brick PC £800/1000		
500 mm high	nr	55.00
750 mm high	nr	69.00
1.00 m high	nr	84.00
1.25 m high	nr	98.00
1.50 m high	nr	110.00
1.80 m high	nr	130.00

8.4.2 WALLS AND SCREENS

Item Excluding site overheads and profit	Unit	Total rate £
Pier 337.5 × 337.5 mm; one and a half brick thick; brick PC £300/1000		
500 mm high	nr	78.00
750 mm high	nr	100.00
1.00 m high	nr	125.00
1.25 m high	nr	150.00
1.50 m high	nr	170.00
1.80 m high	nr	200.00
Pier 337.5 × 337.5 mm; one and a half brick thick; brick PC £800/1000		
500 mm high	nr	240.00
750 mm high	nr	330.00
1.00 m high	nr	410.00
1.25 m high	nr	500.00
1.50 m high	nr	590.00
1.80 m high	nr	690.00
Brick piers on concrete footings 800 × 800 × 600 mm thick; inclusive of excavations and disposal off site; coping of Haddonstone S150C classically moulded corbelled piercap PC £114.00; 120 mm thick overall		
Pier 450 × 450 mm; two bricks thick; brick PC £300/1000; height of brickwork above ground		
500 mm high	nr	290.00
750 mm high	nr	325.00
1.00 m high	nr	360.00
1.25 m high	nr	395.00
1.50 m high	nr	430.00
1.80 m high	nr	470.00
2.00 m high	nr	500.00
Pier 450 × 450 mm; two bricks thick; brick PC £800/1000; height of brickwork above ground		
500 mm high	nr	330.00
750 mm high	nr	375.00
1.00 m high	nr	425.00
1.25 m high	nr	475.00
1.50 m high	nr	530.00
1.80 m high	nr	580.00
2.00 m high	nr	620.00
Brick piers on concrete footings 1000 × 1000 × 600 mm thick; inclusive of excavations and disposal off site; coping Haddonstone S215C classically moulded corbelled pyramidal piercap PC £214.00		
Pier 600 × 600 mm; three bricks thick; brick PC £800/1000; height of brickwork above ground		
1.50 m high	nr	1025.00
1.80 m high	nr	1150.00
2.00 m high	nr	1225.00

Approximate Estimates

8.4.2 WALLS AND SCREENS

Item	Unit	Total rate £
Excluding site overheads and profit		
IN SITU CONCRETE WALLS		
Excavate foundation trench mechanically; remove spoil off site; fix reinforcement starter bars 12 mm at 200 mm centres; lay 1:3:6 site mixed concrete foundations; distance from mixer 25 m; depth of trench to be 225 mm deeper than foundation; cast in situ concrete walls inclusive of bar reinforcement 12 mm; footings cast to blinded exposed ground		
Wall height 500 mm above ground; on foundation 400 mm × 200 mm thick		
150 thick wall; site mixed concrete 20 tonne aggregate loads	m	82.00
150 thick wall; ready mixed concrete ST2	m	81.00
250 thick wall; site mixed concrete; 20 tonne aggregate loads	m	120.00
250 thick wall; ready mixed concrete ST2	m	125.00
Wall 1.00 mm high on foundation 500 mm wide × 300 mm deep		
150 thick wall; site mixed concrete; 20 tonne aggregate loads	m	135.00
150 thick wall; ready mixed concrete ST2	m	130.00
250 thick wall; site mixed concrete; 20 tonne aggregate loads	m	150.00
250 thick wall; ready mixed concrete ST2	m	140.00
Wall 1.50 mm high on foundation 500 mm wide × 300 mm deep		
250 thick wall; site mixed concrete; 20 tonne aggregate loads	m	285.00
250 thick wall; ready mixed concrete ST2	m	270.00
Wall 1.80 mm high on foundation 600 mm wide × 300 mm deep		
250 thick wall; site mixed concrete; 20 tonne aggregate loads	m	360.00
250 thick wall; ready mixed concrete ST2	m	345.00
Wall 1.80 mm high on foundation 800 mm wide × 300 mm deep		
250 thick wall; site mixed concrete; 20 tonne aggregate loads	m	370.00
250 thick wall; ready mixed concrete ST2	m	350.00
SITE MIXED CONCRETE		
Mix concrete on site; aggregates delivered in 20 tonne loads; deliver mixed concrete to location by mechanical dumper distance 25 m		
1:3:6	m³	87.00
1:2:4	m³	99.00
As above but ready mixed concrete		
10 N/mm²	m³	110.00
15 N/mm²	m³	110.00
Mix concrete on site; aggregates delivered in 20 tonne loads; deliver mixed concrete to location by barrow distance 25 m		
Aggregates delivered in 20 t loads to site		
1:3:6	m³	130.00
1:2:4	m³	140.00
aggregates delivered in 1 tonne bags		
1:3:6	m³	180.00
1:2:4	m³	185.00
ready mixed concrete		
10 N/mm²	m³	150.00
15 N/mm²	m³	155.00

8.4.3 RETAINING WALLS

Item	Unit	Total rate £
Excluding site overheads and profit		
Footings for retaining walls		
Excavate trench for retaining wall; Cart material off site; pour concrete footings with L shaped T20 starter bars; total length 2.00 m at 200 mm centres; Concrete poured to 225 mm below ground level		
Ready mixed concrete		
600 mm wide × 300 deep	m	48.00
800 mm wide × 300 deep	m	58.00
1.00 mm wide × 300 deep	m	67.00
1.20 mm wide × 300 deep	m	87.00
1.00 mm wide × 400 deep	m	91.00
1.50 mm wide × 300 deep	m	100.00
1.50 mm wide × 400 deep	m	140.00
Block retaining walls		
Note: The following models do not allow for excavation to the existing bank in preparation of the retaining system. Please amalgamate with the Grading and preparation section above.		
Concrete block retaining walls; Stepoc		
Excavate trench 750 mm deep and lay concrete foundation 750 mm wide × 400 mm deep; construct Forticrete precast hollow concrete block wall with 450 mm below ground laid all in accordance with manufacturer's instructions; fix reinforcing bar 12 mm as work proceeds; fill blocks with ready mixed concrete; Allow for two coats of bitumen based waterproofing to back of wall; backfilling against face of wall with drainage trench perforated pipe filled with gravel rejects 600 mm deep		
Walls 1.00 m high		
type 190; 400 × 225 × 200 mm	m	170.00
type 256; 400 × 225 × 256 mm	m	190.00
As above but 1.50 m high		
type 190; 400 × 225 × 200 mm	m	205.00
type 256; 400 × 225 × 256 mm	m	235.00
Walls 1.50 m high as above but with foundation 1.80 m wide × 450 mm deep		
type 190; 400 × 225 × 200 mm	m	300.00
type 256; 400 × 225 × 256 mm	m	330.00
Walls 1.80 m high		
type 190; 400 × 225 × 200 mm	m	320.00
type 256; 400 × 225 × 256 mm	m	360.00
Hollow concrete block retaining wall		
Excavate trench 750 mm deep and lay concrete foundation 750 mm wide × 400 mm deep; construct hollow concrete block wall with 450 mm below ground laid all in accordance with manufacturer's instructions; fix reinforcing bar 12 mm as work proceeds; fill blocks with concrete 1:3:6 as work proceeds; Allow for two coats of bitumen based waterproofing to back of wall; backfilling against face of wall with drainage trench perforated pipe filled with gravel rejects 600 mm deep		
Walls 1.00 m high	m	155.00
Walls 1.25 m high	m	170.00
Walls 1.50 m high	m	190.00

8.4.3 RETAINING WALLS

Item Excluding site overheads and profit	Unit	Total rate £
Block retaining walls – cont		
Hollow concrete block retaining wall – cont		
Excavate trench 750 mm deep and lay concrete foundation 1.80 mm wide × 400 mm deep; construct hollow concrete block wall with 225 mm below ground; fix reinforcing bar 12 mm as work proceeds; fill blocks with concrete 1:3:6 as work proceeds; Allow for two coats of bitumen based waterproofing to back of wall; backfilling against face of wall with drainage trench perforated pipe filled with gravel rejects 600 mm deep		
Walls 1.50 m high	m	290.00
Walls 1.80 m high	m	310.00
Precast concrete retaining walls		
Excavate trench and lay concrete foundation 600 × 300; 1:3:6: plain concrete, supply and install Milton Precast Concrete precast concrete L shaped units, constructed all in accordance with manufacturer's instructions; backfill with approved excavated material compacted as the work proceeds		
1500 mm high × 1000 mm wide	m	220.00
2500 mm high × 1000 mm wide	m	485.00
3000 mm high × 1000 mm wide	m	630.00
Gabion walls		
Gabion walls; excavation costs excluded; see Grading and preparation of banks above		
Gabions 500 mm high; allowance of 1.00 m working space at top of bank; backfill working space and remove balance of arisings off site; lay concrete footing; 200 mm deep × 1.50 m wide		
Height retained		
500 mm	m	95.00
1.00 m	m	150.00
1.50 m	m	205.00
2.00 m	m	260.00
Timber retaining walls		
Excavate trench 300 mm wide to one third of the finished height of the retaining walls below; lay 100 mm hardcore; fix machine rounded logs set in concrete 1:3:6; remove excavated material from site; fix geofabric to rear of timber logs; backfill with previously excavated material set aside in position; all works by machine		
100 mm dia. logs		
500 mm high (constructed from 1.80 m lengths)	m	67.00
1.20 mm high (constructed from 1.80 m lengths)	m	100.00
1.60 mm high (constructed from 2.40 m lengths)	m	150.00
2.00 mm high (constructed from 3.00 m lengths)	m	195.00

8.4.3 RETAINING WALLS

Item Excluding site overheads and profit	Unit	Total rate £
150 mm dia. logs		
1.20 mm high (constructed from 1.80 m lengths)	m	185.00
1.60 mm high (constructed from 2.40 m lengths)	m	220.00
200 mm dia. logs		
2.00 mm high (constructed from 3.00 m lengths)	m	260.00
Timber crib wall		
Excavate trench to receive foundation 300 mm deep; place plain concrete foundation 150 mm thick in 11.50 N/mm² concrete (sulphate-resisting cement); construct timber crib retaining wall and backfill with excavated spoil behind units		
timber crib wall system; average 5.0 m high	m	1075.00
timber crib wall system; average 4.0 m high	m	800.00
timber crib wall system; average 2.0 m high	m	355.00
timber crib wall system; average 1.00 m high	m	180.00
Railway sleeper wall; Excavate and remove spoil offsite; Pour concrete footing 500 × 250 deep; lay sleeper retaining wall;		
Grade 1 hardwood sleeper 2 450 m × 250 mm wide × 150 mm high; Sleepers set to galvanized pins set in the concrete		
150 high	m	32.00
300 high	m	49.00
450 high	m	66.00
600 high	m	82.00
Grade 1 hardwood; 2 galvanized angle iron stakes per sleeper length set into concrete		
750 high	m	130.00
900 high	m	145.00

8.4.4 BARRIERS AND GUARDRAILS

Item Excluding site overheads and profit	Unit	Total rate £
Trip rails		
Steel trip rail Erect trip rail of galvanized steel tube 38 mm internal dia. with sleeved joint fixed to 38 mm dia. steel posts as above 700 mm long; set in C7P concrete at 1.20 m centres; metalwork primed and painted two coats metal preservative paint	m	175.00

8.5 EXTERNAL FIXTURES

Item Excluding site overheads and profit	Unit	Total rate £
8.5.1 SITE/STREET FURNITURE AND EQUIPMENT		
Bollards and access restriction Supply and install 10 nr cast iron Doric bollards 920 mm high above ground × 170 mm dia. bedded in concrete base 400 mm dia. × 400 mm deep	10 nr	1500.00
Benches and seating In grassed area excavate for base 2500 × 1575 mm and lay 100 mm hardcore, 100 mm concrete, brick pavers in stack bond bedded in 25 mm cement: lime: sand mortar; supply and fix where shown on drawing proprietary seat, hardwood slats on black powder coated steel frame, bolted down with 4 nr 24 × 90 mm recessed hex-head stainless steel anchor bolts set into concrete	set	1125.00
Cycle stand Supply and fix cycle stand 1250 m × 550 mm of 60.3 mm black powder coated hollow steel sections, one-piece with rounded top corners; set 250 mm into paving	each	420.00
8.5.2 ORNAMENTAL WATER FEATURES		
FAIRWATER LTD		
Excavate for small pond or lake maximum depth 1.00 m; remove arisings off site; grade and trim to shape; lay 75 mm sharp sand; line with 0.75 mm butyl liner 75 mm sharp sand and geofabric; cover over with sifted topsoil; anchor liner to anchor trench; install balancing tank and automatic top-up system Pond or lake of organic shape		
100 m²; perimeter 50 m	each	5700.00
250 m²; perimeter 90 m	each	8800.00
500 m²; perimeter 130 m	each	21000.00
1000 m²; perimeter 175 m	each	42000.00
Excavate for lake average depth 1.0 m, allow for bringing to specified levels; reserve topsoil; remove spoil to approved dump on site; remove all stones and debris over 75 mm; lay polythene sheet including welding all joints and seams by specialist; screen and replace topsoil 200 mm thick Prices are for lakes of regular shape		
500 micron sheet	1000 m	16000.00
1000 micron sheet	1000 m	20000.00
Extra for removing spoil to tip	m³	23.00
Extra for 25 mm sand blinding to lake bed	100 m²	115.00
Extra for screening topsoil	m²	1.20
Extra for spreading imported topsoil	100 m²	760.00
Plant aquatic plants in lake topsoil		
Aponogeton distachyum PC £312.00/100	100	79.00
Acorus calamus PC £198.00/100	100	79.00
Butomus umbellatus PC £198.00/100	100	79.00
Typha latifolia PC £198.00/100	100	79.00
Nymphaea PC £816.00/100	100	79.00

8.5 EXTERNAL FIXTURES

Item Excluding site overheads and profit	Unit	Total rate £
8.5.2 ORNAMENTAL WATER FEATURES – cont		
Formal water features		
Excavate and construct water feature of regular shape; lay 100 mm hardcore base and 150 mm concrete 1:2:4 site mixed; line base and vertical face with butyl liner 0.75 micron and construct vertical sides of reinforced blockwork; rendering two coats; anchor the liner behind blockwork; install pumps balancing tanks and all connections to mains supply		
1.00 × 1.00 × 1.00 deep	nr	2475.00
2.00 × 1.00 × 1.00 deep	nr	3100.00
Excavate and construct water feature of circular shape; lay 100 mm hardcore base and 150 mm concrete 1:2:4 site mixed; line base and vertical face with butyl liner 0.75 micron and construct vertical sides of reinforced blockwork; rendering two coats; anchor the liner behind blockwork; install pumps balancing tanks and all connections to mains supply		
Surround water body with ornamental fountain surround		
Architectural Heritage; Great Westwood; 6.00 m dia. 350 high	nr	12500.00
Haddonstone; circular pool surround; internal dia. 1780 mm; kerb features continuous moulding enriched with ovolvo and palmette designs; inclusive of plinth and integral conch shell vases flanked by dolphins	nr	6200.00
8.5.2 ORNAMENTAL BUILDINGS		
Excavate and dispose excavated material; lay concrete base 150 thick ; erect garden building		
Pavillion; Cast stone; Haddonstone; floor of yorkstone diamond sawn 6 sides.		
Venetian Folly L9400; Tuscan columns, pedimented arch, quoins and optional balustrading; 4.70 × 3.10	nr	17000.00
Cast stone;Haddonstone temples with domed roof and stepped floor		
Small classical; 2.54 m dia.	nr	12000.00
Large classical 3.19 m dia.	nr	18000.00
Temple; Soild limestone; Architectural Heritage		
Estate temple 2.70 m dia.; floor of yorkstone; diamond sawn 6 sides	nr	21000.00
Pergola; Solid limestone columns with oak beams; Architectural Heritage		
Estate temple 2.70 m dia.; floor of yorkstone; diamond sawn 6 sides	nr	14000.00

8.6 EXTERNAL DRAINAGE

Item Excluding site overheads and profit	Unit	Total rate £
8.6.1 SURFACE WATER AND FOUL WATER DRAINAGE		
ACCESS CHAMBERS		
Excavate inspection chamber by machine; lay base of concrete 150 thick; allow for half section pipework and benchings		
Precast concrete inspection chamber		
600 × 400 × 600 mm deep	nr	355.00
600 × 400 × 900 mm deep	nr	390.00
Polypropylene inspection chamber		
Mini access chamber 600 deep	nr	290.00
475 mm dia. × 900 mm deep; polymer cover	nr	460.00
475 mm dia. × 900 mm deep; ductile iron cover with screw down lid	nr	495.00
GULLIES		
Gullies; vitrified clay; excavate by hand; supply and set gully in concrete (C10P); connect to drainage system with flexible joints; backfilling with excavated material to 250 mm below finished level. Lay 150 mm type 1 to receive surface treatments (not included)		
Trapped mud (dirt) gully; complete with galvanized bucket and cast iron hinged locking grate and frame		
loading to 1 tonne; 100 or 150 mm outlet	each	360.00
loading to 5 tonne; 100 mm outlet	each	410.00
loading to 5 tonne; 150 mm outlet	each	420.00
Trapped mud (dirt) gully with rodding eye; complete with galvanized bucket and cast iron hinged locking grate and frame		
100 outlet; 300 mm internal dia.; 300 mm internal depth	each	355.00
150 outlet; 400 mm internal dia.; 750 mm internal depth	each	380.00
Concrete road gully		
Excavate and lay 100 mm concrete base (1:3:6) 150 x150 mm to suit given invert level of drain; supply and connect trapped precast concrete road gully set in concrete surround; connect to vitrified clay drainage system with flexible joints; supply and fix straight bar dished top cast iron grating and frame; bedded in cement: sand mortar (1:3)		
450 mm dia. × 1.07 m deep with 160 mm outlet	each	210.00
Gullies PVC-u		
Excavate and lay 100 mm concrete (C20P) base 150 × 150 mm to suit given invert level of drain; connect to drainage system; backfill with DoT Type 1 granular fill; install gully; complete with cast iron grate and frame		
yard gully 300 mm dia. × 600 mm deep	each	340.00
trapped PVC-u gully; 110 mm dia. × 215 mm deep; ductile iron frame	each	175.00
bottle gully 228 × 228 × 642 mm deep	each	150.00
bottle gully 228 × 228 × 317 mm deep	each	135.00

8.6 EXTERNAL DRAINAGE

Item Excluding site overheads and profit	Unit	Total rate £
8.6.1 SURFACE WATER AND FOUL WATER DRAINAGE – cont		
LINEAR DRAINAGE		
Linear drainage to design sensitive areas Excavate trench by machine; lay Aco Brickslot channel drain on concrete base and surround to falls; all to manufacturers specifications		
paving surround to both sides of channel	m	205.00
Linear drainage to pedestrian area Excavate trench by machine; lay Aco Multidrain MD polymer concrete channel drain on concrete base and surround to falls; all to manufacturers specifications; paving surround to channel with brick paving PC £300.00/1000		
Brickslot galvanized grating; paving to both sides	m	205.00
Brickslot stainless steel grating; paving to both sides	m	380.00
slotted galvanized steel grating; paving surround to one side of channel	m	145.00
Excavate trench by machine; lay Aco Multidrain PPD recycled polypropylene channel drain on concrete base and surround to falls; all to manufacturers specifications; paving surround to channel with brick paving PC £300.00/1000		
Heelguard composite black; paving surround to both sides of channel	m	145.00
Linear drainage to light vehicular area Excavate trench by machine; lay Aco Multidrain PPD recycled polypropylene channel drain on concrete base and surround to falls; all to manufacturers specifications; paving surround to channel with brick paving PC £300.00/1000		
Heelguard composite black; paving surround to both sides of channel	m	175.00
ductile iron; paving surround to both sides of channel	m	245.00
Accessories for channel drain Sump unit with sediment bucket	nr	180.00
End cap; inlet/outlet	nr	7.30
MANHOLES		
Brick manhole; excavate pit including earthwork support and working space disposal of surplus spoil to dump on site not exceeding 100 m; lay concrete (1:2:4) base 1500 mm dia. × 200 mm thick; reinforced with mesh reinforcement; 110 mm vitrified clay channels; benching in concrete (1:3:6) allowing one outlet and two inlets for 110 mm dia. pipe; construct inspection chamber 1 brick thick walls of engineering brick Class B; backfill with excavated material; complete with 2 nr cast iron step irons 1200 × 1200 × 1200 mm		
cover slab of precast concrete	each	1025.00
access cover; Group 2; 600 × 450 mm	each	1075.00
1200 × 1200 × 1500 mm		
access cover; Group 2; 600 × 450 mm	each	1275.00
recessed cover 5 tonne load; 600 × 450 mm; filled with block paviors	each	1425.00
1200 × 1200 × 2000 mm; Walls 327.5 thick		
access cover; Group 2; 600 × 450 mm	each	2025.00
recessed cover 5 tonne load; 600 × 450 mm; filled with block paviors	each	2175.00

8.6 EXTERNAL DRAINAGE

Item Excluding site overheads and profit	Unit	Total rate £
1500 × 1500 × 2500 mm; Walls 327.5 thick		
access cover; Group 2; 600 × 450 mm	each	3200.00
recessed cover 5 tonne load; 600 × 450 mm; filled with block paviors	each	3350.00
PIPE LAYING		
PVC-u pipe laying; excavate trench 300 mm wide by excavator; lay bedding; backfill to 150 mm above pipe with gravel rejects; backfill with excavated material to ground level; lay surface water drain pipe to depth shown		
Trench 900 mm deep; pipe 640–590 mm deep; bedding on reject sand		
110 mm	m	16.80
110 mm; short lengths	m	18.50
160 mm	m	28.00
160 mm; short lengths	m	45.00
Trench 900 mm deep; pipe 640–590 mm deep; bedding on 20 mm shingle		
110 mm	m	21.50
110 mm; short lengths	m	23.00
160 mm	m	33.00
160 mm; short lengths	m	50.00
Trench 1.20 m deep; pipe 940–890 mm deep; bedding on reject sand		
110 mm	m	29.00
110 mm; short lengths	m	41.00
160 mm	m	51.00
160 mm; short lengths	m	68.00
Trench 1.20 m deep; pipe 940–890 mm deep; bedding on 20 mm shingle		
110 mm	m	29.00
110 mm; short lengths	m	41.00
160 mm	m	51.00
160 mm; short lengths	m	68.00
Vitrified clay pipe laying; excavate trench by excavator; lay bedding; backfill to 150 mm above pipe with gravel rejects; backfill with excavated material to ground level; lay surface water drain pipe to depth shown		
Trench 900 mm deep; pipe 640–590 mm deep; bedding on reject sand		
100 mm	m	20.00
100 mm; short lengths	m	21.00
160 mm	m	31.00
160 mm; short lengths	m	36.50
Trench 900 mm deep; pipe 640–590 mm deep; bedding on 20 mm shingle		
110 mm	m	25.00
110 mm; short lengths	m	26.00
160 mm	m	35.50
160 mm; short lengths	m	41.50
Trench 1.20 m deep; pipe 940–890 mm deep; bedding on reject sand		
110 mm	m	32.00
110 mm; short lengths	m	44.00
160 mm	m	53.00
160 mm; short lengths	m	59.00
Trench 1.20 m deep; pipe 940–890 mm deep; bedding on 20 mm shingle		
110 mm	m	32.00
110 mm; short lengths	m	44.00
160 mm	m	54.00
160 mm; short lengths	m	60.00

8.6 EXTERNAL DRAINAGE

Item Excluding site overheads and profit	Unit	Total rate £
8.6.1 SURFACE WATER AND FOUL WATER DRAINAGE – cont		
SUBSOIL DRAINAGE – BY HAND		
Main drain; remove 150 mm topsoil and deposit alongside trench; excavate drain trench by machine and lay flexible perforated drain; lay bed of gravel rejects 100 mm; backfill with gravel rejects or similar to within 150 mm of finished ground level; complete fill with topsoil; remove surplus spoil to approved dump on site		
Main drain 160 mm in supplied in 35 m lengths		
450 mm deep	100 m	2000.00
600 mm deep	100 m	2500.00
900 mm deep	100 m	3350.00
Extra for couplings	each	2.35
As above but with 100 mm main drain supplied in 100 m lengths		
450 mm deep	100 m	1300.00
600 mm deep	100 m	1675.00
900 mm deep	100 m	2325.00
Extra for couplings	each	1.90
Laterals to mains above; herringbone pattern; excavation and backfilling as above; inclusive of connecting lateral to main drain		
160 mm pipe to 450 mm deep trench		
laterals at 1.0 m centres	100 m²	2000.00
laterals at 2.0 m centres	100 m²	1000.00
laterals at 3.0 m centres	100 m²	660.00
laterals at 5.0 m centres	100 m²	400.00
laterals at 10.0 m centres	100 m²	200.00
160 mm pipe to 600 mm deep trench		
laterals at 1.0 m centres	100 m²	2500.00
laterals at 2.0 m centres	100 m²	1250.00
laterals at 3.0 m centres	100 m²	830.00
laterals at 5.0 m centres	100 m²	500.00
laterals at 10.0 m centres	100 m²	250.00
160 mm pipe to 900 mm deep trench		
laterals at 1.0 m centres	100 m²	3350.00
laterals at 2.0 m centres	100 m²	1675.00
laterals at 3.0 m centres	100 m²	1100.00
laterals at 5.0 m centres	100 m²	670.00
laterals at 10.0 m centres	100 m²	335.00
Extra for 160/160 mm couplings connecting laterals to main drain		
laterals at 1.0 m centres	10 m	95.00
laterals at 2.0 m centres	10 m	47.50
laterals at 3.0 m centres	10 m	31.50
laterals at 5.0 m centres	10 m	19.00
laterals at 10.0 m centres	10 m	9.50
100 mm pipe to 450 mm deep trench		
laterals at 1.0 m centres	100 m²	1300.00
laterals at 2.0 m centres	100 m²	660.00
laterals at 3.0 m centres	100 m²	430.00
laterals at 5.0 m centres	100 m²	260.00
laterals at 10 m centres	100 m²	130.00

8.6 EXTERNAL DRAINAGE

Item Excluding site overheads and profit	Unit	Total rate £
100 mm pipe to 600 mm deep trench		
laterals at 1.0 m centres	100 m²	1675.00
laterals at 2.0 m centres	100 m²	830.00
laterals at 3.0 m centres	100 m²	550.00
laterals at 5.0 m centres	100 m²	330.00
laterals at 10.0 m centres	100 m²	165.00
100 mm pipe to 900 mm deep trench		
laterals at 1.0 m centres	100 m²	2325.00
laterals at 2.0 m centres	100 m²	1175.00
laterals at 3.0 m centres	100 m²	770.00
laterals at 5.0 m centres	100 m²	465.00
laterals at 10.0 m centres	100 m²	230.00
80 mm pipe to 450 mm deep trench		
laterals at 1.0 m centres	100 m²	1275.00
laterals at 2.0 m centres	100 m²	630.00
laterals at 3.0 m centres	100 m²	420.00
laterals at 5.0 m centres	100 m²	250.00
laterals at 10.0 m centres	100 m²	125.00
80 mm pipe to 600 mm deep trench		
laterals at 1.0 m centres	100 m²	1625.00
laterals at 2.0 m centres	100 m²	810.00
laterals at 3.0 m centres	100 m²	530.00
laterals at 5.0 m centres	100 m²	320.00
laterals at 10.0 m centres	100 m²	160.00
80 mm pipe to 900 mm deep trench		
laterals at 1.0 m centres	100 m²	2275.00
laterals at 2.0 m centres	100 m²	1150.00
laterals at 3.0 m centres	100 m²	750.00
laterals at 5.0 m centres	100 m²	455.00
laterals at 10.0 m centres	100 m²	230.00
Extra for 100/80 mm couplings connecting laterals to main drain		
laterals at 1.0 m centres	10 m	44.00
laterals at 2.0 m centres	10 m	22.00
laterals at 3.0 m centres	10 m	14.70
laterals at 5.0 m centres	10 m	8.80
laterals at 10.0 m centres	10 m	4.40

8.6.4 LAND DRAINAGE

SUBSOIL DRAINAGE – BY MACHINE

Main drain; remove 150 mm topsoil and deposit alongside trench, excavate drain trench by machine and lay flexible perforated drain; lay bed of gravel rejects 100 mm; backfill with gravel rejects or similar to within 150 mm of finished ground level; complete fill with topsoil; remove surplus spoil to approved dump on site

Item	Unit	Total rate £
Main drain 160 mm supplied in 35 m lengths		
450 mm deep	100 m	700.00
600 mm deep	100 m	820.00
900 mm deep	100 m	1375.00
Extra for couplings	each	2.35

Approximate Estimates

8.6 EXTERNAL DRAINAGE

Item Excluding site overheads and profit	Unit	Total rate £
8.6.4 LAND DRAINAGE – cont		
Main drain – cont		
100 mm main drain supplied in 100 m lengths		
450 mm deep	100 m	1050.00
600 mm deep	100 m	1375.00
900 mm deep	100 m	2250.00
Extra for couplings	each	1.90
Laterals to mains above; herringbone pattern; excavation and backfilling as above; inclusive of connecting lateral to main drain		
160 mm pipe to 450 mm deep trench		
laterals at 1.0 m centres	100 m²	700.00
laterals at 2.0 m centres	100 m²	350.00
laterals at 3.0 m centres	100 m²	230.00
laterals at 5.0 m centres	100 m²	140.00
laterals at 10.0 m centres	100 m²	70.00
160 mm pipe to 600 mm deep trench		
laterals at 1.0 m centres	100 m²	820.00
laterals at 2.0 m centres	100 m²	410.00
laterals at 3.0 m centres	100 m²	270.00
laterals at 5.0 m centres	100 m²	160.00
laterals at 10.0 m centres	100 m²	82.00
160 mm pipe to 900 mm deep trench		
laterals at 1.0 m centres	100 m²	1375.00
laterals at 2.0 m centres	100 m²	680.00
laterals at 3.0 m centres	100 m²	450.00
laterals at 5.0 m centres	100 m²	270.00
laterals at 10.0 m centres	100 m²	135.00
Extra for 160/160 mm couplings connecting laterals to main drain		
laterals at 1.0 m centres	10 m	95.00
laterals at 2.0 m centres	10 m	47.50
laterals at 3.0 m centres	10 m	31.50
laterals at 5.0 m centres	10 m	19.00
laterals at 10.0 m centres	10 m	9.50
100 mm pipe to 450 mm deep trench		
laterals at 1.0 m centres	100 m²	1050.00
laterals at 2.0 m centres	100 m²	520.00
laterals at 3.0 m centres	100 m²	340.00
laterals at 5.0 m centres	100 m²	210.00
laterals at 10.0 m centres	100 m²	100.00
100 mm pipe to 600 mm deep trench		
laterals at 1.0 m centres	100 m²	1375.00
laterals at 2.0 m centres	100 m²	690.00
laterals at 3.0 m centres	100 m²	450.00
laterals at 5.0 m centres	100 m²	275.00
laterals at 10.0 m centres	100 m²	140.00
100 mm pipe to 900 mm deep trench		
laterals at 1.0 m centres	100 m²	2250.00
laterals at 2.0 m centres	100 m²	1125.00
laterals at 3.0 m centres	100 m²	740.00
laterals at 5.0 m centres	100 m²	450.00
laterals at 10.0 m centres	100 m²	225.00

8.6 EXTERNAL DRAINAGE

Item Excluding site overheads and profit	Unit	Total rate £
80 mm pipe to 450 mm deep trench		
laterals at 1.0 m centres	100 m²	990.00
laterals at 2.0 m centres	100 m²	495.00
laterals at 3.0 m centres	100 m²	330.00
laterals at 5.0 m centres	100 m²	200.00
laterals at 10.0 m centres	100 m²	99.00
80 mm pipe to 600 mm deep trench		
laterals at 1.0 m centres	100 m²	1325.00
laterals at 2.0 m centres	100 m²	660.00
laterals at 3.0 m centres	100 m²	440.00
laterals at 5.0 m centres	100 m²	265.00
laterals at 10.0 m centres	100 m²	130.00
80 mm pipe to 900 mm deep trench		
laterals at 1.0 m centres	100 m²	2200.00
laterals at 2.0 m centres	100 m²	1100.00
laterals at 3.0 m centres	100 m²	730.00
laterals at 5.0 m centres	100 m²	440.00
laterals at 10.0 m centres	100 m²	220.00
Extra for 100/80 mm junctions connecting laterals to main drain		
laterals at 1.0 m centres	10 m	44.00
laterals at 2.0 m centres	10 m	22.00
laterals at 3.0 m centres	10 m	14.70
laterals at 5.0 m centres	10 m	8.80
laterals at 10.0 m centres	10 m	4.40
Clay land drain		
Excavate trench by excavator to 450 mm deep; lay 100 mm vitrified clay drain with butt joints, bedding Class B; backfill with excavated material screened to remove stones over 40 mm; backfill to be laid in layers not exceeding 150 mm; top with 150 mm topsoil remove surplus material to approved dump on site not exceeding 100 m; final level of fill to allow for settlement	100 m	1725.00
DRAINAGE DITCHES		
Excavate and form ditch and bank with 45 degree sides in light to medium soils; all widths taken at bottom of ditch		
300 mm wide × 600 mm deep	100 m	105.00
600 mm wide × 900 mm deep	100 m	100.00
1.20 m wide × 900 mm deep	100 m	920.00
1.50 m wide × 1.20 m deep	100 m	1550.00
Clear and bottom existing ditch average 1.50 m deep, trim back vegetation and remove debris to licensed tip, lay jointed concrete pipes; including bedding, haunching and topping with 150 mm concrete; 11.50 N/mm²–40 mm aggregate; backfill with approved spoil from site		
Pipes 300 mm dia.	100 m	6800.00
Pipes 450 mm dia.	100 m	8500.00
Pipes 600 mm dia.	100 m	11000.00

8.6 EXTERNAL DRAINAGE

Item Excluding site overheads and profit	Unit	Total rate £
8.6.4 LAND DRAINAGE – cont		
LAND DRAINS		
Excavate trench by excavator; lay Type 2 bedding; backfill to 150 mm above pipe with gravel rejects; lay non woven geofabric and fill with topsoil to ground level		
Trench 600 deep		
160 mm PVC-u drainpipe	100 m	2600.00
110 mm PVC-u drainpipe	100 m	1500.00
150 mm vitrified clay	100 m	2900.00
100 mm vitrified clay	100 m	1825.00
SOAKAWAYS		
Construct soakaway from perforated concrete rings; excavation, casting in situ concrete ring beam base; filling with gravel 250 mm deep; placing perforated concrete rings; surrounding with geofabric and backfilling to external surround of soakaway with 250 mm wide granular surround and excavated material; step irons and cover slab; inclusive of all earthwork retention and disposal off site of surplus material		
900 mm dia.		
1.00 m deep	nr	630.00
2.00 m deep	nr	1050.00
1200 mm dia.		
1.00 m deep	nr	840.00
2.00 m deep	nr	1375.00
2400 mm dia.		
1.00 m deep	nr	2375.00
2.00 m deep	nr	3850.00
Aquacell Infiltration unit soakaway; 1.00 m × 500 × 400 mm; 200 litre volume each in trench		
Excavate for new soakaway; grade bottom of excavation; Lay bedding layer of 100 mm sharp sand; Install geotextile to line excavation and bedding; Install Aquacell units in bonded formation; wrap entire installation in geotextile; back filling to surround in MOT Type 2 150 mm thick. Place sharp sand over infiltration units; inclusive of connection only of in-flow pipework; Cover over 500 deep with Type 2 granular material.		
8 units in trench; 1520 litres	nr	790.00
12 units in trench; 2280 litres	nr	1275.00
16 units in trench; 3040 litres	nr	1575.00
20 units in trench; 3800 litres	nr	1950.00
30 units; in trench 5700 litres	nr	3250.00
60 units; in trench 11400 litres	nr	5900.00

PART 4

Prices for Measured Works

INTRODUCTION

Typical Project Profile

Contract value	£100,000.00–£900,000.00
Labour rate (see page 0)	£19.75 per hour
Labour rate for maintenance contracts	£17.25 per hour
Number of site staff	35
Project area	6000 m²
Project location	Outer London
Project components	50% hard landscape 50% soft landscape and planting
Access to works areas	Very good
Contract	Main contract
Delivery of materials	Full loads
Profit and site overheads	Excluded

Prices for Measured Works

NEW ITEMS 2014

Item Excluding site overheads and profit	PC £	Labour hours	Labour £	Plant £	Material £	Unit	Total rate £
Tree planting in Urban Environments							
Treepits for urban planting							
Excavating tree pits; depositing soil							
alongside pits; by machine							
1.75 × 1.75 × 1.00deep	–	2.09	41.24	56.01	–	nr	97.25
2.00 × 2.00 × 1.00deep	–	2.73	53.86	72.96	–	nr	126.82
2.50 × 2.50 × 1.50deep	–	3.28	64.78	175.96	–	nr	240.74
Earthwork retention to tree pits							
bracing of deep treepits by plywood							
boards braced by timber stays							
Treepit 1.00 m × 1.00 m × 1.00 m	–	0.75	29.63	–	14.73	nr	44.36
Treepit 1.50 m × 1.50 m × 1.00 m	–	0.90	35.55	–	20.51	nr	56.06
Treepit 2.00 m × 2.00 m × 1.50 m	–	1.00	39.50	–	26.29	nr	65.79
Excavate tree break-out zone for tree							
roots to develop in restricted and							
compacted urban environments;							
depositing soil alongside pits by							
machine;							
Trenches 1.50 m wide × 1.50 m deep							
average	–	0.17	3.29	8.94	–	m³	12.23
Bases to treepits; levelling base							
breaking up base by hand and placing of							
granular drainage material 200 mm thick							
Treepit 1.00 m × 1.00 m	–	0.25	4.94	13.41	7.09	nr	25.44
Treepit 1.50 m × 1.50 m	–	0.45	8.89	24.14	15.96	nr	48.99
Treepit 2.00 m × 2.00 m	–	0.75	14.81	40.23	28.37	nr	83.41
Treepit 2.50 m × 2.50 m	–	0.75	14.81	40.23	44.33	nr	99.37
StrataCell; Greenleaf Ltd; Tree							
planting accessories for Urban tree							
planting; Load bearing tree planting							
modules for load support in tree							
environments; Placed to excavated							
treepit;							
single layer; 500 mm × 500 mm × 250 mm							
deep; interlocking load bearing cells;							
backfilling with imported topsoil							
7.5 tonne load limit	53.32	0.25	4.94	3.31	63.36	m²	71.61
40 tonne load limit	75.72	0.25	4.94	3.31	85.75	m²	94.00
double layer;							
500 mm × 500 mm × 500 mm deep;							
interlocking load bearing cells; backfilling							
with imported topsoil							
7.5 tonne load limit	106.64	0.50	9.88	6.63	126.71	m²	143.22
40 tonne load limit	151.44	0.50	9.88	6.63	171.51	m²	188.02

NEW ITEMS 2014

Item Excluding site overheads and profit	PC £	Labour hours	Labour £	Plant £	Material £	Unit	Total rate £
Root director; Greenleaf Ltd; preformed root barrier system with integral root deflecting ribs; Placed over tree rootball; under pavements; overall size;							
510 × 595 × 310 mm	41.59	0.25	9.88	–	41.59	nr	**51.47**
650 × 855 × 455 mm	68.60	0.33	13.04	–	68.60	nr	**81.64**
975 × 1370 × 545 mm	105.63	0.45	17.77	–	105.63	nr	**123.40**
1300 × 1805 × 500 mm	185.90	0.60	23.70	–	185.90	nr	**209.60**
Wild flower meadows							
Note for this section; Prices exclude for delivery and collection of specialized machinery where used to the site							
Topsoil stripping for wild flower planting Stripping to subsoil layer 300 deep							
moving to stockpile	–	–	–	1.58	–	m²	**1.58**
spreading locally	–	–	–	2.40	–	m²	**2.40**
Soil testing for wildflower meadows Collection of samples and submission to soil laboratory for analysis and recommendation for wildflower meadow establishment							
3 nr random samples	–	–	–	–	–	nr	**500.00**
Cultivation and preparation of seedbeds for wildflower meadows Cultivate stripped , filled or existing surface prior to application of herbicides and leave fallow for seedbank eradication period;							
by tractor drawn implements including rough grading to medium tilth	–	–	–	1.38	–	100 m²	**1.38**
by hand drawn rotavator including hand raking, and hand grading, to medium tilth	–	0.72	14.22	2.52	–	100 m²	**16.74**

NEW ITEMS 2014

Item Excluding site overheads and profit	PC £	Labour hours	Labour £	Plant £	Material £	Unit	Total rate £
Wild flower meadows – cont							
Seedbank eradication for wildflower meadows							
Treat stripped intended meadow area to eradicate inherent seedbank; allow optimum 2 years of repeat eradication by herbicide returning regularly to reapply herbicides							
Existing or reclaimed brownfield area by tractor drawn application; 4 applications	–	0.04	0.79	0.60	0.31	100 m²	**1.70**
Existing or reclaimed brownfield area by tractor drawn application; 3 applications	–	0.03	0.59	0.45	0.23	100 m²	**1.27**
Existing paddocks or agricultural area; knapsack; 4 applications	–	2.12	41.87	–	0.31	100 m²	**42.18**
Existing or reclaimed brownfield area by backpack spray application; 3 applications	–	1.59	31.40	–	0.23	100 m²	**31.63**
Subsoil filling to wildflower meadows							
Average thickness 300 mm; filling of poor nutrient value soil to form seedbed for wildflower area; grading to level and preparing to medium tilth							
from stockpile on site	–	–	–	27.70	–	100 m²	**27.70**
imported subsoil fill	–	–	–	27.70	0.72	100 m²	**28.42**
Seeding to wildflower meadows; Spreading of selected wildflower seed to manufacturers specifications; by tractor drawn seeder							
General areas							
DLF Trifolium; Pro Flora 8 Old English Country Meadow Mix; 5 g/m²	–	–	–	1.20	12.50	100 m²	**13.70**
Acid soils							
British Seed Houses; WFG2 (Annual Meadow); 5 g/m²	–	–	–	1.20	22.95	100 m²	**24.15**
Neutral soils							
British Seed Houses; WFG4 (Neutral Meadow); 5 g/m²	–	–	–	1.20	25.19	100 m²	**26.39**
DLF Trifolium; Pro Flora 3 Damp loamy soils; 5 g/m²	–	–	–	1.20	15.75	100 m²	**16.95**
Heavy clay soils							
British Seed Houses; WFG6 (Clay Soils); 5 g/m²	–	–	–	1.20	28.30	100 m²	**29.50**

NEW ITEMS 2014

Item Excluding site overheads and profit	PC £	Labour hours	Labour £	Plant £	Material £	Unit	Total rate £
Seeding to wildflower meadows; **Spreading of selected wildflower seed** **to manufacturers specifications; by** **pedestrian drawn seeder**							
General areas							
DLF Trifolium; Pro Flora 8 Old English Country Meadow Mix; 5 g/m^2	–	0.25	4.94	–	12.50	100 m^2	17.44
Acid soils							
British Seed Houses; WFG2 (Annual Meadow); 5 g/m^2	–	0.25	4.94	–	22.95	100 m^2	27.89
Neutral soils							
British Seed Houses; WFG4 (Neutral Meadow); 5 g/m^2	–	0.25	4.94	–	25.19	100 m^2	30.13
DLF Trifolium; Pro Flora 3 Damp loamy soils; 5 g/m^2	–	0.25	4.94	–	15.75	100 m^2	20.69
Heavy clay soils							
British Seed Houses; WFG6 (Clay Soils); 5 g/m^2	–	0.25	4.94	–	28.30	100 m^2	33.24
Maintenance operations on term **contract sites**							
Note: These new items supplement **items for long term landscape** **maintenance already featured in** **section Q35 of this publication**							
Path or hard surface maintenance							
Cyclic maintenance to surfaces based on annual requirements for long term maintenance contractors							
Check or clear daily for litter or dog faeces	–	12.00	210.00	–	–	100 m^2	210.00
Sweeping monthly	–	1.00	17.50	–	–	100 m^2	17.50
Jet washing bi-annually	–	0.33	5.78	–	–	100 m^2	5.78
Surface grit application; 6 occasions per annum	–	3.00	52.50	–	24.00	100 m^2	76.50
Snow clearance; per occasion	–	1.25	21.88	–	–	100 m^2	21.88
Childrens play area maintenance							
Visit daily to inspect for safety (Play area as part of larger maintained staffed area)	–	0.10	1.75	–	–	nr	1.75
Visit daily to inspect for safety (Play area as isolated play area)	–	0.50	8.75	–	–	nr	8.75
Monthly inspections of play equipment integrity and safety	–	1.00	17.50	–	–	nr	17.50
Annual repairs and maintenance to equipment in play areas; high density populations	–	4.00	70.00	–	–	nr	70.00
Annual repairs and maintenance to equipment in play areas; low density populations	–	1.00	17.50	–	–	nr	17.50

NEW ITEMS 2014

Item Excluding site overheads and profit	PC £	Labour hours	Labour £	Plant £	Material £	Unit	Total rate £
Maintenance operations on term contract sites – cont							
Path or hard surface maintenance – cont							
Street furniture maintenance – annual cost							
Clean benches weekly	–	2.60	51.35	–	–	nr	**51.35**
Sand timber bench bi-annually; remove splinters	–	0.50	9.88	–	–	nr	**9.88**
Maintain lighting bollard inclusive of cleaning and bulb replacement	–	0.50	9.88	–	2.48	nr	**12.36**
Maintain lighting standard inclusive of cleaning and bulb replacement	–	0.75	14.81	–	2.48	nr	**17.29**
Maintain clean exterior to litter and dog bins	–	3.00	52.50	–	–	nr	**52.50**
Litter maintenance							
Check and clear litter bins daily (high density) based on 240 days per annum	–	20.00	350.00	256.62	14.40	nr	**621.02**
Check and clear litter bins weekly	–	4.33	75.83	171.57	3.12	nr	**250.52**
Clear dog litter bins daily (in conjunction with bin clearance on site)	–	6.00	105.00	240.00	3.60	nr	**348.60**
Clear dog litter bins weekly (in conjunction with bin clearance on site)	–	1.00	17.50	52.00	14.40	nr	**83.90**
Sports maintenance							
Take football posts from store and erect	–	2.00	79.00	–	–	set	**79.00**
Remove football posts and remove to store	–	0.75	29.63	–	–	set	**29.63**
Mark football/rugby field; initial mark	19.50	5.00	87.50	–	19.50	nr	**107.00**
Mark football/rugby field; remark per occasion	4.50	1.00	17.50	–	4.50	nr	**22.00**
Graffiti maintenance per occasion based on term graffiti removal contract							
Remove graffiti from street furniture;	–	–	–	–	45.00	nr	**45.00**
Remove graffiti from masonry	–	–	–	–	45.00	nr	**45.00**

A PRELIMINARIES

Item Excluding site overheads and profit	PC £	Labour hours	Labour £	Plant £	Material £	Unit	Total rate £
A DESIGN COSTS FOR LANDSCAPE WORKS							
Landscape design; Craft Pegg (landscape architects); mixed hard and soft landscape; indicative prices for design project stages; brief; planning; detailed design and supervision during construction stage; the complexity ratings shown refer to the Landscape Institute tables published in this book							
Commercial urban development; complexity rating 2; landscape construction value							
£100,000.00	–	–	–	–	–	nr	10000.00
£250,000.00	–	–	–	–	–	nr	20000.00
£500,000.00	–	–	–	–	–	nr	35000.00
£1,000,000.00	–	–	–	–	–	nr	65000.00
Hospitals and education; complexity rating 3; landscape construction value							
£100,000.00	–	–	–	–	–	nr	11000.00
£250,000.00	–	–	–	–	–	nr	23000.00
£500,000.00	–	–	–	–	–	nr	40000.00
£1,000,000.00	–	–	–	–	–	nr	74000.00
Public urban open space; complexity rating 4; landscape construction value							
£100,000.00	–	–	–	–	–	nr	12000.00
£250,000.00	–	–	–	–	–	nr	26000.00
£500,000.00	–	–	–	–	–	nr	47500.00
£1,000,000.00	–	–	–	–	–	nr	85000.00
Countryside or leisure open space; complexity rating 1							
£100,000.00	–	–	–	–	–	nr	9000.00
£250,000.00	–	–	–	–	–	nr	18000.00
£500,000.00	–	–	–	–	–	nr	32000.00
£1,000,000.00	–	–	–	–	–	nr	60000.00
Reclamation project; complexity rating 4							
£100,000.00	–	–	–	–	–	nr	12000.00
£250,000.00	–	–	–	–	–	nr	26000.00
£500,000.00	–	–	–	–	–	nr	45000.00
£1,000,000.00	–	–	–	–	–	nr	80000.00
Housing; complexity rating 3							
£100,000.00	–	–	–	–	–	nr	11000.00
£250,000.00	–	–	–	–	–	nr	23000.00
£500,000.00	–	–	–	–	–	nr	40000.00
£1,000,000.00	–	–	–	–	–	nr	74000.00

A PRELIMINARIES

Item Excluding site overheads and profit	PC £	Labour hours	Labour £	Plant £	Material £	Unit	Total rate £
A11 TENDER AND CONTRACT DOCUMENTS							
Note: The preliminary requirements are different for all projects. Many are time related and many are set costs. Others vary depending on the project requirements. We have attempted to provide a guide to the cost related items in Section A of the National Building Specification. These costs are a guide only. Users of the book should evaluate each project separately. There are other preliminary cost items which may be added in to the preliminary costs of any project which may not be featured here.							
Method statements							
Provide detailed method statements on all aspects of the works; project value							
£30,000	–	2.50	49.38	–	–	nr	49.38
£50,000	–	3.50	69.13	–	–	nr	69.13
£75,000	–	4.00	79.00	–	–	nr	79.00
£100,000	–	5.00	98.75	–	–	nr	98.75
£200,000	–	6.00	118.50	–	–	nr	118.50
Tender and contract documents							
Drawing and plan printing and distributioncosts; plans issued by employer on CD; project value							
£30,000	–	–	–	–	19.00	nr	19.00
£30,000 to £80,000	–	–	–	–	47.50	nr	47.50
£80,000 to £150,000	–	–	–	–	95.00	nr	95.00
£150,000 to £300,000	–	–	–	–	133.00	nr	133.00
£300,000 to £1,000,000	–	–	–	–	228.00	nr	228.00
Contract drawings for project management and distribution to suppliers, subcontractors and site staff							
£30,000	–	–	–	–	38.00	nr	38.00
£30,000 to £80,000	–	–	–	–	95.00	nr	95.00
£80,000 to £150,000	–	–	–	–	190.00	nr	190.00
£150,000 to £300,000	–	–	–	–	266.00	nr	266.00
£300,000 to £1,000,000	–	–	–	–	456.00	nr	456.00

A PRELIMINARIES

Item Excluding site overheads and profit	PC £	Labour hours	Labour £	Plant £	Material £	Unit	Total rate £
A12 THE SITE/EXISTING BUILDINGS							
A12/140 – Existing mains and services; mark positions of existing mains and services; locate and mark; site area							
less than 500 m²	–	1.50	29.63	–	–	nr	29.63
up to 1000 m²	–	3.00	59.25	–	–	nr	59.25
up to 2000 m²	–	8.00	158.00	–	–	nr	158.00
up to 4000 m²	–	16.00	316.00	–	–	nr	316.00
A12/200 – Access to the site							
For pedestrians							
security kiosk	–	–	–	71.80	–	week	71.80
security guard	–	40.00	400.00	–	–	week	400.00
protected walkways; Heras fencing on 2 sides	–	–	–	4.98	–	m	4.98
A12/210 – Parking							
Parking expenses where vehicles do not park on the site area; per vehicle							
metropolitan area; city centre	–	–	–	–	–	week	200.00
metropolitan area; outer areas	–	–	–	–	–	week	160.00
suburban restricted parking areas	–	–	–	–	–	week	40.00
A12/210 – Congestion charging							
London only	–	–	–	–	–	week	40.00
A12/250 – Site visit; pre-tender for purposes of understanding site and tender requirements; prices below are based on a single senior manager attending site							
City location; average distance of travel 20 miles inclusive of travel and parking costs							
less than 500 m²	–	4.00	79.00	–	36.00	nr	115.00
up to 1000 m²	–	5.00	98.75	–	38.00	nr	136.75
up to 2000 m²	–	6.00	118.50	–	44.00	nr	162.50
up to 4000 m²	–	8.00	158.00	–	52.00	nr	210.00
Town or rural location; average distance of travel 20 miles inclusive of travel and parking costs							
less than 500 m²	–	3.00	59.25	–	22.00	nr	81.25
up to 1000 m²	–	4.00	79.00	–	23.00	nr	102.00
up to 2000 m²	–	5.00	98.75	–	24.00	nr	122.75
up to 4000 m²	–	6.00	118.50	–	24.00	nr	142.50

A PRELIMINARIES

Item Excluding site overheads and profit	PC £	Labour hours	Labour £	Plant £	Material £	Unit	Total rate £
A20 THE CONTRACT/SUBCONTRACT							
Contract/subcontract evaluation							
Due diligence on evaluation or							
examination of clauses in contract							
documents; evaluation and report by a							
suitably qualified quantity surveyor or							
legal advisor where necessary							
minor works contract	–	1.00	19.75	–	–	nr	19.75
JCLI	–	–	–	–	70.00	nr	70.00
JCT intermediate contract or							
subcontract	–	1.00	19.75	–	140.00	nr	159.75
A30 EMPLOYER'S REQUIREMENTS:							
TENDERING/SUBLETTING/SUPPLY							
Tendering costs for an employed							
estimator for the acquisition of a							
landscape main or subcontract;							
inclusive of measurement, sourcing							
of materials and suppliers, cost and							
area calculations, pre and post tender							
meetings, bid compliance method							
statements and submission							
documents							
Remeasureable contract from prepared							
bills; contract value							
£25,000	–	–	–	–	–	nr	216.00
£50,000	–	–	–	–	–	nr	360.00
£100,000	–	–	–	–	–	nr	864.00
£200,000	–	–	–	–	–	nr	1440.00
£500,000	–	–	–	–	–	nr	1872.00
£1,000,000	–	–	–	–	–	nr	2880.00
Lump sum contract from specifications							
and drawings only; contract value							
£25,000	–	–	–	–	–	nr	576.00
£50,000	–	–	–	–	–	nr	720.00
£100,000	–	–	–	–	–	nr	1440.00
£200,000	–	–	–	–	–	nr	2592.00
£500,000	–	–	–	–	–	nr	3600.00
£1,000,000	–	–	–	–	–	nr	5760.00
Material costs for tendering							
Drawing and plan printing and							
distributioncosts; plans issued by							
employer on CD; project value							
£30,000	–	–	–	–	19.00	nr	19.00
£30,000 to £80,000	–	–	–	–	47.50	nr	47.50
£80,000 to £150,000	–	–	–	–	95.00	nr	95.00
£150,000 to £300,000	–	–	–	–	133.00	nr	133.00
£300,000 to £1,000,000	–	–	–	–	228.00	nr	228.00

A PRELIMINARIES

Item Excluding site overheads and profit	PC £	Labour hours	Labour £	Plant £	Material £	Unit	Total rate £
A30/480 – Programmes; tender stage programmes							
Allow for production of an outline works programmes for submission with the tenders; project value							
£30,000	–	1.50	29.63	–	–	nr	29.63
£50,000	–	2.00	39.50	–	–	nr	39.50
£75,000	–	2.50	49.38	–	–	nr	49.38
£100,000	–	3.00	59.25	–	–	nr	59.25
£200,000	–	4.00	79.00	–	–	nr	79.00
£500,000	–	4.50	88.88	–	–	nr	88.88
£1,000,000	–	6.00	118.50	–	–	nr	118.50
Method statements for private or commercial projects where award is not points based; provide detailed method statements on all aspects of the works; project value							
£30,000	–	2.50	49.38	–	–	nr	49.38
£50,000	–	3.50	69.13	–	–	nr	69.13
£75,000	–	4.00	79.00	–	–	nr	79.00
£100,000	–	5.00	98.75	–	–	nr	98.75
£200,000	–	6.00	118.50	–	–	nr	118.50
£500,000	–	7.00	138.25	–	–	nr	138.25
£1,000,000	–	8.00	158.00	–	–	nr	158.00
A30/500 – Method statements for local authority type projects where award is points rated; provide detailed method statements on all aspects of the works; project value							
£30,000	–	5.00	98.75	–	–	nr	98.75
£50,000	–	5.00	98.75	–	–	nr	98.75
£75,000	–	6.00	118.50	–	–	nr	118.50
£100,000	–	7.00	138.25	–	–	nr	138.25
£200,000	–	8.00	158.00	–	–	nr	158.00
£500,000	–	9.00	177.75	–	–	nr	177.75
£1,000,000	–	16.00	316.00	–	–	nr	316.00
A30/550–570 – Health and safety							
Produce health and safety file including preliminary meeting and subsequent progress meetings with external planning officer in connection with health and safety; project value							
£35,000	–	8.00	158.00	–	–	nr	158.00
£75,000	–	12.00	237.00	–	–	nr	237.00
£100,000	–	16.00	316.00	–	–	nr	316.00
£200,000 to £500,000	–	40.00	790.00	–	–	nr	790.00

A PRELIMINARIES

Item Excluding site overheads and profit	PC £	Labour hours	Labour £	Plant £	Material £	Unit	Total rate £
A30 EMPLOYER'S REQUIREMENTS: TENDERING/SUBLETTING/SUPPLY – cont							
A30/550–570 – Health and safety – cont							
Maintain health and safety file for project duration; project value							
£35,000	–	4.00	79.00	–	–	week	**79.00**
£75,000	–	4.00	79.00	–	–	week	**79.00**
£100,000	–	8.00	158.00	–	–	week	**158.00**
£200,000 to £500,000	–	8.00	158.00	–	–	week	**158.00**
Produce written risk assessments on all areas of operations within the scope of works of the contract; project value							
£35,000	–	2.00	39.50	–	–	nr	**39.50**
£75,000	–	3.00	59.25	–	–	nr	**59.25**
£100,000	–	5.00	98.75	–	–	nr	**98.75**
£200,000 to £500,000	–	8.00	158.00	–	–	nr	**158.00**
Produce Coshh assessments on all substances to be used in connection with the contract; project value							
£35,000	–	1.50	29.63	–	–	nr	**29.63**
£75,000	–	2.00	39.50	–	–	nr	**39.50**
£100,000	–	3.00	59.25	–	–	nr	**59.25**
£200,000 to £500,000	–	3.00	59.25	–	–	nr	**59.25**
A31 EMPLOYERS REQUIREMENTS: PROVISION CONTENT AND USE OF DOCUMENTS							
Provision, content and use of documents							
Supply as built drawings for elements of the project that may have carried from the original design drawings							
£35,000	–	2.00	70.00	–	–	nr	**70.00**
£75,000	–	2.00	70.00	–	–	nr	**70.00**
£100,000	–	4.00	140.00	–	–	nr	**140.00**
£200,000 to £500,000	–	5.00	175.00	–	–	nr	**175.00**
A31/155 – Fees for Considerate Contractors Scheme (CCS)							
Project value							
up to £100,000	–	–	–	–	100.00	nr	**100.00**
£100,000-£500000	–	–	–	–	200.00	nr	**200.00**
£500,000-£5,000,000	–	–	–	–	400.00	nr	**400.00**
over £5,000,000	–	–	–	–	600.00	nr	**600.00**

A PRELIMINARIES

Item Excluding site overheads and profit	PC £	Labour hours	Labour £	Plant £	Material £	Unit	Total rate £
Climatic conditions – keep records of temperature and rainfall; delays due to weather including descriptions of weather							
daily cost	–	0.08	1.65	–	–	day	1.65
weekly cost	–	0.42	8.23	–	–	day	8.23
A31/211 – Programmes; master programme for the contract works							
Allow for production of works programmes prior to the start of the works; project value							
£30,000	–	3.00	59.25	–	–	nr	59.25
£50,000	–	6.00	118.50	–	–	nr	118.50
£75,000	–	8.00	158.00	–	–	nr	158.00
£100,000	–	10.00	197.50	–	–	nr	197.50
£200,000	–	14.00	276.50	–	–	nr	276.50
£500,000	–	15.00	296.25	–	–	nr	296.25
£1,000,000	–	18.00	355.50	–	–	nr	355.50
A31/212 – Indicative staff resource chart							
Project value							
up to £100000	–	1.00	19.75	–	–	nr	19.75
up to £500000	–	1.50	29.63	–	–	nr	29.63
up to £1,000,000	–	2.00	39.50	–	–	nr	39.50
A31/213 – Curriculum vitae of staff							
Prepare and submit curriculum vitae of pre-construction and construction phase staff							
per staff member	–	0.75	14.81	–	–	nr	14.81
A32 EMPLOYER'S REQUIREMENTS: MANAGEMENT OF THE WORKS							
Allow for updating the works programme during the course of the works; project value							
£30,000	–	1.00	19.75	–	–	nr	19.75
£50,000	–	1.50	29.63	–	–	nr	29.63
£75,000	–	2.00	39.50	–	–	nr	39.50
£100,000	–	3.00	59.25	–	–	nr	59.25
£200,000	–	5.00	98.75	–	–	nr	98.75

A PRELIMINARIES

Item Excluding site overheads and profit	PC £	Labour hours	Labour £	Plant £	Material £	Unit	Total rate £
A32 EMPLOYER'S REQUIREMENTS: MANAGEMENT OF THE WORKS – cont							
Setting out							
Setting out for external works operations comprising hard and soft works elements; placing of pegs and string lines to Landscape Architect's drawings; surveying levels and placing level pegs; obtaining approval from the Landscape Architect to commence works; areas of entire site							
1,000 m²	–	5.00	98.75	–	7.80	nr	106.55
2,500 m²	–	8.00	158.00	–	15.60	nr	173.60
5,000 m²	–	8.00	158.00	–	15.60	nr	173.60
10,000 m²	–	32.00	632.00	–	39.00	nr	671.00
A33 SUBMISSION OF SAMPLES							
Set up sample panels for the works inclusive of arranging delivery of sample materials; paving samples on 100 mm base							
Costs of labours only							
brick paving panel (pointed)	–	6.00	118.50	–	–	m²	118.50
brick paving panel (butt jointed)	–	4.00	79.00	–	–	m²	79.00
block paving panel	–	4.00	79.00	–	–	m²	79.00
stone slab paving panel	–	6.00	118.50	–	–	m²	118.50
cladding panel	–	7.00	138.25	–	–	m²	138.25
brick wall panel	–	5.00	98.75	–	–	m²	98.75
render panel	–	3.50	69.13	–	–	m²	69.13
paint panel	–	3.00	59.25	–	–	m²	59.25
A34 EMPLOYER'S REQUIREMENTS: SECURITY/SAFETY/PROTECTION							
Temporary security fence; HSS Hire; mesh framed unclimbable fencing; including precast concrete supports and couplings							
Weekly hire; 2.85 × 2.00 high							
weekly hire rate	–	–	–	2.49	–	m	2.49
erection of fencing; labour only	–	0.10	1.98	–	–	m	1.98
removal of fencing loading to collection vehicle	–	0.07	1.32	–	–	m	1.32
delivery charge	–	–	–	0.80	–	m	0.80
return haulage charge	–	–	–	0.60	–	m	0.60

A PRELIMINARIES

Item Excluding site overheads and profit	PC £	Labour hours	Labour £	Plant £	Material £	Unit	Total rate £
A41 CONTRACTOR'S GENERAL COST ITEMS: SITE ACCOMMODATION							
General							
The following items are instances of the commonly found preliminary costs associated with external works contracts. The assumption is made that the external works contractor is subcontracted to a main contractor.							
Elliot Hire; erect temporary accommodation and storage on concrete base measured separately							
Prefabricated office hire; jackleg; open plan							
3.6 × 2.4 m	–	–	–	23.00	–	week	**23.00**
4.8 × 2.4 m	–	–	–	26.50	–	week	**26.50**
Armoured store; erect temporary secure storage container for tools and equipment							
3.0 × 2.4 m	–	–	–	11.00	–	week	**11.00**
3.6 × 2.4 m	–	–	–	12.50	–	week	**12.50**
Delivery and collection charges on site offices							
delivery charge	–	–	–	150.00	–	load	**150.00**
collection charge	–	–	–	150.00	–	load	**150.00**
Toilet facilities							
HSS Hire; serviced self-contained toilet delivered to and collected from site; maintained by toilet supply company							
single chemical toilet including wash hand basin and water	–	–	–	43.00	–	week	**43.00**
delivery and collection; each way	–	–	–	40.00	–	nr	**40.00**

A PRELIMINARIES

Item Excluding site overheads and profit	PC £	Labour hours	Labour £	Plant £	Material £	Unit	Total rate £
A43 CONTRACTOR'S GENERAL COST ITEMS: MECHANICAL PLANT							
Southern Conveyors; moving only of granular material by belt conveyor; conveyors fitted with troughed belts, receiving hopper, front and rear undercarriages and driven by electrical and air motors; support work for conveyor installation (scaffolding), delivery, collection and installation all excluded							
Conveyor belt width 400 mm; mechanically loaded and removed at offload point; conveyor length							
up to 5 m	–	1.50	29.63	19.51	–	m³	**49.14**
10 m	–	1.50	29.63	19.97	–	m³	**49.60**
12.5 m	–	1.50	29.63	20.52	–	m³	**50.15**
15 m	–	1.50	29.63	20.91	–	m³	**50.54**
20 m	–	1.50	29.63	21.85	–	m³	**51.48**
25 m	–	1.50	29.63	23.02	–	m³	**52.65**
30 m	–	1.50	29.63	24.19	–	m³	**53.82**
Conveyor belt width 600 mm; mechanically loaded and removed at offload point; conveyor length							
up to 5 m	–	1.00	19.75	13.27	–	m³	**33.02**
10 m	–	1.00	19.75	13.79	–	m³	**33.54**
12.5 m	–	1.00	19.75	14.16	–	m³	**33.91**
15 m	–	1.00	19.75	14.57	–	m³	**34.32**
20 m	–	1.00	19.75	15.35	–	m³	**35.10**
25 m	–	1.00	19.75	16.65	–	m³	**36.40**
30 m	–	1.00	19.75	17.70	–	m³	**37.45**
Conveyor belt width 400 mm; mechanically loaded and removed by hand at offload point; conveyor length							
up to 5 m	–	3.19	63.01	5.62	–	m³	**68.63**
10 m	–	3.19	63.01	6.25	–	m³	**69.26**
12.5 m	–	3.19	63.01	6.98	–	m³	**69.99**
15 m	–	3.19	63.01	7.50	–	m³	**70.51**
20 m	–	3.19	63.01	8.75	–	m³	**71.76**
25 m	–	3.19	63.01	10.31	–	m³	**73.32**
30 m	–	3.19	63.01	11.87	–	m³	**74.88**

A PRELIMINARIES

Item Excluding site overheads and profit	PC £	Labour hours	Labour £	Plant £	Material £	Unit	Total rate £
Terranova Cranes Ltd; crane hire;							
materials handling and lifting;							
telescopic cranes supply and							
management; exclusive of roadway							
management or planning							
applications; prices below illustrate							
lifts of 1 tonne at maximum crane							
reach; lift cycle of 0.25 hours;							
mechanical filling of material skip if							
appropriate							
35 tonne mobile crane; lifting capacity of							
1 tonne at 26 m; C.P.A hire (all							
management by hirer)							
granular materials including concrete	–	0.75	14.81	15.81	–	tonne	**30.62**
palletized or packed materials	–	0.50	9.88	13.59	–	tonne	**23.47**
35 tonne mobile crane; lifting capacity of							
1 tonne at 26 m; contract lift							
granular materials or concrete	–	0.75	14.81	44.28	–	tonne	**59.09**
palletized or packed materials	–	0.50	9.88	42.06	–	tonne	**51.94**
50 tonne mobile crane; lifting capacity of							
1 tonne at 34 m; C.P.A hire (all							
management by hirer)							
granular materials including concrete	–	0.75	14.81	20.02	–	tonne	**34.83**
palletized or packed materials	–	1.00	19.75	17.80	–	tonne	**37.55**
50 tonne mobile crane; lifting capacity of							
1 tonne at 34 m; contract lift							
granular materials or concrete	–	0.75	14.81	50.69	–	tonne	**65.50**
palletized or packed materials	–	0.50	9.88	48.47	–	tonne	**58.35**
80 tonne mobile crane; lifting capacity of							
1 tonne at 44 m; C.P.A. hire (all							
management by hirer)							
granular materials including concrete	–	0.75	14.81	28.11	–	tonne	**42.92**
palletized or packed materials	–	0.50	9.88	25.88	–	tonne	**35.76**
80 tonne mobile crane; lifting capacity of							
1 tonne at 44 m; contract lift							
granular materials or concrete	–	0.75	14.81	57.23	–	tonne	**72.04**
palletized or packed materials	–	0.50	9.88	55.00	–	tonne	**64.88**

A PRELIMINARIES

Item Excluding site overheads and profit	PC £	Labour hours	Labour £	Plant £	Material £	Unit	Total rate £
A44 CONTRACTOR'S GENERAL COST ITEMS: TEMPORARY WORKS							
Eve Trakway; portable roadway systems Temporary roadway system laid directly onto existing surface or onto PVC matting to protect existing surface; most systems are based on a weekly hire charge with transportation, installation and recovery charges included							
Heavy Duty Trakpanel; per panel (3.05 × 2.59 per week	–	–	–	–	–	m²	5.32
Outrigger Mats for use in conjunction with Heavy Duty Trakpanels; per set of 4 mats per week	–	–	–	–	–	set	85.00
Medium Duty Trakpanel; per panel (2.44 × 3.00 per week	–	–	–	–	–	m²	5.74
LD20 Eveolution; Light Duty Trakway; roll out system minimum delivery 50 m	–	–	–	–	–	m²	5.25
Terraplas Walkways; turf protection system; per section (1 × 1 m); per week	–	–	–	–	–	m²	5.50

B COMPLETE BUILDINGS/STRUCTURES/UNITS

Item Excluding site overheads and profit	PC £	Labour hours	Labour £	Plant £	Material £	Unit	Total rate £
B10 PREFABRICATED BUILDINGS/ STRUCTURES							
Cast stone buildings; Haddonstone Ltd; ornamental garden buildings in Portland Bath or Terracotta finished cast stone; prices for stonework and facades only; excavations, foundations, reinforcement, concrete infill, roofing and floors all priced separately							
Pavilion Venetian Folly L9400; Tuscan columns, pedimented arch, quoins and optional balustrading							
4184 mm high × 4728 mm wide × 3147 mm deep	12060.00	175.00	3456.25	256.90	12223.86	nr	**15937.01**
Pavilion L9300; Tuscan columns							
3496 mm high × 3634 mm wide	7320.00	144.00	2844.00	220.20	7483.86	nr	**10548.06**
Small Classical Temple L9250; 6 column with fibreglass lead effect finish dome roof							
overall height 3610 mm; dia. 2540 mm	6330.00	130.00	2567.50	183.50	6493.86	nr	**9244.86**
Large Classical Temple L9100; 8 column with fibreglass lead effect finish dome roof							
overall height 4664 mm; dia. 3190 mm	11820.00	165.00	3258.75	183.50	11983.86	nr	**15426.11**
Stepped floors to temples							
single step; Small Classical Temple	1160.00	24.00	474.00	–	1224.49	nr	**1698.49**
single step; Large Classical Temple	1638.00	26.00	513.50	–	1715.38	nr	**2228.88**
Stone structures; Architectural Heritage Ltd; hand carved from solid natural limestone with wrought iron domed roof, decorated frieze and base and integral seats; supply and erect only; excavations and concrete bases priced separately							
The Park Temple; 5 columns; 3500 mm high × 1650 mm dia.	10000.00	96.00	1896.00	75.00	10000.00	nr	**11971.00**
The Estate Temple; 6 columns; 4000 mm high × 2700 mm dia.	17600.00	120.00	2370.00	150.00	17600.00	nr	**20120.00**
Stone structures; Architectural Heritage Ltd; hand carved from solid natural limestone with solid oak trelliage; prices for stonework and facades only; supply and erect only; excavations and concrete bases priced separately							
The Pergola; 2240 mm high × 2640 mm wide × 6990 mm long	10000.00	96.00	1896.00	300.00	10032.00	nr	**12228.00**

B COMPLETE BUILDINGS/STRUCTURES/UNITS

Item Excluding site overheads and profit	PC £	Labour hours	Labour £	Plant £	Material £	Unit	Total rate £
B10 PREFABRICATED BUILDINGS/ STRUCTURES – cont							
Ornamental stone structures; Architectural Heritage Ltd; setting to bases or plinths (not included)							
The Obelisk; classic natural stone obelisk; tapering square form on panelled square base; 1860 mm high × 360 mm square	1500.00	2.00	39.50	–	1500.00	nr	**1539.50**
The Strigil Sundial Pedestal with Greenwich Armillary Sphere; overall height 2.10 m with on double stepped base	5400.00	2.00	39.50	–	5400.00	nr	**5439.50**
Large obelisk; 2.40 m high × 460 mm square; hand carved natural limestone obelisk of tapering square form raised upon a panelled square base supported by four spheres. Reproduction.	2000.00	1.00	19.75	–	2000.00	nr	**2019.75**
B10 GARDEN STUDIOS							
Garden rooms; 3rdSpace; Sawhorse Ltd; fully insulated pent roof modular garden buildings installed on granular base with jacking legs							
Douglas fir framed with western red cedar cladding; glazed with fully glazed doors; integral floor and internal finishes including birch ply cladding							
12.5 m² ; 5.0 × 2.5 × 2.5 m high	–	–	–	–	–	nr	**16000.00**
9.25 m² ; 3.7 × 2.5 × 2.5 m high	–	–	–	–	–	nr	**14000.00**
6.25 m² ; 2.5 × 2.5 × 2.5 m high	–	–	–	–	–	nr	**11500.00**
15.25 m² ; 6.5 × 2.5 × 2.5 m high	–	–	–	–	–	nr	**18330.00**

C EXISTING SITE/BUILDINGS/SERVICES

Item Excluding site overheads and profit	PC £	Labour hours	Labour £	Plant £	Material £	Unit	Total rate £
C12 UNDERGROUND SERVICES **SURVEY**							
Ground penetrating radar surveys for **artefacts, services and cable** **avoidance**							
Areas not exceeding 1000 m²	–	–	–	–	–	nr	**4900.00**
C13 BUILDING FABRIC SURVEY							
Surveys							
Asbestos surveys							
type 3 asbestos survey (destructive testing) to single storey free-standing buildings such as park buildings, storage sheds and the like; total floor area less than 100 m²	–	–	–	–	–	nr	**950.00**
extra over for the removal only of asbestos off site to tip (tipping charges only)	–	–	–	–	–	tonne	**90.00**
C20 DEMOLITION							
Demolish existing structures; **disposal off site; mechanical** **demolition; with 3 tonne excavator** **and dumper**							
Brick wall							
112.5 mm thick	–	0.07	1.32	2.33	3.55	m²	**7.20**
225 mm thick	–	0.08	1.65	2.91	7.10	m²	**11.66**
337.5 mm thick	–	–	–	9.22	11.92	m²	**21.14**
450 mm thick	–	–	–	12.30	14.21	m²	**26.51**
Demolish existing structures; **disposal off site mechanically loaded;** **all other works by hand**							
Brick wall							
112.5 mm thick	–	0.33	6.58	4.12	3.55	m²	**14.25**
225 mm thick	–	0.50	9.88	–	7.10	m²	**16.98**
337.5 mm thick	–	0.67	13.17	–	11.92	m²	**25.09**
450 mm thick	–	1.00	19.75	–	14.21	m²	**33.96**
Demolish existing structures; **disposal off site mechanically loaded;** **by diesel or electric breaker; all other** **works by hand**							
Brick wall							
112.5 mm thick	–	0.17	3.29	0.55	3.55	m²	**7.39**
225 mm thick	–	0.20	3.95	0.66	7.10	m²	**11.71**
337.5 mm thick	–	0.25	4.94	0.82	10.53	m²	**16.29**
450 mm thick	–	0.33	6.58	1.10	14.21	m²	**21.89**

C EXISTING SITE/BUILDINGS/SERVICES

Item Excluding site overheads and profit	PC £	Labour hours	Labour £	Plant £	Material £	Unit	Total rate £
C20 DEMOLITION – cont							
Break out concrete footings **associated with free-standing walls;** **inclusive of all excavation and** **backfilling with excavated material;** **disposal off site mechanically loaded**							
By mechanical breaker; diesel or electric							
plain concrete	–	1.50	29.62	12.86	18.14	m³	60.62
reinforced concrete	–	2.50	49.38	18.66	31.57	m³	99.61
Remove existing free-standing **buildings; demolition by hand**							
Timber building with suspended timber floor; hardstanding or concrete base not included; disposal off site							
shed 6.0 m²	–	2.00	39.50	–	32.89	nr	72.39
shed 10.0 m²	–	3.00	59.26	–	59.20	nr	118.46
shed 15.0 m²	–	3.50	69.13	–	92.08	nr	161.21
Timber building; insulated; with timber or concrete posts set in concrete, felt covered timber or tiled roof; internal walls cladding with timber or plasterboard; load arisings to skip							
timber structure 6.0 m²	–	2.50	49.38	–	142.07	nr	191.45
timber structure 12.0 m²	–	3.50	69.13	–	198.63	nr	267.76
timber structure 20.0 m²	–	8.00	158.00	374.84	210.47	nr	743.31
Demolition of free-standing brick **buildings with tiled or sheet roof;** **concrete foundations measured** **separately; mechanical demolition;** **maximum distance to stockpile 25 m;** **inclusive for all access scaffolding** **and the like; maximum height of roof** **4.0 m; inclusive of all doors, windows,** **guttering and down pipes; including** **disposal by grab**							
Half brick thick							
10 m²	–	8.00	158.00	168.06	158.73	nr	484.79
20 m²	–	16.00	316.00	299.38	204.14	nr	819.52
1 brick thick							
10 m²	–	10.00	197.50	262.67	317.46	nr	777.63
20 m²	–	16.00	316.00	394.10	408.28	nr	1118.38
Cavity wall with blockwork inner skin and brick outer skin; insulated							
10 m²	–	12.00	237.09	357.39	440.67	nr	1035.15
20 m²	–	20.00	395.00	488.87	567.72	nr	1451.59

C EXISTING SITE/BUILDINGS/SERVICES

Item Excluding site overheads and profit	PC £	Labour hours	Labour £	Plant £	Material £	Unit	Total rate £
Extra over to the above for **disconnection of services**							
Electrical							
disconnection	–	2.00	39.50	–	–	nr	**39.50**
grub out cables and dispose; backfilling; by machine	–	–	–	0.59	0.53	m	**1.12**
grub out cables and dispose; backfilling; by hand	–	0.50	9.88	–	0.53	m	**10.41**
Water supply, foul or surface water drainage							
disconnection; capping off	–	1.00	19.75	–	39.50	nr	**59.25**
grub out pipes and dispose; backfilling; by machine	–	–	–	0.59	0.53	m	**1.12**
grub out pipes and dispose; backfilling; by hand	–	0.50	9.88	–	0.53	m	**10.41**
Demolish existing fence; remove **stakes or grub out posts as** **appropriate; remove to stock pile for** **removal off site; (not included)**							
Mechanical demolition; clear fence line							
chain link fence 1.20–1.50 m high	–	0.01	0.40	0.68	–	m	**1.08**
closeboard or timber panel fence 1.80 m high	–	0.03	0.99	1.29	–	m	**2.28**
Mechanical demolition; light shrubs or creepers in chainlinks							
chain link fence 1.20–1.50 m high	–	0.01	0.49	0.87	–	m	**1.36**
closeboard or timber panel fence 1.80 m high	–	0.03	1.32	1.73	–	m	**3.05**
Mechanical demolition; heavy shrubs or creepers bramble and the like requiring clearance to enable removal							
fence 1.20–1.50 m high	–	0.05	1.98	2.12	–	m	**4.10**
closeboard or timber panel fence 1.80 m high	–	0.05	1.98	2.79	–	m	**4.77**
Demolition and site transport by hand; clear fence lines							
fence 1.20 –1.50 m high	–	0.03	1.32	–	–	m	**1.32**
closeboard or timber panel fence 1.80 m high	–	0.07	2.63	–	–	m	**2.63**
Demolition and site transport by hand; light vegetation and creepers							
fence 1.20 –1.50 m high	–	0.05	1.98	–	–	m	**1.98**
closeboard or timber panel fence 1.80 m high	–	0.13	4.94	–	–	m	**4.94**
Demolition and site transport by hand; heavy shrubs or creepers bramble and the like requiring clearance to enable removal							
fence 1.20 –1.50 m high	–	0.13	4.94	–	–	m	**4.94**
closeboard or timber panel fence 1.80 m high	–	0.17	6.58	–	–	m	**6.58**

C EXISTING SITE/BUILDINGS/SERVICES

Item Excluding site overheads and profit	PC £	Labour hours	Labour £	Plant £	Material £	Unit	Total rate £
C20 DEMOLITION – cont							
Demolition of posts and straining posts							
Break out straining post; grub out concrete footings; remove to stockpile on site							
single straining post by machine	–	0.25	4.94	16.32	–	nr	**21.26**
double straining post by machine	–	0.30	5.92	19.00	–	nr	**24.92**
single straining post by hand	–	1.00	39.50	–	–	nr	**39.50**
double straining post by hand	–	1.25	49.38	–	–	nr	**49.38**
Break out gatepost; concrete steel or timber; girth not exceeding 150 mm; grub out concrete footings							
by machine	–	0.25	4.94	16.32	–	nr	**21.26**
by hand	–	1.00	19.75	–	–	nr	**19.75**
C50 REPAIRING/RENOVATING/ CONSERVING METAL							
Eura Conservation Ltd; conservation of metal railings; works to heritage conservation standards							
Taking down and transporting existing metalwork to an off site workshop for conservation							
pair of gates; maximum overall width 4.00 m	–	–	–	–	–	pair	**596.00**
side screens to gates	–	–	–	–	–	pair	**485.00**
railings; 1.20 m high; plain	–	–	–	–	–	m	**52.00**
railings; 1.80 m high; with finials	–	–	–	–	–	m	**98.00**
ornate cast or wrought iron railings	–	–	–	–	–	m	**108.00**
Conserving metalwork off site; inclusive of repairs to metalwork; rust removal, rubbing down and preparing for re-erection on site							
gates; 2.50 m high; width not exceeding 4.00 m	–	–	–	–	–	m	**1100.00**
railings; 1.20 m high; plain	–	–	–	–	–	m	**145.00**
railings; 1.80 m high; with finials	–	–	–	–	–	m	**220.00**
ornate cast or wrought iron railings	–	–	–	–	–	m	**435.00**
Transporting from store and re-erection of existing gates into locations recorded with position orientation and features of the original installation							
gates; 2.50 m high; width not exceeding 4.00 m	–	–	–	–	–	nr	**1018.00**

C EXISTING SITE/BUILDINGS/SERVICES

Item Excluding site overheads and profit	PC £	Labour hours	Labour £	Plant £	Material £	Unit	Total rate £
Refixing existing railings previously taken down and repaired on site							
railings; 1.20 m high; plain	–	–	–	–	–	m	104.00
railings; 1.80 m high; with finials	–	–	–	–	–	m	200.00
ornate cast or wrought iron railings	–	–	–	–	–	m	248.00
Eura Conservation Ltd; supply of conservation grade railings to match existing materials on site; inclusive of all surveys, measurements and analysis of materials, installation methods and the like							
Wavy bar railings							
1.20 m high; plain	–	–	–	–	–	m	97.00
1.70 m high; straight	–	–	–	–	–	m	310.00
1.70 m high; curved	–	–	–	–	–	m	375.00
1.80 m high; with finials	–	–	–	–	–	m	300.00
ornate cast or wrought iron railings	–	–	–	–	–	m	825.00
Painting of railings: see section M60							

D11 SOIL STABILIZATION

Item Excluding site overheads and profit	PC £	Labour hours	Labour £	Plant £	Material £	Unit	Total rate £
Clarification notes on labour costs in this section							
General groundworks team Generally a three man team is used in this section; The column Labour hours reports team hours. The column Labour £ reports the total cost of the team for the unit of work shown							
3 man team	–	1.00	59.25	–	–	hr	**59.25**
GENERALLY							
Soil stabilization – General Preamble: Earth-retaining and stabilizing materials are often specified as part of the earth-forming work in landscape contracts, therefore this section lists a number of products specially designed for large-scale earth control. There are two types: rigid units for structural retention of earth on steep slopes; and flexible meshes and sheets for control of soil erosion where structural strength is not required. Prices for these items depend on quantity, difficulty of access to the site and availability of suitable filling material; estimates should be obtained from the manufacturer when the site conditions have been determined.							
RETAINING WALLS							
Preamble Earth retaining and stabilization can take numerous forms depending on the height, soil type and loadings. The main types include gravity walls, reinforced soil and soil nailing. Prices for these items depend on quantity, access for the installation and the strength of the soils. Estimates should be obtained from the manufacturer when site conditions and layout have been determined.							

D11 SOIL STABILIZATION

Item Excluding site overheads and profit	PC £	Labour hours	Labour £	Plant £	Material £	Unit	Total rate £
Retaining walls; Milton Precast Concrete Retaining walls of units with plain concrete finish; prices based on 24 tonne loads but other quantities available (excavation, temporary shoring, foundations and backfilling not included)							
1000 mm wide × 1000 mm high	111.00	0.50	29.63	5.59	149.29	m	184.51
1000 mm wide × 1500 mm high	135.00	0.42	25.06	5.59	182.04	m	212.69
1000 mm wide × 2000 mm high	245.00	0.42	25.09	5.59	310.39	m	341.07
1000 mm wide × 2500 mm high	348.00	0.42	25.09	5.59	438.34	m	469.02
1000 mm wide × 3000 mm high	432.00	0.42	25.06	52.50	533.86	m	611.42
Retaining walls; Tensar International Tensartech TW1 retaining wall system; modular dry laid concrete blocks; 220 × 400 mm long × 150 mm high connected to Tensar RE geogrid with proprietary connectors; geogrid laid horizontally within the fill at 300 mm centres; on 150 × 450 mm concrete foundation; filling with imported granular material							
1.00 m high	84.05	1.00	59.25	10.31	90.84	m²	160.40
2.00 m high	84.05	1.33	79.00	10.31	90.84	m²	180.15
3.00 m high	84.05	1.50	88.88	10.31	90.84	m²	190.03
Retaining walls; Grass Concrete Ltd Betoflor precast concrete landscape retaining walls including soil filling to pockets (excavation, concrete foundations, backfilling stones to rear of walls and planting not included)							
Betoflor interlocking units; 250 mm long × 250 × 200 mm modular deep; in walls 250 mm wide	–	–	–	–	–	m²	87.60
extra over Betoflor interlocking units for colours	–	–	–	–	–	m²	7.99
Betoatlas earth retaining walls; 250 mm long × 500 mm wide × 200 mm modular deep; in walls 500 mm wide	–	–	–	–	–	m²	140.65
extra over Betoatlas interlocking units for colours	–	–	–	–	–	m²	15.38

D11 SOIL STABILIZATION

Item Excluding site overheads and profit	PC £	Labour hours	Labour £	Plant £	Material £	Unit	Total rate £
RETAINING WALLS – cont							
Retaining walls; Forticrete Ltd							
Keystone precast concrete block							
retaining wall; geogrid included for walls							
over 1.00 m high; excavation, concrete							
foundation, stone backfill to rear of wall							
all measured separately; includes design							
1.0 m high	95.33	0.80	47.40	–	95.33	m²	**142.73**
2.0 m high	95.33	0.80	47.40	9.18	108.45	m²	**165.03**
3.0 m high	95.33	0.80	47.40	12.23	117.56	m²	**177.19**
4.0 m high	95.33	0.80	47.40	14.68	124.63	m²	**186.71**
Retaining walls; Forticrete Ltd							
Stepoc Blocks; interlocking blocks; 10 mm							
reinforcing laid loose horizontally to							
preformed notches and vertical reinforcing							
nominal size 10 mm fixed to starter bars;							
infilling with concrete; foundations and							
starter bars measured separately							
type 325; 400 × 225 × 325 mm	49.51	0.47	27.65	–	69.19	m²	**96.84**
type 256; 400 × 225 × 256 mm	45.24	0.40	23.70	–	63.26	m²	**86.96**
type 190; 400 × 225 × 200 mm	36.45	0.33	19.75	–	51.13	m²	**70.88**
Timber log retaining walls; AVS Fencing							
Supplies Ltd; timber posts are kiln							
dried redwood with 15 year guarantee							
Machine rounded softwood logs to							
trenches priced separately; disposal of							
excavated material priced separately;							
inclusive of 75 mm hardcore blinding to							
trench and backfilling trench with site							
mixed concrete 1:3:6; geofabric pinned to							
rear of logs; heights of logs above ground							
500 mm (constructed from 1.80 m							
lengths)	26.55	0.50	29.63	–	36.12	m	**65.75**
1.20 m (constructed from 1.80 m							
lengths)	53.10	0.43	25.67	–	75.45	m	**101.12**
1.60 m (constructed from 2.40 m							
lengths)	70.80	0.83	49.36	–	98.68	m	**148.04**
2.00 m (constructed from 3.00 m							
lengths)	88.50	1.17	69.12	–	121.91	m	**191.03**
As above but with 150 mm machine							
rounded timbers							
500 mm	38.34	0.83	49.36	–	47.91	m	**97.27**
1.20 m	127.67	0.58	34.56	–	150.03	m	**184.59**
1.60 m	127.80	1.00	59.25	–	155.68	m	**214.93**
As above but with 200 mm machine							
rounded timbers							
1.80 m (constructed from 2.40 m							
lengths)	143.00	1.33	79.00	–	176.41	m	**255.41**

D11 SOIL STABILIZATION

Item Excluding site overheads and profit	PC £	Labour hours	Labour £	Plant £	Material £	Unit	Total rate £
Railway sleeper walls; AVS Fencing Supplies Ltd; retaining wall from railway sleepers; fixed with steel galvanized pins 12 mm driven into the ground; sleepers laid flat							
Grade 1 hardwood; 2590 × 250 × 150 mm							
150 mm; 1 sleeper high	6.67	0.17	9.87	–	7.10	m	**16.97**
300 mm; 2 sleepers high	13.35	0.33	19.75	–	14.19	m	**33.94**
450 mm; 3 sleepers high	19.86	0.50	29.63	–	20.97	m	**50.60**
600 mm; 4 sleepers high	26.75	0.67	39.50	–	27.87	m	**67.37**
Grade 1 hardwood as above but with 2 nr galvanized angle iron stakes set into concrete internally and screwed to the inside face of the sleepers							
750 mm; 5 sleepers high	33.10	0.83	49.37	–	66.23	m²	**115.60**
900 mm; 6 sleepers high	39.62	0.92	54.31	–	75.79	m²	**130.10**
New pine softwood;							
2400 × 250 × 125 mm							
120 mm; 1 sleeper high	7.77	0.17	9.88	–	8.20	m	**18.08**
240 mm; 2 sleepers high	15.54	0.33	19.75	–	16.39	m	**36.14**
360 mm; 3 sleepers high	23.32	0.50	29.63	–	24.43	m	**54.06**
480 mm; 4 sleepers high	31.09	0.67	39.50	–	32.20	m	**71.70**
New pine softwood as above but with 2 nr galvanized angle iron stakes set into concrete internally and screwed to the inside face of the sleepers							
600 mm; 5 sleepers high	38.86	0.83	49.37	–	71.99	m²	**121.36**
720 mm; 6 sleepers high	46.63	0.92	54.31	–	82.80	m²	**137.11**
New oak hardwood 2600 × 220 × 130 mm							
130 mm; 1 sleeper high	10.58	0.17	9.87	–	11.00	m	**20.87**
260 mm; 2 sleepers high	21.16	0.33	19.75	–	22.01	m	**41.76**
390 mm; 3 sleepers high	31.74	5.00	296.25	–	32.85	m	**329.10**
520 mm; 4 sleepers high	42.32	0.67	39.50	–	43.43	m	**82.93**
New oak hardwood as above but with 2 nr galvanized angle iron stakes set into concrete internally and screwed to the inside face of the sleepers							
640 mm; 5 sleepers high	52.90	0.83	49.37	–	86.03	m²	**135.40**
760 mm; 6 sleepers high	63.48	9.00	533.25	–	99.65	m²	**632.90**
Excavate foundation trench; set railway sleepers vertically on end in concrete 1:3:6 continuous foundation to 33.3% of their length to form retaining wall							
Grade 1 hardwood; finished height above ground level							
300 mm	17.75	1.00	59.25	1.88	22.74	m	**83.87**
500 mm	25.73	1.00	59.25	1.88	30.71	m	**91.84**
600 mm	35.76	1.00	59.25	1.88	44.92	m	**106.05**
750 mm	38.59	1.17	69.13	1.88	47.75	m	**118.76**
1.00 m	43.95	1.25	74.06	1.88	53.94	m	**129.88**

D11 SOIL STABILIZATION

Item Excluding site overheads and profit	PC £	Labour hours	Labour £	Plant £	Material £	Unit	Total rate £
RETAINING WALLS – cont							
Excavate and place vertical steel universal beams 165 mm wide in concrete base at 2.59 m centres; fix railway sleepers set horizontally between beams to form horizontal fence or retaining wall							
Grade 1 hardwood; bay length 2.59 m							
500 mm high; 2 sleepers	14.59	0.83	49.37	–	51.92	bay	**101.29**
750 mm high; 3 sleepers	21.44	0.87	51.35	–	77.10	bay	**128.45**
1.00 m high; 4 sleepers	28.73	1.00	59.25	–	99.60	bay	**158.85**
1.25 m high; 5 sleepers	36.69	1.17	69.13	–	128.33	bay	**197.46**
1.50 m high; 6 sleepers	43.09	1.00	59.25	–	152.73	bay	**211.98**
1.75 m high; 7 sleepers	50.17	1.00	59.25	–	159.80	bay	**219.05**
EMBANKMENTS							
Embankments; Grass Concrete Ltd							
Grasscrete; in situ reinforced concrete surfacing; to 20 mm thick sand blinding layer (not included); including soiling and seeding							
GC1; 100 mm thick	–	–	–	–	42.48	m²	**42.48**
GC2; 150 mm thick	–	–	–	–	54.80	m²	**54.80**
Grassblock 103; solid matrix precast concrete blocks; to 20 mm thick sand blinding layer; excluding edge restraint; including soiling and seeding							
406 × 406 × 103 mm; fully interlocking	–	–	–	–	38.82	m²	**38.82**
Grass reinforcement; Farmura Environmental Ltd							
Matrix grass paver; recycled polyethylene and polypropylene mixed interlocking erosion control and grass reinforcement system laid to rootzone prepared separately and filled with screened topsoil and seeded with grass seed; green							
Golpla; 640 × 330 × 38 mm	13.50	0.04	2.47	–	15.96	m²	**18.43**
Embankments; Cooper Clarke Civils and Lintels							
Ecoblock polyethylene; 925 × 310 × 50 mm; heavy duty for car parking and fire paths							
to firm sub-soil (not included)	19.57	0.01	0.79	–	30.07	m²	**30.86**
to 100 mm granular fill and geotextile	19.57	0.03	1.58	0.13	34.82	m²	**36.53**
to 250 mm granular fill and geotextile	19.57	0.07	3.95	0.22	40.84	m²	**45.01**
Extra for filling Ecoblock with topsoil; seeding with rye grass at 50 g/m²	1.70	0.07	3.97	0.18	1.70	m²	**5.85**

D11 SOIL STABILIZATION

Item Excluding site overheads and profit	PC £	Labour hours	Labour £	Plant £	Material £	Unit	Total rate £
Neoweb polyethylene soil stabilizing panels; to soil surfaces brought to grade (not included); filling with excavated material							
panels; 2.5 × 8.0 m × 100 mm deep	8.22	0.03	1.97	2.56	8.22	m²	**12.75**
panels; 2.5 × 8.0 m × 200 mm deep	16.48	0.04	2.63	3.06	16.48	m²	**22.17**
Neoweb polyethylene soil stabilizing panels; to soil surfaces brought to grade (not included); filling with ballast							
panels; 2.5 × 8.0 m × 100 mm deep	8.22	0.04	2.20	2.56	10.17	m²	**14.93**
panels; 2.5 × 8.0 m × 200 mm deep	16.48	0.05	3.04	3.06	20.38	m²	**26.48**
Neoweb polyethylene soil stabilizing panels; to soil surfaces brought to grade (not included); filling with ST2 10 N/mm² concrete							
panels; 2.5 × 8.0 m × 100 mm deep	8.22	0.05	3.16	–	15.72	m²	**18.88**
panels; 2.5 × 8.0 m × 200 mm deep	16.48	0.07	3.91	–	31.48	m²	**35.39**
Extra over for filling Neoweb with imported topsoil; seeding with rye grass at 50 g/m²	3.53	0.02	1.32	0.30	3.53	m²	**5.15**
Flexible sheet materials; Tensar International							
Tensar Mat 400 erosion mats; 3.0–4.5 m wide; securing with Tensar pegs; lap rolls 100 mm; anchors at top and bottom of slopes; in trenches	4.40	0.01	0.33	–	4.40	m²	**4.73**
Topsoil filling to Tensar Mat; including brushing and raking	0.88	0.01	0.33	0.08	0.88	m²	**1.29**
Tensar Bi-axial Geogrid; to graded compacted base; filling with 200 mm granular fill; compacting (turf or paving to surfaces not included); 400 mm laps							
TriAx 150; 39 × 39 mm mesh	1.40	–	0.27	0.11	6.93	m²	**7.31**
TriAx 160; 39 × 39 mm mesh	1.82	–	0.27	0.11	7.35	m²	**7.73**
TriAx 170; 39 × 39 mm mesh	2.15	–	0.27	0.11	7.68	m²	**8.06**
Flexible sheet materials; Terram Ltd							
Terram synthetic fibre filter fabric; to graded base (not included)							
Terram 1000; 0.70 mm thick; mean water flow 50 l/m²/s	0.41	0.01	0.33	–	0.41	m²	**0.74**
Terram 2000; 1.00 mm thick; mean water flow 33 l/m²/s	0.62	0.01	0.33	–	0.62	m²	**0.95**
Terram Minipack	0.71	–	0.07	–	0.71	m²	**0.78**

D11 SOIL STABILIZATION

Item Excluding site overheads and profit	PC £	Labour hours	Labour £	Plant £	Material £	Unit	Total rate £
EMBANKMENTS – cont							
Flexible sheet materials; Greenfix Ltd							
Greenfix; erosion control mats;							
10–15 mm thick; fixing with 4 nr crimped							
pins in accordance with manufacturer's							
instructions; to graded surface (not							
included)							
unseeded Eromat 1; 2.40 m wide	103.47	0.67	39.50	–	167.47	100 m^2	**206.97**
unseeded Eromat 2; 2.40 m wide	130.69	0.67	39.50	–	194.69	100 m^2	**234.19**
unseeded Eromat 3; 2.40 m wide	147.03	0.67	39.50	–	211.03	100 m^2	**250.53**
seeded Covamat 1; 2.40 m wide	166.25	0.67	39.50	–	230.25	100 m^2	**269.75**
seeded Covamat 2; 2.40 m wide	210.00	0.67	39.50	–	274.00	100 m^2	**313.50**
seeded Covamat 3; 2.40 m wide	227.50	0.67	39.50	–	291.50	100 m^2	**331.00**
Bioroll; 300 mm dia.; to river banks and							
revetments	19.00	0.04	2.47	–	21.90	m	**24.37**
Extra over Greenfix erosion control mats							
for fertilizer applied at 70 g/m^2	9.28	0.07	3.95	–	9.28	100 m^2	**13.23**
Extra over Greenfix erosion control mats							
for Geojute; fixing with steel pins	1.32	0.01	0.44	–	1.32	m^2	**1.76**
Extra over Greenfix erosion control mats							
for laying to slopes exceeding 30°	–	–	–	–	–	25%	–
Extra for the following operations							
Spreading 25 mm approved topsoil							
by machine	0.84	–	0.12	0.13	0.84	m^2	**1.09**
by hand	0.84	0.01	0.69	–	0.84	m^2	**1.53**
Grass seed; PC £4.50/kg; spreading in							
two operations; by hand							
35 g/m^2	15.75	0.06	3.29	–	15.75	100 m^2	**19.04**
50 g/m^2	22.50	0.06	3.29	–	22.50	100 m^2	**25.79**
70 g/m^2	31.50	0.06	3.29	–	31.50	100 m^2	**34.79**
100 g/m^2	45.00	0.07	3.95	–	45.00	100 m^2	**48.95**
125 g/m^2	56.25	0.67	39.50	–	56.25	100 m^2	**95.75**
Extra over seeding by hand for slopes							
over 30° (allowing for the actual area but							
measured in plan)							
35 g/m^2	2.34	–	0.04	–	2.34	100 m^2	**2.38**
50 g/m^2	3.38	–	0.04	–	3.38	100 m^2	**3.42**
70 g/m^2	4.72	–	0.04	–	4.72	100 m^2	**4.76**
100 g/m^2	6.75	–	0.04	–	6.75	100 m^2	**6.79**
125 g/m^2	8.41	–	0.04	–	8.41	100 m^2	**8.45**
Grass seed; PC £4.50/kg; spreading in							
two operations; by machine							
35 g/m^2	15.75	–	–	0.51	15.75	100 m^2	**16.26**
50 g/m^2	22.50	–	–	0.51	22.50	100 m^2	**23.01**
70 g/m^2	31.50	–	–	0.51	31.50	100 m^2	**32.01**
100 g/m^2	45.00	–	–	0.51	45.00	100 m^2	**45.51**
125 kg/ha	562.50	–	–	50.95	562.50	ha	**613.45**
150 kg/ha	675.00	–	–	50.95	675.00	ha	**725.95**
200 kg/ha	900.00	–	–	50.95	900.00	ha	**950.95**

D11 SOIL STABILIZATION

Item Excluding site overheads and profit	PC £	Labour hours	Labour £	Plant £	Material £	Unit	Total rate £
250 kg/ha	1125.00	–	–	50.95	1125.00	ha	1175.95
300 kg/ha	1350.00	–	–	50.95	1350.00	ha	1400.95
350 kg/ha	1575.00	–	–	50.95	1575.00	ha	1625.95
400 kg/ha	1800.00	–	–	50.95	1800.00	ha	1850.95
500 kg/ha	2250.00	–	–	50.95	2250.00	ha	2300.95
700 kg/ha	3150.00	–	–	50.95	3150.00	ha	3200.95
1400 kg/ha	6300.00	–	–	50.95	6300.00	ha	6350.95
Extra over seeding by machine for slopes over 30° (allowing for the actual area but measured in plan)							
35 g/m^2	2.36	–	–	0.08	2.36	100 m^2	2.44
50 g/m^2	3.38	–	–	0.08	3.38	100 m^2	3.46
70 g/m^2	4.72	–	–	0.08	4.72	100 m^2	4.80
100 g/m^2	6.75	–	–	0.08	6.75	100 m^2	6.83
125 kg/ha	84.38	–	–	7.64	84.38	ha	92.02
150 kg/ha	101.25	–	–	7.64	101.25	ha	108.89
200 kg/ha	135.00	–	–	7.64	135.00	ha	142.64
250 kg/ha	168.75	–	–	7.64	168.75	ha	176.39
300 kg/ha	202.50	–	–	7.64	202.50	ha	210.14
350 kg/ha	236.25	–	–	7.64	236.25	ha	243.89
400 kg/ha	270.00	–	–	7.64	270.00	ha	277.64
500 kg/ha	337.50	–	–	7.64	337.50	ha	345.14
700 kg/ha	472.50	–	–	7.64	472.50	ha	480.14
1400 kg/ha	945.00	–	–	7.64	945.00	ha	952.64
REVETMENTS							
Willow walling to riverbanks; LDC Ltd Woven willow walling as retention to riverbanks; driving or concreting posts in at 2 m centres; intermediate posts at 500 mm centres							
1.20 m high	–	–	–	–	–	m	132.67
1.50 m high	–	–	–	–	–	m	169.13
Waterside revetments; Willowbank Services Ltd; natural engineering solutions; ecological sustainable waterside stabilization systems Live willow spiling; 1 m high uprights at 0.50 m centres; woven geotextile backed; included filling with site based topsoil behind spiling to top of bank							
less than 50 m	–	–	–	–	–	m	161.83
50–100 m	–	–	–	–	–	m	126.42

D11 SOIL STABILIZATION

| Item | PC | Labour | Labour | Plant | Material | Unit | Total |
Excluding site overheads and profit	£	hours	£	£	£		rate £
REVETMENTS – cont							
Waterside revetments – cont							
Brushwood faggot and pre-planted coir revetments; 900 mm high revetment; 300 mm dia. faggot; 300 mm dia. pre-planted coir roll; staked in position with 1.8 m tanalized posts; 2 posts per m; faggots and coir fixed to posts with wire							
less than 50 m	–	–	–	–	–	m	**145.02**
50–100 m	–	–	–	–	–	m	**124.84**
Brushwood faggot revetments; 900 mm high revetment; 2 nr 300 mm dia. faggot staked in position with 1.8 m tanalized posts; 2 posts per m; faggots and coir fixed to posts with wire							
less than 50 m	–	–	–	–	–	m	**124.48**
50–100 m	–	–	–	–	–	m	**94.64**

D20 EXCAVATION AND FILLING

Item Excluding site overheads and profit	PC £	Labour hours	Labour £	Plant £	Material £	Unit	Total rate £
MACHINE SELECTION							
MACHINE SELECTION TABLE							
Road Equipment Ltd; machine volumes for excavating/filling only and placing excavated material alongside or to a dumper; no bulkages are allowed for in the material volumes; these rates should be increased by user-preferred percentages to suit prevailing site conditions; the figures in the next section for 'Excavation mechanical' and filling allow for the use of banksmen within the rates shown below							
1.5 tonne excavators: digging volume							
1 cycle/minute; 0.04 m^3	–	0.42	8.23	2.34	–	m^3	**10.57**
2 cycles/minute; 0.08 m^3	–	0.21	4.11	1.62	–	m^3	**5.73**
3 cycles/minute; 0.12 m^3	–	0.14	2.74	1.40	–	m^3	**4.14**
3 tonne excavators; digging volume							
1 cycle/minute; 0.13 m^3	–	0.13	2.53	1.61	–	m^3	**4.14**
2 cycles/minute; 0.26 m^3	–	0.06	1.27	2.90	–	m^3	**4.17**
3 cycles/minute; 0.39 m^3	–	0.04	0.84	1.22	–	m^3	**2.06**
5 tonne excavators; digging volume							
1 cycle/minute; 0.28 m^3	–	0.06	1.18	2.31	–	m^3	**3.49**
2 cycles/minute; 0.56 m^3	–	0.03	0.59	2.18	–	m^3	**2.77**
3 cycles/minute; 0.84 m^3	–	0.02	0.39	2.30	–	m^3	**2.69**
7 tonne excavators; supplied with operator; digging volume							
1 cycle/minute; 0.28 m^3	–	0.06	1.18	3.75	–	m^3	**4.93**
2 cycles/minute; 0.56 m^3	–	0.03	0.59	1.83	–	m^3	**2.42**
3 cycles/minute; 0.84 m^3	–	0.02	0.39	2.23	–	m^3	**2.62**
21 tonne excavators; supplied with operator; digging volume							
1 cycle/minute; 1.21 m^3	–	–	–	0.83	–	m^3	**0.83**
2 cycles/minute; 2.42 m^3	–	–	–	0.34	–	m^3	**0.34**
3 cycles/minute; 3.63 m^3	–	–	–	0.23	–	m^3	**0.23**
Backhoe loader; excavating; JCB 3 CX rear bucket capacity 0.28 m^3							
1 cycle/minute; 0.28 m^3	–	–	–	2.23	–	m^3	**2.23**
2 cycles/minute; 0.56 m^3	–	–	–	1.12	–	m^3	**1.12**
3 cycles/minute; 0.84 m^3	–	–	–	0.74	–	m^3	**0.74**
Backhoe loader; loading from stockpile; JCB 3 CX front bucket capacity 1.00 m^3							
1 cycle/minute; 1.00 m^3	–	–	–	0.63	–	m^3	**0.63**
2 cycles/minute; 2.00 m^3	–	–	–	0.31	–	m^3	**0.31**

D20 EXCAVATION AND FILLING

Item Excluding site overheads and profit	PC £	Labour hours	Labour £	Plant £	Material £	Unit	Total rate £
MACHINE SELECTION – cont							
Note: All volumes below are based on excavated 'earth' moist at 1,997 kg/m³ solid or 1,598 kg/m³ loose; a 25% bulkage factor has been used; the weight capacities below exceed the volume capacities of the machine in most cases; see the memorandum section at the back of this book for further weights of materials.							
Dumpers; Road Equipment Ltd							
1 tonne high tip skip loader; volume 0.485 m³ (775 kg)							
5 loads per hour	–	0.41	8.14	1.59	–	m³	**9.73**
7 loads per hour	–	0.29	5.82	1.20	–	m³	**7.02**
10 loads per hour	–	0.21	4.07	0.92	–	m³	**4.99**
2.5 tonne dumper; excavated material volume 1.3 m³							
4 loads per hour	–	–	–	5.27	–	m³	**5.27**
5 loads per hour	–	–	–	4.21	–	m³	**4.21**
7 loads per hour	–	–	–	3.75	–	m³	**3.75**
10 loads per hour	–	–	–	2.05	–	m³	**2.05**
6 tonne dumper; maximum volume 3.40 m³ (5.4 t); available volume 3.77 m³							
4 loads per hour	–	0.07	1.31	0.33	–	m³	**1.64**
5 loads per hour	–	0.05	1.05	0.27	–	m³	**1.32**
7 loads per hour	–	0.04	0.75	0.21	–	m³	**0.96**
10 loads per hour	–	0.03	0.52	0.18	–	m³	**0.70**
Clarification notes on labour costs in this section							
General groundworks team							
Generally a three man team is used in this section; The column 'Labour hours' reports team hours. The column 'Labour £' reports the total cost of the team for the unit of work shown							
3 man team	–	1.00	59.25	–	–	hr	**59.25**
banksman	–	1.00	19.75	–	–	hr	**19.75**

D20 EXCAVATION AND FILLING

Item Excluding site overheads and profit	PC £	Labour hours	Labour £	Plant £	Material £	Unit	Total rate £
SITE PREPARATION AND CLEARANCE							
Site preparation							
Felling and removing trees off site							
girth 600 mm–1.50 m (95–240 mm trunk dia.)	–	2.50	148.13	19.04	–	nr	167.17
girth 1.50–3.00 m (240–475 mm trunk dia.)	–	10.00	592.50	76.15	–	nr	668.65
girth 3.00–4.00 m (475–630 mm trunk dia.)	–	16.00	948.00	106.62	–	nr	1054.62
Removing tree stumps							
girth 600 mm–1.50 m	–	2.00	39.50	82.50	–	nr	122.00
girth 1.50–3.00 m	–	7.00	138.25	78.20	–	nr	216.45
girth over 3.00 m	–	12.00	237.00	134.06	–	nr	371.06
Stump grinding; disposing to spoil heaps							
girth 600 mm–1.50 m	–	2.00	39.50	32.20	–	nr	71.70
girth 1.50–3.00 m	–	2.50	49.38	64.41	–	nr	113.79
girth over 3.00 m	–	4.00	79.00	56.35	–	nr	135.35
Clearing site vegetation							
mechanical clearance	–	0.25	4.94	13.59	–	100 m²	18.53
hand clearance	–	2.00	39.50	–	–	100 m²	39.50
Lifting turf for preservation							
sod cutter machine lift and stack	–	0.75	14.81	9.96	–	100 m²	24.77
hand lift and stack	–	8.33	164.58	–	–	100 m²	164.58
Lifting turf for disposal (disposal not included)							
by excavator; 5 tonne	–	–	–	40.52	–	100 m²	40.52
by excavator; 7 tonne	–	–	–	30.99	–	100 m²	30.99
by excavator; 7 tonne; working with two dumpers	–	–	–	29.19	–	100 m²	29.19
by excavator; 7 tonne; working with two dumpers	–	–	–	24.96	–	100 m²	24.96
hand lift and stack	–	8.33	164.58	–	–	100 m²	164.58
Site clearance; by machine; clear site of mature shrubs from existing cultivated beds; dig out roots by machine							
Mixed shrubs in beds; planting centres 500 mm average							
height less than 1 m	–	0.03	0.49	0.79	–	m²	1.28
1.00–1.50 m	–	0.04	0.79	1.26	–	m²	2.05
1.50–2.00 m; pruning to ground level by hand	–	0.10	1.98	3.15	–	m²	5.13
2.00–3.00 m; pruning to ground level by hand	–	0.10	1.98	6.29	–	m²	8.27
3.00–4.00 m; pruning to ground level by hand	–	0.20	3.95	10.48	–	m²	14.43

D20 EXCAVATION AND FILLING

Item Excluding site overheads and profit	PC £	Labour hours	Labour £	Plant £	Material £	Unit	Total rate £
SITE PREPARATION AND CLEARANCE – cont							
Site clearance; by hand; clear site of mature shrubs from existing cultivated beds; dig out roots							
Mixed shrubs in beds; planting centres 500 mm average							
height less than 1 m	–	0.33	6.58	–	–	m²	6.58
1.00–1.50 m	–	0.50	9.88	–	–	m²	9.88
1.50–2.00 m	–	1.00	19.75	–	–	m²	19.75
2.00–3.00 m	–	2.00	39.50	–	–	m²	39.50
3.00–4.00 m	–	3.00	59.26	–	–	m²	59.26
Mechanical ground clearance to large areas; Cleartrack (Evl) Limited; vegetation clearance of high capacity mobile mulching machines; 500 HP; 2.50 m head width; tracked; wheeled; skid steer and 360 degree operating							
Site clearance and in situ mulching and incorporating to soil surface to existing area of scrub and small trees up to 150 mm dia.; cleared mulched material left on surface for clearance to stockpile (not included)							
Areas of existing grass, bramble and weed; average height not exceeding 1.00 m	–	–	–	–	–	100 m²	10.25
Areas of existing small shrubs and weed; average height not exceeding 1.80 m	–	–	–	–	–	100 m²	15.38
Areas of existing grass, bramble, weed and regular shrubs; varying heights to 3.00 m; max stem dia. 100 mm	–	–	–	–	–	100 m²	21.52
Areas of existing grass, bramble, weed, regular shrubs and small trees; varying heights up and over 3.00 m; max stem dia. 150 mm	–	–	–	–	–	100 m²	55.35
Spraying of vegetation; Glyphosate at 5l/ha by tractor drawn spray equipment	–	–	–	–	–	100 m²	2.56
Excavated material; on site							
In spoil heaps							
average 25 m distance	–	–	–	2.80	–	m³	2.80
average 50 m distance	–	–	–	3.23	–	m³	3.23
average 100 m distance (1 dumper)	–	–	–	4.05	–	m³	4.05
average 100 m distance (2 dumpers)	–	–	–	4.18	–	m³	4.18
average 200 m distance	–	–	–	4.89	–	m³	4.89
average 200 m distance (2 dumpers)	–	–	–	5.02	–	m³	5.02

D20 EXCAVATION AND FILLING

Item Excluding site overheads and profit	PC £	Labour hours	Labour £	Plant £	Material £	Unit	Total rate £
EXCAVATING							
Market prices of topsoil; prices shown include for 20% settlement							
Multiple source screened topsoil	–	–	–	–	33.60	m³	**33.60**
Single source topsoil; British Sugar PLC	–	–	–	–	33.60	m³	**33.60**
High grade topsoil for planting	–	–	–	–	52.80	m³	**52.80**
Note: The figures in this section relate to the machine capacities shown earlier in this section. The figures below however allow for dig efficiency based on depth. A banksman is allowed for in all excavation build and disposal costs shown. Bulkages of 25% allowed; adjustments should be made for different soil types.							
Excavating; mechanical; topsoil for preservation							
3 tonne tracked excavator (bucket volume 0.13 m³)							
average depth 100 mm	–	0.02	0.33	1.08	–	m²	**1.41**
average depth 150 mm	–	0.02	0.38	1.22	–	m²	**1.60**
average depth 200 mm	–	–	–	1.62	–	m²	**1.62**
average depth 250 mm	–	0.03	0.55	1.80	–	m²	**2.35**
average depth 300 mm	–	0.03	0.66	2.16	–	m²	**2.82**
JCB Sitemaster 3CX (bucket volume 0.28 m³)							
average depth 100 mm	–	1.00	19.75	41.25	–	100 m²	**61.00**
average depth 150 mm	–	1.25	24.69	51.56	–	100 m²	**76.25**
average depth 200 mm	–	1.78	35.16	73.42	–	100 m²	**108.58**
average depth 250 mm	–	1.90	37.52	78.38	–	100 m²	**115.90**
average depth 300 mm	–	2.00	39.50	82.50	–	100 m²	**122.00**
Excavating; mechanical; to reduce levels							
5 tonne excavator (bucket volume 0.28 m³)							
maximum depth not exceeding 0.25 m	–	0.07	1.38	3.76	–	m³	**5.14**
maximum depth not exceeding 1.00 m	–	0.05	0.94	2.55	–	m³	**3.49**
maximum depth not exceeding 2.00 m	–	0.06	1.18	3.19	–	m³	**4.37**
JCB Sitemaster 3CX (bucket volume 0.28 m³)							
maximum depth not exceeding 0.25 m	–	0.07	1.38	2.89	–	m³	**4.27**
maximum depth not exceeding 1.00 m	–	0.06	1.18	2.45	–	m³	**3.63**
7 tonne tracked excavator (bucket volume 0.28 m³)							
maximum depth not exceeding 1.00 m	–	0.06	1.23	4.10	–	m³	**5.33**
maximum depth not exceeding 2.00 m	–	0.07	1.41	4.68	–	m³	**6.09**
maximum depth not exceeding 3.00 m	–	0.09	1.80	5.96	–	m³	**7.76**

Prices for Measured Works

D20 EXCAVATION AND FILLING

Item Excluding site overheads and profit	PC £	Labour hours	Labour £	Plant £	Material £	Unit	Total rate £
EXCAVATING – cont							
Excavating; mechanical; to reduce levels – cont							
21 tonne 360 tracked excavator (bucket volume 1.21 m³)							
maximum depth not exceeding 1.00 m	–	0.01	0.22	0.55	–	m³	0.77
maximum depth not exceeding 2.00 m	–	0.02	0.33	0.82	–	m³	1.15
maximum depth not exceeding 3.00 m	–	0.03	0.66	1.67	–	m³	2.33
Pits; 3 tonne tracked excavator							
maximum depth not exceeding 0.25 m	–	0.33	6.58	15.03	–	m³	21.61
maximum depth not exceeding 1.00 m	–	0.25	4.94	11.28	–	m³	16.22
maximum depth not exceeding 2.00 m	–	0.40	7.90	18.04	–	m³	25.94
Trenches; width not exceeding 0.30 m; 3 tonne excavator							
maximum depth not exceeding 0.25 m	–	1.00	19.75	4.50	–	m³	24.25
maximum depth not exceeding 1.00 m	–	0.69	13.53	3.08	–	m³	16.61
maximum depth not exceeding 2.00 m	–	0.60	11.85	2.70	–	m³	14.55
Trenches; width exceeding 0.30 m; 3 tonne excavator							
maximum depth not exceeding 0.25 m	–	0.60	11.85	2.70	–	m³	14.55
maximum depth not exceeding 1.00 m	–	0.50	9.88	2.25	–	m³	12.13
maximum depth not exceeding 2.00 m	–	0.38	7.60	1.73	–	m³	9.33
Extra over any types of excavating irrespective of depth for breaking out existing materials; JCB with breaker attachment							
hard rock	–	0.50	9.88	78.13	–	m³	88.01
concrete	–	0.50	9.88	29.30	–	m³	39.18
reinforced concrete	–	1.00	19.75	49.08	–	m³	68.83
brickwork, blockwork or stonework	–	0.25	4.94	29.30	–	m³	34.24
Extra over any types of excavating irrespective of depth for breaking out existing hard pavings; JCB with breaker attachment							
concrete; 100 mm thick	–	–	–	2.15	–	m²	2.15
concrete; 150 mm thick	–	–	–	3.58	–	m²	3.58
concrete; 200 mm thick	–	–	–	4.30	–	m²	4.30
concrete; 300 mm thick	–	–	–	6.45	–	m²	6.45
reinforced concrete; 100 mm thick	–	0.08	1.65	2.58	–	m²	4.23
reinforced concrete; 150 mm thick	–	0.08	1.48	3.49	–	m²	4.97
reinforced concrete; 200 mm thick	–	0.10	1.98	4.66	–	m²	6.64
reinforced concrete; 300 mm thick	–	0.15	2.96	6.99	–	m²	9.95
tarmacadam; 75 mm thick	–	–	–	2.15	–	m²	2.15
tarmacadam and hardcore; 150 mm thick	–	–	–	3.44	–	m²	3.44
Extra over any types of excavating irrespective of depth for taking up							
precast concrete paving slabs	–	0.07	1.32	0.71	–	m²	2.03
natural stone paving	–	0.10	1.98	1.06	–	m²	3.04
cobbles	–	0.13	2.47	1.32	–	m²	3.79
brick paviors	–	0.13	2.47	1.32	–	m²	3.79

D20 EXCAVATION AND FILLING

Item Excluding site overheads and profit	PC £	Labour hours	Labour £	Plant £	Material £	Unit	Total rate £
Excavating; hand; for preservation							
Topsoil for preservation; loading to barrows							
average depth 100 mm	–	0.08	4.69	–	–	m²	4.69
average depth 150 mm	–	0.12	7.04	–	–	m²	7.04
average depth 200 mm	–	0.19	11.26	–	–	m²	11.26
average depth 250 mm	–	0.24	14.08	–	–	m²	14.08
average depth 300 mm	–	0.29	16.89	–	–	m²	16.89
Excavating; hand; to reduce							
Topsoil to reduce levels							
maximum depth not exceeding 0.25 m	–	0.79	46.93	–	–	m³	46.93
maximum depth not exceeding 1.00 m	–	1.03	61.00	–	–	m³	61.00
Pits							
maximum depth not exceeding 0.25 m	–	0.88	52.14	–	–	m³	52.14
maximum depth not exceeding 1.00 m	–	1.14	67.78	–	–	m³	67.78
maximum depth not exceeding 2.00 m (includes earthwork support)	–	2.29	135.56	–	–	m³	189.49
Trenches; width not exceeding 0.30 m							
maximum depth not exceeding 0.25 m	–	0.94	55.86	–	–	m³	55.86
maximum depth not exceeding 1.00 m	–	1.23	72.75	–	–	m³	72.75
maximum depth not exceeding 2.00 m (includes earthwork support)	–	1.23	72.75	–	–	m³	99.71
Trenches; width exceeding 0.30 m wide							
maximum depth not exceeding 0.25 m	–	0.94	55.86	–	–	m³	55.86
maximum depth not exceeding 1.00 m	–	1.32	78.21	–	–	m³	78.21
maximum depth not exceeding 2.00 m (includes earthwork support)	–	1.98	117.31	–	–	m³	171.24
Extra over any types of excavating irrespective of depth for breaking out existing materials; hand held pneumatic breaker							
rock	–	1.65	97.76	35.30	–	m³	133.06
concrete	–	0.82	48.88	17.65	–	m³	66.53
reinforced concrete	–	1.32	78.21	33.25	–	m³	111.46
brickwork, blockwork or stonework	–	0.50	29.33	10.59	–	m³	39.92
FILLING							
Filling to make up levels; mechanical							
Arising from the excavations							
average thickness not exceeding 0.25 m	–	0.07	1.45	4.82	–	m³	6.27
average thickness less than 500 mm	–	0.06	1.23	4.10	–	m³	5.33
average thickness 1.00 m	–	0.05	0.91	3.04	–	m³	3.95
Obtained from on site spoil heaps; average 25 m distance; multiple handling							
average thickness less than 250 mm	–	–	–	11.23	–	m³	11.23
average thickness less than 500 mm	–	–	–	9.25	–	m³	9.25
average thickness 1.00 m	–	–	–	7.86	–	m³	7.86

D20 EXCAVATION AND FILLING

Item Excluding site overheads and profit	PC £	Labour hours	Labour £	Plant £	Material £	Unit	Total rate £
FILLING – cont							
Filling to make up levels – cont							
Obtained off site; planting quality topsoil							
PC £33.60/m^3							
average thickness less than 250 mm	–	–	–	11.23	33.60	m^3	**44.83**
average thickness less than 500 mm	–	–	–	9.25	33.60	m^3	**42.85**
average thickness 1.00 m	–	–	–	7.86	33.60	m^3	**41.46**
Obtained off site; hardcore; PC £23.05/m^3							
average thickness less than 250 mm	–	–	–	13.10	23.95	m^3	**37.05**
average thickness less than 500 mm	–	–	–	9.83	23.95	m^3	**33.78**
average thickness 1.00 m	–	–	–	8.28	23.95	m^3	**32.23**
Filling to make up levels; hand							
Arising from the excavations							
average thickness exceeding 0.25 m;							
depositing in layers 150 mm							
maximum thickness	–	0.60	11.85	–	–	m^3	**11.85**
Obtained from on site spoil heaps;							
average 25 m distance; multiple handling							
average thickness exceeding 0.25 m							
thick; depositing in layers 150 mm							
maximum thickness	–	1.00	19.75	–	–	m^3	**19.75**
DISPOSAL							
Disposal							
Note: Most commercial site disposal is							
carried out by 20 tonne, 8 wheeled							
vehicles. It has been customary to							
calculate disposal from construction sites							
in terms of full 15 m^3 loads. Spon's							
research has found that based on							
weights of common materials such as							
clean hardcore and topsoil, vehicles							
could not load more than 12 m^3 at a							
time. Most hauliers do not make it							
apparent that their loads are calculated							
by weight and not by volume. The rates							
below reflect these lesser volumes which							
are limited by the 20 tonne limit.							
The volumes shown below are based on							
volumes 'in the solid'. Weight and							
bulking factors have been applied. For							
further information please see the							
weights of typical materials in the							
Earthworks section of the Memoranda at							
the back of this book.							

D20 EXCAVATION AND FILLING

Item Excluding site overheads and profit	PC £	Labour hours	Labour £	Plant £	Material £	Unit	Total rate £
Disposal; mechanical (all rates include banksman)							
Excavated material; off site; to tip; mechanically loaded (360 Excavator)							
inert (clean landfill, rubble concrete); loose	–	0.03	0.55	–	18.14	m³	**18.69**
inert (clean excavated material); compacted	–	0.03	0.55	–	22.67	m³	**23.22**
hardcore	–	0.02	0.41	0.78	15.30	m³	**16.49**
macadam	–	0.02	0.41	0.78	18.90	m³	**20.09**
clean concrete	–	0.03	0.55	1.04	11.25	m³	**12.84**
broken out compacted materials such as roadbases and the like	–	0.03	0.55	1.04	27.21	m³	**28.80**
Light soils and loams (bulking factor of 1.25); 40 tonnes (2 loads per hour)							
soil (sandy and loam); dry	–	0.03	0.55	–	21.77	m³	**22.32**
soil (sandy and loam); wet	–	0.03	0.55	–	23.58	m³	**24.13**
soil (clay); dry	–	0.03	0.55	–	25.39	m³	**25.94**
soil (clay); wet	–	0.03	0.55	1.04	27.21	m³	**28.80**
Other materials							
rubbish (mixed loads)	–	0.03	0.66	1.25	25.00	m³	**26.91**
green waste	–	0.03	0.66	1.25	24.67	m³	**26.58**
As above but allowing for 3 loads (36 m³, 60 tonne) per hour removed							
inert material	–	0.03	0.55	1.04	35.91	m³	**37.50**
Excavated material; off site to tip; mechanically loaded by grab; capacity of load 13 m³ (18 tonne)							
inert material	–	–	–	–	–	m³	**30.80**
hardcore	–	–	–	–	–	m³	**41.53**
macadam	–	–	–	–	–	m³	**24.00**
clean concrete	–	–	–	–	–	m³	**28.00**
soil (sandy and loam); dry	–	–	–	–	–	m³	**30.80**
soil (sandy and loam); wet	–	–	–	–	–	m³	**32.67**
broken out compacted materials such as roadbases and the like	–	–	–	–	–	m³	**34.22**
soil (clay); dry	–	–	–	–	–	m³	**39.98**
soil (clay); wet	–	–	–	–	–	m³	**41.53**
rubbish (mixed loads)	–	–	–	–	–	m³	**34.62**
green waste	–	–	–	–	–	m³	**37.50**
Disposal by skip; 9 yd³ (6 m³)							
Excavated material loaded to skip							
by machine	–	–	–	48.99	–	m³	**48.99**
by hand	–	3.00	59.25	45.83	–	m³	**105.08**

D20 EXCAVATION AND FILLING

Item Excluding site overheads and profit	PC £	Labour hours	Labour £	Plant £	Material £	Unit	Total rate £
DISPOSAL – cont							
Excavated material; on site							
In spoil heaps							
average 25 m distance	–	–	–	2.80	–	m³	2.80
average 50 m distance	–	–	–	3.23	–	m³	3.23
average 100 m distance (1 dumper)	–	–	–	4.05	–	m³	4.05
average 100 m distance (2 dumpers)	–	–	–	4.18	–	m³	4.18
average 200 m distance	–	–	–	4.89	–	m³	4.89
average 200 m distance (2 dumpers)	–	–	–	5.02	–	m³	5.02
Spreading on site							
average 25 m distance	–	0.02	0.44	5.03	–	m³	5.47
average 50 m distance	–	0.04	0.77	5.63	–	m³	6.40
average 100 m distance	–	0.06	1.22	6.87	–	m³	8.09
average 200 m distance	–	0.12	2.37	6.41	–	m³	8.78
Disposal; hand							
Excavated material; onsite; in spoil heaps							
average 25 m distance	–	2.40	47.40	–	–	m³	47.40
average 50 m distance	–	2.64	52.14	–	–	m³	52.14
average 100 m distance	–	3.00	59.25	–	–	m³	59.25
average 200 m distance	–	3.60	71.10	–	–	m³	71.10
Excavated material; spreading on site							
average 25 m distance	–	2.64	52.14	–	–	m³	52.14
average 50 m distance	–	3.00	59.25	–	–	m³	59.25
average 100 m distance	–	3.60	71.10	–	–	m³	71.10
average 200 m distance	–	4.20	82.95	–	–	m³	82.95
Disposal of material from site **clearance operations**							
Shrubs and groundcovers less than 1.00 m height; disposal by 15 m³ self loaded truck							
deciduous shrubs; not chipped; winter	–	0.05	0.99	0.63	8.22	m²	9.84
deciduous shrubs; chipped; winter	–	0.05	0.99	1.53	1.23	m²	3.75
evergreen or deciduous shrubs; chipped; summer	–	0.08	1.48	0.91	4.93	m²	7.32
Shrubs 1.00–2.00 m height; disposal by 15 m³ self loaded truck							
deciduous shrubs; not chipped; winter	–	0.17	3.29	0.63	24.67	m²	28.59
deciduous shrubs; chipped; winter	–	0.25	4.94	0.91	4.93	m²	10.78
evergreen or deciduous shrubs; chipped; summer	–	0.30	5.92	1.01	9.87	m²	16.80
Shrubs or hedges 2.00–3.00 m height; disposal by 15 m³ self loaded truck							
deciduous plants non-woody growth; not chipped; winter	–	0.25	4.94	0.63	24.67	m²	30.24
deciduous shrubs; chipped; winter	–	0.50	9.88	0.91	12.33	m²	23.12
evergreen or deciduous shrubs non- woody growth; chipped; summer	–	0.25	4.94	0.91	4.93	m²	10.78
evergreen or deciduous shrubs woody growth; chipped; summer	–	0.67	13.17	1.20	14.80	m²	29.17

D20 EXCAVATION AND FILLING

Item Excluding site overheads and profit	PC £	Labour hours	Labour £	Plant £	Material £	Unit	Total rate £
Shrubs and groundcovers less than 1.00 m height; disposal to spoil heap							
deciduous shrubs; not chipped; winter	–	0.05	0.99	–	–	m²	0.99
deciduous shrubs; chipped; winter	–	0.05	0.99	0.29	–	m²	1.28
evergreen or deciduous shrubs; chipped; summer	–	0.08	1.48	0.29	–	m²	1.77
Shrubs 1.00–2.00 m height; disposal to spoil heap							
deciduous shrubs; not chipped; winter	–	0.17	3.29	–	–	m²	3.29
deciduous shrubs; chipped; winter	–	0.25	4.94	0.29	–	m²	5.23
evergreen or deciduous shrubs; chipped; summer	–	0.30	5.92	0.38	–	m²	6.30
Shrubs or hedges 2.00–3.00 m height; disposal to spoil heaps							
deciduous plants non-woody growth; not chipped; winter	–	0.25	4.94	–	–	m²	4.94
deciduous shrubs; chipped; winter	–	0.50	9.88	0.29	–	m²	10.17
evergreen or deciduous shrubs non-woody growth; chipped; summer	–	0.67	13.17	0.57	–	m²	13.74
evergreen or deciduous shrubs woody growth; chipped; summer	–	1.50	29.62	1.14	–	m²	30.76
GRADING AND SURFACE PREPARATION							
Grading operations; surface previously excavated to reduce levels to prepare to receive subsequent treatments; grading to accurate levels and falls 20 mm tolerances							
Clay or heavy soils or hardcore							
5 tonne excavator	–	0.04	0.79	0.18	–	m²	0.97
JCB Sitemaster 3CX	–	0.02	0.40	0.82	–	m²	1.22
21 tonne 360 tracked excavator	–	0.01	0.20	0.50	–	m²	0.70
by hand	–	0.10	1.98	–	–	m²	1.98
Loamy topsoils							
5 tonne excavator	–	0.03	0.53	0.12	–	m²	0.65
JCB Sitemaster 3CX	–	0.01	0.26	0.55	–	m²	0.81
21 tonne 360 tracked excavator	–	0.01	0.15	0.37	–	m²	0.52
by hand	–	0.05	0.99	–	–	m²	0.99
Sand or graded granular materials							
5 tonne excavator	–	0.02	0.40	0.09	–	m²	0.49
JCB Sitemaster 3CX	–	0.01	0.20	0.41	–	m²	0.61
21 tonne 360 tracked excavator	–	0.01	0.11	0.28	–	m²	0.39
by hand	–	0.03	0.66	–	–	m²	0.66

D20 EXCAVATION AND FILLING

Item Excluding site overheads and profit	PC £	Labour hours	Labour £	Plant £	Material £	Unit	Total rate £
GRADING AND SURFACE PREPARATION – cont							
Grading operations to bottoms of excavations or trenches to receive foundations or bases; surface previously excavated to reduce levels to prepare to receive subsequent treatments; grading to accurate levels and falls 20 mm tolerances							
Trenches							
5 tonne excavator	–	0.32	11.19	0.59	–	m²	**11.78**
by hand	–	0.33	13.17	–	–	m²	**13.17**
Excavated areas to receive bases or formwork							
5 tonne excavator	–	0.13	4.61	0.30	–	m²	**4.91**
by hand	–	0.17	6.58	–	–	m²	**6.58**
Grading operations; surface recently filled to raise levels to prepare to receive subsequent treatments; grading to accurate levels and falls 20 mm tolerances							
Clay or heavy soils or hardcore							
5 tonne excavator	–	0.03	0.59	0.13	–	m²	**0.72**
JCB Sitemaster 3CX	–	0.02	0.30	0.62	–	m²	**0.92**
21 tonne 360 tracked excavator	–	0.01	0.13	0.34	–	m²	**0.47**
by hand	–	0.08	1.65	–	–	m²	**1.65**
Loamy topsoils							
5 tonne excavator	–	0.02	0.41	0.09	–	m²	**0.50**
JCB Sitemaster 3CX	–	0.01	0.20	0.42	–	m²	**0.62**
21 tonne 360 tracked excavator	–	0.01	0.13	0.34	–	m²	**0.47**
by hand	–	0.04	0.79	–	–	m²	**0.79**
Sand or graded granular materials							
5 tonne excavator	–	0.02	0.30	0.07	–	m²	**0.37**
JCB Sitemaster 3CX	–	0.01	0.15	0.31	–	m²	**0.46**
21 tonne 360 tracked excavator	–	0.01	0.10	0.25	–	m²	**0.35**
by hand	–	0.03	0.56	–	–	m²	**0.56**
Surface grading; Agripower Ltd; using laser controlled equipment							
Maximum variation of existing ground level							
75 mm; laser box grader and tractor	–	–	–	–	–	ha	**1200.00**
100 mm; laser controlled low ground pressure bulldozer	–	–	–	–	–	ha	**1900.00**
Fraise mowing; scarification of existing turf layer to rootzone level; Agripower Ltd							
Cart arisings to tip on site							
average thickness 50–75 mm	–	–	–	–	–	ha	**2963.00**

D20 EXCAVATION AND FILLING

Item Excluding site overheads and profit	PC £	Labour hours	Labour £	Plant £	Material £	Unit	Total rate £
Cultivating							
Ripping up subsoil; using approved subsoiling machine; minimum depth 250 mm below topsoil; at 1.20 m centres; in							
gravel or sandy clay	–	–	–	2.16	–	100 m²	**2.16**
soil compacted by machines	–	–	–	2.40	–	100 m²	**2.40**
clay	–	–	–	3.59	–	100 m²	**3.59**
chalk or other soft rock	–	–	–	3.59	–	100 m²	**3.59**
Extra for subsoiling at 1 m centres	–	–	–	1.08	–	100 m²	**1.08**
Breaking up existing ground; using pedestrian operated tine cultivator or rotavator; loam or sandy soil							
100 mm deep	–	0.22	4.34	2.52	–	100 m²	**6.86**
150 mm deep	–	0.28	5.43	3.15	–	100 m²	**8.58**
200 mm deep	–	0.37	7.24	4.20	–	100 m²	**11.44**
As above but in heavy clay or wet soils							
100 mm deep	–	0.44	8.69	5.04	–	100 m²	**13.73**
150 mm deep	–	0.66	13.04	7.56	–	100 m²	**20.60**
200 mm deep	–	0.82	16.29	9.44	–	100 m²	**25.73**
Breaking up existing ground; using tractor drawn tine cultivator or rotavator							
100 mm deep	–	–	–	1.38	–	100 m²	**1.38**
150 mm deep	–	–	–	1.72	–	100 m²	**1.72**
200 mm deep	–	–	–	2.29	–	100 m²	**2.29**
400 mm deep	–	–	–	6.88	–	100 m²	**6.88**
final levelling to ploughed and rotavated ploughed ground; using disc, drag, or chain harrow							
4 passes	–	–	–	1.25	–	100 m²	**1.25**
Rolling cultivated ground lightly; using self-propelled agricultural roller	–	0.06	1.10	0.60	–	100 m²	**1.70**
Importing and storing selected and approved topsoil; 20 tonne load = 11.88 m³ average							
20 tonne loads	33.60	–	–	–	33.60	m³	**33.60**
Surface treatments							
Compacting							
bottoms of excavations	–	0.01	0.10	0.04	–	m²	**0.14**
Surface preparation							
Trimming surfaces of cultivated ground to final levels; removing roots, stones and debris exceeding 50 mm in any direction to tip off site; slopes less than 15°							
clean ground with minimal stone content	–	0.25	4.94	–	–	100 m²	**4.94**
slightly stony; 0.5 kg stones per m²	–	0.33	6.58	–	0.01	100 m²	**6.59**
very stony; 1.0–3.0 kg stones per m²	–	0.50	9.88	–	0.02	100 m²	**9.90**
clearing mixed, slightly contaminated rubble; inclusive of roots and vegetation	–	0.50	9.88	–	0.06	100 m²	**9.94**
clearing brick-bats, stones and clean rubble	–	0.60	11.85	–	0.05	100 m²	**11.90**

D41 CRIB WALLS/ GABIONS/ REINFORCED EARTH

Item Excluding site overheads and profit	PC £	Labour hours	Labour £	Plant £	Material £	Unit	Total rate £
CRIB WALLS							
Crib walls; Phi Group							
Permacrib (timber) or Andacrib							
(concrete) crib walling system; machine							
filled infill with crushed rock; inclusive of							
concrete footing and rear wall drain;							
excluding excavation and backfill							
material							
retaining walls up to 2.0 m	–	–	–	–	–	m²	175.00
retaining walls up to 4.0 m	–	–	–	–	–	m²	195.00
retaining walls up to 6.0 m	–	–	–	–	–	m²	215.00
GABIONS							
Retaining walls; Maccaferri Ltd							
Wire mesh gabions; galvanized mesh							
80 × 100 mm; filling with broken stones							
125–200 mm size; wire down securely to							
manufacturer's instructions; filling front							
face by hand							
2 × 1 × 0.50 m	20.67	0.67	39.50	3.72	73.59	nr	116.81
2 × 1 × 1.0 m	28.97	1.33	79.00	7.45	134.81	nr	221.26
PVC coated gabions							
2 × 1 × 0.5 m	26.13	0.67	39.50	3.72	79.05	nr	122.27
2 × 1 × 1.0 m	36.67	1.33	79.00	7.45	142.51	nr	228.96
Reno mattress gabions							
6 × 2 × 0.17 m	90.38	1.00	59.25	5.59	198.34	nr	263.18
6 × 2 × 0.23 m	98.19	1.50	88.88	7.45	244.25	nr	340.58
6 × 2 × 0.30 m	115.31	2.00	118.50	8.38	305.82	nr	432.70
REINFORCED EARTH							
Embankments; Tensar International							
Embankments; reinforced with Tensar							
Geogrid RE520; geogrid laid horizontally							
within fill to 100% of vertical height of							
slope at 1.00 m centres; filling with							
excavated material							
slopes less than 45°; Tensartech							
Natural Green System	2.36	0.61	36.30	1.60	2.36	m³	40.26
slopes exceeding 45°; Tensartech							
Greenslope System	5.08	0.06	3.66	2.31	5.08	m³	11.05
extra over to slopes exceeding 45° for							
temporary shuttering; shuttering							
measured per m² of vertical elevation	–	–	–	–	12.00	m²	12.00

D41 CRIB WALLS/ GABIONS/ REINFORCED EARTH

Item Excluding site overheads and profit	PC £	Labour hours	Labour £	Plant £	Material £	Unit	Total rate £
Embankments; reinforced with Tensar Geogrid RE540; geogrid laid horizontally within fill to 100% of vertical height of slope at 1.00 m centres; filling with excavated material							
slopes less than 45°; Tensartech Natural Green System	2.92	0.61	36.30	1.60	2.92	m³	**40.82**
slopes exceeding 45°; Tensartech Greenslope System	6.28	0.06	3.67	2.31	6.28	m³	**12.26**
extra over to slopes exceeding 45° for temporary shuttering; shuttering measured per m² of vertical elevation	–	–	–	–	12.00	m²	**12.00**
Extra for Tensar Mat 400; erosion control mats; to faces of slopes of 45° or less; filling with 20 mm fine topsoil; seeding	3.95	0.01	0.39	0.21	5.26	m²	**5.86**
Embankments; reinforced with Tensar Geogrid RE540; geogrid laid horizontally within fill to 50% of horizontal length of slopes at specified centres; filling with imported fill PC £11.50/m³							
300 mm centres; slopes up to 45°; Tensartech Natural Green System	4.85	0.08	4.61	1.80	18.65	m³	**25.06**
600 mm centres; slopes up to 45°; Tensartech Natural Green System	2.42	0.06	3.67	2.31	16.22	m³	**22.20**
extra over to slopes exceeding 45° for temporary shuttering; shuttering measured per m² of vertical elevation	–	–	–	–	12.00	m²	**12.00**
Extra for Tensar Mat 400; erosion control mats; to faces of slopes of 45° or less; filling with 20 mm fine topsoil; seeding	3.95	0.01	0.39	0.21	5.26	m²	**5.86**
Embankments; reinforced with Tensar Geogrid RE540; geogrid laid horizontally within fill; filling with excavated material; wrapping around at faces							
300 mm centres; slopes exceeding 45°; Tensartech Natural Green System	9.72	0.06	3.66	1.60	9.72	m³	**14.98**
600 mm centres; slopes exceeding 45°; Tensartech Greenslope System	4.85	0.06	3.66	1.60	4.85	m³	**10.11**
extra over to slopes exceeding 45° for temporary shuttering; shuttering measured per m² of vertical elevation	–	–	–	–	12.00	m²	**12.00**
Extra over embankments for bagwork face supports for slopes exceeding 45°	–	0.17	9.87	–	14.00	m²	**23.87**
Extra over embankments for seeding of bags	–	0.03	1.97	–	0.32	m²	**2.29**
Extra over embankments for Bodkin joints	–	–	–	–	1.10	m	**1.10**

D41 CRIB WALLS/ GABIONS/ REINFORCED EARTH

Item Excluding site overheads and profit	PC £	Labour hours	Labour £	Plant £	Material £	Unit	Total rate £
REINFORCED EARTH – cont							
Anchoring systems; surface stabilization; Platipus Anchors Ltd; centres and depths shown should be verified with a design engineer; they may vary either way dependent on circumstances							
Concrete revetments; hard anodized stealth anchors with stainless steel accessories installed to a depth of 1 m							
SO4 anchors to a depth of 1.00–1.50 m at 1.00 m centres	26.63	0.17	6.58	1.76	26.63	m²	**34.97**
Loadlocking and crimping tool; for stressing and crimping the anchors							
hire rate per week	72.93	–	–	–	72.93	nr	**72.93**
Surface erosion; geotextile anchoring on slopes; aluminium alloy Stealth anchors							
SO4/SO6 to a depth of 1.0–2.0 m at 2.00 m centres	9.10	1.00	39.50	1.76	9.10	m²	**50.36**
Brick block or in situ concrete distressed retaining walls; anchors installed in two rows at varying tensions between 18–50 kN along the length of the retaining wall; core drilling of wall not included							
SO8/BO6/BO8 bronze Stealth and Bat anchors installed in combination with stainless steel accessories to a depth of 6 m; average cost base on 1.50 m centres; long term solution	382.02	0.25	9.88	1.76	382.02	m²	**393.66**
SO8/BO6/BO8 cast SG iron Stealth and Bat anchors installed in combination with galvanized steel accessories to a depth of 6 m; average cost base on 1.50 m centres; short term solution	163.22	0.25	9.88	1.76	163.22	m²	**174.86**
Timber retaining wall support; anchors installed 19 kN along the length of the retaining wall							
SO6/SO8 bronze Stealth anchors including load plate and wege grip; installed in combination with stainless steel accessories to a depth of 3.0 m; 1.00 m centres; average cost	184.80	–	–	–	–	m²	**–**
Gabions; anchors for support and stability to moving, overturning or rotating gabion retaining walls							
SO8/BO6 anchors including base plate at 1.00 m centres to a depth of 4.00 m	289.41	0.17	9.88	1.76	289.41	m²	**301.05**

D41 CRIB WALLS/ GABIONS/ REINFORCED EARTH

Item Excluding site overheads and profit	PC £	Labour hours	Labour £	Plant £	Material £	Unit	Total rate £
Soil nailing of geofabrics; anchors fixed through surface of fabric or erosion control surface treatment (not included)							
SO8/BO6 anchors including base plate at 1.50 m centres to a depth of 3.00 m	115.76	0.08	4.94	1.18	115.76	m²	121.88
Green faced reinforced soil; Phi Group							
Textomur reinforced soil system embankments; to reinforced soil slopes 55–70° to the horizontal; including facing and geogrid; excluding backfill material							
retaining walls up to 3.0 m	–	–	–	–	–	m²	120.00
retaining walls up to 6.0 m	–	–	–	–	–	m²	125.00
Split face concrete blocks reinforced soil; Phi Group							
Modular blocks reinforced soil system embankments; to reinforced soil slopes near vertical; including geogrid; excluding backfill material							
retaining walls up to 3.0 m	–	–	–	–	–	m²	190.00
retaining walls up to 6.0 m	–	–	–	–	–	m²	220.00

E10 IN SITU CONCRETE

Item Excluding site overheads and profit	PC £	Labour hours	Labour £	Plant £	Material £	Unit	Total rate £
Clarification notes on labour costs in this section							
In situ concrete team Generally a three man team is used in this section; The column 'Labour hours' reports team hours. The column 'Labour £' reports the total cost of the team for the unit of work shown							
3 man team	–	1.00	59.25	–	–	hr	**59.25**
2 man team	–	1.00	39.50	–	–	hr	**39.50**
CONCRETE MIX INFORMATION							
General The concrete mixes used here are referred to as Designed, Standard and Designated mixes.							
Designed mix User specifies the performance of the concrete. Producer is responsible for selecting the appropriate mix. Strength testing is essential.							
Prescribed mix User specifies the mix constituents and is responsible for ensuring that the concrete meets performance requirements. Accurate mix proportions are essential.							
Standard mix Made with a restricted range of materials. Specification to include the proposed use of the material as well as the standard mix reference, type of cement, type and size of aggregate, and slump (workability). Quality assurance is required.							
Designated mix Producer to hold current product conformity certification and quality approval. Quality assurance is essential. The mix must not be modified.							

E10 IN SITU CONCRETE

Item Excluding site overheads and profit	PC £	Labour hours	Labour £	Plant £	Material £	Unit	Total rate £
CONCRETE MIXES							
Concrete mixes; mixed on site; costs for producing concrete; prices for commonly used mixes for various types of work; based on bulk load 20 tonne rates for aggregates							
Roughest type mass concrete such as footings and road haunchings; 300 mm thick							
1:3:6	84.22	–	–	–	84.22	m³	**84.22**
As above but aggregates delivered in 10 tonne loads							
1:3:6	97.76	–	–	–	97.76	m³	**97.76**
As above but aggregates delivered in 850 kg bulk bags							
1:3:6	129.81	–	–	–	129.81	m³	**129.81**
Most ordinary use of concrete such as mass walls above ground, road slabs, etc. and general reinforced concrete work							
1:2:4	–	–	–	–	95.81	m³	**95.81**
As above but aggregates delivered in 10 tonne loads							
1:2:4	106.55	–	–	–	106.55	m³	**106.55**
As above but aggregates delivered in 850 kg bulk bags							
1:2:4	–	–	–	–	138.59	m³	**138.59**
Watertight floors, pavements, walls, tanks, pits, steps, paths, surface of two course roads; reinforced concrete where extra strength is required							
1:1.5:3	90.82	1.20	23.64	–	90.82	m³	**114.46**
As above but aggregates delivered in 10 tonne loads							
1:1.5:3	101.56	1.20	23.64	–	101.56	m³	**125.20**
As above but aggregates delivered in 850 kg bulk bags							
1:1.5:3	133.60	1.20	23.64	–	133.60	m³	**157.24**

Prices for Measured Works

E10 IN SITU CONCRETE

Item Excluding site overheads and profit	PC £	Labour hours	Labour £	Plant £	Material £	Unit	Total rate £
IN SITU CONCRETE							
Plain in situ concrete; site mixed;							
10 N/mm²–40 mm aggregate (1:3:6)							
(aggregate delivery indicated)							
Foundations							
ordinary portland cement; 20 tonne							
ballast loads	86.33	0.33	19.55	–	86.33	m³	**105.88**
ordinary portland cement; 10 tonne							
ballast loads	100.20	0.33	19.55	–	100.20	m³	**119.75**
ordinary portland cement; 850 kg bulk							
bags	133.05	0.33	19.55	–	133.05	m³	**152.60**
Foundations; poured on or against earth							
or unblinded hardcore							
ordinary portland cement; 20 tonne							
ballast loads	88.43	0.33	19.55	–	88.43	m³	**107.98**
ordinary portland cement; 10 tonne							
ballast loads	102.65	0.33	19.55	–	102.65	m³	**122.20**
ordinary portland cement; 850 kg bulk							
bags	136.30	0.33	19.55	–	136.30	m³	**155.85**
Isolated foundations							
ordinary portland cement; 20 tonne							
ballast loads	86.33	0.33	19.55	–	86.33	m³	**105.88**
ordinary portland cement; 850 kg bulk							
bags	133.05	0.33	19.55	–	133.05	m³	**152.60**
Plain in situ concrete; site mixed;							
21 N/mm²–20 mm aggregate (1:2:4)							
Foundations							
ordinary portland cement; 20 tonne							
ballast loads	98.20	0.33	19.55	–	98.20	m³	**117.75**
ordinary portland cement; 850 kg bulk							
bags	142.06	0.33	19.55	–	142.06	m³	**161.61**
Foundations; poured on or against earth							
or unblinded hardcore							
ordinary portland cement; 20 tonne							
ballast loads	100.60	0.33	19.55	–	100.60	m³	**120.15**
ordinary portland cement; 850 kg bulk							
bags	145.52	0.33	19.55	–	145.52	m³	**165.07**
Isolated foundations							
ordinary portland cement; 20 tonne							
ballast loads	98.20	0.33	19.55	–	98.20	m³	**117.75**
ordinary portland cement; 850 kg bulk							
bags	142.06	0.33	19.55	–	142.06	m³	**161.61**

E10 IN SITU CONCRETE

Item Excluding site overheads and profit	PC £	Labour hours	Labour £	Plant £	Material £	Unit	Total rate £
Ready mix concrete; concrete mixed on site; Euromix							
Ready mix concrete mixed on site; placed by barrow not more than 25 m distance (16 barrows/m^3)							
concrete 1:3:6	85.00	0.33	19.55	–	85.00	m^3	**104.55**
concrete 1:2:4	87.50	0.33	19.55	–	87.50	m^3	**107.05**
concrete C30	90.00	0.33	19.55	–	90.00	m^3	**109.55**
Reinforced in situ concrete; site mixed; 21 N/mm^2–20 mm aggregate (1:2:4); aggregates delivered in 10 tonne loads							
Foundations							
ordinary portland cement	98.20	0.67	39.46	–	98.20	m^3	**137.66**
Foundations; poured on or against earth or unblinded hardcore							
ordinary portland cement	98.20	0.67	39.46	–	98.20	m^3	**137.66**
Isolated foundations							
ordinary portland cement	98.20	0.83	49.36	–	98.20	m^3	**147.56**
Plain in situ concrete; ready mixed; Tarmac Southern; 10 N/mm mixes; suitable for mass concrete fill and blinding							
Foundations							
C10; 10 N/mm^2	75.00	0.50	29.63	–	75.00	m^3	**104.63**
ST 2; 10 N/mm^2	78.75	0.50	29.63	–	78.75	m^3	**108.38**
Foundations; poured on or against earth or unblinded hardcore							
C10; 10 N/mm^2	75.00	0.50	29.63	–	75.00	m^3	**104.63**
ST 2; 10 N/mm^2	80.63	0.50	29.63	–	80.63	m^3	**110.26**
Isolated foundations							
C10; 10 N/mm^2	75.00	0.67	39.46	–	75.00	m^3	**114.46**
ST 2; 10 N/mm^2	78.75	0.67	39.46	–	78.75	m^3	**118.21**
Plain in situ concrete; ready mixed; Tarmac Southern; 15 N/mm mixes; suitable for oversite below suspended slabs and strip footings in non aggressive soils							
Foundations							
C15; 15 N/mm^2	78.96	0.50	29.63	–	78.96	m^3	**108.59**
ST 3; 15 N/mm^2	79.28	0.50	29.63	–	79.28	m^3	**108.91**
Foundations; poured on or against earth or unblinded hardcore							
C15; 15 N/mm^2	75.20	0.50	29.63	–	75.20	m^3	**104.83**
ST 3; 15 N/mm^2	75.50	0.50	29.63	–	75.50	m^3	**105.13**
Isolated foundations							
C15; 15 N/mm^2	75.20	0.67	39.46	–	75.20	m^3	**114.66**
ST 3; 15 N/mm^2	75.50	0.67	39.46	–	75.50	m^3	**114.96**

E10 IN SITU CONCRETE

Item Excluding site overheads and profit	PC £	Labour hours	Labour £	Plant £	Material £	Unit	Total rate £
IN SITU CONCRETE – cont							
Plain in situ concrete; ready mixed;							
Tarmac Southern; air entrained mixes							
suitable for paving							
Beds or slabs; house drives, parking and							
external paving							
PAV 1; 35 N/mm²; Designated mix	82.53	0.50	29.63	–	82.53	m³	112.16
Beds or slabs; heavy duty external							
paving							
PAV 2; 40 N/mm²; Designated mix	83.06	0.50	29.63	–	83.06	m³	112.69
Reinforced in situ concrete; ready							
mixed; Tarmac Southern; 35 N/							
mm² mix; suitable for foundations in							
class 2 sulphate conditions							
Foundations							
RC 35; Designated mix	83.37	0.67	39.46	–	83.37	m³	122.83
Foundations; poured on or against earth							
or unblinded hardcore							
RC 35; Designated mix	85.36	0.67	39.46	–	85.36	m³	124.82
Isolated foundations							
RC 35; Designated mix	79.40	0.67	39.46	–	79.40	m³	118.86
Reinforced in situ concrete; site							
mixed; 21 N/mm²–20 mm aggregate							
(1:2:4); aggregates delivered in 20							
tonne loads							
Walls; thickness not exceeding 150 mm							
Site mixed concrete 1:2:4: 21 N/mm²;							
20 mm aggregate	98.20	0.67	39.46	3.09	98.20	m³	140.75
C10 Concrete	75.00	0.33	19.75	3.09	75.00	m³	97.84
C15 Concrete	78.96	0.33	19.75	3.09	78.96	m³	101.80
ST 2; 10 N/mm²	78.75	0.33	19.75	3.09	78.75	m³	101.59
ST 3; 15 N/mm²	79.28	0.33	19.75	3.09	79.28	m³	102.12
RC 35; Designated mix	83.37	0.33	19.75	3.09	83.37	m³	106.21
Walls; thickness not exceeding							
150–450 mm							
Site mixed concrete 1:2:4: 21 N/mm²;							
20 mm aggregate	98.20	0.50	29.63	3.09	98.20	m³	130.92
C10 Concrete	75.00	0.20	11.85	3.09	75.00	m³	89.94
C15 Concrete	78.96	0.20	11.85	3.09	78.96	m³	93.90
ST 2; 10 N/mm²	78.75	0.20	11.85	3.09	78.75	m³	93.69
ST 3; 15 N/mm²	79.28	0.20	11.85	3.09	79.28	m³	94.22
RC 35; Designated mix	83.37	0.20	11.85	3.09	83.37	m³	98.31
Adjustments for site mixed concrete; add							
the following amounts for aggregates							
delivered in different load sizes							
850 kg bulk bags	–	–	–	–	42.78	m³	42.78
10 tonne loads	–	–	–	–	10.74	m³	10.74

E20 FORMWORK FOR IN SITU CONCRETE

Item Excluding site overheads and profit	PC £	Labour hours	Labour £	Plant £	Material £	Unit	Total rate £
PLAIN VERTICAL FORMWORK							
Plain vertical formwork; basic finish							
Sides of foundations							
height not exceeding 250 mm	–	0.25	9.88	–	1.13	m	**11.01**
height 250–500 mm	–	0.25	9.88	–	1.84	m	**11.72**
height 500 mm–1.00 m	–	0.33	13.17	–	5.27	m	**18.44**
height exceeding 1.00 m	–	0.50	19.75	–	6.86	m²	**26.61**
Sides of foundations; left in							
height not exceeding 250 mm	–	0.20	7.90	–	3.90	m	**11.80**
height 250–500 mm	–	0.20	7.90	–	7.39	m	**15.29**
height 500 mm–1.00 m	–	0.25	9.88	–	14.78	m	**24.66**
height over 1.00 m	–	0.40	15.80	–	9.38	m²	**25.18**
Walls							
height not exceeding 250 mm	–	0.31	12.34	–	3.13	m	**15.47**
height 250–500 mm	–	0.33	13.17	–	5.84	m	**19.01**
height 500 mm–1.00 m	–	0.42	16.46	–	13.27	m	**29.73**
height exceeding 1.00 m	–	0.75	29.63	–	18.86	m²	**48.49**
Walls curved to 6.00 radius							
height not exceeding 250 mm	–	0.38	14.81	–	3.13	m	**17.94**
height 250–500 mm	–	0.40	15.80	–	5.84	m	**21.64**
height 500 mm–1.00 m	–	0.50	19.75	–	13.27	m	**33.02**
height exceeding 1.00 m	–	0.90	35.55	–	18.86	m²	**54.41**
Walls curved to 3.00 radius							
height not exceeding 250 mm	–	0.47	18.52	–	3.13	m	**21.65**
height 250–500 mm	–	0.50	19.75	–	5.84	m	**25.59**
height 500 mm–1.00 m	–	0.62	24.68	–	13.27	m	**37.95**
height exceeding 1.00 m	–	1.13	44.44	–	18.86	m²	**63.30**

E30 REINFORCEMENT FOR IN SITU CONCRETE

Item Excluding site overheads and profit	PC £	Labour hours	Labour £	Plant £	Material £	Unit	Total rate £
BAR REINFORCEMENT							
The rates for reinforcement shown for steel bars below are based on prices which would be supplied on a typical landscape contract. The steel prices shown have been priced on a selection of steel delivered to site where the total order quantity is in the region of 2 tonnes. The assumption is that should larger quantities be required, the work would fall outside the scope of the typical landscape contract defined in the front of this book. Keener rates can be obtained for larger orders.							
Reinforcement bars; hot rolled plain round mild steel; straight or bent							
Bars							
8 mm nominal size	530.00	13.50	533.25	–	530.00	tonne	**1063.25**
10 mm nominal size	530.00	13.00	513.50	–	530.00	tonne	**1043.50**
12 mm nominal size	530.00	12.50	493.75	–	530.00	tonne	**1023.75**
16 mm nominal size	530.00	12.00	474.00	–	530.00	tonne	**1004.00**
20 mm nominal size	530.00	11.50	454.25	–	530.00	tonne	**984.25**
Reinforcement bar to concrete formwork							
8 mm bar							
100 ccs	2.07	0.17	6.52	–	2.07	m²	**8.59**
200 ccs	1.01	0.13	4.94	–	1.01	m²	**5.95**
300 ccs	0.69	0.08	2.96	–	0.69	m²	**3.65**
10 mm bar							
100 ccs	3.23	0.17	6.52	–	3.23	m²	**9.75**
200 ccs	1.59	0.13	4.94	–	1.59	m²	**6.53**
300 ccs	1.06	0.08	2.96	–	1.06	m²	**4.02**
12 mm bar							
100 ccs	4.66	0.17	6.52	–	4.66	m²	**11.18**
200 ccs	2.33	0.13	4.94	–	2.33	m²	**7.27**
300 ccs	1.54	0.08	2.96	–	1.54	m²	**4.50**
16 mm bar							
100 ccs	8.32	0.20	7.90	–	8.32	m²	**16.22**
200 ccs	4.19	0.17	6.52	–	4.19	m²	**10.71**
300 ccs	2.81	0.13	4.94	–	2.81	m²	**7.75**
25 mm bar							
100 ccs	20.41	0.20	7.90	–	20.41	m²	**28.31**
200 ccs	10.18	0.13	4.94	–	10.18	m²	**15.12**
300 ccs	6.78	0.13	4.94	–	6.78	m²	**11.72**
32 mm bar							
100 ccs	33.44	0.20	7.90	–	33.44	m²	**41.34**
200 ccs	16.70	0.17	6.52	–	16.70	m²	**23.22**
300 ccs	11.13	0.13	4.94	–	11.13	m²	**16.07**

E30 REINFORCEMENT FOR IN SITU CONCRETE

Item Excluding site overheads and profit	PC £	Labour hours	Labour £	Plant £	Material £	Unit	Total rate £
Reinforcement fabric; lapped; in beds or suspended slabs							
Fabric							
A98 (1.54 kg/m^2)	0.94	0.11	4.34	–	0.94	m^2	**5.28**
A142 (2.22 kg/m^2)	1.38	0.11	4.34	–	1.38	m^2	**5.72**
A193 (3.02 kg/m^2)	1.89	0.11	4.34	–	1.89	m^2	**6.23**
A252 (3.95 kg/m^2)	2.44	0.12	4.74	–	2.44	m^2	**7.18**
A393 (6.16 kg/m^2)	3.83	0.14	5.53	–	3.83	m^2	**9.36**

F10 BRICK/BLOCK WALLING

Item Excluding site overheads and profit	PC £	Labour hours	Labour £	Plant £	Material £	Unit	Total rate £
MARKET PRICES OF MATERIALS							
Cement; Builder Centre							
Portland cement	–	–	–	–	3.76	25 kg	**3.76**
Sulphate-resistant cement	–	–	–	–	5.20	25 kg	**5.20**
White cement	–	–	–	–	8.25	25 kg	**8.25**
Sand; Builder Centre							
Building sand							
loose	35.46	–	–	–	35.46	m³	**35.46**
850 kg bulk bags	–	–	–	–	35.59	nr	**35.59**
850 kg bulk bags	–	–	–	–	69.40	m³	**69.40**
Sharp sand							
loose	–	–	–	–	34.02	m³	**34.02**
850 kg bulk bags	–	–	–	–	35.59	nr	**35.59**
850 kg bulk bags	–	–	–	–	69.40	m³	**69.40**
Sand; Yeoman Aggregates Ltd							
Sharp sand	–	–	–	–	18.90	tonne	**18.90**
Sharp sand	–	–	–	–	30.24	m³	**30.24**
Soft sand	–	–	–	–	19.70	tonne	**19.70**
Soft sand	–	–	–	–	31.52	m³	**31.52**
BRICK/BLOCK WALLING DATA							
Note: Batching quantities for these mortar mixes may be found in the Tables and Memoranda section of this book.							
Mortar mixes for brickwork and pavings; all mortars mixed on site and priced at delivery to waiting wheelbarrows at the cement mixer; cement and lime in 25 kg bags; aggregates as described							
Mortar grade 1:3; aggregates delivered in 20 tonne load; cement: sand volumes shown below (bags: litres)							
85 litre mixer (1.84:93.5)	–	0.75	14.81	–	115.89	m³	**130.70**
150 litre mixer (3.24:165)	–	0.50	9.88	–	115.89	m³	**125.77**
Mortar grade 1:3; aggregates delivered in 10 tonne load; cement: sand volumes shown below (bags: litres)							
85 litre mixer (1.84:93.5)	–	0.75	14.81	–	130.72	m³	**145.53**
150 litre mixer (3.24:165)	–	0.50	9.88	–	130.72	m³	**140.60**
Mortar grade 1:3; aggregates delivered in 850 kg bulk bags; cement: sand volumes shown below (bags: litres)							
85 litre mixer (1.84:93.5)	–	0.75	14.81	–	154.89	m³	**169.70**
150 litre mixer (3.24:165)	–	0.50	9.88	–	154.89	m³	**164.77**

F10 BRICK/BLOCK WALLING

Item Excluding site overheads and profit	PC £	Labour hours	Labour £	Plant £	Material £	Unit	Total rate £
Mortar grade 1:4; aggregates delivered in 20 tonne load; cement: sand volumes shown below (bags: litres)							
85 litre mixer (1.36:102)	–	0.75	14.81	–	97.98	m³	**112.79**
150 litre mixer (2.4:180)	–	0.50	9.88	–	97.98	m³	**107.86**
Mortar grade 1:4; aggregates delivered in 10 tonne load; cement: sand volumes shown below (bags: litres)							
85 litre mixer (1.36:102)	–	0.75	14.81	–	114.16	m³	**128.97**
150 litre mixer (2.4:180)	–	0.50	9.88	–	114.16	m³	**124.04**
Mortar grade 1:4; aggregates delivered in 850 kg bulk bags; cement: sand volumes shown below (bags: litres)							
85 litre mixer (1.36:102)	–	0.75	14.81	–	140.59	m³	**155.40**
150 litre mixer (2.4:180)	–	0.50	9.88	–	140.59	m³	**150.47**
Mortar grade 1:1:6; aggregates delivered in 20 tonne load; cement: lime: sand volumes shown below (bags: bags: litres)							
85 litre mixer; (0.9:0.44:93.5)	–	0.75	14.81	–	129.72	m³	**144.53**
150 litre mixer; (1.6: 0.8: 165)	–	0.50	9.88	–	129.72	m³	**139.60**
Mortar grade 1:1:6; aggregates delivered in 10 tonne load; cement: lime: sand volumes shown below (bags: bags: litres)							
85 litre mixer (0.9:0.44:93.5)	–	0.75	14.81	–	149.05	m³	**163.86**
150 litre mixer (1.6:0.8:165)	–	0.50	9.88	–	149.05	m³	**158.93**
Mortar grade 1:1:6; aggregates delivered in 850 kg bulk bags; cement: lime: sand volumes shown below (bags: bags: litres)							
85 litre mixer (0.9:0.44:93.5)	–	0.75	14.81	–	168.72	m³	**183.53**
150 litre mixer (1.6:0.8:165)	–	0.50	9.88	–	168.72	m³	**178.60**
Mortar grade 1:2:9; aggregates delivered in 20 tonne load; cement: lime: sand volumes shown below (bags: bags: litres)							
85 litre mixer (0.7:0.5:102)	–	0.75	14.81	–	130.72	m³	**145.53**
150 litre mixer (1.2:0.9:180)	–	0.50	9.88	–	127.57	m³	**137.45**
Mortar grade 1:2:9; aggregates delivered in 10 tonne load; cement: lime: sand volumes shown below (bags: bags: litres)							
85 litre mixer (0.7:0.5:102)	–	0.75	14.81	–	146.90	m³	**161.71**
150 litre mixer (1.2:0.9:180)	–	0.50	9.88	–	146.90	m³	**156.78**
Mortar grade 1:2:9; aggregates delivered in 850 kg bulk bags; cement: lime: sand volumes shown below (bags: bags: litres)							
85 litre mixer (0.7:0.5:102)	–	0.75	14.81	–	173.33	m³	**188.14**
150 litre mixer (1.2:0.9:180)	–	0.50	9.88	–	172.98	m³	**182.86**

F10 BRICK/BLOCK WALLING

Item Excluding site overheads and profit	PC £	Labour hours	Labour £	Plant £	Material £	Unit	Total rate £
BRICK/BLOCK WALLING DATA – cont							
Variation in brick prices (area × price difference × value shown below)							
Add or subtract the following amounts for every £1.00/1000 difference in the PC price of the measured items below							
half brick thick	–	–	–	–	6.30	m²	**6.30**
one brick thick	–	–	–	–	12.60	m²	**12.60**
one and a half brick thick	–	–	–	–	18.90	m²	**18.90**
two brick thick	–	–	–	–	25.20	m²	**25.20**
Mortar (1:3) required per m² of brickwork; brick size 215 × 102.5 × 65 mm							
Half brick wall (103 mm); 58 bricks/m²							
no frog	–	–	–	–	2.91	m²	**2.91**
single frog	–	–	–	–	3.36	m²	**3.36**
double frog	–	–	–	–	3.97	m²	**3.97**
2 × half brick cavity wall (270 mm); 116 bricks/ m²							
no frog	–	–	–	–	5.81	m²	**5.81**
single frog	–	–	–	–	6.88	m²	**6.88**
double frog	–	–	–	–	8.41	m²	**8.41**
One brick wall (215 mm); 116 bricks/m²							
no frog	–	–	–	–	7.03	m²	**7.03**
single frog	–	–	–	–	8.41	m²	**8.41**
double frog	–	–	–	–	9.78	m²	**9.78**
One and a half brick wall (328 mm) 174 bricks/m²							
no frog	–	–	–	–	9.63	m²	**9.63**
single frog	–	–	–	–	11.31	m²	**11.31**
double frog	–	–	–	–	13.45	m²	**13.45**
Mortar (1:3) required per m² of blockwork; blocks 440 × 215 mm							
Block thickness							
100 mm	–	–	–	–	1.08	m²	**1.08**
140 mm	–	–	–	–	1.38	m²	**1.38**
hollow blocks 440 × 215 mm	–	–	–	–	0.35	m²	**0.35**
Movement of materials							
Loading to wheelbarrows and transporting to location; per 215 mm thick walls; maximum distance 25 m	–	0.42	8.23	–	–	m²	**8.23**

F10 BRICK/BLOCK WALLING

Item Excluding site overheads and profit	PC £	Labour hours	Labour £	Plant £	Material £	Unit	Total rate £
Adjustments for mortars in the brick prices below Difference for cement mortar (1:3) in lieu of gauged mortar							
half brick thick	–	–	–	–	–0.41	m²	**–0.41**
one brick thick	–	–	–	–	–1.03	m²	**–1.03**
one and a half brick thick	–	–	–	–	–1.39	m²	**–1.39**
two brick thick	–	–	–	–	–2.07	m²	**–2.07**
Clarification notes on labour costs in this section							
Brick/Block walling team Generally a three man team is used in this section; The column 'Labour hours' reports team hours. The column 'Labour £' reports the total cost of the team for the unit of work shown							
3 man team	–	1.00	59.25	–	–	hr	**59.25**
2 man team	–	1.00	39.50	–	–	hr	**39.50**
BRICK WALLING							
Note: frog types and mortar quantities used in the calculations below are based on bricks with single frogs. Please make adjustments for mortar volumes used with varying brick frog formats. See data above for mortar quantities for various frog types. Delivery volumes of aggregates for mortars are based on 10 tonne deliveries. The figures below show the approximate areas of walling which would be constructed mixed with building sand from a 10 tonne (6.25 m³) load of building sand.							
half brick wall = 256 m²							
one brick wall = 124 m²							
Class B engineering bricks; PC £276.00/1000; double Flemish bond in cement mortar (1:3) Offloading mechanically; loading to wheelbarrows; transporting to location maximum 25 m distance per 215 mm thick walls	–	0.42	8.23	–	–	m²	**8.23**

F10 BRICK/BLOCK WALLING

Item Excluding site overheads and profit	PC £	Labour hours	Labour £	Plant £	Material £	Unit	Total rate £
BRICK WALLING – cont							
Class B engineering bricks – cont							
Construct walls							
half brick thick	–	0.60	35.55	–	20.59	m²	**56.14**
one brick thick	–	1.20	71.10	–	42.78	m²	**113.88**
one and a half brick thick	–	1.80	106.65	–	68.82	m²	**175.47**
two brick thick	–	2.40	142.14	–	85.56	m²	**227.70**
Construct walls; curved; mean radius 6 m							
half brick thick	–	0.90	53.33	–	20.55	m²	**73.88**
one brick thick	–	1.80	106.65	–	42.78	m²	**149.43**
Construct walls; curved; mean radius 1.50 m							
half brick thick	–	1.20	71.09	–	20.59	m²	**91.68**
one brick thick	–	3.60	142.20	–	42.78	m²	**184.98**
Construct walls; tapering; one face battering; average							
one and a half brick thick	–	2.22	131.32	–	62.93	m²	**194.25**
two brick thick	–	2.96	175.08	–	85.56	m²	**260.64**
Construct walls; battering (retaining)							
one and a half brick thick	–	2.22	131.32	–	62.93	m²	**194.25**
two brick thick	–	2.96	175.08	–	85.56	m²	**260.64**
Projections; vertical							
one brick × half brick	–	2.33	138.23	–	4.37	m	**142.60**
one brick × one brick	–	0.47	27.61	–	8.74	m	**36.35**
one and a half brick × one brick	–	0.70	41.48	–	13.12	m	**54.60**
two brick by one brick	–	0.77	45.39	–	17.49	m	**62.88**
Walls; half brick thick							
in honeycomb bond	–	0.60	35.55	–	14.23	m²	**49.78**
in quarter bond	–	0.56	32.88	–	19.16	m²	**52.04**
Facing bricks; PC £300.00/1000; English garden wall bond; in gauged mortar (1:1:6); facework one side							
Mechanically offloading; maximum 25 m distance; loading to wheelbarrows and transporting to location; walls	–	0.42	8.23	–	–	m²	**8.23**
Construct walls							
half brick thick	–	0.60	35.49	–	22.50	m²	**57.99**
half brick thick (using site cut snap headers to form bond)	–	0.80	47.64	–	22.50	m²	**70.14**
one brick thick	–	1.20	71.10	–	127.93	m²	**199.03**
one and a half brick thick	–	1.80	106.65	–	68.83	m²	**175.48**
two brick thick	–	2.40	142.20	–	93.63	m²	**235.83**
Walls; curved; mean radius 6 m							
half brick thick	–	0.90	53.33	–	23.41	m²	**76.74**
one brick thick	–	1.80	106.65	–	129.73	m²	**236.38**

F10 BRICK/BLOCK WALLING

Item Excluding site overheads and profit	PC £	Labour hours	Labour £	Plant £	Material £	Unit	Total rate £
Walls; curved; mean radius 1.50 m							
half brick thick	–	1.20	71.10	–	22.96	m²	94.06
one brick thick	–	2.40	142.20	–	128.83	m²	271.03
Walls; tapering; one face battering; average							
one and a half brick thick	–	2.20	130.35	–	71.53	m²	201.88
two brick thick	–	3.00	177.75	–	97.23	m²	274.98
Walls; battering (retaining)							
one and a half brick thick	–	2.00	118.50	–	71.53	m²	190.03
two brick thick	–	2.70	159.97	–	89.03	m²	249.00
Projections; vertical							
one brick × half brick	–	0.23	13.63	–	4.77	m	18.40
one brick × one brick	–	0.50	29.63	–	9.54	m	39.17
one and a half brick × one brick	–	0.70	41.48	–	14.32	m	55.80
two brick × one brick	–	0.80	47.40	–	20.08	m	67.48
Brickwork fair faced both sides; facing bricks in gauged mortar (1:1:6)							
extra for fair face both sides; flush, struck, weathered, or bucket-handle pointing	–	0.67	13.17	–	–	m²	13.17
Bricks; PC £800.00/1000; English garden wall bond; in gauged mortar (1:1:6)							
Walls							
half brick thick (stretcher bond)	–	0.60	35.55	–	54.00	m²	89.55
half brick thick (using site cut snap headers to form bond)	–	0.80	47.40	–	54.00	m²	101.40
one brick thick	–	1.20	71.10	–	107.89	m²	178.99
one and a half brick thick	–	1.80	106.65	–	160.45	m²	267.10
two brick thick	–	2.40	142.20	–	207.59	m²	349.79
Walls; curved; mean radius 6 m							
half brick thick	–	0.90	53.33	–	56.40	m²	109.73
one brick thick	–	2.40	142.20	–	114.61	m²	256.81
Walls; curved; mean radius 1.50 m							
half brick thick	–	0.53	31.40	–	58.80	m²	90.20
one brick thick	–	1.07	63.22	–	119.41	m²	182.63
Walls; stretcher bond; wall ties at 450 mm centres vertically and horizontally							
one brick thick	0.96	1.09	64.58	–	108.85	m²	173.43
two brick thick	2.89	1.64	97.17	–	210.48	m²	307.65
Brickwork fair faced both sides; facing bricks in gauged mortar (1:1:6)							
extra for fair face both sides; flush, struck, weathered or bucket-handle pointing	–	0.67	13.17	–	–	m²	13.17

F10 BRICK/BLOCK WALLING

Item Excluding site overheads and profit	PC £	Labour hours	Labour £	Plant £	Material £	Unit	Total rate £
BRICK WALLING – cont							
Brick copings							
Copings; all brick headers-on-edge; two							
angles rounded 53 mm radius; flush							
pointing top and both sides as work							
proceeds; one brick wide; horizontal							
machine-made specials	48.05	0.16	9.72	–	49.04	m	**58.76**
handmade specials	61.80	0.16	9.72	–	62.78	m	**72.50**
Extra over copings for two courses							
machine-made tile creasings; projecting							
25 mm each side; 260 mm wide copings;							
horizontal	6.11	0.17	10.07	–	6.60	m	**16.67**
Copings; all brick headers-on-edge; flush							
pointing top and both sides as work							
proceeds; one brick wide; horizontal							
facing bricks PC £300.00/1000	4.30	0.16	9.72	–	4.30	m	**14.02**
engineering bricks PC £276.00/1000	3.86	0.16	9.72	–	3.86	m	**13.58**
BLOCK WALLING							
Dense aggregate concrete blocks;							
Tarmac Topblock or other equal and							
approved; in gauged mortar (1:2:9);							
one course underground							
Walls							
Solid blocks 7 N/mm²							
440 × 215 × 100 mm thick	8.99	0.40	23.70	–	10.15	m²	**33.85**
440 × 215 × 140 mm thick	13.75	0.44	26.25	–	15.06	m²	**41.31**
Solid blocks 7 N/mm² laid flat							
440 × 100 × 215 mm thick	18.34	1.07	63.40	–	19.65	m²	**83.05**
Hollow concrete blocks							
440 × 215 × 215 mm thick	23.44	0.43	25.66	–	24.75	m²	**50.41**
Filling of hollow concrete blocks with							
concrete as work proceeds; tamping and							
compacting							
440 × 215 × 215 mm thick	17.51	0.07	3.91	–	17.51	m²	**21.42**
BRICK PIERS							
Isolated brick piers in English bond;							
in gauged mortar 1:1:6							
Engineering brick PC £276.00/1000							
one brick thick (225 mm)	–	2.33	138.23	–	44.61	m²	**182.84**
one and a half brick thick (337.5 mm)	–	3.00	177.75	–	66.91	m²	**244.66**
two brick thick (450 mm)	–	3.33	197.48	–	90.85	m²	**288.33**
three brick thick (675 mm)	–	4.13	244.88	–	133.82	m²	**378.70**

F10 BRICK/BLOCK WALLING

Item Excluding site overheads and profit	PC £	Labour hours	Labour £	Plant £	Material £	Unit	Total rate £
Engineering brick PC £276.00/1000 (not SMM)							
one brick (225 mm)	–	0.54	32.27	–	10.21	m	42.48
one and a half brick thick (337.5 mm)	–	1.00	59.25	–	31.12	m	90.37
two brick thick (450 mm)	–	1.50	88.88	–	49.32	m	138.20
three brick thick (675 mm)	–	2.79	165.31	–	99.92	m	265.23
Brick PC £300.00/1000							
one brick thick (225 mm)	–	2.33	138.23	–	47.63	m²	185.86
one and a half brick thick (337.5 mm)	–	3.00	177.75	–	71.45	m²	249.20
two brick thick (450 mm)	–	3.33	197.48	–	96.90	m²	294.38
three brick thick (675 mm)	–	4.13	244.88	–	142.90	m²	387.78
Brick PC £300.00/1000 (not SMM)							
one brick (225 mm)	–	0.54	32.23	–	10.92	m	43.15
one and a half brick thick (337.5 mm)	–	1.00	59.25	–	32.66	m	91.91
two brick thick (450 mm)	–	1.50	88.88	–	52.05	m	140.93
three brick thick (675 mm)	–	2.79	165.31	–	106.04	m	271.35
Brick PC £800.00/1000							
one brick thick (225 mm)	–	2.33	138.23	–	110.63	m²	248.86
one and a half brick thick (337.5 mm)	–	3.00	177.75	–	165.95	m²	343.70
two brick thick (450 mm)	–	3.33	197.48	–	222.90	m²	420.38
three brick thick (675 mm)	–	4.13	244.88	–	331.90	m²	576.78
Brick PC £800.00/1000 (not SMM)							
one brick (225 mm)	–	0.54	32.23	–	25.63	m	57.86
one and a half brick thick (337.5 mm)	–	1.00	59.25	–	64.55	m	123.80
two brick thick (450 mm)	–	1.50	88.88	–	108.75	m	197.63
three brick thick (675 mm)	–	2.79	165.31	–	233.62	m	398.93
BLOCK PIERS							
Isolated blockwork piers; to receive facing treatments (not included)							
Solid blocks; 440 × 100 × 215 mm thick; laid on-flat; 7 N/mm²							
450 × 450 mm (SMM measurement)	35.92	2.00	79.00	–	36.94	m²	115.94
450 × 450 mm (not SMM)	16.18	0.90	35.55	–	16.65	m	52.20

F20 NATURAL STONE WALLING

Item Excluding site overheads and profit	PC £	Labour hours	Labour £	Plant £	Material £	Unit	Total rate £
Clarification notes on labour costs in this section							
Expert stonework team Generally a three man team is used in this section; The column Labour hours reports team hours. The column Labour £ reports the total cost of the team for the unit of work shown							
2 man team	–	1.00	39.50	–	–	hr	**39.50**
GRANITE WALLS							
Granite walls Granite random rubble walls; laid dry							
200 mm thick; single faced	49.29	3.33	131.53	–	49.29	m²	**180.82**
Granite walls; one face battering to 50°; pointing faces							
450 mm (average) thick	119.23	2.50	98.75	–	120.14	m²	**218.89**
DRYSTONE WALLING							
Dry stone walling – General Preamble: In rural areas where natural stone is a traditional material, it may be possible to use dry stone walling or dyking as an alternative to fences or brick walls. Many local authorities are willing to meet the extra cost of stone walling in areas of high landscape value, and they may hold lists of available craftsmen. DSWA Office, Westmorland County Showground, Lane Farm, Crooklands, Milnthorpe, Cumbria, LA7 7NH; Tel: 01539 567953; E-mail: information@dswa.org.uk Note: Traditional walls are not built on concrete foundations. Dry stone wall; wall on concrete foundation (not included); dry stone coursed wall inclusive of locking stones and filling to wall with broken stone or rubble; walls up to 1.20 m high; battered; 2 sides fair faced							
Yorkstone	–	3.25	128.38	–	66.73	m²	**195.11**
Cotswold stone	–	3.25	128.38	–	81.57	m²	**209.95**
Purbeck	–	3.25	128.38	–	88.98	m²	**217.36**

F22 CAST STONE/ASHLAR WALLING/DRESSINGS

Item Excluding site overheads and profit	PC £	Labour hours	Labour £	Plant £	Material £	Unit	Total rate £
RECONSTITUTED STONE WALLING							
Haddonstone Ltd; cast stone piers; **ornamental masonry in Portland Bath** **or Terracotta finished cast stone**							
Gate Pier S120; to foundations and underground work measured separately; concrete infill							
S120G base unit to pier;							
699 × 699 × 172 mm	178.80	0.25	14.81	–	183.48	nr	**198.29**
S120F/F shaft base unit;							
533 × 533 × 280 mm	141.60	0.33	19.73	–	146.47	nr	**166.20**
S120E/E main shaft unit;							
533 × 533 × 280 mm; nr of units required dependent on height of pier	141.60	0.33	19.73	–	146.47	nr	**166.20**
S120D/D top shaft unit;							
33 × 533 × 280 mm	141.60	0.33	19.73	–	146.47	nr	**166.20**
S120C pier cap unit;							
737 × 737 × 114 mm	168.00	0.17	9.84	–	172.87	nr	**182.71**
S120B pier block unit; base for finial;							
533 × 533 × 64 mm	72.00	0.11	6.52	–	72.20	nr	**78.72**
Pier blocks; flat to receive gate finial							
S100B; 440 × 440 × 63 mm	46.80	0.11	6.52	–	47.26	nr	**53.78**
S120B; 546 × 546 × 64 mm	72.00	0.11	6.52	–	72.46	nr	**78.98**
S150B; 330 × 330 × 51 mm	25.20	0.11	6.52	–	25.66	nr	**32.18**
Pier caps; part weathered							
S100C; 915 × 915 × 150 mm	399.60	0.25	14.81	–	400.52	nr	**415.33**
S120C; 737 × 737 × 114 mm	168.00	0.17	9.84	–	168.46	nr	**178.30**
S150C; 584 × 584 × 120 mm	122.40	0.17	9.84	–	122.86	nr	**132.70**
Pier caps; weathered							
S230C; 1029 × 1029 × 175 mm	524.40	0.17	9.84	–	524.86	nr	**534.70**
S215C; 687 × 687 × 175 mm	230.40	0.17	9.84	–	230.86	nr	**240.70**
S210C; 584 × 584 × 175 mm	144.00	0.17	9.84	–	144.46	nr	**154.30**
Pier strings							
S100S; 800 × 800 × 55 mm	136.80	1.66	98.36	–	137.08	nr	**235.44**
S120S; 555 × 555 × 44 mm	62.40	0.17	9.84	–	62.68	nr	**72.52**
S150S; 457 × 457 × 48 mm	38.40	0.17	9.84	–	38.68	nr	**48.52**
Balls and bases							
E150 A Ball 535 mm and E150C collared base	381.00	0.17	9.84	–	381.28	nr	**391.12**
E120 A Ball 330 mm and E120C collared base	144.00	0.17	9.84	–	144.28	nr	**154.12**
E110 A Ball 230 mm and E110C collared base	101.00	0.17	9.84	–	101.28	nr	**111.12**
E100 A Ball 170 mm and E100B plain base	64.00	0.17	9.84	–	64.28	nr	**74.12**
Haddonstone Ltd; cast stone copings; **ornamental masonry in Portland Bath** **or Terracotta finished cast stone**							

F22 CAST STONE/ASHLAR WALLING/DRESSINGS

Item Excluding site overheads and profit	PC £	Labour hours	Labour £	Plant £	Material £	Unit	Total rate £
RECONSTITUTED STONE WALLING – cont							
Copings for walls; bedded, jointed and pointed in approved coloured cement-lime mortar 1:2:9							
T100 weathered coping; 102 mm high × 178 mm wide × 914 mm	41.86	0.11	6.58	–	42.25	m	**48.83**
T140 weathered coping; 102 mm high × 337 mm wide × 914 mm	69.32	0.11	6.58	–	69.72	m	**76.30**
T200 weathered coping; 127 mm high × 508 mm wide × 750 mm	104.64	0.11	6.58	–	105.03	m	**111.61**
T170 weathered coping; 108 mm high × 483 mm wide × 914 mm	112.49	0.11	6.58	–	112.88	m	**119.46**
T340 raked coping; 75–100 mm high × 290 mm wide × 900 mm	64.09	0.11	6.58	–	64.49	m	**71.07**
T310 raked coping; 76–89 mm high × 381 mm wide × 914 mm	74.56	0.11	6.58	–	74.95	m	**81.53**
Bordeaux walling; Forticrete Ltd Dry stacked random sized units 150–400 mm long × 150 mm high cast stone wall mechanically interlocked with fibreglass pins; constructed to levelling pad of coarse compacted granular material back filled behind the elevation with 300 mm wide granular drainage material; walls to 5.00 m high (retaining walls over heights shown below require individual design)							
gravity wall; near vertical; 250 mm thick	88.10	0.33	19.73	–	88.10	m²	**107.83**
gravity wall; 9.5°; battered	88.10	0.67	39.46	–	88.10	m²	**127.56**
copings to Bordeaux wall; 70 mm thick; random lengths	14.30	0.08	4.94	–	14.30	m	**19.24**
Retaining wall; as above but reinforced with Tensar Geogrid RE520 laid between every two courses horizontally into the face of the excavation (excavation not included) Near vertical; 250 mm thick; 1.50 m of geogrid length							
up to 1.20 m high; 2 layers of geogrid	88.10	0.50	29.63	–	94.85	m²	**124.48**
1.20–1.50 m high; 3 layers of geogrid	88.10	0.67	39.46	–	98.22	m²	**137.68**
1.50–1.80 m high; 4 layers of geogrid	88.10	0.75	44.44	–	101.60	m²	**146.04**
Battered; maximum 1:3 slope							
up to 1.20 m high; 3 layers of geogrid	88.10	0.83	49.36	–	98.22	m²	**147.58**
1.20–1.50 m high; 4 layers of geogrid	88.10	0.83	49.36	–	101.60	m²	**150.96**
1.50–1.80 m high; 5 layers of geogrid	88.10	1.00	59.25	–	104.41	m²	**163.66**

F30 ACCESSORIES/SUNDRY ITEMS FOR BRICK/BLOCK/STONE WALLING

Item Excluding site overheads and profit	PC £	Labour hours	Labour £	Plant £	Material £	Unit	Total rate £
DAMP-PROOF COURSES							
Damp-proof courses; pitch polymer; **150 mm laps**							
Horizontal							
width not exceeding 225 mm	4.86	1.14	22.52	–	6.46	m²	28.98
width exceeding 225 mm	5.42	0.58	11.46	–	7.02	m²	18.48
Vertical							
width not exceeding 225 mm	4.86	1.72	33.97	–	6.46	m²	40.43
Two courses slates in cement mortar **(1:3)**							
Horizontal							
width exceeding 225 mm	9.31	3.46	68.33	–	13.67	m²	82.00
Vertical							
width exceeding 225 mm	9.31	5.18	102.31	–	13.67	m²	115.98

F31 PRECAST CONCRETE SILLS/LINTELS/COPINGS/FEATURES

Item Excluding site overheads and profit	PC £	Labour hours	Labour £	Plant £	Material £	Unit	Total rate £
COPINGS AND PIERCAPS							
Precast concrete coping							
Copings; once weathered; twice grooved							
152 × 75 mm	5.62	0.40	7.90	–	5.88	m	**13.78**
178 × 65 mm	7.77	0.40	7.90	–	8.13	m	**16.03**
305 × 75 mm	11.03	0.50	9.88	–	11.61	m	**21.49**
Pier caps; four sides weathered							
305 × 305 mm	7.04	1.00	19.75	–	7.21	nr	**26.96**
381 × 381 mm	7.46	1.00	19.75	–	7.63	nr	**27.38**
533 × 533 mm	7.25	1.20	23.70	–	7.42	nr	**31.12**

H CLADDING/COVERING

Item Excluding site overheads and profit	PC £	Labour hours	Labour £	Plant £	Material £	Unit	Total rate £
H51 NATURAL STONE SLAB CLADDING/FEATURES							
Sawn Yorkstone cladding; Johnsons Wellfield Quarries Ltd							
Six sides sawn stone; rubbed face; sawn and jointed edges; fixed to blockwork (not included) with stainless steel fixings; Ancon Ltd; grade 304 stainless steel frame cramp and dowel 7 mm; cladding units drilled 4 × to receive dowels							
440 × 200 × 50 mm thick	89.95	1.87	36.93	–	93.35	m²	**130.28**
H52 CAST STONE SLAB CLADDING/ FEATURES							
Cast stone cladding; Haddonstone Ltd							
Reconstituted stone in Portland Bath or Terracotta; fixed to blockwork or concrete (not included) with stainless steel fixings; stainless steel frame cramp and dowel M6 mm; cladding units drilled 4 × to receive dowels							
1000 × 300 × 50 mm thick	45.00	1.87	36.93	–	48.40	m²	**85.33**

J10 SPECIALIST WATERPROOF RENDERING

Item Excluding site overheads and profit	PC £	Labour hours	Labour £	Plant £	Material £	Unit	Total rate £
SIKA							
Sika 1 waterproof rendering; steel trowelled							
Walls; 18 mm thick; three coats; to concrete base							
width exceeding 300 mm	–	–	–	–	–	m²	**45.42**
width not exceeding 300 mm	–	–	–	–	–	m²	**47.92**
Walls; 18 mm thick; three coats; to concrete base							
width exceeding 300 mm	–	–	–	–	–	m²	**42.43**
width not exceeding 300 mm	–	–	–	–	–	m²	**44.93**

J20 MASTIC ASPHALT TANKING/DAMP-PROOFING

Item Excluding site overheads and profit	PC £	Labour hours	Labour £	Plant £	Material £	Unit	Total rate £
TANKING AND DAMP-PROOFING							
Tanking and damp-proofing; mastic asphalt; type T1097; Bituchem							
13 mm thick; one coat covering; to concrete base; flat; work subsequently covered							
width exceeding 300 mm	–	–	–	–	–	m²	**11.24**
20 mm thick; two coat coverings; to concrete base; flat; work subsequently covered							
width exceeding 300 mm	–	–	–	–	–	m²	**14.09**
30 mm thick; three coat coverings; to concrete base; flat; work subsequently covered							
width exceeding 300 mm	–	–	–	–	–	m²	**19.13**
13 mm thick; two coat coverings; to brickwork base; vertical; work subsequently covered							
width exceeding 300 mm	–	–	–	–	–	m²	**37.34**
20 mm thick; three coat coverings; to brickwork base; vertical; work subsequently covered							
width exceeding 300 mm	–	–	–	–	–	m²	**50.98**
Internal angle fillets; work subsequently covered	–	–	–	–	–	m	**3.97**
Turning asphalt nibs into grooves; 20 mm deep	–	–	–	–	–	m	**2.52**

J30 LIQUID APPLIED TANKING/DAMP-PROOFING

Item Excluding site overheads and profit	PC £	Labour hours	Labour £	Plant £	Material £	Unit	Total rate £
COLD APPLIED BITUMINOUS EMULSION							
Tanking and damp-proofing; Ruberoid Building Products; Synthaprufe cold applied bituminous emulsion waterproof coating							
Synthaprufe; to smooth finished concrete or screeded slabs; flat; blinding with sand							
two coats	6.34	0.22	4.39	–	6.49	m²	**10.88**
three coats	9.50	0.31	6.12	–	9.66	m²	**15.78**
Synthaprufe; to fair faced brickwork with flush joints, rendered brickwork or smooth finished concrete walls; vertical							
two coats	7.13	0.29	5.64	–	7.28	m²	**12.92**
three coats	10.45	0.40	7.90	–	10.61	m²	**18.51**
LIQUID ASPHALT							
Tanking and damp-proofing; RIW Ltd							
Liquid asphaltic composition; to smooth finished concrete screeded slabs or screeded slabs; flat							
two coats	6.41	0.33	6.58	–	6.41	m²	**12.99**
Liquid asphaltic composition; fair faced brickwork with flush joints, rendered brickwork or smooth finished concrete walls; vertical							
two coats	6.41	0.50	9.88	–	6.41	m²	**16.29**
Heviseal; to smooth finished concrete or screeded slabs; to surfaces of ponds, tanks or planters; flat							
two coats	9.11	0.33	6.58	–	9.11	m²	**15.69**
Heviseal; to fair faced brickwork with flush joints, rendered brickwork or smooth finished concrete walls; to surfaces of retaining walls, ponds, tanks or planters; vertical							
two coats	9.11	0.50	9.88	–	9.11	m²	**18.99**

J40 FLEXIBLE SHEET TANKING/DAMP-PROOFING

Item Excluding site overheads and profit	PC £	Labour hours	Labour £	Plant £	Material £	Unit	Total rate £
SELF ADHESIVE							
Tanking and damp-proofing; Grace Construction Products Bitu-thene 3000; 1.50 mm thick; overlapping and bonding; including sealing all edges							
to concrete slabs; flat	7.04	0.25	4.94	–	7.78	m²	**12.72**
to brick/concrete walls; vertical	7.04	0.40	7.90	–	7.78	m²	**15.68**

M SURFACE FINISHES

Item Excluding site overheads and profit	PC £	Labour hours	Labour £	Plant £	Material £	Unit	Total rate £
M20 PLASTERED/RENDERED/ **ROUGHCAST COATINGS**							
Cement: lime: sand (1:1:6); 19 mm **thick; two coats; wood floated finish** Walls							
width exceeding 300 mm; to brickwork or blockwork base	–	–	–	–	–	m²	**14.25**
Extra over cement: sand: lime (1:1:6) coatings for decorative texture finish with water repellent cement							
combed or floated finish	–	–	–	–	–	m²	**3.80**
M40 STONE/CONCRETE/QUARRY/ **CERAMIC TILING/MOSAIC**							
Ceramic tiles; unglazed slip-resistant; **various colours and textures; jointing** Floors							
level or to falls only not exceeding 15° from horizontal; 150 × 150 × 8 mm thick	21.00	1.00	19.75	–	23.91	m²	**43.66**
level or to falls only not exceeding 15° from horizontal; 150 × 150 × 12 mm thick	26.25	1.00	19.75	–	29.16	m²	**48.91**

Clay tiles – General
Preamble: Typical specification – Clay
tiles shall be reasonably true to shape,
flat, free from flaws, frost resistant and
true to sample approved by the
Landscape Architect prior to laying.
Quarry tiles (or semi-vitrified tiles) shall
be of external quality, either heather
brown or blue, to size specified, laid on
1:2:4 concrete, with 20 mm maximum
aggregate 100 mm thick, on 100 mm
hardcore. The hardened concrete should
be well wetted and the surplus water
taken off. Clay tiles shall be thoroughly
wetted immediately before laying and
then drained and shall be bedded to
19 mm thick cement: sand (1:3) screed.
Joints should be approximately 4 mm (or
3 mm for vitrified tiles) grouted in
cement: sand (1:2) and cleaned off
immediately.

M SURFACE FINISHES

Item Excluding site overheads and profit	PC £	Labour hours	Labour £	Plant £	Material £	Unit	Total rate £
Quarry tiles; external quality;							
including bedding; jointing							
Floors							
level or to falls only not exceeding 15° from horizontal; 150 × 150 × 12.5 mm thick; heather brown; PC £15.00 / m²	–	0.80	15.80	–	18.66	m²	34.46
level or to falls only not exceeding 15° from horizontal; 225 × 225 × 29 mm thick; heather brown PC £25.00 /m²	–	0.67	13.17	–	29.16	m²	42.33
level or to falls only not exceeding 15° from horizontal; 150 × 150 × 12.5 mm thick; blue/black PC £16.00 / m²	–	1.00	19.75	–	19.71	m²	39.46
level or to falls only not exceeding 15° from horizontal; 194 × 194 × 12.5 mm thick; heather brown PC £15.00 / m²	–	0.80	15.80	–	18.66	m²	34.46
M60 PAINTING/CLEAR FINISHING							
Eura Conservation Ltd; restoration of							
railings; works carried out off site;							
excludes removal of railings from site							
Shotblast railings; remove all traces of paint and corrosion; apply temporary protective primer; measured overall 2 sides							
to plain railings	–	–	–	–	–	m²	42.00
to decorative railings	–	–	–	–	–	m²	84.00
Paint new galvanized railings in situ with 3 coats system							
vertical bar railings or gates	–	–	–	–	–	m²	11.00
vertical bar railings or gates with dog mesh panels	–	–	–	–	–	m²	12.00
gate posts not exceeding 300 mm girth	–	–	–	–	–	m²	6.00
Paint previously painted railings; shotblasted off site							
vertical bar railings or gates	–	–	–	–	–	m²	14.00
gate posts not exceeding 300 mm girth	–	–	–	–	–	m²	6.00
Prepare; touch up primer; two							
undercoats and one finishing coat of							
gloss oil paint; on metal surfaces							
General surfaces							
girth exceeding 300 mm	5.31	0.67	13.17	–	5.31	m²	18.48
isolated surfaces; girth not exceeding 300 mm	16.09	0.40	7.90	–	16.09	m	23.99
isolated areas not exceeding 0.50 m² irrespective of girth	10.62	0.67	13.17	–	10.62	nr	23.79

M SURFACE FINISHES

Item Excluding site overheads and profit	PC £	Labour hours	Labour £	Plant £	Material £	Unit	Total rate £
M60 PAINTING/CLEAR FINISHING – cont							
Prepare – cont Ornamental railings; each side measured separately							
girth exceeding 300 mm	5.31	0.57	11.29	–	5.31	m²	**16.60**
Prepare; one coat primer; two finishing coats of gloss paint; on wood surfaces							
New wood surfaces							
girth exceeding 300 mm	2.28	0.80	15.80	–	2.28	m²	**18.08**
isolated surfaces; girth not exceeding 300 mm	0.76	1.00	19.75	–	0.76	m	**20.51**
isolated areas not exceeding 0.50 m² irrespective of girth	3.36	0.50	9.88	–	3.36	nr	**13.24**
Previously painted wood surfaces							
girth exceeding 300 mm	0.80	0.50	9.88	–	0.80	m²	**10.68**
isolated surfaces; girth not exceeding 300 mm	0.26	0.67	13.17	–	0.26	m	**13.43**
isolated areas not exceeding 0.50 m² irrespective of girth	1.59	0.40	7.90	–	1.59	nr	**9.49**
Fences and sheds; prepare; two coats of Protek wood preserver on wood surfaces							
Planed surfaces							
girth exceeding 300 mm	0.60	0.07	1.32	–	0.60	m²	**1.92**
isolated surfaces; girth not exceeding 300 mm	0.20	0.17	3.29	–	0.20	m	**3.49**
isolated areas not exceeding 0.50 m² irrespective of girth	0.60	0.25	4.94	–	0.60	nr	**5.54**
Sawn surfaces							
girth exceeding 300 mm	0.22	0.10	1.98	–	0.22	m²	**2.20**
isolated surfaces; girth not exceeding 300 mm	2.97	0.17	3.29	–	2.97	m	**6.26**
isolated areas not exceeding 0.50 m² irrespective of girth	0.89	0.25	4.94	–	0.89	nr	**5.83**
Fences and sheds; prepare; two coats of Sadolin wood preserver on wood surfaces							
Planed surfaces							
girth exceeding 300 mm	1.89	0.07	1.32	–	1.89	m²	**3.21**
isolated surfaces; girth not exceeding 300 mm	0.63	0.17	3.29	–	0.63	m	**3.92**
isolated areas not exceeding 0.50 m² irrespective of girth	0.95	0.25	4.94	–	0.95	nr	**5.89**

M SURFACE FINISHES

Item Excluding site overheads and profit	PC £	Labour hours	Labour £	Plant £	Material £	Unit	Total rate £
Sawn surfaces							
girth exceeding 300 mm	1.43	0.10	1.98	–	1.43	m²	**3.41**
isolated surfaces; girth not exceeding 300 mm	1.01	0.17	3.29	–	1.01	m	**4.30**
isolated areas not exceeding 0.50 m² irrespective of girth	1.51	0.25	4.94	–	1.51	nr	**6.45**
Prepare; proprietary solution primer; two coats of dark stain; on wood surfaces							
General surfaces							
girth exceeding 300 mm	1.80	0.17	3.29	–	1.80	m²	**5.09**
isolated surfaces; girth not exceeding 300 mm	0.60	0.13	2.47	–	0.60	m	**3.07**
Two coats Weathershield; to clean, dry surfaces; in accordance with manufacturer's instructions							
Brick or block walls							
girth exceeding 300 mm	1.20	0.40	7.90	–	1.20	m²	**9.10**
Cement render or concrete walls							
girth exceeding 300 mm	1.20	0.33	6.58	–	1.20	m²	**7.78**

P BUILDING FABRIC SUNDRIES

Item Excluding site overheads and profit	PC £	Labour hours	Labour £	Plant £	Material £	Unit	Total rate £
P30 TRENCHES/PIPEWAYS/PITS FOR BURIED ENGINEERING SERVICES							
Excavating trenches; using 3 tonne tracked excavator; to receive pipes; grading bottoms; earthwork support; filling with excavated material to within 150 mm of finished surfaces and compacting; completing fill with topsoil; disposal of surplus soil							
Services not exceeding 200 mm nominal size							
average depth of run not exceeding 0.50 m	1.51	0.12	2.37	1.06	1.51	m	**4.94**
average depth of run not exceeding 0.75 m	1.51	0.16	3.22	1.44	1.51	m	**6.17**
average depth of run not exceeding 1.00 m	1.51	0.28	5.60	2.51	1.51	m	**9.62**
average depth of run not exceeding 1.25 m	1.51	0.38	7.57	3.40	1.51	m	**12.48**
Excavating trenches; using 3 tonne tracked excavator; to receive pipes; grading bottoms; earthwork support; filling with imported granular material and compacting; disposal of surplus soil							
Services not exceeding 200 mm nominal size							
average depth of run not exceeding 0.50 m	1.50	0.09	1.71	0.76	4.22	m	**6.69**
average depth of run not exceeding 0.75 m	2.25	0.11	2.14	0.95	6.33	m	**9.42**
average depth of run not exceeding 1.00 m	3.00	0.14	2.76	1.23	8.44	m	**12.43**
average depth of run not exceeding 1.25 m	3.74	0.23	4.51	2.03	10.54	m	**17.08**
Excavating trenches; using 3 tonne tracked excavator; to receive pipes; grading bottoms; earthwork support; filling with lean mix concrete; disposal of surplus soil							
Services not exceeding 200 mm nominal size							
average depth of run not exceeding 0.50 m	11.25	0.11	2.11	0.36	14.52	m	**16.99**
average depth of run not exceeding 0.75 m	16.88	0.13	2.57	0.45	21.78	m	**24.80**

P BUILDING FABRIC SUNDRIES

Item Excluding site overheads and profit	PC £	Labour hours	Labour £	Plant £	Material £	Unit	Total rate £
average depth of run not exceeding 1.00 m	22.50	0.17	3.29	0.60	29.03	m	**32.92**
average depth of run not exceeding 1.25 m	28.13	0.23	4.44	0.90	36.29	m	**41.63**
Earthwork support; providing support to opposing faces of excavation; moving along as work proceeds							
Maximum depth not exceeding 2.00 m							
trenchbox; distance between opposing faces not exceeding 2.00 m	–	0.80	15.80	21.00	–	m	**36.80**
timber; distance between opposing faces not exceeding 500 mm	–	0.20	7.90	–	0.02	m	**7.92**

Q10 KERBS/EDGINGS/CHANNELS/PAVING ACCESSORIES

Item Excluding site overheads and profit	PC £	Labour hours	Labour £	Plant £	Material £	Unit	Total rate £
Clarification notes on labour costs in this section							
General groundworks team							
Generally a three man team is used in this section; The column Labour hours reports team hours. The column Labour £ reports the total cost of the team for the unit of work shown							
3 man team	–	1.00	59.25	–	–	hr	**59.25**
2 man team	–	1.00	39.50	–	–	hr	**39.50**
FOUNDATIONS TO KERBS							
Foundations to kerbs							
Excavating trenches; width 300 mm; 3 tonne excavator; disposal off site							
depth 200 mm	–	–	–	2.71	1.09	m	**3.80**
depth 300 mm	–	–	–	3.93	1.63	m	**5.56**
depth 400 mm	–	–	–	4.29	2.18	m	**6.47**
By hand; barrowing to spoil heap and disposal off site							
depth 300	–	0.06	3.56	–	18.14	m	**21.70**
depth 400	–	0.08	4.74	–	18.14	m	**22.88**
Excavating trenches; width 450 mm; 3 tonne excavator; disposal off site							
depth 300 mm	–	–	–	4.90	2.45	m	**7.35**
depth 400 mm	–	–	–	5.61	3.27	m	**8.88**
Foundations; to kerbs, edgings or channels; in situ concrete; 21 N/mm²– 20 mm aggregate ((1:2:4) site mixed); one side against earth face, other against formwork (not included); site mixed concrete							
Site mixed concrete							
150 mm wide × 100 mm deep	–	0.04	2.64	–	1.44	m	**4.08**
150 mm wide × 150 mm deep	–	0.06	3.29	–	2.16	m	**5.45**
200 mm wide × 150 mm deep	–	0.07	3.95	–	2.87	m	**6.82**
300 mm wide × 150 mm deep	–	0.06	3.46	–	4.31	m	**7.77**
600 mm wide × 200 mm deep	–	0.06	3.45	–	11.50	m	**14.95**
Ready mixed concrete							
150 mm wide × 100 mm deep	–	0.04	2.64	–	1.14	m	**3.78**
150 mm wide × 150 mm deep	–	0.06	3.29	–	1.71	m	**5.00**
200 mm wide × 150 mm deep	–	0.07	3.95	–	2.28	m	**6.23**
300 mm wide × 150 mm deep	–	0.08	4.61	–	3.42	m	**8.03**
600 mm wide × 200 mm deep	–	0.10	5.64	–	9.12	m	**14.76**

Q10 KERBS/EDGINGS/CHANNELS/PAVING ACCESSORIES

Item Excluding site overheads and profit	PC £	Labour hours	Labour £	Plant £	Material £	Unit	Total rate £
Formwork; sides of foundations (this will usually be required to one side of each kerb foundation adjacent to road subbases)							
100 mm deep	–	0.01	0.82	–	0.18	m	**1.00**
150 mm deep	–	0.01	0.82	–	0.27	m	**1.09**
PRECAST CONCRETE KERBS							
Precast concrete kerbs, channels, edgings etc.; Marshalls Mono; bedding, jointing and pointing in cement mortar (1:3); including haunching with in situ concrete; 11.50 N/mm²–40 mm aggregate one side							
Kerbs; straight							
150 × 305 mm; HB1	8.79	0.17	9.88	–	11.31	m	**21.19**
125 × 255 mm; HB2; SP	4.20	0.15	8.77	–	6.73	m	**15.50**
125 × 150 mm; BN	2.85	0.13	7.90	–	5.38	m	**13.28**
Dropper kerbs; left and right handed							
125 × 255–150 mm; DL1 or DR1	6.36	0.17	9.88	–	8.89	m	**18.77**
125 × 255–150 mm; DL2 or DR2	6.36	0.17	9.88	–	8.89	m	**18.77**
Quadrant kerbs							
305 mm radius	10.69	0.17	9.88	–	12.37	nr	**22.25**
455 mm radius	11.51	0.17	9.88	–	14.04	nr	**23.92**
Straight kerbs or channels; to radius; 125 × 255 mm							
0.90 m radius (2 units per quarter circle)	8.04	0.27	15.80	–	10.65	m	**26.45**
1.80 m radius (4 units per quarter circle)	8.04	0.24	14.36	–	10.65	m	**25.01**
2.40 m radius (5 units per quarter circle)	8.04	0.23	13.39	–	10.65	m	**24.04**
3.00 m radius (5 units per quarter circle)	8.04	0.22	13.17	–	10.65	m	**23.82**
4.50 m radius (2 units per quarter circle)	8.04	0.20	11.86	–	10.65	m	**22.51**
6.10 m radius (11 units per quarter circle)	8.04	0.19	11.45	–	10.65	m	**22.10**
7.60 m radius (14 units per quarter circle)	8.04	0.19	11.29	–	10.65	m	**21.94**
9.15 m radius (17 units per quarter circle)	8.04	0.19	10.97	–	10.65	m	**21.62**
10.70 m radius (20 units per quarter circle)	8.04	0.19	10.97	–	10.65	m	**21.62**
12.20 m radius (22 units per quarter circle)	8.04	0.18	10.40	–	10.65	m	**21.05**

Q10 KERBS/EDGINGS/CHANNELS/PAVING ACCESSORIES

Item Excluding site overheads and profit	PC £	Labour hours	Labour £	Plant £	Material £	Unit	Total rate £
PRECAST CONCRETE KERBS – cont							
Precast concrete kerbs, channels, **edgings etc. – cont**							
Kerbs; Conservation Kerb units; to simulate natural granite kerbs							
255 × 150 × 914 mm; laid flat	22.40	0.19	11.29	–	26.94	m	**38.23**
150 × 255 × 914 mm; laid vertical	22.40	0.19	11.29	–	26.19	m	**37.48**
145 × 255 mm; radius internal 3.25 m	27.77	0.28	16.46	–	31.56	m	**48.02**
150 × 255 mm; radius external 3.40 m	26.52	0.28	16.46	–	30.31	m	**46.77**
145 × 255 mm; radius internal 6.50 m	24.81	0.22	13.17	–	28.60	m	**41.77**
145 × 255 mm; radius external 6.70 m	24.67	0.22	13.17	–	28.46	m	**41.63**
150 × 255 mm; radius internal 9.80 m	24.54	0.22	13.17	–	28.33	m	**41.50**
150 × 255 mm; radius external 10.00 m	23.26	0.22	13.17	–	27.05	m	**40.22**
305 × 305 × 255 mm; solid quadrants	32.14	0.19	11.29	–	35.93	nr	**47.22**
Marshalls Mono; small element **precast concrete kerb system;** **Keykerb Large (KL) upstand of** **100–125 mm; on 150 mm concrete** **foundation including haunching with** **in situ concrete (1:3:6) 1 side**							
Bullnosed or half battered; 100 × 127 × 200 mm							
laid straight	12.77	0.27	15.80	–	13.65	m	**29.45**
radial blocks laid to curve; 8 blocks/1/ 4 circle – 500 mm radius	9.83	0.33	19.75	–	10.71	m	**30.46**
radial blocks laid to curve; 8 radial blocks, alternating 8 standard blocks/ 1/4 circle – 1000 mm radius	16.21	0.42	24.69	–	17.10	m	**41.79**
radial blocks, alternating 16 standard blocks/1/4 circle – 1500 mm radius	14.92	0.50	29.63	–	15.80	m	**45.43**
internal angle 90 degree	5.40	0.07	3.95	–	5.40	nr	**9.35**
external angle	5.40	0.07	3.95	–	5.40	nr	**9.35**
drop crossing kerbs; KL half battered to KL Splay; LH and RH	22.56	0.33	19.75	–	23.44	pair	**43.19**
drop crossing kerbs; KL half battered to KS bullnosed	21.94	0.33	19.75	–	22.82	pair	**42.57**

Q10 KERBS/EDGINGS/CHANNELS/PAVING ACCESSORIES

Item Excluding site overheads and profit	PC £	Labour hours	Labour £	Plant £	Material £	Unit	Total rate £
Marshalls Mono; small element precast concrete kerb system; Keykerb Small (KS) upstand of 25–50 mm; on 150 mm concrete foundation including haunching with in situ concrete (1:3:6) 1 side							
Half battered							
laid straight	8.97	0.27	15.80	–	9.85	m	**25.65**
radial blocks laid to curve; 8 blocks/1/ 4 circle; 500 mm radius	7.36	0.33	19.75	–	8.24	m	**27.99**
radial blocks laid to curve; 8 radial blocks, alternating 8 standard blocks/ 1/4 circle; 1000 mm radius	11.85	0.42	24.69	–	12.73	m	**37.42**
radial blocks, alternating 16 standard blocks/1/4 circle; 1500 mm radius	12.68	0.50	29.63	–	13.56	m	**43.19**
internal angle 90 degree	5.40	0.07	3.95	–	5.40	nr	**9.35**
external angle	5.40	0.07	3.95	–	5.40	nr	**9.35**
PRECAST CONCRETE EDGINGS							
Precast concrete edging units; including haunching with in situ concrete; 11.50 N/mm^2–40 mm aggregate both sides							
Edgings; rectangular, bullnosed or chamfered							
50 × 150 mm	1.93	0.11	6.58	–	6.14	m	**12.72**
125 × 150 mm bullnosed	2.10	0.11	6.58	–	6.31	m	**12.89**
50 × 200 mm	2.76	0.11	6.58	–	6.97	m	**13.55**
50 × 250 mm	3.05	0.11	6.58	–	7.26	m	**13.84**
50 × 250 mm flat top	3.39	0.11	6.58	–	7.60	m	**14.18**

Q10 KERBS/EDGINGS/CHANNELS/PAVING ACCESSORIES

Item Excluding site overheads and profit	PC £	Labour hours	Labour £	Plant £	Material £	Unit	Total rate £
STONE KERBS							
Dressed natural stone kerbs – General Preamble: The kerbs are to be good, sound and uniform in texture and free from defects; worked straight or to radius, square and out of wind, with the top front and back edges parallel or concentric to the dimensions specified. All drill and pick holes shall be removed from dressed faces. Standard dressings shall be designated as either fine picked, single axed or nidged or rough punched.							
Dressed natural stone kerbs; CED Ltd; on concrete foundations (not included); including haunching with in situ concrete; 11.50 N/mm²–40 mm aggregate one side Granite kerbs; 125 × 250 mm							
special quality; straight; random lengths	22.88	0.27	15.80	–	25.88	m	41.68
Granite kerbs; 125 × 250 mm; curved to mean radius 3 m							
special quality; random lengths	32.89	0.30	17.97	–	35.89	m	53.86
STONE EDGINGS							
New granite sett edgings; CED Ltd; 100 × 100 × 100 mm; bedding in cement mortar (1:4); including haunching with in situ concrete; 11.50 N/mm²–40 mm aggregate one side; concrete foundations (not included) Edgings 100 × 100 × 100 mm; ref 1R 'better quality' with 20 mm pointing gaps							
100 mm wide	3.14	0.44	26.27	–	6.13	m	32.40
2 rows; 220 mm wide	6.28	0.74	43.84	–	10.15	m	53.99
3 rows; 340 mm wide	9.42	1.00	59.25	–	12.46	m	71.71
Edgings 100 × 100 × 200 mm; ref 1R 'better quality' with 20 mm pointing gaps							
100 mm wide	3.20	0.25	14.81	–	6.20	m	21.01
2 rows; 220 mm wide	6.41	0.50	29.63	–	10.28	m	39.91
3 rows; 340 mm wide	9.61	0.75	44.44	–	12.65	m	57.09
Edgings 100 × 100 × 100 mm; ref 8R 'standard quality' with 20 mm pointing gaps							
100 mm wide	2.58	0.44	26.27	–	5.57	m	31.84
2 rows; 220 mm wide	5.15	0.74	43.84	–	9.02	m	52.86
3 rows; 340 mm wide	7.73	1.00	59.25	–	10.77	m	70.02

Q10 KERBS/EDGINGS/CHANNELS/PAVING ACCESSORIES

Item Excluding site overheads and profit	PC £	Labour hours	Labour £	Plant £	Material £	Unit	Total rate £
Edgings 100 × 100 × 200 mm; ref 8R 'standard quality' with 20 mm pointing gaps							
100 mm wide	2.83	0.25	14.81	–	5.83	m	**20.64**
2 rows; 220 mm wide	5.67	0.50	29.63	–	9.54	m	**39.17**
3 rows; 340 mm wide	8.50	0.75	44.44	–	11.54	m	**55.98**
Reclaimed granite setts edgings; CED Ltd; 100 × 100 mm; bedding in cement mortar (1:4); including haunching with in situ concrete; 11.50 N/mm^2–40 mm aggregate one side; on concrete foundations (not included)							
Edgings; with 20 mm pointing gaps							
100 mm wide	3.58	0.44	26.27	–	6.58	m	**32.85**
2 rows; 220 mm wide	7.17	0.74	43.84	–	11.04	m	**54.88**
3 rows; 340 mm wide	10.75	1.00	59.25	–	13.80	m	**73.05**
Natural yorkstone edgings; Johnsons Wellfield Quarries; sawn 6 sides; 50 mm thick; on prepared base measured separately; bedding on 25 mm cement: sand (1:3); cement: sand (1:3) joints							
Edgings							
100 mm wide × random lengths	10.50	0.17	9.88	–	10.86	m	**20.74**
100 × 100 mm	7.98	0.17	9.88	–	11.62	m	**21.50**
100 × 200 mm	15.96	0.17	9.88	–	19.59	m	**29.47**
250 mm wide × random lengths	15.75	0.13	7.90	–	19.39	m	**27.29**
500 mm wide × random lengths	31.53	0.11	6.58	–	35.17	m	**41.75**
Yorkstone edgings; 600 mm long × 250 mm wide; cut to radius							
1.00 m to 3.00	50.18	0.17	9.88	–	50.55	m	**60.43**
3.00 m to 5.00 m	50.18	0.15	8.77	–	50.55	m	**59.32**
exceeding 5.00 m	50.18	0.13	7.90	–	50.55	m	**58.45**
Natural yorkstone edgings; Johnsons Wellfield Quarries; sawn 6 sides; 75 mm thick; on prepared base measured separately; bedding on 25 mm cement: sand (1:3); cement: sand (1:3) joints							
Edgings							
100 mm wide × random lengths	9.40	0.20	11.85	–	9.76	m	**21.61**
100 × 100 mm	0.89	0.20	11.85	–	4.53	m	**16.38**
100 × 200 mm	3.57	0.17	9.88	–	7.21	m	**17.09**
250 mm wide × random lengths	24.60	0.17	9.88	–	28.24	m	**38.12**
500 mm wide × random lengths	49.21	0.13	7.90	–	52.85	m	**60.75**

Q10 KERBS/EDGINGS/CHANNELS/PAVING ACCESSORIES

Item Excluding site overheads and profit	PC £	Labour hours	Labour £	Plant £	Material £	Unit	Total rate £
STONE EDGINGS – cont							
Natural yorkstone edgings – cont							
Edgings; 600 mm long × 250 mm wide;							
cut to radius							
1.00 m to 3.00	59.06	0.20	11.85	–	59.43	m	**71.28**
3.00 m to 5.00 m	59.06	0.17	9.88	–	59.43	m	**69.31**
exceeding 5.00 m	56.44	0.15	8.77	–	56.80	m	**65.57**
Natural stone, slate or granite flag							
edgings; CED Ltd; on prepared base							
(not included); bedding on 25 mm							
cement: sand (1:3); cement: sand							
(1:3) joints							
Edgings; silver grey							
100 mm wide × random lengths	6.77	0.33	19.75	–	7.13	m	**26.88**
100 × 100 mm	7.74	0.50	29.63	–	8.10	m	**37.73**
100 mm long × 200 mm wide	12.57	0.20	11.85	–	12.94	m	**24.79**
250 mm wide × random lengths	13.30	0.25	14.81	–	16.94	m	**31.75**
300 mm wide × random lengths	14.51	0.25	14.81	–	18.15	m	**32.96**
BRICK EDGINGS							
Brick or block stretchers; bedding in							
cement mortar (1:4); on 150 mm deep							
concrete foundations (not included);							
Single course							
concrete paving blocks; PC £9.67/m²;							
200 × 100 × 60 mm	1.93	0.10	6.07	–	2.66	m	**8.73**
engineering bricks; PC £276.00/1000;							
215 × 102.5 × 65 mm	1.29	0.13	7.90	–	2.01	m	**9.91**
paving bricks; PC £450.00/1000;							
215 × 102.5 × 65 mm	2.00	0.13	7.90	–	2.72	m	**10.62**
Two courses							
concrete paving blocks; PC £9,67/m²;							
200 × 100 × 60 mm	3.87	0.13	7.90	–	5.17	m	**13.07**
engineering bricks; PC £276.00/1000;							
215 × 102.5 × 65 mm	2.58	0.19	11.29	–	3.88	m	**15.17**
paving bricks; PC £450.00/1000;							
215 × 102.5 × 65 mm	4.00	0.19	11.29	–	5.30	m	**16.59**
Three courses							
concrete paving blocks; PC £9.67/m²;							
200 × 100 × 60 mm	5.80	0.15	8.77	–	7.10	m	**15.87**
engineering bricks; PC £276.00/1000;							
215 × 102.5 × 65 mm	3.86	0.22	13.17	–	5.17	m	**18.34**
paving bricks; PC £450.00/1000;							
215 × 102.5 × 65 mm	6.00	0.22	13.17	–	7.30	m	**20.47**

Q10 KERBS/EDGINGS/CHANNELS/PAVING ACCESSORIES

Item Excluding site overheads and profit	PC £	Labour hours	Labour £	Plant £	Material £	Unit	Total rate £
Bricks on edge; bedding in cement mortar (1:4); on 150 mm deep concrete foundations;							
One brick wide							
engineering bricks;							
215 × 102.5 × 65 mm	3.86	0.19	11.29	–	5.31	m	**16.60**
paving bricks; 215 × 102.5 × 65 mm	6.00	0.19	11.29	–	7.44	m	**18.73**
Two courses; stretchers laid on edge; 225 mm wide							
engineering bricks;							
215 × 102.5 × 65 mm	7.73	0.40	23.70	–	10.46	m	**34.16**
paving bricks; 215 × 102.5 × 65 mm	12.00	0.40	23.70	–	14.73	m	**38.43**
Edge restraints; to brick paving; on prepared base (not included); 65 mm thick bricks; PC £300.00/1000							
Header course							
200 × 100 mm; butt joints	3.00	0.09	5.27	–	3.34	m	**8.61**
200x 100 mm; mortar joints	2.80	0.17	9.88	–	4.98	m	**14.86**
200 × 100 mm × 50; on edge; mortar joints	4.19	0.23	13.83	–	4.89	m	**18.72**
215 × 102.5 mm mortar joints	2.67	0.17	9.88	–	3.36	m	**13.24**
215 × 102.5 mm × 65; on edge; mortar joints	4.00	0.25	14.82	–	4.70	m	**19.52**
Stretcher course; on flat or on edge							
200 × 100 mm; butt joints	1.50	0.06	3.29	–	1.67	m	**4.96**
210 × 105 mm; mortar joints	1.36	0.11	6.58	–	1.70	m	**8.28**
Precast concrete block edgings; PC £8.99/m²; 200 × 100 × 60 mm; on prepared base (not included); haunching one side							
Edgings; butt joints							
stretcher course	0.90	0.06	3.29	–	1.07	m	**4.36**
header course	1.80	0.09	5.27	–	2.14	m	**7.41**
Haunching							
Haunching with in situ concrete; 11.50 N/mm²–40 mm aggregate one side							
one side	–	0.13	4.94	–	2.88	m	**7.82**
Variation in brick prices; add or subtract the following amounts for every £100/1000 difference in the PC price							
Edgings; mortar jointed							
100 mm wide	0.50	–	–	–	0.50	m	**0.50**
200 mm wide	0.95	–	–	–	0.95	m	**0.95**
102.5 mm wide	0.47	–	–	–	0.47	m	**0.47**
215 mm wide	0.93	–	–	–	0.93	m	**0.93**

Q10 KERBS/EDGINGS/CHANNELS/PAVING ACCESSORIES

Item Excluding site overheads and profit	PC £	Labour hours	Labour £	Plant £	Material £	Unit	Total rate £
TIMBER EDGINGS							
Timber edging boards; fixed with 50 × 50 × 750 mm timber pegs at 1000 mm centres (excavations and hardcore under edgings not included)							
Straight							
38 × 150 mm treated softwood edge boards	2.21	0.07	2.63	–	2.21	m	4.84
50 × 150 mm treated softwood edge boards	2.70	0.10	3.95	–	2.70	m	6.65
Curved							
38 × 150 mm treated softwood edge boards	2.21	0.13	4.94	–	2.21	m	7.15
50 × 150 mm treated softwood edge boards	2.70	0.14	5.64	–	2.70	m	8.34
METAL EDGINGS							
Permaloc AshphaltEdge; Kinley Systems; extruded aluminium alloy L shaped edging with 5.33 mm exposed upper lip; edging fixed to roadway base and edge profile with 250 mm steel fixing spike; laid to straight or curvilinear road edge; subsequently filled with macadam (not included)							
Depth of macadam							
38 mm	8.65	0.02	0.66	–	8.65	m	9.31
51 mm	9.50	0.06	3.40	–	9.50	m	12.90
64 mm	9.73	0.01	0.36	–	9.73	m	10.09
76 mm	11.41	0.01	0.40	–	11.41	m	11.81
102 mm	13.36	0.01	0.55	–	13.36	m	13.91
Permaloc L-shaped aluminium edging; Kinley Systems; heavy duty straight profile edging; for edgings to soft landscape beds or turf areas; 3.2 mm × 100 mm high; 3.2 mm thick with 4.75 mm exposed upper lip; fixed to form straight or curvilinear edge with 305 mm fixing spike							
Milled aluminium							
100 mm deep	–	0.10	3.95	–	13.03	m	16.98

Q10 KERBS/EDGINGS/CHANNELS/PAVING ACCESSORIES

Item Excluding site overheads and profit	PC £	Labour hours	Labour £	Plant £	Material £	Unit	Total rate £
Permaloc Permastrip; Kinley Systems; heavy duty L-shaped profile maintenance strip; 3.2 mm × 89 mm high with 5.2 mm exposed top lip; for straight or gentle curves on paths or bed turf interfaces; fixed to form straight or curvilinear edge with standard 305 mm stake; other stake lengths available							
Milled aluminium							
75 mm deep	11.41	0.10	3.95	–	11.41	m	**15.36**
Permaloc Proline; Kinley Systems; medium duty straight profiled maintenance strip; 3.2 mm × 102 mm high with 3.18 mm exposed top lip; for straight or gentle curves on paths or bed turf interfaces; fixed to form straight or curvilinear edge with standard 305 mm stake; other stake lengths available							
Milled aluminium							
50 mm deep	–	0.10	3.95	–	9.27	m	**13.22**
CHANNELS							
Channels; bedding in cement mortar (1:3); joints pointed flush; on concrete foundations (not included)							
Three courses stretchers; 350 mm wide; quarter bond to form dished channels							
engineering bricks; PC £276.00/1000;							
215 × 102.5 × 65 mm	3.86	0.33	19.75	–	5.17	m	**24.92**
paving bricks; PC £450.00/1000;							
215 × 102.5 × 65 mm	6.00	0.33	19.75	–	7.30	m	**27.05**
Three courses granite setts; 340 mm wide; to form dished channels							
340 mm wide	12.03	0.67	39.50	–	15.02	m	**54.52**
Precast concrete channels etc.; Marshalls Mono; bedding, jointing and pointing in cement mortar (1:3); including haunching with in situ concrete; 11.50 N/mm²–40 mm aggregate one side							
Channels; square							
125 × 225 × 915 mm long; CS1	4.01	0.13	7.90	–	13.15	m	**21.05**
125 × 150 × 915 mm long; CS2	3.25	0.13	7.90	–	10.87	m	**18.77**
Channels; dished							
305 × 150 × 915 mm long; CD	11.34	0.13	7.90	–	21.20	m	**29.10**
150 × 100 × 915 mm	6.66	0.13	7.90	–	12.28	m	**20.18**

Q20 GRANULAR SUBBASES TO ROADS AND PAVINGS

Item Excluding site overheads and profit	PC £	Labour hours	Labour £	Plant £	Material £	Unit	Total rate £
Clarification notes on labour costs in this section							
General groundworks team Generally a three man team is used in this section; The column Labour hours reports team hours. The column Labour £ reports the total cost of the team for the unit of work shown							
3 man team	–	1.00	59.25	–	–	hr	**59.25**
HARDCORE BASES							
Hardcore bases; obtained off site; PC £23.05/m³							
By machine							
100 mm thick	2.31	0.05	0.99	1.67	2.31	m²	**4.97**
150 mm thick	3.46	0.07	1.32	2.16	3.46	m²	**6.94**
200 mm thick	4.61	0.08	1.58	2.28	4.61	m²	**8.47**
300 mm thick	6.92	0.08	1.65	2.41	6.92	m²	**10.98**
exceeding 300 mm thick	27.66	0.17	3.29	5.57	27.66	m³	**36.52**
By hand							
100 mm thick	2.31	0.20	3.95	0.20	2.31	m²	**6.46**
150 mm thick	4.15	0.30	5.93	0.20	4.15	m²	**10.28**
200 mm thick	5.53	0.40	7.90	0.32	5.53	m²	**13.75**
300 mm thick	8.30	0.60	11.85	0.20	8.30	m²	**20.35**
exceeding 300 mm thick	27.66	2.00	39.50	0.66	27.66	m³	**67.82**
Hardcore; difference for each £1.00 increase/decrease in PC price per m³; price will vary with type and source of hardcore							
average 75 mm thick	–	–	–	–	0.07	m²	**0.07**
average 100 mm thick	–	–	–	–	0.10	m²	**0.10**
average 150 mm thick	–	–	–	–	0.14	m²	**0.14**
average 200 mm thick	–	–	–	–	0.19	m²	**0.19**
average 250 mm thick	–	–	–	–	0.24	m²	**0.24**
average 300 mm thick	–	–	–	–	0.28	m²	**0.28**
exceeding 300 mm thick	–	–	–	–	1.04	m³	**1.04**
TYPE 1 BASES							
Type 1 granular fill base; PC £17.95/ tonne (£39.49 / m³ compacted)							
By machine							
100 mm thick	3.95	0.02	1.23	0.84	3.95	m²	**6.02**
150 mm thick	5.92	0.03	1.48	0.84	5.92	m²	**8.24**
250 mm thick	9.87	0.03	1.69	1.67	9.87	m²	**13.23**
over 250 mm thick	39.49	0.20	11.85	5.08	39.49	m³	**56.42**

Q20 GRANULAR SUBBASES TO ROADS AND PAVINGS

Item Excluding site overheads and profit	PC £	Labour hours	Labour £	Plant £	Material £	Unit	Total rate £
By hand (mechanical compaction)							
100 mm thick	2.63	0.07	3.95	0.07	2.63	m²	6.65
150 mm thick	5.92	0.08	4.94	0.10	5.92	m²	10.96
250 mm thick	9.87	0.13	7.41	0.17	9.87	m²	17.45
over 250 mm thick	19.75	0.20	11.85	0.17	19.75	m³	31.77
Type 1 (crushed concrete) granular fill base; PC £11.80/tonne (£21.24/m³ compacted)							
By machine							
100 mm thick	2.60	0.02	1.23	0.55	2.60	m²	4.38
150 mm thick	3.89	0.03	1.48	0.84	3.89	m²	6.21
250 mm thick	6.49	0.03	1.69	1.39	6.49	m²	9.57
over 250 mm thick	11.80	–	–	–	–	m³	–
By hand (mechanical compaction)							
100 mm thick	2.60	0.07	3.95	0.07	2.60	m²	6.62
150 mm thick	3.89	0.08	4.94	0.10	3.89	m²	8.93
250 mm thick	6.49	0.13	7.41	0.17	6.49	m²	14.07
over 250 mm thick	11.80	0.20	11.85	0.17	11.80	m³	23.82
SURFACE TREATMENTS TO GRANULAR BASES							
Surface treatments							
Sand blinding; to hardcore base (not included); 25 mm thick	0.85	0.03	0.66	–	0.85	m²	1.51
Sand blinding; to hardcore base (not included); 50 mm thick	1.70	0.05	0.99	–	1.70	m²	2.69
Filter fabrics; to hardcore base (not included)	0.43	0.01	0.20	–	0.43	m²	0.63
CELLULAR CONFINED BASES							
Cellweb; Geosynthetics Ltd; cellular confinement system for load support or permeable bases below pavings							
To graded and compacted substrate (not included); on 50 mm sharp sand base filled with angular drainage aggregate 20–5 mm							
100 mm deep	9.62	0.05	2.96	2.03	17.65	m²	22.64
Nidagravel; Cedar Nursery; honeycomb shaped cellular polypropylene containment grids for retaining filled decorative aggregate surfaces (not included) for driveways, cycle paths and footpaths; to graded compacted substrate (not included) and 50 mm sharp sand base; integrated geofabric base							

Q20 GRANULAR SUBBASES TO ROADS AND PAVINGS

Item Excluding site overheads and profit	PC £	Labour hours	Labour £	Plant £	Material £	Unit	Total rate £
CELLULAR CONFINED BASES – cont							
Foot traffic, wheelchairs, bikes and pushchairs							
Nidagravel 125; 1200 × 800 × 25 mm deep	1.70	0.08	2.47	–	13.70	m²	**16.17**
Driveways: vehicles, wheelchairs, bikes and pushchairs (requires compacted granular base to support vehicle loadings)							
Nidagravel 140; 2400 × 1200 × 40 mm deep	1.70	0.08	2.68	–	16.65	m²	**19.33**
Aggregate filling to Nidagravel 125							
shingle	0.89	0.03	1.97	1.98	0.89	m²	**4.84**
decorative aggregate; CED Ltd; PC £48.00 / tonne	2.16	0.03	1.97	1.98	2.16	m²	**6.11**
Aggregate filling to Nidagravel 140							
shingle	1.42	0.03	1.66	2.40	1.42	m²	**5.48**
decorative aggregate; CED Ltd; PC £48.00 / tonne	3.46	0.03	1.66	2.40	3.46	m²	**7.52**
StableDRIVE; Cedar Nursery; honeycomb shaped polypropylene cellular containment system for retaining filled decorative aggregate surfaces for driveways, cyclepaths and footpaths; on compacted subgrade and subbase (both not included)							
Driveways: vehicles, wheelchairs, bikes and pushchairs (requires compacted granular base to support vehicle loadings)							
StableDRIVE; 600 × 500 × 32 mm deep	1.70	0.08	2.84	–	17.65	m²	**20.49**
Aggregate filling to StableDRIVE							
shingle	1.13	0.03	1.85	2.94	1.13	m²	**5.92**
decorative aggregate; CED Ltd; PC £48.00 / tonne	2.76	0.03	1.85	2.94	2.76	m²	**7.55**

Q21 IN SITU CONCRETE ROADS/PAVINGS

Item Excluding site overheads and profit	PC £	Labour hours	Labour £	Plant £	Material £	Unit	Total rate £
Clarification notes on labour costs in this section							
General groundworks team							
Generally a three man team is used in this section; The column Labour hours reports team hours. The column Labour £ reports the total cost of the team for the unit of work shown							
3 man team	–	1.00	59.25	–	–	hr	59.25
IN SITU CONCRETE PAVINGS							
Unreinforced concrete; on prepared subbase (not included)							
Roads; 21.00 N/mm²–20 mm aggregate (1:2:4) mechanically mixed on site							
100 mm thick	10.06	0.04	2.47	–	10.06	m²	12.53
150 mm thick	15.09	–	0.12	–	15.09	m²	15.21
Reinforced in situ concrete; mechanically mixed on site; normal Portland cement; on hardcore base (not included); reinforcement (not included)							
Roads; 11.50 N/mm²–40 mm aggregate (1:3:6)							
100 mm thick	8.63	0.03	1.97	–	8.63	m²	10.60
150 mm thick	12.95	0.20	11.85	–	12.95	m²	24.80
200 mm thick	17.69	0.27	15.80	–	17.69	m²	33.49
250 mm thick	21.58	0.33	19.75	0.35	21.58	m²	41.68
300 mm thick	25.90	0.40	23.70	0.35	25.90	m²	49.95
Roads; 21.00 N/mm²–20 mm aggregate (1:2:4)							
100 mm thick	9.82	0.13	7.90	–	9.82	m²	17.72
150 mm thick	14.74	0.20	11.85	–	14.74	m²	26.59
200 mm thick	19.64	0.27	15.80	–	19.64	m²	35.44
250 mm thick	24.55	0.33	19.75	0.35	24.55	m²	44.65
300 mm thick	29.46	0.40	23.70	0.35	29.46	m²	53.51
Roads; 25.00 N/mm²–20 mm aggregate C 25/30 ready mixed							
100 mm thick	7.89	0.13	7.90	–	7.89	m²	15.79
150 mm thick	11.84	0.20	11.85	–	11.84	m²	23.69
200 mm thick	15.79	0.27	15.80	–	15.79	m²	31.59
250 mm thick	19.73	0.33	19.75	0.35	19.73	m²	39.83
300 mm thick	23.68	0.40	23.70	0.35	23.68	m²	47.73

Q21 IN SITU CONCRETE ROADS/PAVINGS

Item Excluding site overheads and profit	PC £	Labour hours	Labour £	Plant £	Material £	Unit	Total rate £
IN SITU CONCRETE PAVINGS – cont							
Reinforced in situ concrete; ready mixed; discharged directly into location from supply vehicle; normal Portland cement; on hardcore base (not included); reinforcement (not included)							
Roads; 11.50 N/mm²–40 mm aggregate (1:3:6)							
100 mm thick	7.50	0.05	3.16	–	7.50	m²	10.66
150 mm thick	11.25	0.08	4.74	–	11.25	m²	15.99
200 mm thick	15.00	0.12	7.11	–	15.00	m²	22.11
250 mm thick	18.75	0.18	10.66	0.35	18.75	m²	29.76
300 mm thick	22.50	0.22	13.04	0.35	22.50	m²	35.89
Roads; 21.00 N/mm²–20 mm aggregate (1:2:4)							
100 mm thick	7.66	2.08	123.44	–	7.66	m²	131.10
150 mm thick	11.49	0.08	4.74	–	11.49	m²	16.23
200 mm thick	15.32	0.12	7.11	–	15.32	m²	22.43
250 mm thick	19.15	0.18	10.66	0.35	19.15	m²	30.16
300 mm thick	22.98	0.22	13.04	0.35	22.98	m²	36.37
Roads; 26.00 N/mm²–20 mm aggregate (1:1.5:3)							
100 mm thick	7.75	0.05	3.16	–	7.75	m²	10.91
150 mm thick	11.63	0.08	4.74	–	11.63	m²	16.37
200 mm thick	15.32	0.12	7.11	–	15.32	m²	22.43
250 mm thick	19.38	0.18	10.66	0.35	19.38	m²	30.39
300 mm thick	23.25	0.22	13.04	0.35	23.25	m²	36.64
Roads; PAV1 concrete – 35 N/mm²; Designated mix							
100 mm thick	8.06	0.05	3.16	–	8.06	m²	11.22
150 mm thick	12.09	0.08	4.74	–	12.09	m²	16.83
200 mm thick	16.11	0.12	7.11	–	16.11	m²	23.22
250 mm thick	20.14	0.18	10.66	0.35	20.14	m²	31.15
300 mm thick	24.17	0.22	13.04	0.35	24.17	m²	37.56
Concrete sundries							
Treating surfaces of unset concrete; grading to cambers; tamping with 75 mm thick steel shod tamper or similar	–	0.13	5.27	–	–	m²	5.27

Q21 IN SITU CONCRETE ROADS/PAVINGS

Item Excluding site overheads and profit	PC £	Labour hours	Labour £	Plant £	Material £	Unit	Total rate £
FORMWORK FOR CONCRETE PAVINGS							
Formwork for in situ concrete							
Sides of foundations							
height not exceeding 250 mm	0.46	0.01	0.66	–	0.55	m	1.21
height 250–500 mm	0.61	0.01	0.79	–	0.83	m	1.62
height 500 mm–1.00 m	0.61	0.02	0.99	–	0.94	m	1.93
height exceeding 1.00 m	1.82	0.67	39.50	–	3.69	m²	43.19
Extra over formwork for curved work 6 m radius	–	0.25	9.88	–	–	m	9.88
Steel road forms; to edges of beds or faces of foundations							
150 mm wide	–	0.07	3.95	0.22	–	m	4.17
EXPANSION JOINTS							
General groundworks team							
Generally a two man team is used in this section; The column Labour hours reports team hours. The column Labour £ reports the total cost of the team for the unit of work shown							
2 man team	–	1.00	39.50	–	–	hr	39.50
Expansion joints							
13 mm thick joint filler; formwork							
width or depth not exceeding 150 mm	1.04	0.04	1.77	–	2.02	m	3.79
width or depth 150–300 mm	2.08	0.06	2.20	–	4.04	m	6.24
width or depth 300–450 mm	3.11	0.07	2.65	–	6.06	m	8.71
25 mm thick joint filler; formwork							
width or depth not exceeding 150 mm	5.00	0.04	1.77	–	5.98	m	7.75
width or depth 150–300 mm	10.00	0.06	2.20	–	11.97	m	14.17
width or depth 300–450 mm	15.00	0.07	2.65	–	17.95	m	20.60
Sealants; sealing top 25 mm of joint with rubberized bituminous compound	1.25	0.06	2.20	–	1.25	m	3.45
REINFORCEMENT TO CONCRETE PAVINGS							
Reinforcement; fabric; side laps 150 mm; head laps 300 mm; mesh 200 × 200 mm; in roads, footpaths or pavings							
Fabric							
A142 (2.22 kg/m²)	1.38	0.03	1.64	–	1.38	m²	3.02
A193 (3.02 kg/m²)	1.89	0.02	1.05	–	1.89	m²	2.94

Q22 COATED MACADAM ROADS/PAVINGS

Item Excluding site overheads and profit	PC £	Labour hours	Labour £	Plant £	Material £	Unit	Total rate £
MACADAM SURFACING							
Coated macadam/asphalt roads/							
pavings – General							
Preamble: The prices for all in situ							
finishings to roads and footpaths include							
for work to falls, crossfalls or slopes not							
exceeding 15° from horizontal; for laying							
on prepared bases (not included) and for							
rolling with an appropriate roller.							
Users should note the new terminology							
for the surfaces described below which							
is to European standard descriptions.							
The now redundant descriptions for each							
course are shown in brackets.							
Macadam surfacing; Spadeoak							
Construction Co Ltd; surface							
(wearing) course; 20 mm of 6 mm							
dense bitumen macadam							
Machine lay; areas 1000 m² and over							
limestone aggregate	–	–	–	–	–	m²	7.33
granite aggregate	–	–	–	–	–	m²	7.42
red	–	–	–	–	–	m²	13.19
Hand lay; areas 400 m² and over							
limestone aggregate	–	–	–	–	–	m²	9.34
granite aggregate	–	–	–	–	–	m²	9.43
red	–	–	–	–	–	m²	15.53
Macadam surfacing; Spadeoak							
Construction Co Ltd; surface							
(wearing) course; 30 mm of 10 mm							
dense bitumen macadam							
Machine lay; areas 1000 m² and over							
limestone aggregate	–	–	–	–	–	m²	8.86
granite aggregate	–	–	–	–	–	m²	8.98
red	–	–	–	–	–	m²	18.59
Hand lay; areas 400 m² and over							
limestone aggregate	–	–	–	–	–	m²	10.94
granite aggregate	–	–	–	–	–	m²	11.08
red	–	–	–	–	–	m²	21.28
Macadam surfacing; Spadeoak							
Construction Co Ltd; surface							
(wearing) course; 40 mm of 10 mm							
dense bitumen macadam							
Machine lay; areas 1000 m² and over							
limestone aggregate	–	–	–	–	–	m²	11.27
granite aggregate	–	–	–	–	–	m²	11.44
red	–	–	–	–	–	m²	21.07

Q22 COATED MACADAM ROADS/PAVINGS

Item Excluding site overheads and profit	PC £	Labour hours	Labour £	Plant £	Material £	Unit	Total rate £
Hand lay; areas 400 m² and over							
limestone aggregate	–	–	–	–	–	m²	13.53
granite aggregate	–	–	–	–	–	m²	13.69
red	–	–	–	–	–	m²	23.91
Macadam surfacing; Spadeoak Construction Co Ltd; binder (base) course; 50 mm of 20 mm dense bitumen macadam							
Machine lay; areas 1000 m² and over							
limestone aggregate	–	–	–	–	–	m²	11.31
granite aggregate	–	–	–	–	–	m²	11.49
Hand lay; areas 400 m² and over							
limestone aggregate	–	–	–	–	–	m²	13.55
granite aggregate	–	–	–	–	–	m²	13.76
Macadam surfacing; Spadeoak Construction Co Ltd; binder (base) course; 60 mm of 20 mm dense bitumen macadam							
Machine lay; areas 1000 m² and over							
limestone aggregate	–	–	–	–	–	m²	12.43
granite aggregate	–	–	–	–	–	m²	12.65
Hand lay; areas 400 m² and over							
limestone aggregate	–	–	–	–	–	m²	14.76
granite aggregate	–	–	–	–	–	m²	14.97
Macadam surfacing; Spadeoak Construction Co Ltd; base (roadbase) course; 75 mm of 28 mm dense bitumen macadam							
Machine lay; areas 1000 m² and over							
limestone aggregate	–	–	–	–	–	m²	15.26
granite aggregate	–	–	–	–	–	m²	15.53
Hand lay; areas 400 m² and over							
limestone aggregate	–	–	–	–	–	m²	17.76
granite aggregate	–	–	–	–	–	m²	18.07
Macadam surfacing; Spadeoak Construction Co Ltd; base (roadbase) course; 100 mm of 28 mm dense bitumen macadam							
Machine lay; areas 1000 m² and over							
limestone aggregate	–	–	–	–	–	m²	18.90
granite aggregate	–	–	–	–	–	m²	19.27
Hand lay; areas 400 m² and over							
limestone aggregate	–	–	–	–	–	m²	21.66
granite aggregate	–	–	–	–	–	m²	22.04

Q22 COATED MACADAM ROADS/PAVINGS

Item Excluding site overheads and profit	PC £	Labour hours	Labour £	Plant £	Material £	Unit	Total rate £
MACADAM SURFACING – cont							
Macadam surfacing; Spadeoak							
Construction Co Ltd; base (roadbase)							
course; 150 mm of 28 mm dense							
bitumen macadam in two layers							
Machine lay; areas 1000 m² and over							
limestone aggregate	–	–	–	–	–	m²	29.80
granite aggregate	–	–	–	–	–	m²	30.36
Hand lay; areas 400 m² and over							
limestone aggregate	–	–	–	–	–	m²	34.80
granite aggregate	–	–	–	–	–	m²	35.37
Base (roadbase) course; 200 mm of							
28 mm dense bitumen macadam in							
two layers							
Machine lay; areas 1000 m² and over							
limestone aggregate	–	–	–	–	–	m²	37.88
granite aggregate	–	–	–	–	–	m²	38.62
Hand lay; areas 400 m² and over							
limestone aggregate	–	–	–	–	–	m²	29.43
granite aggregate	–	–	–	–	–	m²	44.17
Resin bound macadam pavings;							
Bituchem; to pedestrian or vehicular							
hard landscape areas; laid to base							
course (not included)							
Natratex wearing course; clear resin							
bound macadam							
25 mm thick to pedestrian areas	17.00	0.04	2.47	–	17.00	m²	19.47
30 mm thick to vehicular areas	21.00	0.01	0.66	–	21.00	m²	21.66
Colourtex coloured resin bound							
macadam							
25 mm thick	17.00	0.04	2.47	–	17.00	m²	19.47
RESIN BONDED SURFACING							
Bonded aggregates; Addagrip							
Surface Treatments UK Ltd; natural							
decorative resin bonded surface							
dressing laid to concrete, macadam							
or to plywood panels priced							
separately							
Primer coat to macadam or concrete							
base	–	0.04	2.47	–	4.00	m²	6.47
Golden pea gravel; 1–3 mm							
buff adhesive	–	0.04	2.47	–	20.00	m²	22.47
red adhesive	–	–	–	–	–	m²	20.00
green adhesive	–	–	–	–	–	m²	20.00

Q22 COATED MACADAM ROADS/PAVINGS

Item Excluding site overheads and profit	PC £	Labour hours	Labour £	Plant £	Material £	Unit	Total rate £
Golden pea gravel; 2–5 mm							
buff adhesive	–	–	–	–	–	m²	24.00
Chinese bauxite; 1–3 mm							
buff adhesive	–	–	–	–	–	m²	20.00
Cobalt Blue Glass; 6 mm							
15 mm depth	–	–	–	–	–	m²	75.00
Midnight Grey; 6 mm							
15 mm depth	–	–	–	–	–	m²	45.00
Chocolate; 6 mm							
15 mm depth	–	–	–	–	–	m²	45.00
18 mm depth	–	–	–	–	–	m²	55.00
RESIN BOUND SURFACING							
Resin bound paving; Sureset Ltd; fully mixed permeable decorative aggregate paving; laid to macadam binder and base course (not included); thickness dependent on loading application and aggregate size							
18 mm thick – 6 mm aggregate							
Natural gravel							
areas 100–300 m²	–	–	–	–	–	m²	52.00
areas 300–500 m²	–	–	–	–	–	m²	49.50
areas over 500 m²	–	–	–	–	–	m²	44.00
Crushed rock							
areas 100–300 m²	–	–	–	–	–	m²	54.50
areas 300–500 m²	–	–	–	–	–	m²	51.50
areas over 500 m²	–	–	–	–	–	m²	46.00
Marble							
areas 100–300 m²	–	–	–	–	–	m²	56.50
areas 300–500 m²	–	–	–	–	–	m²	54.50
areas over 500 m²	–	–	–	–	–	m²	47.50
Recycled glass							
areas 100–300 m²	–	–	–	–	–	m²	67.00
areas 300–500 m²	–	–	–	–	–	m²	67.00
areas over 500 m²	–	–	–	–	–	m²	63.00
Spectrum							
areas 100–300 m²	–	–	–	–	–	m²	80.50
areas 300–500 m²	–	–	–	–	–	m²	76.00
areas over 500 m²	–	–	–	–	–	m²	73.50

Q22 COATED MACADAM ROADS/PAVINGS

Item Excluding site overheads and profit	PC £	Labour hours	Labour £	Plant £	Material £	Unit	Total rate £
RESIN BOUND SURFACING – cont							
Resin bound macadam pavings; **Bituchem; to pedestrian or vehicular** **hard landscape areas; laid to base** **course (not included)**							
Natratex wearing course; clear resin bound macadam							
25 mm thick to pedestrian areas	17.00	0.04	2.47	–	17.00	m²	**19.47**
30 mm thick to vehicular areas	21.00	0.01	0.66	–	21.00	m²	**21.66**
Colourtex coloured resin bound macadam							
25 mm thick	17.00	0.04	2.47	–	17.00	m²	**19.47**
Resin bound paving to individual **treepits; SureSet Ltd; fully mixed** **permeable decorative aggregate** **paving; laid to well compacted stone** **Type 3 (not included)**							
30–40 mm thick – average tree pit size 1.50 × 1.50 tree pit aggregate							
Natural gravel							
areas 100–300 m²	–	–	–	–	–	nr	**98.00**
MARKING CAR PARKS							
Marking car parks							
Car parking space division strips; laid hot at 115° C; on bitumen macadam surfacing							
minimum daily rate	–	–	–	–	–	item	**500.00**
Stainless metal road studs							
100 × 100 mm	5.78	0.08	4.94	–	5.78	nr	**10.72**

Q23 GRAVEL/HOGGIN/WOODCHIP ROADS/PAVINGS

Item Excluding site overheads and profit	PC £	Labour hours	Labour £	Plant £	Material £	Unit	Total rate £
Clarification notes on labour costs in this section							
General groundworks team Generally a three man team is used in this section; The column Labour hours reports team hours. The column Labour £ reports the total cost of the team for the unit of work shown							
3 man team	–	1.00	59.25	–	–	hr	**59.25**
EXCAVATION OF PATHWAYS							
Excavation and path preparation Excavating; 300 mm deep; to width of path; depositing excavated material at sides of excavation							
width 1.00 m	–	–	–	2.75	–	m²	**2.75**
width 1.50 m	–	–	–	2.29	–	m²	**2.29**
width 2.00 m	–	–	–	1.97	–	m²	**1.97**
width 3.00 m	–	–	–	1.65	–	m²	**1.65**
Excavating trenches; in centre of pathways; 100 mm flexible drain pipes; filling with clean broken stone or gravel rejects							
300 × 450 mm deep	5.05	0.03	1.97	1.37	5.05	m	**8.39**
Hand trimming and compacting reduced surface of pathway; by machine							
width 1.00 m	–	0.02	0.99	0.20	–	m	**1.19**
width 1.50 m	–	0.01	0.88	0.18	–	m	**1.06**
width 2.00 m	–	0.01	0.79	0.16	–	m	**0.95**
width 3.00 m	–	0.01	0.79	0.16	–	m	**0.95**
Permeable membranes; to trimmed and compacted surface of pathway							
Terram 1000	0.43	0.01	0.40	–	0.43	m²	**0.83**
FILLING TO MAKE UP LEVELS							
Filling to make up levels Obtained off site; hardcore; PC £23.05/m³							
150 mm thick	3.69	0.01	0.79	0.57	3.69	m²	**5.05**
Type 1 granular fill base; PC £17.95/ tonne (£36.30 / m³ compacted)							
100 mm thick	3.95	0.01	0.55	0.41	3.95	m²	**4.91**
150 mm thick	5.92	0.01	0.49	0.62	5.92	m²	**7.03**
Surface treatments Sand blinding; to hardcore (not included)							
50 mm thick	1.87	0.01	0.79	–	1.87	m²	**2.66**
Filter fabric; to hardcore (not included)	0.43	–	0.20	–	0.43	m²	**0.63**

Q23 GRAVEL/HOGGIN/WOODCHIP ROADS/PAVINGS

Item Excluding site overheads and profit	PC £	Labour hours	Labour £	Plant £	Material £	Unit	Total rate £
CELLULAR CONFINED GRANULAR PAVINGS							
Nidagravel; Cedar Nursery; honeycomb shaped cellular polypropylene containment grids for retaining filled decorative aggregate surfaces (not included) for driveways, cycle paths and footpaths; to graded compacted substrate (not included) and 50 mm sharp sand base; integrated geofabric base							
Foot traffic, wheelchairs, bikes and pushchairs							
Nidagravel 125; 1200 × 800 × 25 mm deep	1.70	0.08	2.47	–	13.70	m²	**16.17**
Driveways: vehicles, wheelchairs, bikes and pushchairs (requires compacted granular base to support vehicle loadings)							
Nidagravel 140; 2400 × 1200 × 40 mm deep	1.70	0.08	2.68	–	16.65	m²	**19.33**
Aggregate filling to Nidagravel 125							
shingle	–	0.03	1.97	1.98	0.89	m²	**4.84**
decorative aggregate; CED Ltd; PC £48.00 / tonne	–	0.03	1.97	1.98	2.16	m²	**6.11**
Aggregate filling to Nidagravel 140							
shingle	–	0.03	1.66	2.40	1.42	m²	**5.48**
decorative aggregate; CED Ltd; PC £48.00 / tonne	3.46	0.03	1.66	2.40	3.46	m²	**7.52**
StableDRIVE; Cedar Nursery; honeycomb shaped polypropylene cellular containment system for retaining filled decorative aggregate surfaces for driveways, cyclepaths and footpaths; on compacted subgrade and subbase (both not included)							
Driveways: vehicles, wheelchairs, bikes and pushchairs (requires compacted granular base to support vehicle loadings)							
StableDRIVE; 600 × 500 × 32 mm deep	1.70	0.08	2.84	–	17.65	m²	**20.49**
Aggregate filling to StableDRIVE							
shingle	–	0.03	1.85	2.94	1.13	m²	**5.92**
decorative aggregate; CED Ltd; PC £48.00 / tonne	–	0.03	1.85	2.94	2.76	m²	**7.55**

Q23 GRAVEL/HOGGIN/WOODCHIP ROADS/PAVINGS

Item Excluding site overheads and profit	PC £	Labour hours	Labour £	Plant £	Material £	Unit	Total rate £
GRANULAR PAVINGS							
Footpath gravels; porous self binding gravel							
CED Ltd; Cedec gravel; self-binding; laid to inert (non-limestone) base measured separately; compacting							
red, silver or gold; 50 mm thick	13.01	0.01	0.49	0.42	13.01	m²	**13.92**
Grundon Ltd; Coxwell self-binding path gravels laid and compacted to excavation or base measured separately							
50 mm thick	4.79	0.01	0.49	0.42	4.79	m²	**5.70**
Breedon Special Aggregates; Golden Gravel or equivalent; rolling wet; on hardcore base (not included); for pavements; to falls and crossfalls and to slopes not exceeding 15° from horizontal; over 300 mm wide							
50 mm thick	11.10	0.01	0.49	0.36	11.10	m²	**11.95**
75 mm thick	16.65	0.03	1.97	0.44	16.65	m²	**19.06**
Breedon Special Aggregates; Wayfarer specially formulated fine gravel for use on golf course pathways							
50 mm thick	10.00	0.01	0.49	0.36	10.00	m²	**10.85**
75 mm thick	14.99	0.02	0.99	0.63	14.99	m²	**16.61**
Hoggin (stabilized); PC £26.00/m³ on hardcore base (not included); to falls and crossfalls and to slopes not exceeding 15° from horizontal; over 300 mm wide							
100 mm thick	4.27	0.01	0.66	0.56	4.27	m²	**5.49**
150 mm thick	6.40	0.02	0.99	0.85	6.40	m²	**8.24**
Ballast; as dug; watering; rolling; on hardcore base (not included)							
100 mm thick	4.04	0.01	0.66	0.56	4.04	m²	**5.26**
150 mm thick	6.06	0.02	0.99	0.85	6.06	m²	**7.90**
Footpath gravels; porous loose gravels							
Breedon Special Aggregates; Breedon Buff decorative limestone chippings							
50 mm thick	5.88	–	0.20	0.09	5.88	m²	**6.17**
75 mm thick	14.58	–	0.25	0.11	14.58	m²	**14.94**
Breedon Special Aggregates; Breedon Buff decorative limestone chippings							
50 mm thick	5.88	–	0.20	0.09	5.88	m²	**6.17**
75 mm thick	8.83	–	0.25	0.11	8.83	m²	**9.19**
Breedon Special Aggregates; Brindle or Moorland Black chippings							
50 mm thick	9.19	–	0.20	0.09	9.19	m²	**9.48**
75 mm thick	13.79	–	0.25	0.11	13.79	m²	**14.15**

Q23 GRAVEL/HOGGIN/WOODCHIP ROADS/PAVINGS

Item Excluding site overheads and profit	PC £	Labour hours	Labour £	Plant £	Material £	Unit	Total rate £
GRANULAR PAVINGS – cont							
Footpath gravels – cont							
Breedon Special Aggregates; slate chippings; plum/blue							
50 mm thick	9.32	–	0.20	0.09	9.32	m²	9.61
75 mm thick	13.99	–	0.25	0.11	13.99	m²	14.35
Washed shingle; on prepared base (not included)							
25–50 mm size; 25 mm thick	0.93	0.01	0.49	0.08	0.93	m²	1.50
25–50 mm size; 75 mm thick	2.80	0.03	1.48	0.24	2.80	m²	4.52
50–75 mm size; 25 mm thick	0.93	0.01	0.40	0.09	0.93	m²	1.42
50–75 mm size; 75 mm thick	2.80	0.02	1.32	0.30	2.80	m²	4.42
Pea shingle; on prepared base (not included)							
10–15 mm size; 25 mm thick	0.89	0.01	0.35	0.08	0.89	m²	1.32
5–10 mm size; 75 mm thick	2.67	0.02	1.04	0.24	2.67	m²	3.95
WOODCHIP PATHWAYS							
Wood chip surfaces; Melcourt Industries Ltd (items labelled FSC are Forest Stewardship Council certified)							
Wood chips; to surface of pathways by machine; material delivered in 80 m³ loads; levelling and spreading by hand (excavation and preparation not included)							
Walk Chips; 100 mm thick; FSC (25 m³ loads)	4.06	0.01	0.66	0.34	4.06	m²	5.06
Walk Chips; 100 mm thick; FSC (80 m³ loads)	2.70	0.01	0.66	0.34	2.70	m²	3.70
Woodfibre; 100 mm thick; FSC (80 m³ loads)	2.51	0.01	0.66	0.34	2.51	m²	3.51
PATH EDGINGS							
Permaloc Asphalt edge L-shaped aluminium edging; Kinley Systems; extruded aluminium alloy L shaped edging with 5.33 mm exposed upper lip; edging fixed to roadway base and edge profile with 250 mm steel fixing spike; laid to straight or curvilinear road edge; subsequently filled with macadam (not included)							
Depth of macadam							
40 mm	8.65	0.07	2.63	–	8.65	m	11.28
50 mm	9.50	0.07	2.63	–	9.50	m	12.13
65 mm	9.73	0.07	2.82	–	9.73	m	12.55
75 mm	11.41	0.08	3.04	–	11.41	m	14.45
100 mm	13.36	0.08	3.29	–	13.36	m	16.65

Q23 GRAVEL/HOGGIN/WOODCHIP ROADS/PAVINGS

Item Excluding site overheads and profit	PC £	Labour hours	Labour £	Plant £	Material £	Unit	Total rate £
Permaloc L-shaped aluminium edging; Kinley Systems; heavy duty straight profile edging; for edgings to soft landscape beds or turf areas; 3.2 mm × 100 mm high; 3.2 mm thick with 4.75 mm exposed upper lip; fixed to form straight or curvilinear edge with 305 mm fixing spike Milled aluminium							
100 mm deep	–	0.10	3.95	–	13.03	m	**16.98**
Permaloc Permastrip; Kinley Systems; heavy duty L-shaped profile maintenance strip; 3.2 mm × 89 mm high with 5.2 mm exposed top lip; for straight or gentle curves on paths or bed turf interfaces; fixed to form straight or curvilinear edge with standard 305 mm stake; other stake lengths available Milled aluminium							
75 mm deep	11.41	0.10	3.95	–	11.41	m	**15.36**
Permaloc Proline; Kinley Systems; medium duty straight profiled maintenance strip; 3.2 mm × 102 mm high with 3.18 mm exposed top lip; for straight or gentle curves on paths or bed turf interfaces; fixed to form straight or curvilinear edge with standard 305 mm stake; other stake lengths available Milled aluminium							
50 mm deep	–	0.10	3.95	–	9.27	m	**13.22**

Q24 INTERLOCKING BRICK/BLOCK ROADS/PAVINGS

Item Excluding site overheads and profit	PC £	Labour hours	Labour £	Plant £	Material £	Unit	Total rate £
Clarification notes on labour costs in this section							
General groundworks team							
Generally a three man team is used in this section; The column Labour hours reports team hours. The column Labour £ reports the total cost of the team for the unit of work shown							
3 man team	–	1.00	59.25	–	–	hr	**59.25**
PRECAST CONCRETE BLOCK PAVINGS							
Precast concrete block edgings; PC £8.99/m^2; 200 × 100 × 60 mm; on prepared base (not included); haunching one side							
Edgings; butt joints							
stretcher course	0.90	0.06	3.29	–	3.26	m	**6.55**
header course	1.80	0.09	5.27	–	4.33	m	**9.60**
Precast concrete vehicular paving blocks; Marshalls Plc; on prepared base (not included); on 50 mm compacted sharp sand bed; blocks laid in 7 mm loose sand and vibrated; joints filled with sharp sand and vibrated; level and to falls only							
Trafica paving blocks; 450 × 450 × 70 mm							
Perfecta finish; colour natural	31.68	0.17	9.88	0.11	33.55	m^2	**43.54**
Perfecta finish; colour buff	36.69	0.17	9.88	0.11	38.56	m^2	**48.55**
Saxon finish; colour natural	27.89	0.17	9.88	0.11	29.76	m^2	**39.75**
Saxon finish; colour buff	30.98	0.17	9.88	0.11	32.85	m^2	**42.84**
Precast concrete vehicular paving blocks; Keyblok Marshalls Plc; on prepared base (not included); on 50 mm compacted sharp sand bed; blocks laid in 7 mm loose sand and vibrated; joints filled with sharp sand and vibrated; level and to falls only							
Herringbone bond							
200 × 100 × 60 mm; natural grey	9.44	0.13	7.70	0.11	11.37	m^2	**19.18**
200 × 100 × 60 mm; colours	10.15	0.13	7.90	0.11	12.08	m^2	**20.09**
200 × 100 × 80 mm; natural grey	10.50	0.14	8.47	0.11	12.43	m^2	**21.01**
200 × 100 × 80 mm; colours	11.87	0.14	8.47	0.11	13.79	m^2	**22.37**

Q24 INTERLOCKING BRICK/BLOCK ROADS/PAVINGS

Item Excluding site overheads and profit	PC £	Labour hours	Labour £	Plant £	Material £	Unit	Total rate £
Basketweave bond							
200 × 100 × 60 mm; natural grey	9.44	0.14	8.47	0.11	11.37	m²	**19.95**
200 × 100 × 60 mm; colours	10.15	0.14	8.47	0.11	12.08	m²	**20.66**
200 × 100 × 80 mm; natural grey	10.50	0.15	8.77	0.11	12.43	m²	**21.31**
200 × 100 × 80 mm; colours	11.87	0.15	8.77	0.11	13.79	m²	**22.67**
Precast concrete vehicular paving blocks; Charcon Hard Landscaping; on prepared base (not included); on 50 mm compacted sharp sand bed; blocks laid in 7 mm loose sand and vibrated; joints filled with sharp sand and vibrated; level and to falls only							
Europa concrete blocks							
200 × 100 × 60 mm; natural grey	11.62	0.13	7.90	0.11	13.55	m²	**21.56**
200 × 100 × 60 mm; colours	12.67	0.13	7.90	0.11	14.60	m²	**22.61**
Parliament concrete blocks							
200 × 100 × 65 mm; natural grey	31.31	0.14	8.47	0.11	33.24	m²	**41.82**
200 × 100 × 65 mm; colours	31.31	0.14	8.47	0.11	33.24	m²	**41.82**

Q24 INTERLOCKING BRICK/BLOCK ROADS/PAVINGS

Item Excluding site overheads and profit	PC £	Labour hours	Labour £	Plant £	Material £	Unit	Total rate £
SUDS PERMEABLE BLOCK PAVINGS							
SUDS paving; Concrete Block Permeable Paving (CBPP) Note: Sustainable Drainage Systems (SUDS) aim to reduce flood risk by controlling the rate and volume of surface water run off from developments. Permeable Paving is a SUDS technique. In addition to managing the quantity of surface water run-off more effectively, one of the primary benefits of CBPP is the potential enhancement of water quality. Permeable pavements improve water quality by mirroring nature in providing filtration and allowing for natural biodegradation of hydrocarbons and the dilution of other contaminants as water passes through the system.							
Bases for Sustainable Urban Drainage System (SUDS) subbase for use below impervious paving; Aggregate Industries Ltd; Bardon DrainAgg 20 4/20 mm open graded material laid loose on prepared sub-grade; within edge restraints to paving area (not included); graded to levels and falls as per paving manufacturer's instructions							
150 mm thick	6.02	0.01	0.49	0.62	6.02	m²	7.13
200 mm thick	8.03	–	0.04	0.82	8.03	m²	8.89
250 mm thick	10.02	0.01	0.41	1.02	10.02	m²	11.45
350 mm thick	14.02	0.01	0.56	1.41	14.02	m²	15.99
450 mm thick	18.04	0.01	0.79	1.97	18.04	m²	20.80

Q24 INTERLOCKING BRICK/BLOCK ROADS/PAVINGS

Item Excluding site overheads and profit	PC £	Labour hours	Labour £	Plant £	Material £	Unit	Total rate £
Subbase replacement system for SUDS paving Charcon Permavoid; Aggregate Industries Ltd; geocellular, interlocking, high strength; high void capacity (95%); typically a 4:1 ratio when compared with alternative granular systems; allows for water storage to be confined at shallow depth within the subbase layer; inclusive of connection, ties, pins; does not include for encapsulation in geotextile/geomembranes							
708 × 354 × 150 mm							
150 mm thick	60.87	0.06	3.30	–	60.87	m²	**64.17**
300 mm thick	126.30	0.11	6.58	–	126.30	m²	**132.88**
Encapsulation of Permavoid; 150 mm thick systems; geomembrane geofabric or combination of both for filtration or attenuation of water; please see the manufacturers design guide							
Charcon geomembrane, heat welded	14.96	0.02	0.99	–	14.96	m²	**15.95**
Charcon non reinforced geotextile	9.27	0.11	6.58	–	9.27	m²	**15.85**
Charcon reinforced geotextile	4.97	0.01	0.66	–	4.97	m²	**5.63**
Encapsulation of Permavoid; 300 mm thick systems; geomembrane geofabric or combination of both for filtration or attenuation of water; please see the manufacturers design guide							
Charcon geomembrane, heat welded	16.91	0.02	1.10	–	16.91	m²	**18.01**
Charcon non reinforced geotextile	10.48	0.01	0.66	–	10.48	m²	**11.14**
Charcon reinforced geotextile	4.68	0.01	0.71	–	4.68	m²	**5.39**
Pervious surfacing materials; Aggregate Industries Ltd; Charcon Infilta CBPP							
Rectangular permeable block paving system; incorporating a 5 mm spacer design that provides a 5 mm void allowing ingress of water through to the subbase storage system (priced separately); various colours; bedded on 50 mm thick 2–6.3 mm clean, angular, free draining uncompacted aggregate; joints infilled with 3 mm clean grit							
Infilta; 200 × 100 × 80 mm thick; natural grey	16.80	0.15	8.77	0.11	19.74	m²	**28.62**
Infilta; 200 × 100 × 80 mm thick; coloured	17.38	0.15	8.77	0.11	20.32	m²	**29.20**

Q24 INTERLOCKING BRICK/BLOCK ROADS/PAVINGS

Item Excluding site overheads and profit	PC £	Labour hours	Labour £	Plant £	Material £	Unit	Total rate £
CELLULAR PAVINGS							
Recycled polyethylene grassblocks;							
Fiberweb Reinforcement Solutions;							
interlocking units laid to prepared							
base or rootzone (not included)							
BodPave85; load bearing <400 tonnes							
per m²; 500 × 500 × 50 mm deep; 35 mm							
ground spike							
1–50 m²	16.50	0.07	3.95	0.18	19.69	m²	**23.82**
51–500 m²	13.75	0.07	3.95	0.18	16.94	m²	**21.07**
501–1000 m²	13.20	0.07	3.95	0.18	16.39	m²	**20.52**
1001–1559 m²	12.50	0.07	3.95	0.18	15.69	m²	**19.82**
1560 m² or over	12.00	0.07	3.95	0.18	15.19	m²	**19.32**
GrassProtecta; extruded expanded							
polyethylene flexible mesh laid to							
existing grass surface or newly seeded							
areas to provide heavy surface							
protection from traffic and pedestrians							
Standard; 1.2 kg/m²; 20 × 2 m; up to							
320 m²	7.50	–	0.17	–	7.50	m²	**7.67**
Standard; 1.2 kg/m²; 20 × 2 m;							
321–3000 m²	6.50	–	0.17	–	6.50	m²	**6.67**
Standard; 1.2 kg/m²; 20 × 2 m; over							
3000 m²	6.00	–	0.17	–	6.00	m²	**6.17**
Heavy; 2 kg/m²; 20 × 2 m; up to							
320 m²	8.50	–	0.17	–	8.50	m²	**8.67**
Heavy; 2 kg/m²; 20 × 2 m; 321–3000 m²	8.00	–	0.17	–	8.00	m²	**8.17**
Heavy; 2 kg/m²; 20 × 2 m; over							
3000 m²	7.50	–	0.17	–	7.50	m²	**7.67**
TurfProtecta; extruded polyethylene							
flexible mesh laid to existing grass							
surface or newly seeded areas to							
provide surface protection from traffic							
including vehicle or animal wear and tear							
Standard; 30 × 2 m; up to 300 m²	2.90	–	0.17	–	2.90	m²	**3.07**
Standard; 30 × 2 m; 301–600 m²	2.49	–	0.17	–	2.49	m²	**2.66**
Standard; 30 × 2 m; 601–1440 m²	2.41	–	0.17	–	2.41	m²	**2.58**
Standard; 30 × 2 m; over 1440 m²	2.24	–	0.17	–	2.24	m²	**2.41**
Premium; 30 × 2 m; up to 300 m²	2.99	–	0.17	–	2.99	m²	**3.16**
Premium; 30 × 2 m; 301–600 m²	2.66	–	0.17	–	2.66	m²	**2.83**
Premium; 30 × 2 m; 601–1440 m²	2.57	–	0.17	–	2.57	m²	**2.74**
Premium; 30 × 2 m; over 1440 m²	2.41	–	0.17	–	2.41	m²	**2.58**

Q24 INTERLOCKING BRICK/BLOCK ROADS/PAVINGS

Item Excluding site overheads and profit	PC £	Labour hours	Labour £	Plant £	Material £	Unit	Total rate £
Grassroad; Cooper Clarke Civils and Lintels; heavy duty for car parking and fire paths verge hardening and shallow embankments Honeycomb cellular polyproylene interconnecting paviors with integral downstead anti-shear cleats including topsoil but excluding edge restraints; to granular subbase (not included)							
635 × 330 × 42 mm overall laid to a module of 622 × 311 × 32 mm	–	–	–	–	–	m²	30.33

Q25 SLAB/BRICK/STONE/TIMBER PAVINGS

Item Excluding site overheads and profit	PC £	Labour hours	Labour £	Plant £	Material £	Unit	Total rate £
Clarification notes on labour costs in this section							
General and Expert paving team Generally a three man team is used in this section; The column 'Labour hours' reports team hours. The column 'Labour £' reports the total cost of the team for the unit of work shown							
3 man team	–	1.00	59.25	–	–	hr	**59.25**
BRICK PAVINGS							
Bricks – General Preamble: Bricks shall be hard, well burnt, non-dusting, resistant to frost and sulphate attack and true to shape, size and sample.							
Movement of materials Mechanically offloading bricks; loading wheelbarrows; transporting maximum 25 m distance	–	0.07	3.95	–	–	m²	**3.95**
Edge restraints; to brick paving; on prepared base (not included); 65 mm thick bricks; PC £300.00/1000; haunching one side Header course							
200 × 100 mm; butt joints	3.00	0.09	5.27	–	5.53	m	**10.80**
210 × 105 mm; mortar joints	2.67	0.17	9.88	–	5.55	m	**15.43**
Stretcher course							
200 × 100 mm; butt joints	1.50	0.06	3.29	–	3.86	m	**7.15**
210 × 105 mm; mortar joints	1.36	0.11	6.58	–	3.89	m	**10.47**
Variation in brick prices; add or subtract the following amounts for every £100/1000 difference in the PC price Edgings; mortar jointed							
100 mm wide	0.50	–	–	–	0.50	m	**0.50**
200 mm wide	0.95	–	–	–	0.95	m	**0.95**
102.5 mm wide	0.47	–	–	–	0.47	m	**0.47**
215 mm wide	0.93	–	–	–	0.93	m	**0.93**
Clay brick pavings; on prepared base (not included); bedding on 50 mm sharp sand; kiln dried sand joints Pavings; 200 × 100 × 65 mm wirecut chamfered paviors							
bricks; PC £450.00/1000	23.06	0.46	27.36	0.23	25.18	m²	**52.77**

Q25 SLAB/BRICK/STONE/TIMBER PAVINGS

Item Excluding site overheads and profit	PC £	Labour hours	Labour £	Plant £	Material £	Unit	Total rate £
Clay brick pavings; 200 × 100 × 50 mm; laid to running stretcher or stack bond only; on prepared base (not included); bedding on cement: sand (1:4) pointing mortar as work proceeds							
PC £600.00/1000							
laid on edge	48.81	1.59	94.05	–	56.62	m²	**150.67**
laid on edge but pavior 65 mm thick	41.00	1.27	75.24	–	48.81	m²	**124.05**
laid flat	26.62	0.73	43.45	–	31.28	m²	**74.73**
PC £500.00/1000							
laid on edge	40.67	1.59	94.05	–	48.49	m²	**142.54**
laid on edge but pavior 65 mm thick	34.16	1.27	75.24	–	41.98	m²	**117.22**
laid flat	22.19	0.73	43.45	–	26.84	m²	**70.29**
PC £400.00/1000							
laid flat	17.75	0.73	43.45	–	22.41	m²	**65.86**
laid on edge	32.54	1.59	94.05	–	40.35	m²	**134.40**
laid on edge but pavior 65 mm thick	27.33	1.27	75.24	–	35.15	m²	**110.39**
PC £300.00/1000							
laid on edge	24.40	1.59	94.05	–	32.22	m²	**126.27**
laid on edge but pavior 65 mm thick	20.50	1.27	75.24	–	28.31	m²	**103.55**
laid flat	13.31	0.73	43.45	–	17.97	m²	**61.42**
Clay brick pavings; 200 × 100 × 50 mm; butt jointed laid herringbone or basketweave pattern only; on prepared base (not included); bedding on 50 mm sharp sand							
PC £600.00/1000							
laid flat	30.75	0.46	27.36	0.30	32.87	m²	**60.53**
PC £500.00/1000							
laid flat	25.63	0.46	27.36	0.30	27.74	m²	**55.40**
PC £400.00/1000							
laid flat	20.50	0.46	27.36	0.30	22.62	m²	**50.28**
PC £300.00/1000							
laid flat	15.38	0.46	27.36	0.30	17.49	m²	**45.15**
Clay brick pavings; 215 × 102.5 × 65 mm; on prepared base (not included); bedding on cement: sand (1:4) pointing mortar as work proceeds							
Paving bricks; PC £600.00/1000; herringbone bond							
laid on edge	36.44	1.19	70.21	–	43.31	m²	**113.52**
laid flat	23.70	0.79	46.84	–	30.57	m²	**77.41**
Paving bricks; PC £600.00/1000; basketweave bond							
laid on edge	36.44	0.79	46.81	–	43.31	m²	**90.12**
laid flat	23.70	0.53	31.21	–	30.57	m²	**61.78**

Q25 SLAB/BRICK/STONE/TIMBER PAVINGS

Item Excluding site overheads and profit	PC £	Labour hours	Labour £	Plant £	Material £	Unit	Total rate £
BRICK PAVINGS – cont							
Clay brick pavings – cont							
Paving bricks; PC £600.00/1000;							
running or stack bond							
laid on edge	36.44	0.63	37.46	–	43.31	m²	**80.77**
laid flat	23.70	0.42	24.97	–	30.57	m²	**55.54**
Paving bricks; PC £500.00/1000;							
herringbone bond							
laid on edge	30.37	1.19	70.21	–	34.31	m²	**104.52**
laid flat	19.75	0.79	46.81	–	26.62	m²	**73.43**
Paving bricks; PC £500.00/1000;							
basketweave bond							
laid on edge	30.37	0.79	46.81	–	34.31	m²	**81.12**
laid flat	19.75	0.53	31.21	–	26.62	m²	**57.83**
Paving bricks; PC £500.00/1000;							
running or stack bond							
laid on edge	30.37	0.63	37.46	–	34.31	m²	**71.77**
laid flat	19.75	0.42	24.97	–	26.62	m²	**51.59**
Paving bricks; PC £400.00/1000;							
herringbone bond							
laid on edge	24.29	1.19	70.21	–	28.24	m²	**98.45**
laid flat	15.80	0.79	46.81	–	22.55	m²	**69.36**
Paving bricks; PC £400.00/1000;							
basketweave bond							
laid on edge	24.29	0.79	46.81	–	28.24	m²	**75.05**
laid flat	15.80	0.53	31.21	–	22.55	m²	**53.76**
Paving bricks; PC £400.00/1000;							
running or stack bond							
laid on edge	24.29	0.63	37.46	–	28.24	m²	**65.70**
laid flat	15.80	0.42	24.97	–	22.55	m²	**47.52**
Paving bricks; PC £300.00/1000;							
herringbone bond							
laid on edge	17.77	1.19	70.21	–	21.72	m²	**91.93**
laid flat	11.85	0.79	46.81	–	18.72	m²	**65.53**
Paving bricks; PC £300.00/1000;							
basketweave bond							
laid on edge	17.77	0.79	46.81	–	21.72	m²	**68.53**
laid flat	11.85	0.53	31.21	–	18.72	m²	**49.93**
Paving bricks; PC £300.00/1000;							
running or stack bond							
laid on edge	17.77	0.63	37.45	–	21.72	m²	**59.17**
laid flat	11.85	0.42	24.97	–	18.72	m²	**43.69**
Cutting							
curved cutting	–	0.15	8.77	6.53	–	m	**15.30**
raking cutting	–	0.11	6.58	5.11	–	m	**11.69**

Q25 SLAB/BRICK/STONE/TIMBER PAVINGS

Item Excluding site overheads and profit	PC £	Labour hours	Labour £	Plant £	Material £	Unit	Total rate £
Add or subtract the following amounts for every £10.00/1000 difference in the prime cost of bricks							
Butt joints							
200 × 100 mm	–	–	–	–	0.50	m²	0.50
215 × 102.5 mm	–	–	–	–	0.45	m²	0.45
10 mm mortar joints							
200 × 100 mm	–	–	–	–	0.43	m²	0.43
215 × 102.5 mm	–	–	–	–	0.40	m²	0.40
PRECAST CONCRETE SLAB PAVINGS							
Slab paving; precast concrete pavings; Charcon Hard Landscaping; on prepared subbase (not included); bedding on 25 mm thick cement: sand mortar (1:4); butt joints; straight both ways; on 50 mm thick sharp sand base							
Pavings; natural grey							
450 × 450 × 70 mm chamfered	18.91	0.15	8.77	–	23.84	m²	**32.61**
450 × 450 × 50 mm chamfered	14.77	0.15	8.77	–	19.69	m²	**28.46**
600 × 300 × 50 mm	12.89	0.15	8.77	–	17.81	m²	**26.58**
400 × 400 × 65 mm chamfered	25.69	0.13	7.90	–	30.61	m²	**38.51**
450 × 600 × 50 mm	12.33	0.15	8.77	–	17.26	m²	**26.03**
600 × 600 × 50 mm	10.11	0.13	7.90	–	15.04	m²	**22.94**
750 × 600 × 50 mm	9.67	0.13	7.90	–	14.59	m²	**22.49**
900 × 600 × 50 mm	8.24	0.13	7.90	–	13.17	m²	**21.07**
Pavings; coloured							
450 × 450 × 70 mm chamfered	25.18	0.15	8.77	–	30.11	m²	**38.88**
450 × 600 × 50 mm	15.22	0.15	8.77	–	20.15	m²	**28.92**
400 × 400 × 65 mm chamfered	25.69	0.13	7.90	–	30.61	m²	**38.51**
600 × 600 × 50 mm	12.25	0.13	7.90	–	17.18	m²	**25.08**
750 × 600 × 50 mm	11.33	0.13	7.90	–	16.26	m²	**24.16**
900 × 600 × 50 mm	9.75	0.13	7.90	–	14.67	m²	**22.57**
Precast concrete pavings; Charcon Hard Landscaping; on prepared subbase (not included); bedding on 25 mm thick cement: sand mortar (1:4); butt joints; straight both ways; jointing in cement: sand (1:3) brushed in; on 50 mm thick sharp sand base							
Appalacian rough textured exposed aggregate pebble paving							
600 × 600 × 65 mm	32.27	0.17	9.88	–	35.49	m²	**45.37**

Q25 SLAB/BRICK/STONE/TIMBER PAVINGS

Item Excluding site overheads and profit	PC £	Labour hours	Labour £	Plant £	Material £	Unit	Total rate £
PRECAST CONCRETE SLAB PAVINGS – cont							
Pavings; Marshalls Plc; spot bedding on 5 nr pads of cement: sand mortar (1:4); on sharp sand							
Blister Tactile pavings; specially textured slabs for guidance of blind pedestrians; red or buff							
400 × 400 × 50 mm	33.57	0.17	9.88	–	35.33	m²	45.21
450 × 450 × 50 mm	27.33	0.17	9.88	–	29.10	m²	38.98
Metric Four Square pavings							
496 × 496 × 50 mm; exposed river							
gravel aggregate	104.71	0.17	9.88	–	106.47	m²	116.35
Metric Four Square cycle blocks							
496 × 496 × 50 mm; exposed							
aggregate	104.71	0.08	4.94	–	107.22	m²	112.16
Precast concrete pavings; Marshalls Plc; Heritage imitation riven yorkstone paving; on prepared subbase measured separately; bedding on 25 mm thick cement: sand mortar (1:4); pointed straight both ways cement: sand (1:3)							
Square and rectangular paving							
450 × 300 × 38 mm	38.27	0.33	19.75	–	42.41	m²	62.16
450 × 450 × 38 mm	23.57	0.25	14.81	–	27.70	m²	42.51
600 × 300 × 38 mm	28.25	0.27	15.80	–	32.38	m²	48.18
600 × 450 × 38 mm	26.01	0.25	14.81	–	30.14	m²	44.95
600 × 600 × 38 mm	23.52	0.17	9.88	–	27.73	m²	37.61
Extra labours for laying the a selection of the above sizes to random rectangular pattern	–	0.33	6.58	–	–	m²	6.58
Radial paving for circles							
circle with centre stone and first ring (8 slabs); 450 × 230/560 × 38 mm; dia. 1.54 m (total area 1.86 m²)	63.52	0.50	29.63	–	69.48	nr	99.11
circle with second ring (16 slabs); 450 × 300/460 × 38 mm; dia. 2.48 m (total area 4.83 m²)	63.52	1.33	79.00	–	178.99	nr	257.99
circle with third ring (16 slabs); 450 × 470/625 × 38 mm; dia. 3.42 m (total area 9.18 m²)	63.52	2.67	158.00	–	326.76	nr	484.76

Q25 SLAB/BRICK/STONE/TIMBER PAVINGS

Item Excluding site overheads and profit	PC £	Labour hours	Labour £	Plant £	Material £	Unit	Total rate £
Marshalls Plc; La Linia Paving; fine textured exposed aggregate 80 mm thick pavings in various sizes to designed laying patterns; laid to 50 mm sharp sand bed on Type 1 base all priced separately							
Bonded laying patterns; 300 × 300 mm							
light granite/anthracite basalt	35.29	0.25	14.81	–	35.29	m²	**50.10**
Random scatter pattern incorporating 100 × 200 mm, 200 × 200 mm and 300 × 200 mm units							
light granite/anthracite basalt	34.43	0.33	19.75	–	34.43	m²	**54.18**
DETERRENT SLAB PAVINGS							
Pedestrian deterrent pavings; Marshalls Plc; on prepared base (not included); bedding on 25 mm cement: sand (1:3); cement: sand (1:3) joints							
Lambeth pyramidal pavings							
600 × 600 × 75 mm	19.93	0.08	4.94	–	24.06	m²	**29.00**
Thaxted pavings; granite sett appearance							
600 × 600 × 75 mm	23.50	0.11	6.58	–	27.63	m²	**34.21**
Pedestrian deterrent pavings; Townscape Products Ltd; on prepared base (not included); bedding on 25 mm cement: sand (1:3); cement: sand (1:3) joints							
Strata striated textured slab pavings; giving bonded appearance; grey							
600 × 600 × 60 mm	14.46	0.17	9.88	–	18.59	m²	**28.47**
Geoset raised chamfered studs pavings; grey							
600 × 600 × 60 mm	14.46	0.17	9.88	–	18.59	m²	**28.47**
Abbey square cobble pattern pavings; reinforced							
600 × 600 × 65 mm	13.68	0.27	15.80	–	17.81	m²	**33.61**
Edge restraints; to block paving; on prepared base (not included); 200 × 100 × 80 mm; PC £10.00/m²; haunching one side							
Header course							
200 × 100 mm; butt joints	2.00	0.09	5.27	–	4.53	m	**9.80**
Stretcher course							
200 × 100 mm; butt joints	1.00	0.06	3.29	–	3.36	m	**6.65**

Q25 SLAB/BRICK/STONE/TIMBER PAVINGS

Item Excluding site overheads and profit	PC £	Labour hours	Labour £	Plant £	Material £	Unit	Total rate £
PRECAST CONCRETE BLOCK PAVINGS							
Concrete cobble paviors; Charcon Hard Landscaping; Concrete Products; on prepared base (not included); bedding on 50 mm sand; kiln dried sand joints swept in							
Paviors							
Woburn blocks; 100–201 × 134 × 80 mm; random sizes	32.91	0.22	13.17	0.11	34.84	m²	**48.12**
Woburn blocks; 100–201 × 134 × 80 mm; single size	32.91	0.17	9.88	0.11	34.84	m²	**44.83**
Woburn blocks; 100–201 × 134 × 60 mm; random sizes	27.12	0.22	13.17	0.11	29.05	m²	**42.33**
Woburn blocks; 100–201 × 134 × 60 mm; single size	27.09	0.17	9.88	0.11	29.02	m²	**39.01**
Concrete setts; on 25 mm sand; compacted; vibrated; joints filled with sand; natural or coloured; well rammed hardcore base (not included)							
Marshalls Plc; Tegula cobble paving							
60 mm thick; random sizes	23.08	0.19	11.29	0.11	25.18	m²	**36.58**
60 mm thick; single size	23.08	0.15	8.98	0.11	25.18	m²	**34.27**
80 mm thick; random sizes	26.60	0.19	11.29	0.11	28.69	m²	**40.09**
80 mm thick; single size	26.60	0.15	8.98	0.11	28.69	m²	**37.78**
cobbles; 80 × 80 × 60 mm thick; traditional	39.35	0.19	10.97	0.11	41.45	m²	**52.53**
Cobbles							
Charcon Hard Landscaping; Country setts							
100 mm thick; random sizes	43.84	0.33	19.75	0.11	45.94	m²	**65.80**
100 mm thick; single size	44.94	0.22	13.17	0.11	47.03	m²	**60.31**
PRECAST CONCRETE CYCLE BLOCKS							
Concrete cycle blocks; Marshalls Plc; bedding in cement: sand (1:4)							
Metric 4 Square cycle stand blocks; smooth grey concrete							
496 × 496 × 100 mm	45.00	0.25	4.94	–	45.61	nr	**50.55**
Concrete cycle blocks; Townscape Products Ltd; on 100 mm concrete (1:2:4); on 150 mm hardcore; bedding in cement: sand (1:4)							
Cycle blocks							
Cycle Bloc; in white concrete	34.50	0.25	4.94	–	35.31	nr	**40.25**
Mountain Cycle Bloc; in white concrete	46.50	0.25	4.94	–	47.31	nr	**52.25**

Q25 SLAB/BRICK/STONE/TIMBER PAVINGS

Item Excluding site overheads and profit	PC £	Labour hours	Labour £	Plant £	Material £	Unit	Total rate £
GRASS CONCRETE							
Grass concrete – General							
Preamble: Grass seed should be a							
perennial ryegrass mixture, with the							
proportion depending on expected traffic.							
Hardwearing winter sportsground							
mixtures are suitable for public areas.							
Loose gravel, shingle or sand is liable to							
be kicked out of the blocks; rammed							
hoggin or other stabilized material							
should be specified.							
Grass concrete; Grass Concrete Ltd;							
on blinded granular Type 1 subbase							
(not included)							
Grasscrete in situ concrete continuously							
reinforced surfacing; including expansion							
joints at 10 m centres; soiling; seeding							
GC2; 150 mm thick; traffic up to 40.00							
tonnes	–	–	–	–	–	m²	**48.55**
GC1; 100 mm thick; traffic up to 13.30							
tonnes	–	–	–	–	–	m²	**38.26**
GC3; 76 mm thick; traffic up to 4.30							
tonnes	–	–	–	–	–	m²	**34.61**
Grass concrete; Grass Concrete Ltd;							
406 × 406 mm blocks; on 20 mm sharp							
sand; on blinded MOT type 1 subbase							
(not included); level and to falls only;							
including filling with topsoil; seeding							
with dwarf rye grass at PC £4.50/kg							
Pavings							
GB103; 103 mm thick	18.68	0.40	7.90	0.18	21.32	m²	**29.40**
GB83; 83 mm thick	17.85	0.36	7.18	0.18	20.49	m²	**27.85**
Grass concrete; Charcon Hard							
Landscaping; on 25 mm sharp sand;							
including filling with topsoil; seeding							
with dwarf rye grass at PC £4.50/kg							
Grassgrid grass/concrete paving blocks							
366 × 274 × 100 mm thick	19.72	0.38	7.41	0.18	22.36	m²	**29.95**

Q25 SLAB/BRICK/STONE/TIMBER PAVINGS

Item Excluding site overheads and profit	PC £	Labour hours	Labour £	Plant £	Material £	Unit	Total rate £
GRASS CONCRETE – cont							
Grass concrete; Marshalls Plc; on 25 mm sharp sand; including filling with topsoil; seeding with dwarf rye grass at PC £4.50/kg							
Concrete grass pavings							
Grassguard 130; for light duty applications (80 mm prepared base not included)	19.48	0.38	7.41	0.18	22.30	m²	29.89
Grassguard 160; for medium duty applications (80–150 mm prepared base not included)	22.96	0.46	9.11	0.18	25.78	m²	35.07
Grassguard 180; for heavy duty applications (150 mm prepared base not included)	25.70	0.60	11.85	0.18	28.52	m²	40.55
Full mortar bedding							
Extra over pavings for bedding on 25 mm cement: sand (1:4); in lieu of spot bedding on sharp sand	–	0.03	0.49	–	1.65	m²	2.14
NATURAL STONE SLAB PAVINGS							
Natural stone, slab or granite paving – General							
Preamble: Provide paving slabs of the specified thickness in random sizes but not less than 25 slabs per 10 m² of surface area, to be laid in parallel courses with joints alternately broken and laid to falls.							
Natural stone, slate or granite flag pavings; CED Ltd; on prepared base (not included); bedding on 25 mm cement: sand (1:3); cement: sand (1:3) joints							
Yorkstone; riven laid random rectangular							
new slabs; 40–60 mm thick	82.50	0.57	33.86	–	86.33	m²	120.19
reclaimed slabs; Cathedral grade; 50–75 mm thick	102.96	0.93	55.30	–	106.79	m²	162.09
Donegal quartzite slabs; standard tiles							
200 mm × random lengths × 15–25 mm	89.18	1.17	69.13	–	93.01	m²	162.14
250 mm × random lengths × 15–25 mm	89.18	1.10	65.17	–	93.01	m²	158.18
300 mm × random lengths × 15–25 mm	89.18	1.03	61.22	–	93.01	m²	154.23
350 mm × random lengths × 15–25 mm	89.18	0.93	55.30	–	93.01	m²	148.31
400 mm × random lengths × 15–25 mm	89.18	0.83	49.37	–	93.01	m²	142.38
450 mm × random lengths × 15–25 mm	89.18	0.78	46.08	–	93.01	m²	139.09

Q25 SLAB/BRICK/STONE/TIMBER PAVINGS

Item Excluding site overheads and profit	PC £	Labour hours	Labour £	Plant £	Material £	Unit	Total rate £
Natural yorkstone pavings; Johnsons Wellfield Quarries; sawn 6 sides; 50 mm thick; on prepared base measured separately; bedding on 25 mm cement: sand (1:3); cement: sand (1:3) joints							
Paving							
laid to random rectangular pattern	58.25	0.57	33.86	–	61.89	m²	**95.75**
laid to coursed laying pattern; 3 sizes	62.84	0.57	34.05	–	66.48	m²	**100.53**
Paving; single size							
600 × 600 mm	68.20	0.28	16.79	–	71.84	m²	**88.63**
600 × 400 mm	68.20	0.33	19.75	–	71.84	m²	**91.59**
300 × 200 mm	77.21	0.67	39.50	–	80.84	m²	**120.34**
215 × 102.5 mm	79.78	0.83	49.37	–	83.42	m²	**132.79**
Paving; cut to template off site; 600 × 600 mm; radius							
1.00 m	166.66	1.11	65.83	–	170.30	m²	**236.13**
2.50 m	166.66	0.67	39.50	–	170.30	m²	**209.80**
5.00 m	166.66	0.67	39.50	–	170.30	m²	**209.80**
Natural yorkstone pavings; Johnsons Wellfield Quarries; sawn 6 sides; 75 mm thick; on prepared base measured separately; bedding on 25 mm cement: sand (1:3); cement: sand (1:3) joints							
Paving							
laid to random rectangular pattern	68.20	0.32	18.76	–	71.84	m²	**90.60**
laid to coursed laying pattern; 3 sizes	76.13	0.32	18.76	–	79.76	m²	**98.52**
Paving; single size							
600 × 600 mm	78.70	0.32	18.76	–	82.34	m²	**101.10**
600 × 400 mm	78.70	0.32	18.76	–	82.34	m²	**101.10**
300 × 200 mm	88.53	0.25	14.81	–	92.16	m²	**106.97**
215 × 102.5 mm	94.24	0.83	49.37	–	97.88	m²	**147.25**
Paving; cut to template off site; 600 × 600 mm; radius							
1.00 m	235.00	1.33	79.00	–	238.64	m²	**317.64**
2.50 m	225.00	0.83	49.37	–	228.64	m²	**278.01**
5.00 m	215.00	0.83	49.37	–	218.64	m²	**268.01**
CED Ltd; Indian sandstone, riven pavings or edgings; calibrated +/−26 mm thick; on prepared base measured separately; bedding on 25 mm cement: sand (1:3); cement: sand (1:3) joints							
Paving							
laid to random rectangular pattern	25.12	0.80	47.41	–	28.75	m²	**76.16**
laid to coursed laying pattern; 3 sizes	25.12	0.67	39.50	–	28.75	m²	**68.25**
Paving; single size							
600 × 600 mm	25.12	0.33	19.75	–	28.75	m²	**48.50**
600 × 400 mm	25.12	0.42	24.69	–	28.75	m²	**53.44**
400 × 400 mm	25.12	0.56	32.92	–	28.75	m²	**61.67**

Q25 SLAB/BRICK/STONE/TIMBER PAVINGS

Item Excluding site overheads and profit	PC £	Labour hours	Labour £	Plant £	Material £	Unit	Total rate £
NATURAL STONE SLAB PAVINGS – cont							
Natural stone, slate or granite flag pavings; CED Ltd; on prepared base (not included); bedding on 25 mm cement: sand (1:3); cement: sand (1:3) joints							
Granite paving; sawn 6 sides; textured top							
new slabs; silver grey; 50 mm thick	32.84	0.57	33.86	–	36.67	m²	**70.53**
new slabs; blue grey; 50 mm thick	42.84	0.57	33.86	–	46.67	m²	**80.53**
new slabs; yellow grey; 50 mm thick	37.73	0.57	33.86	–	41.55	m²	**75.41**
new slabs; black; 50 mm thick	57.70	0.57	33.86	–	61.52	m²	**95.38**
Edgings; silver grey							
100 mm wide × random lengths	6.77	0.33	19.75	–	7.13	m	**26.88**
100 × 100 mm	7.74	0.50	29.63	–	8.10	m	**37.73**
100 mm long × 200 mm wide	12.57	0.20	11.85	–	12.94	m	**24.79**
250 mm wide × random lengths	13.30	0.25	14.81	–	16.94	m	**31.75**
300 mm wide × random lengths	14.51	0.25	14.81	–	18.15	m	**32.96**
RECONSTITUTED STONE SLAB PAVINGS							
Reconstituted yorkstone aggregate pavings; Marshalls Plc; Saxon on prepared subbase measured separately; bedding on 25 mm thick cement: sand mortar (1:4) ;on 50 mm thick sharp sand base							
Square and rectangular paving in buff; butt joints straight both ways							
300 × 300 × 35 mm	37.19	0.27	15.80	–	42.77	m²	**58.57**
600 × 300 × 35 mm	24.58	0.22	12.84	–	30.17	m²	**43.01**
450 × 450 × 50 mm	26.86	0.25	14.81	–	32.45	m²	**47.26**
600 × 600 × 35 mm	18.40	0.17	9.88	–	23.98	m²	**33.86**
600 × 600 × 50 mm	23.13	0.18	10.86	–	28.72	m²	**39.58**
Square and rectangular paving in natural; butt joints straight both ways							
300 × 300 × 35 mm	30.41	0.27	15.80	–	36.00	m²	**51.80**
450 × 450 × 50 mm	22.88	0.23	13.82	–	28.47	m²	**42.29**
600 × 300 × 35 mm	21.62	0.25	14.81	–	27.20	m²	**42.01**
600 × 600 × 35 mm	15.88	0.17	9.88	–	21.46	m²	**31.34**
600 × 600 × 50 mm	19.26	0.20	11.85	–	24.85	m²	**36.70**
Radial paving for circles; 20 mm joints							
circle with centre stone and first ring (8 slabs); 450 × 230/560 × 35 mm; dia. 1.54 m (total area 1.86 m²)	73.22	0.50	29.63	–	79.51	nr	**109.14**
circle with second ring (16 slabs); 450 × 300/460 × 35 mm; dia. 2.48 m (total area 4.83 m²)	178.34	1.33	79.00	–	194.59	nr	**273.59**
circle with third ring (24 slabs); 450 × 310/430 × 35 mm; dia. 3.42 m (total area 9.18 m²)	336.02	2.67	158.00	–	367.10	nr	**525.10**

Q25 SLAB/BRICK/STONE/TIMBER PAVINGS

Item Excluding site overheads and profit	PC £	Labour hours	Labour £	Plant £	Material £	Unit	Total rate £
NATURAL STONE SETTS							
Granite setts; bedding on 25 mm **cement: sand (1:3)** Natural granite setts; 100 × 100 mm to 125 × 150 mm × 150–250 mm length; riven surface; silver grey							
new; standard grade	28.11	0.67	39.50	–	36.84	m²	76.34
new; high grade	31.77	0.67	39.50	–	40.50	m²	80.00
reclaimed; cleaned	36.26	0.67	39.50	–	44.99	m²	84.49
Granite setts; bedding on **30 mmSteinTec mortar; surfaces** **primed with SteinTec primer** Natural granite setts; 100 × 100 mm to 125 × 150 mm × 150–250 mm length; riven surface; silver grey							
new; standard grade	36.60	0.85	43.19	–	36.60	m²	79.79
new; high grade	40.26	0.85	43.19	–	40.26	m²	83.45
reclaimed; cleaned	44.75	0.85	43.19	–	44.75	m²	87.94
COBBLE PAVINGS							
Cobble pavings – General Cobbles should be embedded by hand, tight-butted, endwise to a depth of 60% of their length. A dry grout of rapid-hardening cement: sand (1:2) shall be brushed over the cobbles until the interstices are filled to the level of the adjoining paving. Surplus grout shall then be brushed off and a light, fine spray of water applied over the area.							
Cobble pavings Cobbles; to present a uniform colour in panels; or varied in colour as required							
Scottish Beach Cobbles; 50–75 mm	24.16	1.11	65.83	–	30.41	m²	96.24
Scottish Beach Cobbles; 75–100 mm	34.41	0.83	49.37	–	40.66	m²	90.03
Scottish Beach Cobbles; 100–200 mm	35.41	2.67	79.00	–	41.66	m²	120.66
TIMBER DECKING							
Market prices of timber decking **materials; Exterior Decking Ltd** Hardwood decking; stainless steel screw fixings included in rates shown							
Ipe; pre-drilled and countersunk	76.22	–	–	–	–	m²	–

Q25 SLAB/BRICK/STONE/TIMBER PAVINGS

Item Excluding site overheads and profit	PC £	Labour hours	Labour £	Plant £	Material £	Unit	Total rate £
TIMBER DECKING – cont							
Market prices of timber decking **materials – cont**							
Hardwood decking; 'Exterpark' invisible fixings included in rates shown							
Ipe; 21 mm thick	83.42	–	–	–	–	m²	–
Ipe; 28 mm thick	126.74	–	–	–	–	m²	–
Ipe; 35 mm thick	172.55	–	–	–	–	m²	–
Teak Rustic FSC 21 mm thick	94.71	–	–	–	–	m²	–
Merbau 21 mm	67.64	–	–	–	–	m²	–
Merbau 28 mm	88.66	–	–	–	–	m²	–
Merbau 35 mm	103.84	–	–	–	–	m²	–
Composite decking; stainless steel screw fixings included in rates shown							
Millboard 176 mm wide	66.99	–	–	–	–	m²	–
Millboard 196 mm wide	60.28	–	–	–	60.28	m²	**60.28**
Softwood decking							
Pine smooth; 145 × 38 mm	15.39	–	–	–	–	m²	–
Timber decking; Exterior Decking Ltd; **timber decking substructure/** **carcassing costs**							
Roof decks exclusive of beams; timber craned (not included); joists at 400 mm centres							
47 × 47 mm	–	–	–	–	–	m²	**34.63**
47 × 100 mm	–	–	–	–	–	m²	**37.48**
47 × 150 mm	–	–	–	–	–	m²	**40.39**
pedestals up to 150 mm	–	–	–	–	–	m²	**60.37**
Commercial applications; decks with double beams at 300 mm centres							
47 × 100 mm	–	–	–	–	–	m²	**81.64**
47 × 150 mm	–	–	–	–	–	m²	**87.74**
Domestic garden applications; decks with double beams at 300 mm centres							
47 × 100 mm	–	–	–	–	–	m²	**71.24**
47 × 150 mm	–	–	–	–	–	m²	**75.53**
Structural posts to decks; tanalized treated softwood; rebated including excavation and setting in concrete							
100 × 100 mm; single beam; single rebate	–	–	–	–	–	m	**10.43**
150 × 150 mm; double beam; double rebate	–	–	–	–	–	m	**14.25**

Q25 SLAB/BRICK/STONE/TIMBER PAVINGS

Item Excluding site overheads and profit	PC £	Labour hours	Labour £	Plant £	Material £	Unit	Total rate £
Timber decking; Exterior Decking Ltd; **timber decking surfaces; fixed to** **substructure (not included)**							
Softwood deck timbers; screw fixed pine; smooth; 145 × 38 mm fixed with 65 × 5 mm coated deck screws	–	–	–	–	–	m²	40.33
Hardwood deck timbers; screw fixed Ipe; S4S E4E; 145 × 21 mm; pre-drilled	–	–	–	–	–	m²	115.30
Hardwood deck timbers; invisibly fixed							
Ipe; 21 mm thick	–	–	–	–	–	m²	140.59
Ipe; 28 mm thick	–	–	–	–	–	m²	200.58
Ipe; 35 mm thick	–	–	–	–	–	m²	263.81
Teak Rustic FSC 21 mm	–	–	–	–	–	m²	155.41
Merbau 21 mm	–	–	–	–	–	m²	119.87
Merbau 28 mm	–	–	–	–	–	m²	150.58
Merbau 35 mm	–	–	–	–	–	m²	173.62
Composite deck surface; Millboard 146 mm wide; fixed with stainless screws 4.5 × 70 mm	–	–	–	–	–	m²	87.94
196 mm wide; fixed with stainless screws 4.5 × 70 mm	–	–	–	–	–	m²	87.94
Surface treatments to timber decks; 'Owatrol' decking oil							
Seasonite	–	–	–	–	–	m²	6.40
Textrol	–	–	–	–	–	m²	4.67
Timber decking							
Supports for timber decking; softwood joists to receive decking boards; joists at 400 mm centres; Southern Yellow pine							
38 × 88 mm	15.84	1.00	19.75	–	17.58	m²	37.33
47 × 100 mm	9.96	1.00	19.75	–	11.70	m²	31.45
47 × 150 mm	14.90	1.00	19.75	–	16.64	m²	36.39
Hardwood decking; Yellow Balau; grooved or smooth; 6 mm joints							
deck boards; 90 mm wide × 19 mm thick	57.05	1.00	19.75	–	59.03	m²	78.78
deck boards; 145 mm wide × 21 mm thick	62.77	1.00	19.75	–	64.75	m²	84.50
deck boards; 145 mm wide × 28 mm thick	39.10	1.00	19.75	–	41.08	m²	60.83
Hardwood decking; Ipe; smooth; 6 mm joints							
deck boards; 90 mm wide × 19 mm thick	49.23	1.00	19.75	–	51.21	m²	70.96
deck boards; 145 mm wide × 19 mm thick	50.24	1.00	19.75	–	52.22	m²	71.97
AVS; Profile; Western Red Cedar; 6 mm joints smooth or reeded							
ex 125 mm wide × 38 mm thick	52.45	1.00	19.75	–	54.43	m²	74.18
ex 150 mm wide × 32 mm thick	16.60	1.00	19.75	–	18.58	m²	38.33

Q25 SLAB/BRICK/STONE/TIMBER PAVINGS

Item Excluding site overheads and profit	PC £	Labour hours	Labour £	Plant £	Material £	Unit	Total rate £
TIMBER DECKING – cont							
Timber decking – cont							
Handrails and base rail; fixed to posts at 2.00 m centres							
posts; 100 × 100 × 1370 mm high	9.42	1.00	19.75	–	9.88	m	**29.63**
posts; turned; 1220 mm high	15.63	1.00	19.75	–	16.08	m	**35.83**
Handrails; balusters							
square balusters at 100 mm centres	34.70	0.50	9.88	–	35.16	m	**45.04**
square balusters at 300 mm centres	11.56	0.33	6.58	–	11.92	m	**18.50**
turned balusters at 100 mm centres	54.80	0.50	9.88	–	55.25	m	**65.13**
turned balusters at 300 mm centres	18.25	0.33	6.52	–	18.61	m	**25.13**
STEPS							
In situ concrete steps; ready mixed concrete; Gen 1; on 100 mm hardcore base; base concrete thickness 100 mm treads and risers cast simultaneously; inclusive of A142 mesh within the concrete; step riser 170 mm high; inclusive of all shuttering and concrete labours; floated finish; to graded ground (not included)							
Treads 250 mm wide							
one riser	5.06	1.32	78.21	–	7.77	m	**85.98**
two risers	8.11	1.98	117.31	–	13.20	m	**130.51**
three risers	11.74	2.64	156.42	–	22.52	m	**178.94**
four risers	15.77	3.96	234.63	–	27.60	m	**262.23**
six risers	19.96	5.28	312.84	–	31.39	m	**344.23**
Lay surface to concrete in situ steps above							
granite treads; pre-cut off site; 305 mm wide × 50 mm thick bullnose finish inclusive of 20 mm overhang; granite riser 20 mm thick all laid on 15 mm mortar bed	56.00	0.66	39.10	–	57.09	m	**96.19**
riven yorkstone tread and riser 50 mm thick; cut on site	34.50	0.82	48.88	–	35.59	m	**84.47**
riven yorkstone treads 225 mm wide × 50 mm thick cut on site; two courses brick risers laid on flat stretcher bond; bricks PC £500.00/1000	16.88	0.99	58.66	–	22.73	m	**81.39**
Marshalls Saxon paving slab 450 × 450 mm cut to 225 mm wide × 38 mm thick; cut on site; two courses brick risers laid on flat stretcher bond; bricks PC £500.00/1000	5.18	0.99	58.66	–	11.04	m	**69.70**
Treads 450 mm wide							
one riser	9.11	1.65	97.76	–	12.70	m	**110.46**
two risers	18.61	2.31	136.87	–	25.97	m	**162.84**
three risers	26.31	2.97	175.97	–	37.62	m	**213.59**
four risers	45.51	4.29	254.18	–	60.19	m	**314.37**
six risers	57.84	5.94	351.94	–	70.89	m	**422.83**

Q25 SLAB/BRICK/STONE/TIMBER PAVINGS

Item Excluding site overheads and profit	PC £	Labour hours	Labour £	Plant £	Material £	Unit	Total rate £
Lay surface to concrete in situ steps							
granite treads; pre-cut off site;							
470 mm wide × 50 mm thick bullnose							
finish inclusive of 20 mm overhang;							
granite riser 20 mm thick all laid on							
15 mm mortar bed	0.82	1.32	78.21	–	2.19	m	80.40
riven yorkstone tread and riser 50 mm							
thick; cut on site	46.88	1.49	87.99	–	48.24	m	136.23
riven yorkstone treads 460 mm wide							
50 mm thick cut on site; two courses							
brick risers laid on flat stretcher bond;							
bricks PC £500.00/1000	34.50	1.65	97.76	–	40.58	m	138.34
Marshalls Saxon paving slab							
450 × 450 × 38 mm thick cut on site;							
two courses brick risers laid on flat							
stretcher bond; bricks PC £500.00/1000	10.37	0.91	53.77	–	16.23	m	70.00
SPECIALISED MORTARS FOR PAVINGS							
SteinTec Ltd							
Primer for specialized mortar; SteinTec							
Tuffbond priming mortar immediately							
prior to paving material being placed on							
bedding mortar							
maximum thickness 1.5 mm (1.5 kg/m^2)	0.86	0.02	0.40	–	0.86	m^2	1.26
Tuffbed; 2 pack hydraulic mortar for							
paving applications							
30 mm thick	7.63	0.17	3.29	–	7.63	m^2	10.92
40 mm thick	10.17	0.20	3.95	–	10.17	m^2	14.12
50 mm thick	12.71	0.25	4.94	–	12.71	m^2	17.65
SteinTec jointing mortar; joint size							
100 × 100 × 100 mm; 10 mm joints;							
31.24 kg/m^2	15.15	0.50	9.88	–	15.15	m^2	25.03
100 × 100 × 100 mm; 20 mm joints;							
31.24 kg/m^2	26.66	1.00	19.75	–	26.66	m^2	46.41
200 × 100 × 65 mm; 10 mm joints;							
15.7 kg/m^2	7.61	0.10	1.98	–	7.61	m^2	9.59
215 × 112.5 × 65 mm; 10 mm joints;							
14.3 kg/m^2	6.93	0.08	1.65	–	6.93	m^2	8.58
300 × 450 × 65 mm; 10 mm joints;							
6.24 kg/m^2	3.03	0.04	0.79	–	3.03	m^2	3.82
300 × 450 × 65 mm; 20 mm joints;							
11.98 kg/m^2	5.81	0.10	1.98	–	5.81	m^2	7.79
450 × 450 × 65 mm; 20 mm joints;							
9.75 kg/m^2	4.73	0.11	2.19	–	4.73	m^2	6.92
450 × 600 × 65 mm; 20 mm joints;							
8.6 kg/m^2	4.17	0.09	1.80	–	4.17	m^2	5.97
600 × 600 × 65 mm; 20 mm joints;							
7.43 kg/m^2	3.60	0.07	1.41	–	3.60	m^2	5.01

Q26 SPECIAL SURFACINGS/PAVINGS FOR SPORT/GENERAL AMENITY

Item Excluding site overheads and profit	PC £	Labour hours	Labour £	Plant £	Material £	Unit	Total rate £
MARKET PRICES OF SPORTS AND PLAY SURFACING MATERIALS							
Market prices of surfacing materials							
Surfacings; Melcourt Industries Ltd							
(items labelled FSC are Forest							
Stewardship Council certified)							
Playbark® 10/50; per 25 m³ load	–	–	–	–	63.60	m³	**63.60**
Playbark® 10/50; per 80 m³ load	–	–	–	–	50.60	m³	**50.60**
Playbark® 8/25; per 25 m³ load	–	–	–	–	62.10	m³	**62.10**
Playbark® 8/25; per 80 m³ load	–	–	–	–	49.10	m³	**49.10**
Playchips®; per 25 m³ load; FSC	–	–	–	–	41.75	m³	**41.75**
Playchips®; per 80 m³ load; FSC	–	–	–	–	28.75	m³	**28.75**
Kushyfall; per 25 m³ load; FSC	–	–	–	–	39.70	m³	**39.70**
Kushyfall; per 80 m³ load; FSC	–	–	–	–	26.70	m³	**26.70**
Softfall; per 25 m³ load	–	–	–	–	31.40	m³	**31.40**
Softfall; per 80 m³ load	–	–	–	–	18.40	m³	**18.40**
Playsand; per 10 t load	–	–	–	–	101.16	m³	**101.16**
Playsand; per 20 t load	–	–	–	–	88.92	m³	**88.92**
Walk Chips; per 25 m³ load; FSC	–	–	–	–	38.70	m³	**38.70**
Walk Chips; per 80 m³ load; FSC	–	–	–	–	25.70	m³	**25.70**
Woodfibre; per 25 m³ load; FSC	–	–	–	–	36.95	m³	**36.95**
Woodfibre; per 80 m³ load; FSC	–	–	–	–	23.95	m³	**23.95**
NATURAL SPORTS PITCHES							
Natural sports pitches; Agripower Ltd; sports pitches; gravel raft construction							
Professional standard							
football pitch; 6500 m²	–	–	–	–	–	nr	**150000.00**
cricket wicket; county standard	–	–	–	–	–	nr	**30000.00**
Playing fields; Sport England Compliant							
football pitch; 6500 m²	–	–	–	–	–	nr	**65000.00**
cricket wicket; playing field standard	–	–	–	–	–	nr	**15000.00**
track and infield	–	–	–	–	–	nr	**92000.00**
ARTIFICIAL SPORTS PITCHES							
Artificial surfaces and finishes – General							
Preamble: Advice should also be sought from the Technical Unit for Sport, the appropriate regional office of the Sports Council or the National Playing Fields Association. Some of the following prices include base work whereas others are for a specialist surface only on to a base prepared and costed separately.							

Q26 SPECIAL SURFACINGS/PAVINGS FOR SPORT/GENERAL AMENITY

Item Excluding site overheads and profit	PC £	Labour hours	Labour £	Plant £	Material £	Unit	Total rate £
Artificial sports pitches; Agripower Ltd; third generation rubber crumb Sports pitches to respective National governing body standards; inclusive of all excavation, drainage, lighting, fencing, etc.							
football pitch; 6500 m²	–	–	–	–	–	nr	465000.00
rugby union multi-use pitch; 120 × 75 m	–	–	–	–	–	nr	590000.00
hockey pitch; water-based; 101.4 × 63 m	–	–	–	–	–	nr	550000.00
athletic track; 8 lane; International amateur athletic federation	–	–	–	–	–	nr	550000.00
Multi-sport pitches; inclusive of all excavation, drainage, lighting, fencing, etc.							
sand dressed; 101.4 × 63 m	–	–	–	–	–	nr	390000.00
sand filled; 101.4 × 63 m	–	–	–	–	–	nr	340000.00
polymeric (synthetic bound rubber)	–	–	–	–	–	m²	130.00
macadam	–	–	–	–	–	m²	80.00
Sports areas; Agripower Ltd Sports tracks; polyurethane rubber surfacing; on bitumen-macadam (not included); prices for 5500 m² minimum							
International	–	–	–	–	–	m²	40.00
Club Grade	–	–	–	–	–	m²	31.00
Sports areas; multi-component polyurethane rubber surfacing; on bitumen-macadam; finished with polyurethane or acrylic coat; green or red							
Permaprene	–	–	–	–	–	m²	41.00
Cricketweave artificial wicket system; including proprietary subbase (all by specialist subcontractor)							
28 × 2.74 m; hard porous base	–	–	–	–	–	nr	6702.00
28 × 2.74 m; tarmac base	–	–	–	–	–	nr	8470.00
Cricketweave practice batting end							
11 × 2 .74 m; hard porous base	–	–	–	–	–	nr	3515.00
11 × 2 .74 m; tarmac base	–	–	–	–	–	nr	4378.00
11 × 3.65 m; hard porous base	–	–	–	–	–	nr	4584.00
11 × 3.65 m; tarmac base	–	–	–	–	–	nr	5732.00
Cricketweave practice bowling end							
8 × 2.74 m; hard porous base	–	–	–	–	–	nr	2639.00
8 × 2.74 m; tarmac base	–	–	–	–	–	nr	–
Supply and erection of single bay cricket cage							
18.3 × 3.65 × 3.6 m high	–	–	–	–	–	nr	2599.00

Q26 SPECIAL SURFACINGS/PAVINGS FOR SPORT/GENERAL AMENITY

Item Excluding site overheads and profit	PC £	Labour hours	Labour £	Plant £	Material £	Unit	Total rate £
TENNIS COURTS							
Tennis courts; Spadeoak Ltd							
Hard playing surfaces to SAPCA Code							
of Practice minimum requirements; laid							
on 65 mm thick macadam base on							
150 mm thick stone foundation; to							
include lines, nets, posts, brick edging,							
2.75 m high fence and perimeter							
drainage (excluding excavation, levelling							
or additional foundation); based on court							
size 36.6 × 18.3 m (670 m²)							
Durapore; all weather porous							
macadam and acrylic colour coating	–	–	–	–	–	m²	**63.47**
Tiger Turf Advantage Pro; sand filled							
artificial grass	–	–	–	–	–	m²	**87.04**
Tiger Turf Baseline; short pile sand							
filled artificial grass	–	–	–	–	–	m²	**86.55**
DecoColour; impervious acrylic							
hardcourt (200 mm foundation)	–	–	–	–	–	m²	**80.78**
DecoTurf; cushioned impervious							
acrylic tournament surface (200 mm							
foundation)	–	–	–	–	–	m²	**92.32**
Porous Kushion Kourt; porous							
cushioned acrylic surface	–	–	–	–	–	m²	**90.39**
EasiClay; synthetic clay system	–	–	–	–	–	m²	**94.23**
Canada Tenn; American green clay;							
fast dry surface (no macadam but							
including irrigation)	–	–	–	–	–	m²	**96.16**
PLAY SURFACES							
Playgrounds; Rubaflex in situ							
playground surfacing, porous; on							
prepared stone/granular base Type 1							
(not included) and macadam base							
course (not included)							
Black							
15 mm thick (0.50 m critical fall height)	–	–	–	–	–	m²	**27.15**
35 mm thick (1.00 m critical fall height)	–	–	–	–	–	m²	**41.85**
60 mm thick (1.50 m critical fall height)	–	–	–	–	–	m²	**55.45**
Playgrounds; Wetpour safety							
surfacing in situ playground surfacing,							
porous; on prepared stone/granular							
base Type 1 (not included) and							
macadam base course (not included)							
Coloured							
15 mm thick (0.50 m critical fall height)	–	–	–	–	–	m²	**64.48**
35 mm thick (1.00 m critical fall height)	–	–	–	–	–	m²	**70.15**
60 mm thick (1.50 m critical fall height)	–	–	–	–	–	m²	**79.19**

Q26 SPECIAL SURFACINGS/PAVINGS FOR SPORT/GENERAL AMENITY

Item Excluding site overheads and profit	PC £	Labour hours	Labour £	Plant £	Material £	Unit	Total rate £
Playgrounds; Matta Ltd; Play Matta **safety tiles; porous; on prepared Type** **1 base (not included)**							
Critical fall height 500 mm							
natural colours	–	–	–	–	–	m²	67.00
bright colours	–	–	–	–	–	m²	79.00
Critical fall height 1.70 m							
natural colours	–	–	–	–	–	m²	72.00
bright colours	–	–	–	–	–	m²	82.00
Critical fall height 3.20 m							
natural colours	–	–	–	–	–	m²	88.00
bright colours	–	–	–	–	–	m²	98.00
Critical fall height 4.20 m							
natural colours	–	–	–	–	–	m²	99.00
bright colours	–	–	–	–	–	m²	106.00
Playgrounds; Matta Ltd; Safety Matta **tiles; supplied and fitted on to grass** **(turf) or seeded soil base (not** **included)**							
Green or black							
critical fall height 2.70 m	–	–	–	–	–	m²	66.95
critical fall height 2.90 m	–	–	–	–	–	m²	75.00
Playgrounds; Melcourt Industries Ltd; **specifiers and users should contact** **the supplier for performance** **specifications of the materials below**							
Play surfaces; on drainage layer (not included); minimum 300 mm settled depth; prices for 80 m³ loads unless specified							
Playbark® 8/25; 8–25 mm particles; red/brown; 25 m³ loads	18.63	0.35	6.93	–	18.63	m²	25.56
Playbark® 8/25; 8–25 mm particles; red/brown	16.20	0.35	6.93	–	16.20	m²	23.13
Playbark® 10/50; 10–50 mm particles; red/brown	18.55	0.35	6.93	–	18.55	m²	25.48
Playchips®; FSC graded woodchips	9.58	0.35	6.93	–	9.58	m²	16.51
Kushyfall; fibreised woodchips	8.90	0.35	6.93	–	8.90	m²	15.83
Softfall; conifer shavings	6.13	0.35	6.93	–	6.13	m²	13.06

Prices for Measured Works

Q30 SEEDING/TURFING

Item Excluding site overheads and profit	PC £	Labour hours	Labour £	Plant £	Material £	Unit	Total rate £
GENERALLY							
Seeding/turfing – General							
Preamble: The following market prices generally reflect the manufacturer's recommended retail prices. Trade and bulk discounts are often available on the prices shown. The manufacturers of these products generally recommend application rates. Note: the following rates reflect the average rate for each product.							
MARKET PRICES OF SEEDING MATERIALS							
Market prices of pre-seeding materials							
Rigby Taylor Ltd							
turf fertilizer; Mascot Outfield 16+6+6	–	–	–	–	3.09	100 m²	**3.09**
Boughton Loam							
screened topsoil; 100 mm	–	–	–	–	50.40	m³	**50.40**
screened Kettering loam; 3 mm	–	–	–	–	117.00	m³	**117.00**
screened Kettering loam; sterilised; 3 mm	–	–	–	–	120.60	m³	**120.60**
top dressing; sand soil mixtures; 90/10 to 50/50	–	–	–	–	93.60	m³	**93.60**
Market prices of turf fertilizers; recommended average application rates							
Everris; turf fertilizers							
grass fertilizer; slow release; Sierraform GT Preseeder 18+22+05	–	–	–	–	5.19	100 m²	**5.19**
grass fertilizer; slow release; Sierraform GT All Season 18+06+18	–	–	–	–	4.34	100 m²	**4.34**
grass fertilizer; slow release; Sierraform GT Anti-Stress 15+00+26	–	–	–	–	5.19	100 m²	**5.19**
grass fertilizer; slow release; Sierraform GT Momentum 22+05+11	–	–	–	–	3.46	100 m²	**3.46**
grass fertilizer; slow release; Sierraform GT Spring Start 16+00+16	–	–	–	–	4.33	100 m²	**4.33**
grass fertilizer; controlled release; Sierrablen 27+05+05 (8–9 months)	–	–	–	–	6.02	100 m²	**6.02**
grass fertilizer; controlled release; Sierrablen 28+05+05 (5–6 months)	–	–	–	–	5.75	100 m²	**5.75**
grass fertilizer; controlled release; Sierrablen 15+00+22 (5–6 months)	–	–	–	–	7.19	100 m²	**7.19**
grass fertilizer; Sierrablen Fine 38+00+00 (2–3 months)	–	–	–	–	2.06	100 m²	**2.06**

Q30 SEEDING/TURFING

Item Excluding site overheads and profit	PC £	Labour hours	Labour £	Plant £	Material £	Unit	Total rate £
grass fertilizer; Sierrablen Mini 00+00 +37 (5–6 months)	–	–	–	–	3.92	100 m²	3.92
grass fertilizer; Greenmaster Pro-Lite Turf Tonic 08+00+00	–	–	–	–	2.51	100 m²	2.51
grass fertilizer; Greenmaster Pro-Lite Spring & Summer 14+05+10	–	–	–	–	3.59	100 m²	3.59
grass fertilizer; Greenmaster Pro-Lite Zero Phosphate 14+00+10	–	–	–	–	3.59	100 m²	3.59
grass fertilizer; Greenmaster Pro-Lite Mosskiller 14+00+00	–	–	–	–	3.87	100 m²	3.87
grass fertilizer; Greenmaster Pro-Lite Invigorator 04+00+08	–	–	–	–	2.72	100 m²	2.72
grass fertilizer; Greenmaster Pro-Lite Autumn 06+05+10	–	–	–	–	3.87	100 m²	3.87
grass fertilizer; Greenmaster Pro-Lite Double K 07+00+14	–	–	–	–	3.87	100 m²	3.87
grass fertilizer; Greenmaster Pro-Lite NK 12+00+12	–	–	–	–	3.87	100 m²	3.87
outfield turf fertilizer; Sportsmaster Standard Spring & Summer 09+07+07	–	–	–	–	3.14	100 m²	3.14
outfield turf fertilizer; Sportsmaster Standard Autumn 04+12+12	–	–	–	–	3.43	100 m²	3.43
outfield turf fertilizer; Sportsmaster Standard Zero Phosphate 12+00+09	–	–	–	–	3.58	100 m²	3.58
TPMC; tree planting and mulching compost	–	–	–	–	5.62	75 l	5.62
Rigby Taylor Ltd; 35 g/m²							
grass fertilizer; Mascot Microfine 12+0 +10 + 2% Mg + 2% Fe	–	–	–	–	6.53	100 m²	6.53
grass fertilizer; Mascot Microfine 8+0 +6 + 2% Mg + 4% Fe	–	–	–	–	5.29	100 m²	5.29
grass fertilizer; Mascot Microfine Organic OC1 8+0+0 + 2% Fe	–	–	–	–	5.53	100 m²	5.53
grass fertilizer; Mascot Microfine Organic OC2 5+2+10	–	–	–	–	6.15	100 m²	6.15
grass fertilizer; Mascot Guardian 6+1 +12 + 2% Mg + 2% Fe + seaweed	–	–	–	–	7.88	100 m²	7.88
grass fertilizer; Mascot Delta Sport 12 +4+8 + 0.5% Fe	–	–	–	–	5.00	100 m²	5.00
grass fertilizer; Mascot Fine Turf 12+0 +9 + 1% Mg + 1% Fe + seaweed	–	–	–	–	6.01	100 m²	6.01
outfield fertilizer; Mascot Outfield 16 +6+6	–	–	–	–	4.33	100 m²	4.33
outfield fertilizer; Mascot Outfield 4 +10+10	–	–	–	–	4.27	100 m²	4.27
outfield fertilizer; Mascot Outfield 9+5 +5	–	–	–	–	3.67	100 m²	3.67
outfield fertilizer; Mascot Outfield 12 +4+4	–	–	–	–	3.70	100 m²	3.70

Q30 SEEDING/TURFING

Item Excluding site overheads and profit	PC £	Labour hours	Labour £	Plant £	Material £	Unit	Total rate £
MARKET PRICES OF SEEDING MATERIALS – cont							
Market prices of turf fertilizers – cont							
Rigby Taylor Ltd – cont							
liquid fertilizer; Mascot Microflow-C 25 +0+0 + trace elements; 200–1400 ml /100 m²	–	–	–	–	7.14	100 m²	**7.14**
liquid fertilizer; Mascot Microflow-CX 4+2+18 + trace elements; 200–1400 ml/100 m²	–	–	–	–	6.88	100 m²	**6.88**
liquid fertilizer; Mascot Microflow-CX 12+0+8 + trace elements; 200–1400 ml/100 m²	–	–	–	–	5.93	100 m²	**5.93**
liquid fertilizer; Mascot Microflow-CX 17+2+5 + trace elements; 200–1400 ml/100 m²	–	–	–	–	6.66	100 m²	**6.66**
Market prices of grass seed							
Preamble: The prices shown are for supply only at one number 20 kg or 25 kg bag purchase price unless otherwise stated. Rates shown are based on the manufacturer's maximum recommendation for each seed type. Trade and bulk discounts are often available on the prices shown for quantities of more than one bag.							
Bowling greens; fine lawns; ornamental turf; croquet lawns							
British Seed Houses; A1 Green; 35 g/m²	–	–	–	–	25.93	100 m²	**25.93**
DLF Trifolium; J Green; 34–50 g/m²	–	–	–	–	41.55	100 m²	**41.55**
DLF Trifolium; J Premier Green; 34–50 g/m²	–	–	–	–	68.50	100 m²	**68.50**
DLF Trifolium; Promaster 20; 35–50 g/m²	–	–	–	–	22.20	100 m²	**22.20**
DLF Trifolium; Promaster 10; 35–50 g/m²	–	–	–	–	34.70	100 m²	**34.70**
Tennis courts; cricket squares							
British Seed Houses; A2 Lawns; 35 g/m²	–	–	–	–	19.35	100 m²	**19.35**
British Seed Houses; A5 Cricket; 35 g/m²	–	–	–	–	29.43	100 m²	**29.43**
DLF Trifolium; J Premier Green; 34–50 g/m²	–	–	–	–	68.50	100 m²	**68.50**
DLF Trifolium; J Court; 18–25 g/m²	–	–	–	–	12.75	100 m²	**12.75**
DLF Trifolium; Promaster 35; 35 g/m²	–	–	–	–	17.15	100 m²	**17.15**

Q30 SEEDING/TURFING

Item Excluding site overheads and profit	PC £	Labour hours	Labour £	Plant £	Material £	Unit	Total rate £
Amenity grassed areas; general purpose lawns							
British Seed Houses; A3 Banks; 25–50 g/m²	–	–	–	–	23.77	100 m²	**23.77**
DLF Trifolium; J Rye Fairway; 18–25 g/m²	–	–	–	–	12.82	100 m²	**12.82**
DLF Trifolium; J Court; 18–25 g/m²	–	–	–	–	12.75	100 m²	**12.75**
DLF Trifolium; Promaster 50; 25–35 g/m²	–	–	–	–	16.48	100 m²	**16.48**
DLF Trifolium; Promaster 120; 25–35 g/m²	–	–	–	–	14.98	100 m²	**14.98**
Conservation; country parks; slopes and banks							
British Seed Houses; A4 Parkland; 17–35 g/m²	–	–	–	–	17.79	100 m²	**17.79**
British Seed Houses; A16 Country Park; 8–19 g/m²	–	–	–	–	12.68	100 m²	**12.68**
British Seed Houses; A17 Legume; 2 g/m²	–	–	–	–	24.62	100 m²	**24.62**
Shaded areas							
British Seed Houses; A6 Shade; 50 g/m²	–	–	–	–	35.65	100 m²	**35.65**
DLF Trifolium; J Green; 34–50 g/m²	–	–	–	–	41.55	100 m²	**41.55**
DLF Trifolium; Promaster 60; 35–50 g/m²	–	–	–	–	28.30	100 m²	**28.30**
Sports pitches; rugby; soccer pitches							
British Seed Houses; A7 Sportsground; 20 g/m²	–	–	–	–	13.19	100 m²	**13.19**
DLF Trifolium; J Pitch; 18–30 g/m²	–	–	–	–	14.67	100 m²	**14.67**
DLF Trifolium; Promaster 70; 15–35 g/m²	–	–	–	–	13.93	100 m²	**13.93**
DLF Trifolium; Promaster 75; 15–35 g/m²	–	–	–	–	18.45	100 m²	**18.45**
DLF Trifolium; Promaster 80; 17–35 g/m²	–	–	–	–	13.58	100 m²	**13.58**
Rigby Taylor; Mascot R11 Football & Rugby; 35 g/m²	–	–	–	–	18.52	100 m²	**18.52**
Rigby Taylor; Mascot R12 General Playing Fields; 35 g/m²	–	–	–	–	28.00	100 m²	**28.00**
Outfields							
British Seed Houses; A7 Sportsground; 20 g/m²	–	–	–	–	13.19	100 m²	**13.19**
British Seed Houses; A9 Outfield; 17–35 g/m²	–	–	–	–	19.37	100 m²	**19.37**
DLF Trifolium; J Fairway; 12–25 g/m²	–	–	–	–	12.82	100 m²	**12.82**
DLF Trifolium; Promaster 40; 35 g/m²	–	–	–	–	16.14	100 m²	**16.14**
DLF Trifolium; Promaster 70; 15–35 g/m²	–	–	–	–	13.93	100 m²	**13.93**
Rigby Taylor; Mascot R4 Cricket Outfields; 35 g/m²	–	–	–	–	25.20	100 m²	**25.20**

Q30 SEEDING/TURFING

Item Excluding site overheads and profit	PC £	Labour hours	Labour £	Plant £	Material £	Unit	Total rate £
MARKET PRICES OF SEEDING MATERIALS – cont							
Market prices of grass seed – cont							
Hockey pitches							
DLF Trifolium; J Fairway; 12–25 g/m²	–	–	–	–	12.82	100 m²	**12.82**
DLF Trifolium; Promaster 70; 15–35 g/m²	–	–	–	–	13.93	100 m²	**13.93**
Rigby Taylor; Mascot R10 Cricket & Hockey Outfield; 35 g/m²	–	–	–	–	25.20	100 m²	**25.20**
Parks							
British Seed Houses; A7 Sportsground 20 g/m²	–	–	–	–	13.19	100 m²	**13.19**
British Seed Houses; A9 Outfield; 17–35 g/m²	–	–	–	–	19.37	100 m²	**19.37**
DLF Trifolium; Promaster 120; 25–35 g/m²	–	–	–	–	14.98	100 m²	**14.98**
Informal playing fields							
DLF Trifolium; J Rye Fairway; 18–25 g/m²	–	–	–	–	12.82	100 m²	**12.82**
DLF Trifolium; Promaster 45; 35 g/m²	–	–	–	–	15.68	100 m²	**15.68**
Caravan sites							
British Seed Houses; A9 Outfield; 17–35 g/m²	–	–	–	–	19.37	100 m²	**19.37**
Sports pitch re-seeding and repair							
British Seed Houses; A8 Pitch Renovator; 20–35 g/m²	–	–	–	–	18.26	100 m²	**18.26**
British Seed Houses; A20 Ryesport; 20–35 g/m²	–	–	–	–	20.90	100 m²	**20.90**
DLF Trifolium; J Pitch; 18–30 g/m²	–	–	–	–	14.67	100 m²	**14.67**
DLF Trifolium; Promaster 80; 17–35 g/m²	–	–	–	–	13.58	100 m²	**13.58**
DLF Trifolium; Promaster 81; 17–35 g/m²	–	–	–	–	15.86	100 m²	**15.86**
Rigby Taylor; Mascot R14 Premier Winter Games Renovation + ESP coating; 35 g/m²	–	–	–	–	21.84	100 m²	**21.84**
Racecourses; gallops; polo grounds; horse rides							
British Seed Houses; Racecourse; 25–30 g/m²	–	–	–	–	17.84	100 m²	**17.84**
DLF Trifolium; J Court; 18–25 g/m²	–	–	–	–	12.75	100 m²	**12.75**
DLF Trifolium; Promaster 65; 17–35 g/m²	–	–	–	–	19.00	100 m²	**19.00**
Motorway and road verges							
British Seed Houses; A18 Road Verge; 6–15 g/m²	–	–	–	–	9.14	100 m²	**9.14**
DLF Trifolium; Promaster 85; 10 g/m²	–	–	–	–	4.17	100 m²	**4.17**
DLF Trifolium; Promaster 120; 25–35 g/m²	–	–	–	–	14.98	100 m²	**14.98**

Q30 SEEDING/TURFING

Item Excluding site overheads and profit	PC £	Labour hours	Labour £	Plant £	Material £	Unit	**Total rate £**
Golf courses; tees							
British Seed Houses; A10 Tees, 35–50 g/m^2	–	–	–	–	33.52	100 m^2	**33.52**
DLF Trifolium; J Premier Fairway; 18–30 g/m^2	–	–	–	–	17.70	100 m^2	**17.70**
DLF Trifolium; J Court; 18–25 g/m^2	–	–	–	–	12.75	100 m^2	**12.75**
DLF Trifolium; Promaster 40; 35 g/m^2	–	–	–	–	16.14	100 m^2	**16.14**
DLF Trifolium; Promaster 45; 35 g/m^2	–	–	–	–	15.68	100 m^2	**15.68**
Golf courses; greens							
British Seed Houses; A11 Golf Greens; 35 g/m^2	–	–	–	–	26.45	100 m^2	**26.45**
British Seed Houses; A13 Roughs; 8 g/m^2	–	–	–	–	4.09	100 m^2	**4.09**
DLF Trifolium; J All Bent; 8 g/m^2	–	–	–	–	13.86	100 m^2	**13.86**
DLF Trifolium; J Green; 34–50 g/m^2	–	–	–	–	41.55	100 m^2	**41.55**
DLF Trifolium; Promaster 5; 35–50 g/m^2	–	–	–	–	24.70	100 m^2	**24.70**
Rigby Taylor; Mascot R1 Greenkeeper; 35 g/m^2	–	–	–	–	77.63	100 m^2	**77.63**
Golf courses; fairways							
British Seed Houses; A12 Fairways; 15–25 g/m^2	–	–	–	–	13.71	100 m^2	**13.71**
DLF Trifolium; J Rye Fairway; 18–30 g/m^2	–	–	–	–	15.39	100 m^2	**15.39**
DLF Trifolium; J Premier Fairway; 18–30 g/m^2	–	–	–	–	17.70	100 m^2	**17.70**
DLF Trifolium; Promaster 40; 35 g/m^2	–	–	–	–	16.14	100 m^2	**16.14**
DLF Trifolium; Promaster 45; 35 g/m^2	–	–	–	–	15.68	100 m^2	**15.68**
Rigby Taylor; Mascot R6 Fescue Rye Fairway; 35 g/m^2	–	–	–	–	56.00	100 m^2	**56.00**
Golf courses; roughs							
DLF Trifolium; Promaster 25; 17–35 g/m^2	–	–	–	–	14.46	100 m^2	**14.46**
Rigby Taylor; Mascot R127 Golf Links & Rough; 35 g/m^2	–	–	–	–	5.38	100 m^2	**5.38**
Waste land; spoil heaps; quarries							
British Seed Houses; A15 Reclamation; 15–20 g/m^2	–	–	–	–	12.68	100 m^2	**12.68**
DLF Trifolium; Promaster 95; 12–35 g/m^2	–	–	–	–	19.14	100 m^2	**19.14**
DLF Trifolium; Promaster 105; 5 g/m^2	–	–	–	–	5.67	100 m^2	**5.67**
Low maintenance; housing estates; amenity grassed areas							
British Seed Houses; A19 Housing; 25–35 g/m^2	–	–	–	–	17.92	100 m^2	**17.92**
British Seed Houses; A22 Low Maintenance; 25–35 g/m^2	–	–	–	–	22.18	100 m^2	**22.18**
DLF Trifolium; Promaster 120; 25–35 g/m^2	–	–	–	–	14.98	100 m^2	**14.98**

Q30 SEEDING/TURFING

Item Excluding site overheads and profit	PC £	Labour hours	Labour £	Plant £	Material £	Unit	Total rate £
MARKET PRICES OF SEEDING MATERIALS – cont							
Market prices of grass seed – cont							
Saline coastal; roadside areas							
British Seed Houses; A21 Saline; 15–20 g/m²	–	–	–	–	10.97	100 m²	**10.97**
DLF Trifolium; Promaster 90; 15–35 g /m²	–	–	–	–	15.79	100 m²	**15.79**
Turf production							
British Seed Houses; A25 Meadow Ley; 160 kg/ha	–	–	–	–	1189.58	ha	**1189.58**
British Seed Houses; A24 Wear & Tear; 185 kg/ha	–	–	–	–	994.32	ha	**994.32**
Market prices of wild flora seed mixtures							
Acid soils							
British Seed Houses; WF1 (Annual Flowering); 1–2 g/m²	–	–	–	–	20.63	100 m²	**20.63**
Neutral soils							
British Seed Houses; WF3 (Neutral Soils); 0.5–1 g/m²	–	–	–	–	10.31	100 m²	**10.31**
Market prices of wild flora and grass seed mixtures							
General purpose							
DLF Trifolium; Pro Flora 8 Old English Country Meadow Mix; 5 g/m²	–	–	–	–	12.50	100 m²	**12.50**
DLF Trifolium; Pro Flora 9 General purpose; 5 g/m²	–	–	–	–	10.00	100 m²	**10.00**
Acid soils							
British Seed Houses; WFG2 (Annual Meadow); 5 g/m²	–	–	–	–	22.95	100 m²	**22.95**
DLF Trifolium; Pro Flora 2 Acidic soils; 5 g/m²	–	–	–	–	11.50	100 m²	**11.50**
Neutral soils							
British Seed Houses; WFG4 (Neutral Meadow); 5 g/m²	–	–	–	–	25.19	100 m²	**25.19**
British Seed Houses; WFG13 (Scotland); 5 g/m²	–	–	–	–	22.24	100 m²	**22.24**
DLF Trifolium; Pro Flora 3 Damp loamy soils; 5 g/m²	–	–	–	–	15.75	100 m²	**15.75**
Calcareous soils							
British Seed Houses; WFG5 (Calcareous Soils); 5 g/m²	–	–	–	–	23.56	100 m²	**23.56**
DLF Trifolium; Pro Flora 4 Calcareous soils; 5 g/m²	–	–	–	–	13.80	100 m²	**13.80**

Q30 SEEDING/TURFING

Item Excluding site overheads and profit	PC £	Labour hours	Labour £	Plant £	Material £	Unit	Total rate £
Heavy clay soils							
British Seed Houses; WFG6 (Clay Soils); 5 g/m²	–	–	–	–	28.30	100 m²	**28.30**
British Seed Houses; WFG12 (Ireland); 5 g/m²	–	–	–	–	22.31	100 m²	**22.31**
DLF Trifolium; Pro Flora 5 Wet loamy soils; 5 g/m²	–	–	–	–	29.15	100 m²	**29.15**
Sandy soils							
British Seed Houses; WFG7 (Free Draining Soils); 5 g/m²	–	–	–	–	33.15	100 m²	**33.15**
British Seed Houses; WFG11 (Ireland); 5 g/m²	–	–	–	–	23.46	100 m²	**23.46**
British Seed Houses; WFG14 (Scotland); 5 g/m²	–	–	–	–	25.76	100 m²	**25.76**
DLF Trifolium; Pro Flora 6 Dry free draining loamy soils; 5 g/m²	–	–	–	–	29.25	100 m²	**29.25**
Shaded areas							
British Seed Houses; WFG8 (Woodland and Hedgerow); 5 g/m²	–	–	–	–	24.73	100 m²	**24.73**
DLF Trifolium; Pro Flora 7 Hedgerow and light shade; 5 g/m²	–	–	–	–	33.75	100 m²	**33.75**
Educational							
British Seed Houses; WFG15 (Schools and Colleges); 5 g/m²	–	–	–	–	34.68	100 m²	**34.68**
Wetlands							
British Seed Houses; WFG9 (Wetlands and Ponds); 5 g/m²	–	–	–	–	30.60	100 m²	**30.60**
DLF Trifolium; Pro Flora 5 Wet loamy soils; 5 g/m²	–	–	–	–	29.15	100 m²	**29.15**
Scrub and moorland							
British Seed Houses; WFG10 (Cornfield Annuals); 5 g/m²	–	–	–	–	33.71	100 m²	**33.71**
Hedgerow							
DLF Trifolium; Pro Flora 7 Hedgerow and light shade; 5 g/m²	–	–	–	–	33.75	100 m²	**33.75**
Vacant sites							
DLF Trifolium; Pro Flora 1 Cornfield annuals; 5 g/m²	–	–	–	–	30.00	100 m²	**30.00**
Regional Environmental mixes							
British Seed Houses; RE1 (Traditional Hay); 5 g/m²	–	–	–	–	25.45	100 m²	**25.45**
British Seed Houses; RE2 (Lowland Meadow); 5 g/m²	–	–	–	–	33.41	100 m²	**33.41**
British Seed Houses; RE3 (Riverflood Plain/Water Meadow); 5 g/m²	–	–	–	–	32.89	100 m²	**32.89**

Q30 SEEDING/TURFING

Item Excluding site overheads and profit	PC £	Labour hours	Labour £	Plant £	Material £	Unit	Total rate £
MARKET PRICES OF SEEDING MATERIALS – cont							
Market prices of wild flora and grass seed mixtures – cont							
Regional Environmental mixes – cont							
British Seed Houses; RE4 (Lowland Limestone); 5 g/m²	–	–	–	–	32.39	100 m²	**32.39**
British Seed Houses; RE5 (Calcareous Sub-mountain Restoration); 5 g/m²	–	–	–	–	36.82	100 m²	**36.82**
British Seed Houses; RE6 (Upland Limestone); 5 g/m²	–	–	–	–	47.38	100 m²	**47.38**
British Seed Houses; RE7 (Acid Sub-mountain Restoration); 5 g/m²	–	–	–	–	31.88	100 m²	**31.88**
British Seed Houses; RE8 (Coastal Reclamation); 5 g/m²	–	–	–	–	41.26	100 m²	**41.26**
British Seed Houses; RE9 (Farmland Mixture); 5 g/m²	–	–	–	–	28.41	100 m²	**28.41**
British Seed Houses; RE10 (Marginal Land); 5 g/m²	–	–	–	–	42.58	100 m²	**42.58**
British Seed Houses; RE11 (Heath Scrubland); 5 g/m²	–	–	–	–	13.06	100 m²	**13.06**
British Seed Houses; RE12 (Drought Land); 5 g/m²	–	–	–	–	40.29	100 m²	**40.29**
Clarification notes on labour costs in this section							
General landscape team							
Generally a three man team is used in this section; The column Labour hours reports team hours. The column Labour £ reports the total cost of the team for the unit of work shown							
3 man team	–	1.00	59.25	–	–	hr	**59.25**
CULTIVATION AND SOIL PREPARATION							
Cultivation by tractor; Agripower Ltd							
Ripping up subsoil; using approved subsoiling machine; minimum depth 250 mm below topsoil; at 1.20 m centres; in							
gravel or sandy clay	–	–	–	2.16	–	100 m²	**2.16**
soil compacted by machines	–	–	–	2.40	–	100 m²	**2.40**
clay	–	–	–	3.08	–	100 m²	**3.08**
chalk or other soft rock	–	–	–	3.59	–	100 m²	**3.59**
Extra for subsoiling at 1 m centres	–	–	–	0.62	–	100 m²	**0.62**

Q30 SEEDING/TURFING

Item Excluding site overheads and profit	PC £	Labour hours	Labour £	Plant £	Material £	Unit	Total rate £
Breaking up existing ground; using tractor drawn Blec ground preparation equipment in one operation							
Single pass							
100 mm deep	–	–	–	13.76	–	100 m²	**13.76**
150 mm deep	–	–	–	17.19	–	100 m²	**17.19**
200 mm deep	–	–	–	19.65	–	100 m²	**19.65**
Cultivating ploughed ground; using disc, drag or chain harrow							
4 passes	–	–	–	15.63	–	100 m²	**15.63**
Rolling cultivated ground lightly; using self-propelled agricultural roller	–	0.07	1.32	0.72	–	100 m²	**2.04**
Cultivation by pedestrian operated rotavator							
Breaking up existing ground; tine cultivator or rotavator							
100 mm deep	–	0.22	4.34	2.52	–	100 m²	**6.86**
150 mm deep	–	0.28	5.43	3.15	–	100 m²	**8.58**
200 mm deep	–	0.37	7.24	4.20	–	100 m²	**11.44**
As above but in heavy clay or wet soils							
100 mm deep	–	0.44	8.69	5.04	–	100 m²	**13.73**
150 mm deep	–	0.66	13.04	7.56	–	100 m²	**20.60**
200 mm deep	–	0.82	16.29	9.44	–	100 m²	**25.73**
Importing and storing selected and approved topsoil; inclusive of settlement							
small quantities (less than 15 m³)	60.00	–	–	–	60.00	m³	**60.00**
over 15 m³	33.60	–	–	–	33.60	m³	**33.60**
Spreading and lightly consolidating approved topsoil (imported or from spoil heaps); in layers not exceeding 150 mm; travel distance from spoil heaps not exceeding 100 m; by machine (imported topsoil not included)							
minimum depth 100 mm	–	1.55	30.61	45.34	–	100 m²	**75.95**
minimum depth 150 mm	–	2.33	46.08	68.23	–	100 m²	**114.31**
minimum depth 300 mm	–	4.67	92.16	136.45	–	100 m²	**228.61**
minimum depth 450 mm	–	6.99	138.05	204.46	–	100 m²	**342.51**
Spreading and lightly consolidating approved topsoil (imported or from spoil heaps); in layers not exceeding 150 mm; travel distance from spoil heaps not exceeding 100 m; by hand (imported topsoil not included)							
minimum depth 100 mm	–	20.00	395.08	–	–	100 m²	**395.08**
minimum depth 150 mm	–	30.01	592.62	–	–	100 m²	**592.62**
minimum depth 300 mm	–	60.01	1185.24	–	–	100 m²	**1185.24**
minimum depth 450 mm	–	90.02	1777.86	–	–	100 m²	**1777.86**
Extra over for spreading topsoil to slopes 15–30°; by machine or hand	–	–	–	–	–	10%	–
Extra over for spreading topsoil to slopes over 30°; by machine or hand	–	–	–	–	–	25%	–

Q30 SEEDING/TURFING

Item Excluding site overheads and profit	PC £	Labour hours	Labour £	Plant £	Material £	Unit	Total rate £
CULTIVATION AND SOIL PREPARATION – cont							
Cultivation by pedestrian operated rotavator – cont							
Extra over for spreading topsoil from spoil heaps; travel exceeding 100 m; by machine							
100–150 m	–	0.01	0.24	0.06	–	m³	**0.30**
150–200 m	–	0.02	0.37	0.10	–	m³	**0.47**
200–300 m	–	0.03	0.55	0.15	–	m³	**0.70**
Extra over spreading topsoil for travel exceeding 100 m; by hand							
100 m	–	0.83	16.46	–	–	m³	**16.46**
200 m	–	1.67	32.92	–	–	m³	**32.92**
300 m	–	2.50	49.38	–	–	m³	**49.38**
Evenly grading; to general surfaces to bring to finished levels							
by machine (tractor mounted rotavator)	–	–	–	0.06	–	m²	**0.06**
by pedestrian operated rotavator	–	–	0.08	0.05	–	m²	**0.13**
by hand	–	0.01	0.20	–	–	m²	**0.20**
Extra over grading for slopes 15–30°; by machine or hand	–	–	–	–	–	10%	**–**
Extra over grading for slopes over 30°; by machine or hand	–	–	–	–	–	25%	**–**
Apply screened topdressing to grass surfaces; spread using Tru-Lute							
sand soil mixes 90/10 to 50/50	–	–	0.04	0.03	0.14	m²	**0.21**
Spread only existing cultivated soil to final levels using Tru-Lute							
cultivated soil	–	–	0.04	0.03	–	m²	**0.07**
Clearing stones; disposing off site							
by hand; stones not exceeding 50 mm in any direction; loading to skip 4.6 m³	–	0.01	0.20	0.04	–	m²	**0.24**
by mechanical stone rake; stones not exceeding 50 mm in any direction; loading to 15 m³ truck by mechanical loader	–	–	0.04	0.19	–	m²	**0.23**
Lightly cultivating; weeding; to fallow areas; disposing debris off site							
by hand	–	0.01	0.28	–	0.09	m²	**0.37**

Q30 SEEDING/TURFING

Item Excluding site overheads and profit	PC £	Labour hours	Labour £	Plant £	Material £	Unit	Total rate £
SOIL ADDITIVES AND IMPROVEMENTS							
Surface applications; soil additives; pre-seeding; material delivered to a maximum of 25 m from area of application; applied; by machine							
Soil conditioners; to cultivated ground; mushroom compost; delivered in 25 m³ loads; including turning in							
1 m³ per 40 m² = 25 mm thick	0.59	0.02	0.33	0.16	0.59	m²	1.08
1 m³ per 20 m² = 50 mm thick	1.19	0.03	0.61	0.22	1.19	m²	2.02
1 m³ per 13.33 m² = 75 mm thick	1.78	0.04	0.79	0.29	1.78	m²	2.86
1 m³ per 10 m² = 100 mm thick	2.37	0.05	0.99	0.36	2.37	m²	3.72
Soil conditioners; to cultivated ground; mushroom compost; delivered in 35 m³ loads; including turning in							
1 m³ per 40 m² = 25 mm thick	0.47	0.02	0.33	0.16	0.47	m²	0.96
1 m³ per 20 m² = 50 mm thick	0.95	0.03	0.61	0.22	0.95	m²	1.78
1 m³ per 13.33 m² = 75 mm thick	1.43	0.04	0.79	0.29	1.43	m²	2.51
1 m³ per 10 m² = 100 mm thick	1.90	0.05	0.99	0.36	1.90	m²	3.25
Surface applications and soil additives; pre-seeding; material delivered to a maximum of 25 m from area of application; applied; by hand							
Soil conditioners; to cultivated ground; mushroom compost; delivered in 25 m³ loads; including turning in							
1 m³ per 40 m² = 25 mm thick	0.59	0.02	0.44	–	0.59	m²	1.03
1 m³ per 20 m² = 50 mm thick	1.19	0.04	0.88	–	1.19	m²	2.07
1 m³ per 13.33 m² = 75 mm thick	1.78	0.07	1.32	–	1.78	m²	3.10
1 m³ per 10 m² = 100 mm thick	2.37	0.08	1.58	–	2.37	m²	3.95
Soil conditioners; to cultivated ground; mushroom compost; delivered in 60 m³ loads; including turning in							
1 m³ per 40 m² = 25 mm thick	0.30	0.02	0.44	–	0.30	m²	0.74
1 m³ per 20 m² = 50 mm thick	0.53	0.04	0.88	–	0.53	m²	1.41
1 m³ per 13.33 m² = 75 mm thick	0.90	0.07	1.32	–	0.90	m²	2.22
1 m³ per 10 m² = 100 mm thick	1.20	0.08	1.58	–	1.20	m²	2.78
PREPARATION OF SEEDBEDS							
Preparation of seedbeds – General Preamble: For preliminary operations see 'Cultivation' section.							
Preparation of seedbeds; soil preparation							
Lifting selected and approved topsoil from spoil heaps; passing through 6 mm screen; removing debris	–	0.08	1.65	4.92	0.02	m³	6.59

Q30 SEEDING/TURFING

Item Excluding site overheads and profit	PC £	Labour hours	Labour £	Plant £	Material £	Unit	Total rate £
PREPARATION OF SEEDBEDS – cont							
Preparation of seedbeds – cont							
Topsoil; supply only; PC £28.00/m³; allowing for 20% settlement							
25 mm	–	–	–	–	0.84	m²	0.84
50 mm	–	–	–	–	1.68	m²	1.68
100 mm	–	–	–	–	3.36	m²	3.36
150 mm	–	–	–	–	5.04	m²	5.04
200 mm	–	–	–	–	6.72	m²	6.72
250 mm	–	–	–	–	8.40	m²	8.40
300 mm	–	–	–	–	10.08	m²	10.08
400 mm	–	–	–	–	13.44	m²	13.44
450 mm	–	–	–	–	15.12	m²	15.12
Spreading topsoil to form seedbeds (topsoil not included); by machine							
25 mm deep	–	–	0.05	0.12	–	m²	0.17
50 mm deep	–	–	0.07	0.15	–	m²	0.22
75 mm deep	–	–	0.08	0.18	–	m²	0.26
100 mm deep	–	0.01	0.10	0.23	–	m²	0.33
150 mm deep	–	0.01	0.15	0.35	–	m²	0.50
Spreading only topsoil to form seedbeds (topsoil not included); by hand							
25 mm deep	–	0.03	0.49	–	–	m²	0.49
50 mm deep	–	0.03	0.66	–	–	m²	0.66
75 mm deep	–	0.04	0.85	–	–	m²	0.85
100 mm deep	–	0.05	0.99	–	–	m²	0.99
150 mm deep	–	0.08	1.48	–	–	m²	1.48
Bringing existing topsoil to a fine tilth for seeding; by raking or harrowing; stones not to exceed 6 mm; by machine	–	–	0.08	0.11	–	m²	0.19
Bringing existing topsoil to a fine tilth for seeding; by raking or harrowing; stones not to exceed 6 mm; by hand	–	0.01	0.17	–	–	m²	0.17
Preparation of seedbeds; soil treatments							
For the following operations add or subtract the following amounts for every £0.10 difference in the material cost price							
35 g/m²	–	–	–	–	0.35	100 m²	0.35
50 g/m²	–	–	–	–	0.50	100 m²	0.50
70 g/m²	–	–	–	–	0.70	100 m²	0.70
100 g/m²	–	–	–	–	1.00	100 m²	1.00
125 kg/ha	–	–	–	–	12.50	ha	12.50
150 kg/ha	–	–	–	–	15.00	ha	15.00
175 kg/ha	–	–	–	–	17.50	ha	17.50
200 kg/ha	–	–	–	–	20.00	ha	20.00
225 kg/ha	–	–	–	–	22.50	ha	22.50
250 kg/ha	–	–	–	–	25.00	ha	25.00
300 kg/ha	–	–	–	–	30.00	ha	30.00

Q30 SEEDING/TURFING

Item Excluding site overheads and profit	PC £	Labour hours	Labour £	Plant £	Material £	Unit	Total rate £
350 kg/ha	–	–	–	–	35.00	ha	**35.00**
400 kg/ha	–	–	–	–	40.00	ha	**40.00**
500 kg/ha	–	–	–	–	50.00	ha	**50.00**
700 kg/ha	–	–	–	–	70.00	ha	**70.00**
1000 kg/ha	–	–	–	–	100.00	ha	**100.00**
1250 kg/ha	–	–	–	–	125.00	ha	**125.00**
Pre-seeding fertilizers (12+00+09); PC £0.68/kg; to seedbeds; by machine							
35 g/m^2	2.39	–	–	0.19	2.39	100 m^2	**2.58**
50 g/m^2	3.41	–	–	0.19	3.41	100 m^2	**3.60**
70 g/m^2	4.78	–	–	0.19	4.78	100 m^2	**4.97**
100 g/m^2	6.83	–	–	0.19	6.83	100 m^2	**7.02**
Pre-seeding fertilizers (12+00+09); PC £0.68]/kg; to seedbeds; by hand							
35 g/m^2	2.39	0.17	3.29	–	2.39	100 m^2	**5.68**
50 g/m^2	3.41	0.17	3.29	–	3.41	100 m^2	**6.70**
70 g/m^2	4.78	0.17	3.29	–	4.78	100 m^2	**8.07**
100 g/m^2	6.83	0.20	3.95	–	6.83	100 m^2	**10.78**

SEEDING

Seeding

Seeding labours only in two operations; by machine (for seed prices see above)							
35 g/m^2	–	–	–	0.51	–	100 m^2	**0.51**
Grass seed; spreading in two operations; PC £4.50/kg (for changes in material prices please refer to table above); by machine							
35 g/m^2	–	–	–	0.51	15.75	100 m^2	**16.26**
50 g/m^2	–	–	–	0.51	22.50	100 m^2	**23.01**
70 g/m^2	–	–	–	0.51	31.50	100 m^2	**32.01**
100 g/m^2	–	–	–	0.51	45.00	100 m^2	**45.51**
125 kg/ha	–	–	–	50.95	562.50	ha	**613.45**
150 kg/ha	–	–	–	50.95	675.00	ha	**725.95**
200 kg/ha	–	–	–	50.95	900.00	ha	**950.95**
250 kg/ha	–	–	–	50.95	1125.00	ha	**1175.95**
300 kg/ha	–	–	–	50.95	1350.00	ha	**1400.95**
350 kg/ha	–	–	–	50.95	1575.00	ha	**1625.95**
400 kg/ha	–	–	–	50.95	1800.00	ha	**1850.95**
500 kg/ha	–	–	–	50.95	2250.00	ha	**2300.95**
700 kg/ha	–	–	–	50.95	3150.00	ha	**3200.95**
Extra over seeding by machine for slopes over 30° (allowing for the actual area but measured in plan)							
35 g/m^2	–	–	–	0.08	2.36	100 m^2	**2.44**
50 g/m^2	–	–	–	0.08	3.38	100 m^2	**3.46**
70 g/m^2	–	–	–	0.08	4.72	100 m^2	**4.80**
100 g/m^2	–	–	–	0.08	6.75	100 m^2	**6.83**
125 kg/ha	–	–	–	7.64	84.38	ha	**92.02**
150 kg/ha	–	–	–	7.64	101.25	ha	**108.89**
200 kg/ha	–	–	–	7.64	135.00	ha	**142.64**

Q30 SEEDING/TURFING

Item Excluding site overheads and profit	PC £	Labour hours	Labour £	Plant £	Material £	Unit	Total rate £
SEEDING – cont							
Seeding – cont							
Extra over seeding by machine for slopes over 30° (allowing for the actual area but measured in plan) – cont							
250 kg/ha	–	–	–	7.64	168.75	ha	**176.39**
300 kg/ha	–	–	–	7.64	202.50	ha	**210.14**
350 kg/ha	–	–	–	7.64	236.25	ha	**243.89**
400 kg/ha	–	–	–	7.64	270.00	ha	**277.64**
500 kg/ha	–	–	–	7.64	337.50	ha	**345.14**
700 kg/ha	–	–	–	7.64	472.50	ha	**480.14**
Seeding labours only in two operations; by machine (for seed prices see above)							
35 g/m^2	–	0.17	3.29	–	–	100 m^2	**3.29**
Grass seed; spreading in two operations; PC £4.50/kg (for changes in material prices please refer to table above); by hand							
35 g/m^2	–	0.17	3.29	–	15.75	100 m^2	**19.04**
50 g/m^2	–	0.17	3.29	–	22.50	100 m^2	**25.79**
70 g/m^2	–	0.17	3.29	–	31.50	100 m^2	**34.79**
100 g/m^2	–	0.20	3.95	–	45.00	100 m^2	**48.95**
125 g/m^2	–	0.20	3.95	–	56.25	100 m^2	**60.20**
Extra over seeding by hand for slopes over 30° (allowing for the actual area but measured in plan)							
35 g/m^2	2.34	–	0.07	–	2.34	100 m^2	**2.41**
50 g/m^2	3.38	–	0.07	–	3.38	100 m^2	**3.45**
70 g/m^2	4.72	–	0.07	–	4.72	100 m^2	**4.79**
100 g/m^2	6.75	–	0.08	–	6.75	100 m^2	**6.83**
125 g/m^2	8.41	–	0.08	–	8.41	100 m^2	**8.49**
Harrowing seeded areas; light chain harrow	–	–	–	0.09	–	100 m^2	**0.09**
Raking over seeded areas							
by mechanical stone rake	–	–	–	2.18	–	100 m^2	**2.18**
by hand	–	0.80	15.80	–	–	100 m^2	**15.80**
Rolling seeded areas; light roller							
by tractor drawn roller	–	–	–	0.54	–	100 m^2	**0.54**
by pedestrian operated mechanical roller	–	0.08	1.65	0.66	–	100 m^2	**2.31**
by hand drawn roller	–	0.17	3.29	–	–	100 m^2	**3.29**
Extra over harrowing, raking or rolling seeded areas for slopes over 30°; by machine or hand	–	–	–	–	–	25%	**–**
Turf edging; to seeded areas; 300 mm wide	–	0.05	0.94	–	1.85	m^2	**2.79**

Q30 SEEDING/TURFING

Item Excluding site overheads and profit	PC £	Labour hours	Labour £	Plant £	Material £	Unit	Total rate £
PREPARATION OF TURF BEDS							
Preparation of turf beds							
Rolling turf to be lifted; lifting by hand or mechanical turf stripper; stacks to be not more than 1 m high							
cutting only preparing to lift; pedestrian turf cutter	–	0.75	14.81	9.96	–	100 m²	**24.77**
lifting and stacking; by hand	–	8.33	164.58	–	–	100 m²	**164.58**
Rolling up; moving to stacks							
distance not exceeding 100 m	–	2.50	49.38	–	–	100 m²	**49.38**
extra over rolling and moving turf to stacks to transport per additional 100 m	–	0.83	16.46	–	–	100 m²	**16.46**
Lifting selected and approved topsoil from spoil heaps							
passing through 6 mm screen; removing debris	–	0.17	3.29	9.84	–	m³	**13.13**
Extra over lifting topsoil and passing through screen for imported topsoil; plus 20% allowance for settlement	–	–	–	–	33.60	m³	**33.60**
Topsoil; PC £28.00/m³; plus 20% allowance for settlement							
25 mm deep	–	–	–	–	0.84	m²	**0.84**
50 mm deep	–	–	–	–	1.68	m²	**1.68**
100 mm deep	–	–	–	–	3.36	m²	**3.36**
150 mm deep	–	–	–	–	5.04	m²	**5.04**
200 mm deep	–	–	–	–	6.72	m²	**6.72**
250 mm deep	–	–	–	–	8.40	m²	**8.40**
300 mm deep	–	–	–	–	10.08	m²	**10.08**
400 mm deep	–	–	–	–	13.44	m²	**13.44**
450 mm deep	–	–	–	–	15.12	m²	**15.12**
Spreading topsoil to form turf beds (topsoil not included); by machine							
25 mm deep	–	–	0.05	0.12	–	m²	**0.17**
50 mm deep	–	–	0.07	0.15	–	m²	**0.22**
75 mm deep	–	–	0.08	0.18	–	m²	**0.26**
100 mm deep	–	0.01	0.10	0.23	–	m²	**0.33**
150 mm deep	–	0.01	0.15	0.35	–	m²	**0.50**
Spreading topsoil to form turf beds (topsoil not included); by hand							
25 mm deep	–	0.03	0.49	–	–	m²	**0.49**
50 mm deep	–	0.03	0.66	–	–	m²	**0.66**
75 mm deep	–	0.04	0.85	–	–	m²	**0.85**
100 mm deep	–	0.05	0.99	–	–	m²	**0.99**
150 mm deep	–	0.08	1.48	–	–	m²	**1.48**
Bringing existing topsoil to a fine tilth for turfing by raking or harrowing; stones not to exceed 6 mm; by machine	–	–	0.08	0.11	–	m²	**0.19**
Bringing existing topsoil to a fine tilth for turfing by raking or harrowing; stones not to exceed 6 mm; by hand	–	0.01	0.18	–	–	m²	**0.18**

Q30 SEEDING/TURFING

Item Excluding site overheads and profit	PC £	Labour hours	Labour £	Plant £	Material £	Unit	Total rate £
TURFING							
Turfing							
Turfing; laying only; to stretcher bond; butt joints; including providing and working from barrow plank runs where necessary to surfaces not exceeding 30° from horizontal							
specially selected lawn turves from previously lifted stockpile	–	0.08	1.48	–	–	m²	**1.48**
cultivated lawn turves; to large open areas	–	0.06	1.15	–	–	m²	**1.15**
cultivated lawn turves; to domestic or garden areas	–	0.08	1.53	–	–	m²	**1.53**
road verge quality turf	–	0.04	0.79	–	–	m²	**0.79**
Industrially grown turf; PC prices listed represent the general range of industrial turf prices for sportsfields and amenity purposes; prices will vary with quantity and site location							
Rolawn							
RB Medallion; sports fields, domestic lawns, general landscape; full loads 1720 m²	1.69	0.07	1.38	–	1.69	m²	**3.07**
RB Medallion; sports fields, domestic lawns, general landscape; part loads	1.85	0.07	1.38	–	1.85	m²	**3.23**
Tensar Ltd							
Tensar Mat 400 reinforced turf for embankments; laid to embankments 3.3 m² turves	3.85	0.11	2.17	–	3.85	m²	**6.02**
Inturf							
Inturf Masters; formal lawns, golf greens, bowling greens and low maintenance areas	2.68	0.10	1.88	–	2.68	m²	**4.56**
Inturf Classic; golf tees, surrounds, lawns, parks, general purpose landscaping areas and winter sports	2.08	0.05	1.04	–	2.08	m²	**3.12**
Inturf Ornamental; medium fine turf for ornamental lawns, fairways and green surrounds	2.18	0.05	1.07	–	2.18	m²	**3.25**
Inturf Custom Grown Turf; tailor made for any turfgrass specification	7.08	0.08	1.58	–	7.08	m²	**8.66**
RTF Rhizomatous Tall Fescue; deep rooted self repairing for our changing climate	2.68	0.08	1.58	–	2.68	m²	**4.26**
Turf laid in Big roll; Inturf							
Inturf Classic; golf tees, surrounds, lawns, parks, general purpose landscaping areas and winter sports	1.50	0.01	0.47	–	2.23	m²	**2.70**

Q30 SEEDING/TURFING

Item Excluding site overheads and profit	PC £	Labour hours	Labour £	Plant £	Material £	Unit	Total rate £
Reinforced turf; Fiberweb Advanced Turf; blended mesh elements incorporated into root zone; root zone spread and levelled over cultivated, prepared and reduced and levelled ground (not included); compacted with vibratory roller							
Terram ATS 400/B with selected turf and fertilizer 100 mm thick							
100–500 m^2	25.00	0.03	0.66	2.38	25.00	m^2	**28.04**
over 500 m^2	23.50	0.03	0.66	2.38	23.50	m^2	**26.54**
Terram ATS 400/B with selected turf and fertilizer 150 mm thick							
100–500 m^2	30.00	0.04	0.79	2.80	30.00	m^2	**33.59**
over 500 m^2	28.00	0.04	0.79	2.80	28.00	m^2	**31.59**
Terram ATS 400/B with selected turf and fertilizer 200 mm thick							
100–500 m^2	36.00	0.05	0.99	3.42	36.00	m^2	**40.41**
over 500 m^2	34.00	0.05	0.99	3.42	34.00	m^2	**38.41**
Firming turves with wooden beater	–	0.01	0.20	–	–	m^2	**0.20**
Rolling turfed areas; light roller							
by tractor with turf tyres and roller	–	–	–	0.54	–	100 m^2	**0.54**
by pedestrian operated mechanical roller	–	0.08	1.65	0.66	–	100 m^2	**2.31**
by hand drawn roller	–	0.17	3.29	–	–	100 m^2	**3.29**
Dressing with finely sifted topsoil; brushing into joints	0.04	0.05	0.99	–	0.04	m^2	**1.03**
Turfing; laying only							
to slopes over 30°; to diagonal bond (measured as plan area – add 15% to these rates for the incline area of 30 degree slopes)	–	0.12	2.37	–	–	m^2	**2.37**
Extra over laying turfing for pegging down turves							
wooden or galvanized wire pegs; 200 mm long; 2 pegs per 0.50 m^2	1.60	0.01	0.26	–	1.60	m^2	**1.86**
Wildflower meadows							
Note for this section; Prices exclude for delivery and collection of specialized machinery where used to the site							
Topsoil stripping for wild flower planting							
Stripping to subsoil layer 300 deep							
moving to stockpile	–	–	–	1.58	–	m^2	**1.58**
spreading locally	–	–	–	2.40	–	m^2	**2.40**

Q30 SEEDING/TURFING

Item Excluding site overheads and profit	PC £	Labour hours	Labour £	Plant £	Material £	Unit	Total rate £
Wildflower meadows – cont							
Soil testing for wildflower meadows Collection of samples and submission to soil laboratory for analysis and recommendation for wildflower meadow establishment							
3 nr random samples	–	–	–	–	–	nr	**500.00**
Cultivation and preparation of seedbeds for wildflower meadows Cultivate stripped, filled or existing surface prior to application of herbicides and leave fallow for seedbank eradication period;							
by tractor drawn implements including rough grading to medium tilth	–	–	–	1.38	–	100 m²	**1.38**
by hand drawn rotavator including hand raking, and hand grading, to medium tilth	–	0.72	14.22	2.52	–	100 m²	**16.74**
Seedbank eradication for wildflower meadows Treat stripped intended meadow area to eradicate inherent seedbank; allow optimum 2 years of repeat eradication by herbicide returning regularly to reapply herbicides							
Existing or reclaimed brownfield area by tractor drawn application; 4 applications	–	0.04	0.79	0.60	0.31	100 m²	**1.70**
Existing or reclaimed brownfield area by tractor drawn application; 3 applications	–	0.03	0.59	0.45	0.23	100 m²	**1.27**
Existing paddocks or agricultural area; knapsack; 4 applications	–	2.12	41.87	–	0.31	100 m²	**42.18**
Existing or reclaimed brownfield area by backpack spray application; 3 applications	–	1.59	31.40	–	0.23	100 m²	**31.63**
Subsoil filling to wildflower meadows Average thickness 300 mm; filling of poor nutrient value soil to form seedbed for wildflower area; grading to level and preparing to medium tilth							
from stockpile on site	–	–	–	27.70	–	100 m²	**27.70**
imported subsoil fill	–	–	–	27.70	0.72	100 m²	**28.42**

Q30 SEEDING/TURFING

Item Excluding site overheads and profit	PC £	Labour hours	Labour £	Plant £	Material £	Unit	Total rate £
Seeding to wildflower meadows; **Spreading of selected wildflower seed** **to manufacturers specifications; by** **tractor drawn seeder**							
General areas							
DLF Trifolium; Pro Flora 8 Old English Country Meadow Mix; 5 g/m²	–	–	–	1.20	12.50	100 m²	**13.70**
Acid soils							
British Seed Houses; WFG2 (Annual Meadow); 5 g/m²	–	–	–	1.20	22.95	100 m²	**24.15**
Neutral soils							
British Seed Houses; WFG4 (Neutral Meadow); 5 g/m²	–	–	–	1.20	25.19	100 m²	**26.39**
DLF Trifolium; Pro Flora 3 Damp loamy soils; 5 g/m²	–	–	–	1.20	15.75	100 m²	**16.95**
Heavy clay soils							
British Seed Houses; WFG6 (Clay Soils); 5 g/m²	–	–	–	1.20	28.30	100 m²	**29.50**
Seeding to wildflower meadows; **Spreading of selected wildflower seed** **to manufacturers specifications; by** **pedestrian drawn seeder**							
General areas							
DLF Trifolium; Pro Flora 8 Old English Country Meadow Mix; 5 g/m²	–	0.25	4.94	–	12.50	100 m²	**17.44**
Acid soils							
British Seed Houses; WFG2 (Annual Meadow); 5 g/m²	–	0.25	4.94	–	22.95	100 m²	**27.89**
Neutral soils							
British Seed Houses; WFG4 (Neutral Meadow); 5 g/m²	–	0.25	4.94	–	25.19	100 m²	**30.13**
DLF Trifolium; Pro Flora 3 Damp loamy soils; 5 g/m²	–	0.25	4.94	–	15.75	100 m²	**20.69**
Heavy clay soils							
British Seed Houses; WFG6 (Clay Soils); 5 g/m²	–	0.25	4.94	–	28.30	100 m²	**33.24**

Q30 SEEDING/TURFING

Item Excluding site overheads and profit	PC £	Labour hours	Labour £	Plant £	Material £	Unit	Total rate £
ARTIFICIAL TURF							
Artificial grass; Trulawn Limited Artificial lawns laid to cleared and levelled area (not included); laid to 50 mm Type 1 drainage layer on 25 mm sharp sand bed							
Trulawn Value; 16 mm roof; terraces, balconies, general use	–	–	–	–	–	m²	**35.95**
Trulawn Play; 20 mm; entry level; lawn and play areas	–	–	–	–	–	m²	**37.95**
Trulawn Continental; 20 mm realistic appearance; lawns and patios	–	–	–	–	–	m²	**33.85**
Trulawn Optimum; 26 mm; softest pile; lush green; lawns and patios	–	–	–	–	–	m²	**45.50**
Trulawn Luxury; 30 mm; deep pile; realistic lawn	–	–	–	–	–	m²	**47.50**
MAINTENANCE OF SEEDED/TURFED AREAS							
Maintenance operations (Note: the following rates apply to aftercare maintenance executed as part of a landscaping contract only) Initial cutting; to turfed areas							
20 mm high; using pedestrian guided power driven cylinder mower; including boxing off cuttings (stone picking and rolling not included)	–	0.18	3.56	0.31	–	100 m²	**3.87**
Repairing damaged grass areas scraping out; removing slurry; from							
ruts and holes; average 100 mm deep	–	0.13	2.63	–	–	m²	**2.63**
100 mm topsoil	–	0.13	2.63	–	3.36	m²	**5.99**
Repairing damaged grass areas; sowing grass seed to match existing or as specified; to individually prepared worn patches							
35 g/m²	0.18	0.01	0.20	–	0.18	m²	**0.38**
50 g/m²	0.26	0.01	0.20	–	0.26	m²	**0.46**
Sweeping leaves; disposing off site; motorized vacuum sweeper or rotary brush sweeper							
areas of maximum 2500 m² with occasional large tree and established boundary planting; 4.6 m³ (1 skip of material to be removed)	–	0.40	7.90	3.28	–	100 m²	**11.18**

Q30 SEEDING/TURFING

Item Excluding site overheads and profit	PC £	Labour hours	Labour £	Plant £	Material £	Unit	Total rate £
Leaf clearance; clearing grassed area of leaves and other extraneous debris							
Using equipment towed by tractor							
large grassed areas with perimeters of mature trees such as sports fields and amenity areas	–	0.01	0.25	0.05	–	100 m²	0.30
large grassed areas containing ornamental trees and shrub beds	–	0.03	0.49	0.07	–	100 m²	0.56
Using pedestrian operated mechanical equipment and blowers							
grassed areas with perimeters of mature trees such as sports fields and amenity areas	–	0.04	0.79	0.05	–	100 m²	0.84
grassed areas containing ornamental trees and shrub beds	–	0.10	1.98	0.13	–	100 m²	2.11
verges	–	0.07	1.32	0.09	–	100 m²	1.41
By hand							
grassed areas with perimeters of mature trees such as sports fields and amenity areas	–	0.05	0.99	0.10	–	100 m²	1.09
grassed areas containing ornamental trees and shrub beds	–	0.08	1.65	0.17	–	100 m²	1.82
verges	–	1.00	19.75	1.99	–	100 m²	21.74
Removal of arisings							
areas with perimeters of mature trees	–	0.01	0.12	0.09	0.10	100 m²	0.31
areas containing ornamental trees and shrub beds	–	0.02	0.35	0.34	0.26	100 m²	0.95
Cutting grass to specified height; per cut							
multi unit gang mower	–	0.59	11.62	19.35	–	ha	30.97
ride-on triple cylinder mower	–	0.01	0.28	0.14	–	100 m²	0.42
ride-on triple rotary mower	–	0.01	0.28	–	–	100 m²	0.28
pedestrian mower	–	0.18	3.56	0.73	–	100 m²	4.29
Cutting grass to banks; per cut							
side arm cutter bar mower	–	0.02	0.46	0.36	–	100 m²	0.82
Cutting rough grass; per cut							
power flail or scythe cutter	–	0.04	0.69	–	–	100 m²	0.69
Extra over cutting grass for slopes not exceeding 30°	–	–	–	–	–	10%	–
Extra over cutting grass for slopes exceeding 30°	–	–	–	–	–	40%	–
Cutting fine sward							
pedestrian operated seven-blade cylinder lawn mower	–	0.14	2.77	0.23	–	100 m²	3.00
Extra over cutting fine sward for boxing off cuttings							
pedestrian mower	–	0.03	0.55	0.05	–	100 m²	0.60
Cutting areas of rough grass							
scythe	–	1.00	19.75	–	–	100 m²	19.75
sickle	–	2.00	39.50	–	–	100 m²	39.50
petrol operated strimmer	–	0.30	5.93	0.42	–	100 m²	6.35

Q30 SEEDING/TURFING

Item Excluding site overheads and profit	PC £	Labour hours	Labour £	Plant £	Material £	Unit	Total rate £
MAINTENANCE OF SEEDED/TURFED **AREAS – cont**							
Maintenance operations (Note: the **following rates apply to aftercare** **maintenance executed as part of a** **landscaping contract only) – cont**							
Cutting areas of rough grass which contain trees or whips							
petrol operated strimmer	–	0.40	7.90	0.56	–	100 m²	**8.46**
Extra over cutting rough grass for on site raking up and dumping	–	0.33	6.58	–	–	100 m²	**6.58**
Trimming edge of grass areas; edging tool							
with petrol powered strimmer	–	0.13	2.63	0.19	–	100 m	**2.82**
by hand	–	0.67	13.17	–	–	100 m	**13.17**
Marking out pitches using approved line marking compound; including initial setting out and marking							
discus, hammer, javelin or shot putt area	4.68	2.00	39.50	–	4.68	nr	**44.18**
cricket square	3.12	2.00	39.50	–	3.12	nr	**42.62**
cricket boundary	10.92	8.00	158.00	–	10.92	nr	**168.92**
grass tennis court	4.68	4.00	79.00	–	4.68	nr	**83.68**
hockey pitch	15.60	8.00	158.00	–	15.60	nr	**173.60**
football pitch	15.60	8.00	158.00	–	15.60	nr	**173.60**
rugby pitch	15.60	8.00	158.00	–	15.60	nr	**173.60**
eight lane running track; 400 m	31.20	16.00	316.00	–	31.20	nr	**347.20**
Re-marking out pitches using approved line marking compound							
discus, hammer, javelin or shot putt area	3.12	0.50	9.88	–	3.12	nr	**13.00**
cricket square	3.12	0.50	9.88	–	3.12	nr	**13.00**
grass tennis court	10.92	1.00	19.75	–	10.92	nr	**30.67**
hockey pitch	10.92	1.00	19.75	–	10.92	nr	**30.67**
football pitch	10.92	1.00	19.75	–	10.92	nr	**30.67**
rugby pitch	10.92	1.00	19.75	–	10.92	nr	**30.67**
eight lane running track; 400 m	31.20	2.50	49.38	–	31.20	nr	**80.58**
Rolling grass areas; light roller							
by tractor drawn roller	–	–	–	0.54	–	100 m²	**0.54**
by pedestrian operated mechanical roller	–	0.08	1.65	0.66	–	100 m²	**2.31**
by hand drawn roller	–	0.17	3.29	–	–	100 m²	**3.29**
Aerating grass areas; to a depth of 100 mm							
using tractor-drawn aerator	–	0.06	1.15	5.04	–	100 m²	**6.19**

using pedestrian-guided motor
  powered solid or slitting tine turf
  aerator
	–	0.18	3.46	2.96	–	100 m²	**6.42**

using hollow tine aerator; including sweeping up and dumping corings	–	0.50	9.88	5.92	–	100 m²	**15.80**
using hand aerator or fork	–	1.67	32.92	–	–	100 m²	**32.92**

Q30 SEEDING/TURFING

Item Excluding site overheads and profit	PC £	Labour hours	Labour £	Plant £	Material £	Unit	Total rate £
Extra over aerating grass areas for on site sweeping up and dumping corings	–	0.17	3.29	–	–	100 m²	3.29
Switching off dew; from fine turf areas	–	0.20	3.95	–	–	100 m²	3.95
Scarifying grass areas to break up thatch; removing dead grass							
using tractor-drawn scarifier	–	0.07	1.38	0.31	–	100 m²	1.69
using self-propelled scarifier; including removing and disposing of grass on site	–	0.33	6.58	0.13	–	100 m²	6.71
Harrowing grass areas							
using drag mat	–	0.03	0.55	0.30	–	100 m²	0.85
using chain harrow	–	0.04	0.69	0.38	–	100 m²	1.07
using drag mat	–	2.80	55.31	30.25	–	ha	85.56
using chain harrow	–	3.50	69.13	37.81	–	ha	106.94
Extra for scarifying and harrowing grass areas for disposing excavated material off site; to tip; loading by machine							
slightly contaminated	–	–	–	2.06	25.00	m³	27.06
rubbish	–	–	–	2.06	25.00	m³	27.06
inert material	–	–	–	1.37	18.14	m³	19.51
For the following topsoil improvement and seeding operations add or subtract the following amounts for every £0.10 difference in the material cost price							
35 g/m²	–	–	–	–	0.35	100 m²	0.35
50 g/m²	–	–	–	–	0.50	100 m²	0.50
70 g/m²	–	–	–	–	0.70	100 m²	0.70
100 g/m²	–	–	–	–	1.00	100 m²	1.00
125 kg/ha	–	–	–	–	12.50	ha	12.50
150 kg/ha	–	–	–	–	15.00	ha	15.00
175 kg/ha	–	–	–	–	17.50	ha	17.50
200 kg/ha	–	–	–	–	20.00	ha	20.00
225 kg/ha	–	–	–	–	22.50	ha	22.50
250 kg/ha	–	–	–	–	25.00	ha	25.00
300 kg/ha	–	–	–	–	30.00	ha	30.00
350 kg/ha	–	–	–	–	35.00	ha	35.00
400 kg/ha	–	–	–	–	40.00	ha	40.00
500 kg/ha	–	–	–	–	50.00	ha	50.00
700 kg/ha	–	–	–	–	70.00	ha	70.00
1000 kg/ha	–	–	–	–	100.00	ha	100.00
1250 kg/ha	–	–	–	–	125.00	ha	125.00
Top dressing fertilizers (7+7+7); PC £1.33/kg; to seedbeds; by machine							
35 g/m²	4.64	–	–	0.19	4.64	100 m²	4.83
50 g/m²	6.63	–	–	0.19	6.63	100 m²	6.82
300 kg/ha	397.50	–	–	18.76	397.50	ha	416.26
350 kg/ha	463.75	–	–	18.76	463.75	ha	482.51
400 kg/ha	530.00	–	–	18.76	530.00	ha	548.76
500 kg/ha	662.50	–	–	30.01	662.50	ha	692.51

Q30 SEEDING/TURFING

Item Excluding site overheads and profit	PC £	Labour hours	Labour £	Plant £	Material £	Unit	Total rate £
MAINTENANCE OF SEEDED/TURFED AREAS – cont							
Maintenance operations (Note: the following rates apply to aftercare maintenance executed as part of a landscaping contract only) – cont							
Top dressing fertilizers (7+7+7);							
PC £1.33/kg; to seedbeds; by hand							
35 g/m^2	4.64	0.17	3.29	–	4.64	100 m^2	**7.93**
50 g/m^2	6.63	0.17	3.29	–	6.63	100 m^2	**9.92**
70 g/m^2	9.28	0.17	3.29	–	9.28	100 m^2	**12.57**
Watering turf; evenly; at a rate of 5 l/m^2							
using movable spray lines powering 3 nr sprinkler heads with a radius of 15 m and allowing for 60% overlap (irrigation machinery costs not included)	–	0.02	0.31	–	–	100 m^2	**0.31**
using sprinkler equipment and with sufficient water pressure to run 1 nr 15 m radius sprinkler	–	0.02	0.39	–	–	100 m^2	**0.39**
using hand-held watering equipment	–	0.25	4.94	–	–	100 m^2	**4.94**

Q31 PLANTING

Item Excluding site overheads and profit	PC £	Labour hours	Labour £	Plant £	Material £	Unit	Total rate £
PLANTING GENERAL							
Planting – General Preamble: Prices for all planting work are deemed to include carrying out planting in accordance with all good horticultural practice.							
MARKET PRICES OF MATERIALS							
Note: For market prices of landscape chemicals please see sections Q30 or Q35.							
Market prices of planting materials (Note: the rates shown generally reflect the manufacturer's recommended retail prices; trade and bulk discounts are often available on the prices shown)							
Imported topsoil; market prices; British Sugar Topsoil; sustainably sourced graded topsoil to British standards and with independent certification; rates shown allow for 20% settlement Landscape 20; BS3882:2007; sandy loam							
8 wheel delivery; 20 tonne loads (approximately 13.33 m³)	33.45	–	–	–	33.45	m³	**33.45**
articulated delivery	25.20	–	–	–	25.20	m³	**25.20**
Sports 10; for topdressing to sports pitches; application rate 40–80 t/6000 m² pitch							
8 wheel delivery; 20 tonne loads (approximately 13.33 m³)	36.75	–	–	–	36.75	m³	**36.75**
articulated delivery	29.25	–	–	–	29.25	m³	**29.25**
Topsoil prices; Average locally sourced prices Topsoil							
general purpose	28.00	–	–	–	28.00	m³	**28.00**
Topsoil enhanced with organic matter	42.00	–	–	–	42.00	m³	**42.00**

Q31 PLANTING

Item Excluding site overheads and profit	PC £	Labour hours	Labour £	Plant £	Material £	Unit	Total rate £
MARKET PRICES OF MATERIALS – **cont**							
Market prices of mulching materials							
Melcourt Industries Ltd; 25 m³ loads							
(items labelled FSC are Forest							
Stewardship Council certified, items							
marked FT are certified fire tested)							
Ornamental Bark Mulch FT	–	–	–	–	58.40	m³	**58.40**
Bark Nuggets® FT	–	–	–	–	54.70	m³	**54.70**
Amenity Bark Mulch FSC FT	–	–	–	–	39.65	m³	**39.65**
Contract Bark Mulch FSC	–	–	–	–	36.90	m³	**36.90**
Spruce Ornamental FSC FT	–	–	–	–	40.40	m³	**40.40**
Decorative Biomulch®	–	–	–	–	39.55	m³	**39.55**
Rustic Biomulch®	–	–	–	–	43.30	m³	**43.30**
Mulch 2000	–	–	–	–	29.70	m³	**29.70**
Forest BioMulch®	–	–	–	–	37.05	m³	**37.05**
Melcourt Industries Ltd; 50 m³ loads							
Mulch 2000	–	–	–	–	18.95	m³	**18.95**
Melcourt Industries Ltd; 70 m³ loads							
Contract Bark Mulch FSC	–	–	–	–	24.50	m³	**24.50**
Melcourt Industries Ltd; 80 m³ loads							
Ornamental Bark Mulch FT	–	–	–	–	45.40	m³	**45.40**
Bark Nuggets® FT	–	–	–	–	41.70	m³	**41.70**
Amenity Bark Mulch FSC	–	–	–	–	26.65	m³	**26.65**
Spruce Ornamental FSC FT	–	–	–	–	27.40	m³	**27.40**
Decorative Biomulch®	–	–	–	–	26.55	m³	**26.55**
Rustic Biomulch®	–	–	–	–	30.30	m³	**30.30**
Forest BioMulch®	–	–	–	–	24.05	m³	**24.05**
Mulch; Melcourt Industries Ltd; 25 m³ loads							
Composted Fine Bark FSC	–	–	–	–	36.65	m³	**36.65**
Humus 2000	–	–	–	–	30.35	m³	**30.35**
Spent Mushroom Compost	–	–	–	–	23.70	m³	**23.70**
Topgrow	–	–	–	–	36.05	m³	**36.05**
Mulch; Melcourt Industries Ltd; 50 m³ loads							
Humus 2000	–	–	–	–	18.85	m³	**18.85**
Mulch; Melcourt Industries Ltd; 60 m³ loads							
Spent Mushroom Compost	–	–	–	–	12.00	m³	**12.00**
Mulch; Melcourt Industries Ltd; 65 m³ loads							
Composted Fine Bark	–	–	–	–	24.15	m³	**24.15**
Topgrow	–	–	–	–	28.05	m³	**28.05**
Market prices of fertilizers							
Fertilizers; British Seed Houses							
BSH1; 6+9+6	27.79	–	–	–	27.79	kg	**27.79**
BSH4; 11+5+5	27.30	–	–	–	27.30	kg	**27.30**
BSH6; 20+10+10	30.23	–	–	–	30.23	kg	**30.23**
BSH9; 12+2+9+Fe	28.62	–	–	–	28.62	kg	**28.62**

Q31 PLANTING

Item Excluding site overheads and profit	PC £	Labour hours	Labour £	Plant £	Material £	Unit	Total rate £
Fertilizers; Everris							
Enmag; controlled release fertilizer							
(8–9 months); 11+22+09; 70 g/m^2	–	–	–	–	15.76	100 m^2	**15.76**
Fertilizers; Everris; granular Osmocote							
Flora; 15+09+11; controlled release							
fertilizer; costs for recommended							
application rates							
transplant	–	–	–	–	0.09	nr	**0.09**
whip	–	–	–	–	0.14	nr	**0.14**
feathered	–	–	–	–	0.18	nr	**0.18**
light standard	–	–	–	–	0.18	nr	**0.18**
standard	–	–	–	–	0.23	nr	**0.23**
selected standard	–	–	–	–	0.32	nr	**0.32**
heavy standard	–	–	–	–	0.37	nr	**0.37**
extra heavy standard	–	–	–	–	0.46	nr	**0.46**
16–18 cm girth	–	–	–	–	0.50	nr	**0.50**
18–20 cm girth	–	–	–	–	0.55	nr	**0.55**
20–22 cm girth	–	–	–	–	0.64	nr	**0.64**
22–24 cm girth	–	–	–	–	0.68	nr	**0.68**
24–26 cm girth	–	–	–	–	0.73	nr	**0.73**
Fertilizers; Farmura Environmental Ltd;							
Seanure Soilbuilder							
soil amelioration; 70 g/m^2	–	–	–	–	10.16	100 m^2	**10.16**
to plant pits; 300 × 300 × 300 mm	–	–	–	–	0.04	nr	**0.04**
to plant pits; 600 × 600 × 600 mm	–	–	–	–	0.47	nr	**0.47**
to tree pits; 1.00 × 1.00 × 1.00	–	–	–	–	2.18	nr	**2.18**
Fertilizers; Everris							
TPMC tree planting and mulching							
compost	–	–	–	–	5.62	bag	**5.62**
Fertilizers; Rigby Taylor Ltd; fertilizer							
application rates 35 g/m^2 unless							
otherwise shown							
liquid fertilizer; Mascot Microflow-C;							
25+0+0 + trace elements;							
200–1400 ml/100 m^2	–	–	–	–	7.14	100 m^2	**7.14**
liquid fertilizer; Mascot Microflow-CX;							
4+2+18 + trace elements;							
200–1400 ml/100 m^2	–	–	–	–	6.88	100 m^2	**6.88**
liquid fertilizer; Mascot Microflow-CX;							
12+0+8 + trace elements;							
200–1400 ml/100 m^2	–	–	–	–	5.93	100 m^2	**5.93**
liquid fertilizer; Mascot Microflow-CX;							
17+2+5 + trace elements;							
200–1400 ml/100 m^2	–	–	–	–	6.66	100 m^2	**6.66**
Wetting agents; Rigby Taylor Ltd							
wetting agent; Breaker Advance							
Granules; 20 kg	–	–	–	–	16.38	100 m^2	**16.38**

Q31 PLANTING

Item Excluding site overheads and profit	PC £	Labour hours	Labour £	Plant £	Material £	Unit	Total rate £
Clarification notes on labour costs in this section							
General landscape team Generally a three man team is used in this section; The column Labour hours reports team hours. The column Labour £ reports the total cost of the team for the unit of work shown							
3 man team	–	1.00	59.25	–	–	hr	**59.25**
PLANTING PREPARATION AND CULTIVATION							
Site protection; temporary protective fencing Cleft chestnut rolled fencing; to 100 mm dia. chestnut posts; driving into firm ground at 3 m centres; pales at 50 mm centres							
900 mm high	4.06	0.11	2.11	–	4.80	m	**6.91**
1100 mm high	4.95	0.11	2.11	–	5.69	m	**7.80**
1500 mm high	6.40	0.11	2.11	–	7.14	m	**9.25**
Extra over temporary protective fencing for removing and making good (no allowance for reuse of material)	–	0.07	1.32	0.20	–	m	**1.52**
Cultivation by tractor; Agripower Ltd; Specialist subcontract prepared turf and sports area preparation Ripping up subsoil; using approved subsoiling machine; minimum depth 250 mm below topsoil; at 1.20 m centres; in							
gravel or sandy clay	–	–	–	2.16	–	100 m²	**2.16**
soil compacted by machines	–	–	–	2.40	–	100 m²	**2.40**
clay	–	–	–	3.08	–	100 m²	**3.08**
chalk or other soft rock	–	–	–	3.59	–	100 m²	**3.59**
Extra for subsoiling at 1 m centres	–	–	–	0.62	–	100 m²	**0.62**
Breaking up existing ground; using tractor drawn tine cultivator or rotavator with stone burier Single pass							
100 mm deep	–	–	–	1.38	–	100 m²	**1.38**
150 mm deep	–	–	–	1.72	–	100 m²	**1.72**
200 mm deep	–	–	–	2.29	–	100 m²	**2.29**

Q31 PLANTING

Item Excluding site overheads and profit	PC £	Labour hours	Labour £	Plant £	Material £	Unit	Total rate £
Cultivating ploughed ground; using disc, drag or chain harrow							
4 passes	–	–	–	–	–	100 m²	-
Rolling cultivated ground lightly; using self-propelled pedestrian agricultural roller	–	0.10	1.98	0.53	–	100 m²	2.51
Cultivation; pedestrian operated rotavator							
Breaking up existing ground; tine cultivator or rotavator							
100 mm deep	–	0.22	4.34	2.52	–	100 m²	6.86
150 mm deep	–	0.28	5.43	3.15	–	100 m²	8.58
200 mm deep	–	0.37	7.24	4.20	–	100 m²	11.44
As above but in heavy clay or wet soils							
100 mm deep	–	0.44	8.69	5.04	–	100 m²	13.73
150 mm deep	–	0.66	13.04	7.56	–	100 m²	20.60
200 mm deep	–	0.82	16.29	9.44	–	100 m²	25.73
Clearing stones; disposing off site							
by hand; stones not exceeding 50 mm in any direction; loading to skip 4.6 m³	–	0.01	0.20	0.04	–	m²	0.24
by mechanical stone rake; stones not exceeding 50 mm in any direction; loading to 15 m³ truck by mechanical loader	–	–	0.04	0.19	–	m²	0.23
Lightly cultivating; weeding; to fallow areas; disposing debris off site							
by hand	–	0.01	0.28	–	0.09	m²	0.37
TOPSOIL FILLING							
Market prices of topsoil; price shown includes a 20% settlement factor							
topsoil to BS3882	28.00	–	–	–	61.60	m³	61.60
blended topsoil	–	–	–	–	50.40	m³	50.40
small quantities (less than 15 m³)	60.00	–	–	–	60.00	m³	60.00
Spreading and lightly consolidating approved topsoil (imported or from spoil heaps); in layers not exceeding 150 mm; travel distance from spoil heaps not exceeding 100 m; by machine (imported topsoil not included)							
minimum depth 100 mm	–	1.55	30.61	45.34	–	100 m²	75.95
minimum depth 150 mm	–	2.33	46.08	68.23	–	100 m²	114.31
minimum depth 300 mm	–	4.67	92.16	136.45	–	100 m²	228.61
minimum depth 450 mm	–	6.99	138.05	204.46	–	100 m²	342.51

Q31 PLANTING

Item Excluding site overheads and profit	PC £	Labour hours	Labour £	Plant £	Material £	Unit	Total rate £
TOPSOIL FILLING – cont							
Spreading and lightly consolidating approved topsoil (imported or from spoil heaps); in layers not exceeding 150 mm; travel distance from spoil heaps not exceeding 100 m; by hand (imported topsoil not included)							
minimum depth 100 mm	–	20.00	395.08	–	–	100 m²	395.08
minimum depth 150 mm	–	30.01	592.62	–	–	100 m²	592.62
minimum depth 300 mm	–	60.01	1185.24	–	–	100 m²	1185.24
minimum depth 450 mm	–	90.02	1777.86	–	–	100 m²	1777.86
Spreading topsoil; from dump not exceeding 100 m distance; by machine (topsoil not included)							
at 1 m³ per 13 m²; 75 mm thick	–	0.56	10.97	33.86	–	100 m²	44.83
at 1 m³ per 10 m²; 100 mm thick	–	0.89	17.55	34.44	–	100 m²	51.99
at 1 m³ per 6.50 m²; 150 mm thick	–	1.11	21.92	60.77	–	100 m²	82.69
at 1 m³ per 5 m²; 200 mm thick	–	1.48	29.23	80.78	–	100 m²	110.01
Spreading topsoil; from dump not exceeding 100 m distance; by hand (topsoil not included)							
at 1 m³ per 13 m²; 75 mm thick	–	15.00	296.31	–	–	100 m²	296.31
at 1 m³ per 10 m²; 100 mm thick	–	20.00	395.08	–	–	100 m²	395.08
at 1 m³ per 6.50 m²; 150 mm thick	–	30.01	592.62	–	–	100 m²	592.62
at 1 m³ per 5 m²; 200 mm thick	–	36.67	724.31	–	–	100 m²	724.31
Imported topsoil; tipped 100 m from area of application; by machine							
at 1 m³ per 13 m²; 75 mm thick	252.00	0.56	10.97	33.86	252.00	100 m²	296.83
at 1 m³ per 10 m²; 100 mm thick	336.00	0.89	17.55	34.44	336.00	100 m²	387.99
at 1 m³ per 6.50 m²; 150 mm thick	504.00	1.11	21.92	60.77	504.00	100 m²	586.69
at 1 m³ per 5 m²; 200 mm thick	672.00	1.48	29.23	80.78	672.00	100 m²	782.01
Imported topsoil; tipped 100 m from area of application; by hand							
at 1 m³ per 13 m²; 75 mm thick	252.00	15.00	296.31	–	252.00	100 m²	548.31
at 1 m³ per 10 m²; 100 mm thick	336.00	20.00	395.08	–	336.00	100 m²	731.08
at 1 m³ per 6.50 m²; 150 mm thick	504.00	30.01	592.62	–	504.00	100 m²	1096.62
at 1 m³ per 5 m²; 200 mm thick	672.00	36.67	724.31	–	672.00	100 m²	1396.31
Extra over for spreading topsoil to slopes 15–30°; by machine or hand	–	–	–	–	–	10%	–
Extra over for spreading topsoil to slopes over 30°; by machine or hand	–	–	–	–	–	25%	–
Extra over for spreading topsoil from spoil heaps; travel exceeding 100 m; by machine							
100–150 m	–	0.01	0.24	0.06	–	m³	0.30
150–200 m	–	0.02	0.37	0.10	–	m³	0.47
200–300 m	–	0.03	0.55	0.15	–	m³	0.70

Q31 PLANTING

Item Excluding site overheads and profit	PC £	Labour hours	Labour £	Plant £	Material £	Unit	Total rate £
Extra over spreading topsoil for travel exceeding 100 m; by hand							
100 m	–	0.83	16.46	–	–	m³	**16.46**
200 m	–	1.67	32.92	–	–	m³	**32.92**
300 m	–	2.50	49.38	–	–	m³	**49.38**
Evenly grading; to general surfaces to bring to finished levels							
by machine (tractor mounted rotavator)	–	–	–	0.07	–	m²	**0.07**
by pedestrian operated rotavator	–	–	0.08	0.05	–	m²	**0.13**
by hand	–	0.01	0.20	–	–	m²	**0.20**
Extra over grading for slopes 15–30°; by machine or hand	–	–	–	–	–	10%	**–**
Extra over grading for slopes over 30°; by machine or hand	–	–	–	–	–	25%	**–**
Apply screened topdressing to grass surfaces; spread using Tru-Lute							
sand soil mixes 90/10 to 50/50	–	–	0.04	0.03	0.14	m²	**0.21**
Spread only existing cultivated soil to final levels using Tru-Lute							
cultivated soil	–	–	0.04	0.03	–	m²	**0.07**
SOIL IMPROVEMENTS							
Preparation of planting operations							
For the following topsoil improvement and planting operations add or subtract the following amounts for every £0.10 difference in the material cost price							
35 g/m²	–	–	–	–	0.35	100 m²	**0.35**
50 g/m²	–	–	–	–	0.50	100 m²	**0.50**
70 g/m²	–	–	–	–	0.70	100 m²	**0.70**
100 g/m²	–	–	–	–	1.00	100 m²	**1.00**
150 kg/ha	–	–	–	–	15.00	ha	**15.00**
200 kg/ha	–	–	–	–	20.00	ha	**20.00**
250 kg/ha	–	–	–	–	25.00	ha	**25.00**
300 kg/ha	–	–	–	–	30.00	ha	**30.00**
400 kg/ha	–	–	–	–	40.00	ha	**40.00**
500 kg/ha	–	–	–	–	50.00	ha	**50.00**
700 kg/ha	–	–	–	–	70.00	ha	**70.00**
1000 kg/ha	–	–	–	–	100.00	ha	**100.00**
1250 kg/ha	–	–	–	–	125.00	ha	**125.00**
Fertilizers; in top 150 mm of topsoil; at 35 g/m²							
Mascot Microfine; controlled release turf fertilizer; 8+0+6 + 2% Mg + 4% Fe	0.05	–	0.03	–	0.05	m²	**0.08**
Enmag; controlled release fertilizer (8–9 months); 11+22+09	0.08	–	0.03	–	0.08	m²	**0.11**
Mascot Outfield; turf fertilizer; 4+10 +10	0.04	–	0.03	–	0.04	m²	**0.07**
Mascot Outfield; turf fertilizer; 9+5+5	0.04	–	0.03	–	0.04	m²	**0.07**

Q31 PLANTING

Item Excluding site overheads and profit	PC £	Labour hours	Labour £	Plant £	Material £	Unit	Total rate £
SOIL IMPROVEMENTS – cont							
Preparation of planting operations – cont							
Fertilizers; in top 150 mm of topsoil at 70 g/m^2							
Mascot Microfine; controlled release turf fertilizer; 8+0+6 + 2% Mg + 4% Fe	0.11	–	0.03	–	0.11	m^2	0.14
Enmag; controlled release fertilizer (8–9 months); 11+22+09	0.17	–	0.03	–	0.17	m^2	0.20
Mascot Outfield; turf fertilizer; 4+10+10	0.09	–	0.03	–	0.09	m^2	0.12
Mascot Outfield; turf fertilizer; 9+5+5	0.07	–	0.03	–	0.07	m^2	0.10
'Topgrow'; Melcourt; peat free tree and shrub compost (25 m^3 loads); placing on beds by mechanical loader; spreading and rotavating into topsoil by tractor drawn machine							
50 mm thick	180.25	–	–	14.44	180.25	100 m^2	194.69
100 mm thick	378.52	–	–	19.75	378.52	100 m^2	398.27
150 mm thick	567.79	–	–	25.28	567.79	100 m^2	593.07
200 mm thick	757.05	–	–	30.69	757.05	100 m^2	787.74
'Topgrow'; Melcourt; peat free tree and shrub compost (65 m^3 loads); placing on beds by mechanical loader; spreading and rotavating into topsoil by tractor drawn machine							
50 mm thick	140.25	–	–	14.44	140.25	100 m^2	154.69
100 mm thick	294.52	–	–	19.75	294.52	100 m^2	314.27
150 mm thick	441.79	–	–	25.28	441.79	100 m^2	467.07
200 mm thick	589.05	–	–	30.69	589.05	100 m^2	619.74
Mushroom compost; Melcourt; (25 m^3 loads); from not further than 25 m from location; cultivating into topsoil by pedestrian drawn machine							
50 mm thick	118.50	2.86	56.43	2.10	118.50	100 m^2	177.03
100 mm thick	237.00	6.05	119.47	2.10	237.00	100 m^2	358.57
150 mm thick	355.50	8.90	175.87	2.10	355.50	100 m^2	533.47
200 mm thick	474.00	12.90	254.87	2.10	474.00	100 m^2	730.97
Mushroom compost; Melcourt (25 m^3 loads); placing on beds by mechanical loader; spreading and rotavating into topsoil by tractor drawn cultivator							
50 mm thick	118.50	–	–	14.44	118.50	100 m^2	132.94
100 mm thick	237.00	–	–	19.75	237.00	100 m^2	256.75
150 mm thick	355.50	–	–	25.28	355.50	100 m^2	380.78
200 mm thick	474.00	–	–	30.69	474.00	100 m^2	504.69
Manure (60 m^3 loads); from not further than 25 m from location; cultivating into topsoil by pedestrian drawn machine							
50 mm thick	114.90	2.86	56.43	2.10	114.90	100 m^2	173.43
100 mm thick	241.29	6.05	119.47	2.10	241.29	100 m^2	362.86
150 mm thick	361.94	8.90	175.87	2.10	361.94	100 m^2	539.91
200 mm thick	482.58	12.90	254.87	2.10	482.58	100 m^2	739.55

Q31 PLANTING

Item Excluding site overheads and profit	PC £	Labour hours	Labour £	Plant £	Material £	Unit	Total rate £
Surface applications and soil additives; pre-planting; from not further than 25 m from location; by machine							
Medium bark soil conditioner; Melcourt Ltd; including turning in to cultivated ground; delivered in 25 m³ loads							
1 m³ per 40 m² = 25 mm thick	0.92	–	–	0.16	0.92	m²	**1.08**
1 m³ per 20 m² = 50 mm thick	1.83	–	–	0.22	1.83	m²	**2.05**
1 m³ per 13.33 m² = 75 mm thick	2.75	–	–	0.29	2.75	m²	**3.04**
1 m³ per 10 m² = 100 mm thick	3.67	–	–	0.36	3.67	m²	**4.03**
Works by hand; surface applications and soil additives; pre-planting; from not further than 25 m from location; by hand							
Mushroom compost soil conditioner; Melcourt Industries Ltd; including turning in to cultivated ground; delivered in 25 m³ loads							
1 m³ per 40 m² = 25 mm thick	0.59	0.02	0.44	–	0.59	m²	**1.03**
1 m³ per 20 m² = 50 mm thick	1.19	0.04	0.88	–	1.19	m²	**2.07**
1 m³ per 13.33 m² = 75 mm thick	1.78	0.07	1.32	–	1.78	m²	**3.10**
1 m³ per 10 m² = 100 mm thick	2.37	0.08	1.58	–	2.37	m²	**3.95**
Medium bark soil conditioner; Melcourt Ltd Ltd; including turning in to cultivated ground; delivered in 25 m³ loads							
1 m³ per 40 m² = 25 mm thick	0.92	0.02	0.44	–	0.92	m²	**1.36**
1 m³ per 20 m² = 50 mm thick	1.83	0.04	0.88	–	1.83	m²	**2.71**
1 m³ per 13.33 m² = 75 mm thick	2.75	0.07	1.32	–	2.75	m²	**4.07**
1 m³ per 10 m² = 100 mm thick	3.67	0.08	1.58	–	3.67	m²	**5.25**
Manure; 20 m³ loads; delivered not further than 25 m from location; cultivating into topsoil by pedestrian operated machine							
50 mm thick	225.00	2.86	56.43	2.10	225.00	100 m²	**283.53**
100 mm thick	241.29	6.05	119.47	2.10	241.29	100 m²	**362.86**
150 mm thick	361.94	8.90	175.87	2.10	361.94	100 m²	**539.91**
200 mm thick	482.58	12.90	254.87	2.10	482.58	100 m²	**739.55**
Fertilizer (7+7+7); PC £1.33/kg; to beds; by hand							
35 g/m²	4.64	0.17	3.29	–	4.64	100 m²	**7.93**
50 g/m²	6.63	0.17	3.29	–	6.63	100 m²	**9.92**
70 g/m²	9.28	0.17	3.29	–	9.28	100 m²	**12.57**
Fertilizers; Enmag; PC £2.25/kg; controlled release fertilizer; to beds; by hand							
35 g/m²	7.88	0.17	3.29	–	7.88	100 m²	**11.17**
50 g/m²	11.26	0.17	3.29	–	11.26	100 m²	**14.55**
70 g/m²	16.88	0.17	3.29	–	16.88	100 m²	**20.17**
Note: For machine incorporation of fertilizers and soil conditioners see 'Cultivation'.							

Prices for Measured Works

Q31 PLANTING

Item Excluding site overheads and profit	PC £	Labour hours	Labour £	Plant £	Material £	Unit	Total rate £
TREE PLANTING							
Tree planting; pre-planting operations							
Excavating tree pits; depositing soil alongside pits; by machine							
600 × 600 × 600 mm deep	–	0.15	2.91	0.66	–	nr	3.57
900 × 900 × 600 mm deep	–	0.33	6.52	1.49	–	nr	8.01
1.00 × 1.00 × 600 mm deep	–	0.61	12.12	1.84	–	nr	13.96
1.25 × 1.25 × 600 mm deep	–	0.96	18.97	2.88	–	nr	21.85
1.00 × 1.00 × 1.00 deep	–	1.02	20.20	3.07	–	nr	23.27
1.50 × 1.50 × 750 mm deep	–	1.73	34.09	5.18	–	nr	39.27
1.50 × 1.50 × 1.00 deep	–	2.30	45.33	6.89	–	nr	52.22
1.75 × 1.75 × 1.00 deep	–	3.13	61.86	9.40	–	nr	71.26
2.00 × 2.00 × 1.00 deep	–	4.09	80.79	12.27	–	nr	93.06
Excavating tree pits; depositing soil alongside pits; by hand							
600 × 600 × 600 mm deep	–	0.44	8.69	–	–	nr	8.69
900 × 900 × 600 mm deep	–	1.00	19.75	–	–	nr	19.75
1.00 × 1.00 × 600 mm deep	–	1.13	22.22	–	–	nr	22.22
1.25 × 1.25 × 600 mm deep	–	1.93	38.12	–	–	nr	38.12
1.00 × 1.00 × 1.00 deep	–	2.06	40.69	–	–	nr	40.69
1.50 × 1.50 × 750 mm deep	–	3.47	68.53	–	–	nr	68.53
1.75 × 1.50 × 750 mm deep	–	4.05	79.99	–	–	nr	79.99
1.50 × 1.50 × 1.00 deep	–	4.63	91.44	–	–	nr	91.44
2.00 × 2.00 × 750 mm deep	–	6.17	121.86	–	–	nr	121.86
2.00 × 2.00 × 1.00 deep	–	8.23	162.55	–	–	nr	162.55
Breaking up subsoil in tree pits; to a depth of 200 mm	–	0.03	0.66	–	–	m²	0.66
Spreading and lightly consolidating approved topsoil (imported or from spoil heaps); in layers not exceeding 150 mm; distance from spoil heaps not exceeding 100 m (imported topsoil not included); by machine							
minimum depth 100 mm	–	1.55	30.61	45.34	–	100 m²	75.95
minimum depth 150 mm	–	2.33	46.08	68.23	–	100 m²	114.31
minimum depth 300 mm	–	4.67	92.16	136.45	–	100 m²	228.61
minimum depth 450 mm	–	6.99	138.05	204.46	–	100 m²	342.51
Spreading and lightly consolidating approved topsoil (imported or from spoil heaps); in layers not exceeding 150 mm; distance from spoil heaps not exceeding 100 m (imported topsoil not included); by hand							
minimum depth 100 mm	–	20.00	395.08	–	–	100 m²	395.08
minimum depth 150 mm	–	30.01	592.62	–	–	100 m²	592.62
minimum depth 300 mm	–	60.01	1185.24	–	–	100 m²	1185.24
minimum depth 450 mm	–	90.02	1777.86	–	–	100 m²	1777.86

Q31 PLANTING

Item Excluding site overheads and profit	PC £	Labour hours	Labour £	Plant £	Material £	Unit	Total rate £
Extra for filling tree pits with imported topsoil; PC £28.00/m³; plus allowance for 20% settlement							
depth 100 mm	–	–	–	–	4.03	m²	4.03
depth 150 mm	–	–	–	–	6.05	m²	6.05
depth 200 mm	–	–	–	–	8.06	m²	8.06
depth 300 mm	–	–	–	–	12.10	m²	12.10
depth 400 mm	–	–	–	–	16.13	m²	16.13
depth 450 mm	–	–	–	–	18.14	m²	18.14
depth 500 mm	–	–	–	–	20.16	m²	20.16
depth 600 mm	–	–	–	–	24.19	m²	24.19
depth 750 mm	–	–	–	–	25.20	m²	25.20
depth 1.00 m	–	–	–	–	33.60	m²	33.60
Backfilling of tree pits with imported topsoil; compacting in layers							
depth 500 mm	–	–	–	6.71	16.80	m²	23.51
depth 600 mm	–	–	–	7.67	20.16	m²	27.83
depth 750 mm	–	–	–	8.94	25.20	m²	34.14
depth 1.00 m	–	–	–	13.41	33.60	m²	47.01
Add or deduct the following amounts for every £1.00 change in the material price of topsoil							
depth 100 mm	–	–	–	–	0.12	m²	0.12
depth 150 mm	–	–	–	–	0.18	m²	0.18
depth 200 mm	–	–	–	–	0.24	m²	0.24
depth 300 mm	–	–	–	–	0.36	m²	0.36
depth 400 mm	–	–	–	–	0.48	m²	0.48
depth 450 mm	–	–	–	–	0.54	m²	0.54
depth 500 mm	–	–	–	–	0.60	m²	0.60
depth 600 mm	–	–	–	–	0.72	m²	0.72
Structural soils (Amsterdam tree sand); Heicom; non compressive soil mixture for tree planting in areas to receive compressive surface treatments							
Backfilling and lightly compacting in layers; excavation, disposal, moving of material from delivery position and surface treatments not included; by machine							
individual treepits (29 t loads)	49.50	0.50	9.88	2.25	49.50	m³	61.63
individual treepits (20 t loads)	57.75	0.50	9.88	2.25	57.75	m³	69.88
in trenches (29 t loads)	49.50	0.42	8.23	1.88	49.50	m³	59.61
Backfilling and lightly compacting in layers; excavation, disposal, moving of material from delivery position and surface treatments not included; by hand							
individual treepits (29 t loads)	49.50	1.60	31.60	–	49.50	m³	81.10
individual treepits (20 t loads)	57.75	1.60	31.60	–	57.75	m³	89.35
in trenches	49.50	1.33	26.27	–	49.50	m³	75.77

Q31 PLANTING

Item Excluding site overheads and profit	PC £	Labour hours	Labour £	Plant £	Material £	Unit	Total rate £
TREE PLANTING – cont							
Tree staking							
J Toms Ltd; extra over trees for tree stake(s); driving 500 mm into firm ground; trimming to approved height; including two tree ties to approved pattern							
one stake; 1.52 m long × 32 × 32 mm	1.52	0.20	3.95	–	1.52	nr	**5.47**
two stakes; 1.21 m long × 25 × 25 mm	1.64	0.30	5.92	–	1.64	nr	**7.56**
two stakes; 1.52 m long × 32 × 32 mm	2.24	0.30	5.92	–	2.24	nr	**8.16**
three stakes; 1.52 m long × 32 × 32 mm	3.36	0.36	7.11	–	3.36	nr	**10.47**
Tree anchors							
Platipus Anchors Ltd; extra over trees for tree anchors							
RF1P rootball kit; for 75–220 mm girth × 2–4.5 m high; inclusive of Plati-Mat PM1	29.67	1.00	19.75	–	29.67	nr	**49.42**
RF2P; rootball kit; for 220–450 mm girth × 4.5–7.5 m high; inclusive of Plati-Mat PM2	50.51	1.33	26.27	–	50.51	nr	**76.78**
RF3P; rootball kit; for 450–750 mm girth × 7.5 to 12 m high; inclusive of Plati-Mat PM3	106.92	1.50	29.63	–	106.92	nr	**136.55**
CG1; guy fixing kit; 75–220 mm girth × 2–4.5 m high	19.08	1.67	32.92	–	19.08	nr	**52.00**
CG2; guy fixing kit; 220–450 mm girth × 4.5–7.5 m high	33.56	2.00	39.50	–	33.56	nr	**73.06**
installation tools; drive rod for RF1/ CG1 kits	–	–	–	–	61.85	nr	**61.85**
installation tools; drive rod for RF2/ CG2 kits	–	–	–	–	93.20	nr	**93.20**
Extra over trees for land drain to tree pits; 100 mm dia. perforated flexible agricultural drain; including excavating drain trench; laying pipe; backfilling	1.26	1.00	19.75	–	1.26	m	**21.01**
Platipus anchors; deadman anchoring							
Deadman kits; anchoring to concrete kerbs placed in base of tree pit							
trees 12–25 cm girth; 2.5–4.0 m high	26.62	0.75	14.81	–	42.70	nr	**57.51**
trees 25–45 cm girth; 4.0–7.5 m high	45.82	1.50	29.63	–	61.90	nr	**91.53**
trees 45–75 cm girth; 7.5–12.0 m high	100.23	2.00	39.50	–	116.31	nr	**155.81**

Q31 PLANTING

Item Excluding site overheads and profit	PC £	Labour hours	Labour £	Plant £	Material £	Unit	Total rate £
Tree planting; tree pit additives							
Melcourt Industries Ltd; Topgrow;							
incorporating into topsoil at 1 part							
Topgrow to 3 parts excavated topsoil;							
supplied in 75 l bags; pit size							
600 × 600 × 600 mm	1.76	0.02	0.40	–	1.76	nr	2.16
900 × 900 × 900 mm	5.95	0.06	1.19	–	5.95	nr	7.14
1.00 × 1.00 × 1.00	8.16	0.24	4.74	–	8.16	nr	12.90
1.25 × 1.25 × 1.25	15.95	0.40	7.90	–	15.95	nr	23.85
1.50 × 1.50 × 1.50	27.56	0.90	17.77	–	27.56	nr	45.33
Melcourt Industries Ltd; Topgrow;							
incorporating into topsoil at 1 part							
Topgrow to 3 parts excavated topsoil;							
supplied in 60 m³ loose loads; pit size							
600 × 600 × 600 mm	1.51	0.02	0.33	–	1.51	nr	1.84
900 × 900 × 900 mm	5.11	0.05	0.99	–	5.11	nr	6.10
1.00 × 1.00 × 1.00	7.01	0.20	3.95	–	7.01	nr	10.96
1.25 × 1.25 × 1.25	13.70	0.33	6.58	–	13.70	nr	20.28
1.50 × 1.50 × 1.50	23.67	0.75	14.81	–	23.67	nr	38.48
Tree planting; root barriers							
English Woodlands; Root Director;							
one-piece root control planters for							
installation at time of planting to divert							
root growth down away from pavements							
and out for anchorage; excavation							
measured separately							
RD 1050; 1050 × 1050 mm to							
1300 × 1300 mm at base	105.63	0.25	4.94	–	105.63	nr	110.57
RD 640; 640 × 640 mm to							
870 × 870 mm at base	68.60	0.25	4.94	–	68.60	nr	73.54
Greenleaf Horticulture; linear root							
deflection barriers; installed to trench							
measured separately							
Re-Root 2000; 2.0 mm thick × 2 m							
wide	15.14	0.05	0.99	–	15.14	m	16.13
Re-Root 2000; 1.0 mm thick × 1 m							
wide	5.28	0.05	0.99	–	5.28	m	6.27
Re-Root 600; 1.0 mm thick	7.76	0.05	0.99	–	7.76	m	8.75
Greenleaf Horticulture; irrigation							
systems; Root Rain tree pit irrigation							
systems							
Metro; small; 35 mm pipe dia. × 1.25 m							
long; for specimen shrubs and standard							
trees							
plastic	6.87	0.25	4.94	–	6.87	nr	11.81
plastic; with chain	9.00	0.25	4.94	–	9.00	nr	13.94
metal; with chain	11.15	0.25	4.94	–	11.15	nr	16.09

Q31 PLANTING

Item Excluding site overheads and profit	PC £	Labour hours	Labour £	Plant £	Material £	Unit	Total rate £
TREE PLANTING – cont							
Greenleaf Horticulture – cont							
Metro; medium; 35 mm pipe dia. × 1.75 m long; for standard and selected standard trees							
plastic	7.13	0.29	5.64	–	7.13	nr	**12.77**
plastic; with chain	9.44	0.29	5.64	–	9.44	nr	**15.08**
metal; with chain	11.55	0.29	5.64	–	11.55	nr	**17.19**
Metro; large; 35 mm pipe dia. × 2.50 m long; for selected standards and extra heavy standards							
plastic	8.44	0.33	6.58	–	8.44	nr	**15.02**
plastic; with chain	10.57	0.33	6.58	–	10.57	nr	**17.15**
metal; with chain	14.13	0.33	6.58	–	14.13	nr	**20.71**
Urban; for large capacity general purpose irrigation to parkland and street verge planting							
RRUrb1; 3.0 m pipe	–	0.29	5.64	–	16.94	nr	**22.58**
RRUrb2; 5.0 m pipe	19.45	0.33	6.58	–	19.45	nr	**26.03**
RRUrb3; 8.0 m pipe	23.24	0.40	7.90	–	23.24	nr	**31.14**
Civic; heavy cast aluminium inlet; for heavily trafficked locations							
5.0 m pipe	34.45	0.33	6.58	–	34.45	nr	**41.03**
8.0 m pipe	38.27	0.40	7.90	–	38.27	nr	**46.17**
Tree irrigation kits; Platipus; direct water **delivery to the rootball area of trees**							
Piddler tree irrigation system; permeable membrane delivering targeted irrigation to rootball surround							
root balls up to 550 mm dia.; Irrigation Kit PID0	7.00	0.25	4.94	–	7.00	nr	**11.94**
root balls up to 900 mm dia.; Irrigation Kit PID1	12.50	0.30	5.92	–	12.50	nr	**18.42**
root balls up to 1.55 m dia.; Irrigation Kit PID2	16.50	0.40	7.90	–	16.50	nr	**24.40**
root balls up to 2.40 m dia.; Irrigation Kit PID3	20.50	0.45	8.89	–	20.50	nr	**29.39**
root balls up to 3.10 m dia.; Irrigation Kit PID4	25.50	0.50	9.88	–	25.50	nr	**35.38**
Mulching of tree pits; Melcourt **Industries Ltd; FSC (Forest** **Stewardship Council) and FT (fire** **tested) certified**							
Spreading mulch; to individual trees; maximum distance 25 m (mulch not included)							
50 mm thick	–	0.05	0.96	–	–	m²	**0.96**
75 mm thick	–	0.07	1.44	–	–	m²	**1.44**
100 mm thick	–	0.10	1.92	–	–	m²	**1.92**

Q31 PLANTING

Item Excluding site overheads and profit	PC £	Labour hours	Labour £	Plant £	Material £	Unit	Total rate £
Mulch; Bark Nuggets®; to individual trees; delivered in 80 m³ loads; maximum distance 25 m							
50 mm thick	2.19	0.05	0.96	–	2.19	m²	**3.15**
75 mm thick	3.29	0.07	1.44	–	3.29	m²	**4.73**
100 mm thick	4.38	0.10	1.92	–	4.38	m²	**6.30**
Mulch; Bark Nuggets®; to individual trees; delivered in 25 m³ loads; maximum distance 25 m							
50 mm thick	2.87	0.05	0.96	–	2.87	m²	**3.83**
75 mm thick	4.31	0.05	0.99	–	4.31	m²	**5.30**
100 mm thick	5.74	0.07	1.32	–	5.74	m²	**7.06**
Mulch; Amenity Bark Mulch; to individual trees; delivered in 80 m³ loads; maximum distance 25 m							
50 mm thick	1.40	0.05	0.96	–	1.40	m²	**2.36**
75 mm thick	2.10	0.07	1.44	–	2.10	m²	**3.54**
100 mm thick	2.80	0.10	1.92	–	2.80	m²	**4.72**
Mulch; Amenity Bark Mulch; to individual trees; delivered in 25 m³ loads; maximum distance 25 m							
50 mm thick	2.08	0.05	0.96	–	2.08	m²	**3.04**
75 mm thick	3.12	0.07	1.44	–	3.12	m²	**4.56**
100 mm thick	4.16	0.07	1.32	–	4.16	m²	**5.48**
Trees; planting labours only							
Bare root trees; including backfilling with previously excavated material (all other operations and materials not included)							
light standard; 6–8 cm girth	–	0.35	6.91	–	–	nr	**6.91**
standard; 8–10 cm girth	–	0.40	7.90	–	–	nr	**7.90**
selected standard; 10–12 cm girth	–	0.58	11.46	–	–	nr	**11.46**
heavy standard; 12–14 cm girth	–	0.83	16.46	–	–	nr	**16.46**
extra heavy standard; 14–16 cm girth	–	1.00	19.75	–	–	nr	**19.75**
Root balled trees; including backfilling with previously excavated material (all other operations and materials not included)							
standard; 8–10 cm girth	–	0.50	9.88	–	–	nr	**9.88**
selected standard; 10–12 cm girth	–	0.60	11.85	–	–	nr	**11.85**
heavy standard; 12–14 cm girth	–	0.80	15.80	–	–	nr	**15.80**
extra heavy standard;14–16 cm girth	–	1.50	29.63	–	–	nr	**29.63**
16–18 cm girth	–	1.30	25.60	26.73	–	nr	**52.33**
18–20 cm girth	–	1.60	31.60	33.00	–	nr	**64.60**
20–25 cm girth	–	4.50	88.88	103.56	–	nr	**192.44**
25–30 cm girth	–	6.00	118.50	136.65	–	nr	**255.15**
30–35 cm girth	–	11.00	217.25	243.00	–	nr	**460.25**

Q31 PLANTING

Item Excluding site overheads and profit	PC £	Labour hours	Labour £	Plant £	Material £	Unit	Total rate £
TREE PLANTING – cont							
Treepits for urban planting							
Excavating tree pits; depositing soil							
alongside pits; by machine							
1.75 × 1.75 × 1.00deep	–	2.09	41.24	56.01	–	nr	97.25
2.00 × 2.00 × 1.00deep	–	2.73	53.86	72.96	–	nr	126.82
2.50 × 2.50 × 1.50deep	–	3.28	64.78	175.96	–	nr	240.74
Earthwork retention to tree pits							
bracing of deep treepits by plywood							
boards braced by timber stays							
Treepit 1.00 m × 1.00 m × 1.00 m	–	0.75	29.63	–	14.73	nr	44.36
Treepit 1.50 m × 1.50 m × 1.00 m	–	0.90	35.55	–	20.51	nr	56.06
Treepit 2.00 m × 2.00 m × 1.50 m	–	1.00	39.50	–	26.29	nr	65.79
Excavate tree break-out zone for tree							
roots to develop in restricted and							
compacted urban environments;							
depositing soil alongside pits by							
machine;							
Trenches 1.50 m wide × 1.50 m deep							
average	–	0.17	3.29	8.94	–	m³	12.23
Bases to treepits; levelling base							
breaking up base by hand and placing of							
granular drainage material 200 mm thick							
Treepit 1.00 m × 1.00 m	–	0.25	4.94	13.41	7.09	nr	25.44
Treepit 1.50 m × 1.50 m	–	0.45	8.89	24.14	15.96	nr	48.99
Treepit 2.00 m × 2.00 m	–	0.75	14.81	40.23	28.37	nr	83.41
Treepit 2.50 m × 2.50 m	–	0.75	14.81	40.23	44.33	nr	99.37
StrataCell; Greenleaf Ltd; Tree							
planting accessories for Urban tree							
planting; Load bearing tree planting							
modules for load support in tree							
environments; Placed to excavated							
treepit;							
single layer; 500 mm × 500 mm × 250 mm							
deep; interlocking load bearing cells;							
backfilling with imported topsoil							
7.5 tonne load limit	53.32	0.25	4.94	3.31	63.36	m²	71.61
40 tonne load limit	75.72	0.25	4.94	3.31	85.75	m²	94.00
double layer;							
500 mm × 500 mm × 500 mm deep;							
interlocking load bearing cells; backfilling							
with imported topsoil							
7.5 tonne load limit	106.64	0.50	9.88	6.63	126.71	m²	143.22
40 tonne load limit	151.44	0.50	9.88	6.63	171.51	m²	188.02

Q31 PLANTING

Item Excluding site overheads and profit	PC £	Labour hours	Labour £	Plant £	Material £	Unit	Total rate £
Root director; Greenleaf Ltd;							
preformed root barrier system with							
integral root deflecting ribs;							
Placed over tree rootball; under							
pavements; overall size;							
510 × 595 × 310 mm	41.59	0.25	9.88	–	41.59	nr	**51.47**
650 × 855 × 455 mm	68.60	0.33	13.04	–	68.60	nr	**81.64**
975 × 1370 × 545 mm	105.63	0.45	17.77	–	105.63	nr	**123.40**
1300 × 1805 × 500 mm	185.90	0.60	23.70	–	185.90	nr	**209.60**
Tree planting; containerized trees;							
nursery stock; James Coles & Sons							
(Nurseries) Ltd							
Acer platanoides 'Emerald Queen';							
including backfilling with excavated							
material (other operations not included)							
standard; 8–10 cm girth	44.75	0.48	9.48	–	44.75	nr	**54.23**
selected standard; 10–12 cm girth	70.00	0.56	11.07	–	70.00	nr	**81.07**
heavy standard; 12–14 cm girth	91.00	0.76	15.10	–	91.00	nr	**106.10**
extra heavy standard; 14–16 cm girth	105.00	1.20	23.70	–	105.00	nr	**128.70**
Carpinus betulus; including backfilling							
with excavated material (other							
operations not included)							
standard; 8–10 cm girth	44.75	0.48	9.48	–	44.75	nr	**54.23**
selected standard; 10–12 cm girth	70.00	0.56	11.07	–	70.00	nr	**81.07**
heavy standard; 12–14 cm girth	91.00	0.76	15.10	–	91.00	nr	**106.10**
extra heavy standard; 14–16 cm girth	112.00	1.20	23.70	–	112.00	nr	**135.70**
Fraxinus excelsior 'Altena'; including							
backfilling with excavated material							
(other operations not included)							
standard; 8–10 cm girth	42.00	0.48	9.48	–	42.00	nr	**51.48**
selected standard; 10–12 cm girth	67.25	0.56	11.07	–	67.25	nr	**78.32**
heavy standard; 12–14 cm girth	84.00	0.76	15.10	–	84.00	nr	**99.10**
extra heavy standard; 14–16 cm girth	98.00	1.20	23.70	–	98.00	nr	**121.70**
Prunus avium 'Plena'; including							
backfilling with excavated material							
(other operations not included)							
standard; 8–10 cm girth	42.00	0.40	7.90	–	42.00	nr	**49.90**
selected standard; 10–12 cm girth	70.00	0.56	11.07	–	70.00	nr	**81.07**
heavy standard; 12–14 cm girth	84.00	0.76	15.10	–	84.00	nr	**99.10**
extra heavy standard; 14–16 cm girth	98.00	1.20	23.70	–	98.00	nr	**121.70**
Quercus robur; including backfilling with							
excavated material (other operations not							
included)							
standard; 8–10 cm girth	49.00	0.48	9.48	–	49.00	nr	**58.48**
selected standard; 10–12 cm girth	77.00	0.56	11.07	–	77.00	nr	**88.07**
heavy standard; 12–14 cm girth	105.00	0.76	15.10	–	105.00	nr	**120.10**
extra heavy standard; 14–16 cm girth	119.00	1.20	23.70	–	119.00	nr	**142.70**

Q31 PLANTING

Item Excluding site overheads and profit	PC £	Labour hours	Labour £	Plant £	Material £	Unit	Total rate £
TREE PLANTING – cont							
Tree planting – cont							
Betula utilis jaquemontii; multistemmed;							
including backfillling with excavated							
material (other operations not included)							
175/200 mm high	84.00	0.48	9.48	–	84.00	nr	93.48
200/250 mm high	98.00	0.56	11.07	–	98.00	nr	109.07
250/300 mm high	84.00	0.76	15.10	–	84.00	nr	99.10
300/350 mm high	196.00	1.20	23.70	–	196.00	nr	219.70
Tree planting; root balled trees;							
advanced nursery stock and							
semi-mature – General							
Preamble: The cost of planting							
semi-mature trees will depend on the							
size and species and on the access to							
the site for tree handling machines.							
Prices should be obtained for individual							
trees and planting.							
Tree planting; bare root trees; nursery							
stock; James Coles & Sons							
(Nurseries) Ltd							
Acer platanoides; including backfillling							
with excavated material (other							
operations not included)							
light standard; 6–8 cm girth	9.00	0.35	6.91	–	9.00	nr	15.91
standard; 8–10 cm girth	12.00	0.40	7.90	–	12.00	nr	19.90
selected standard; 10–12 cm girth	19.50	0.58	11.46	–	19.50	nr	30.96
heavy standard; 12–14 cm girth	42.00	0.83	16.46	–	42.00	nr	58.46
extra heavy standard; 14–16 cm girth	56.00	1.00	19.75	–	56.00	nr	75.75
Carpinus betulus; including backfillling							
with excavated material (other							
operations not included)							
light standard; 6–8 cm girth	13.25	0.35	6.91	–	13.25	nr	20.16
standard; 8–10 cm girth	19.50	0.40	7.90	–	19.50	nr	27.40
selected standard; 10–12 cm girth	28.00	0.58	11.46	–	28.00	nr	39.46
heavy standard; 12–14 cm girth	39.25	0.83	16.47	–	39.25	nr	55.72
extra heavy standard; 14–16 cm girth	42.00	1.00	19.75	–	42.00	nr	61.75
Fraxinus excelsior; including backfillling							
with excavated material (other							
operations not included)							
light standard; 6–8 cm girth	9.75	0.35	6.91	–	9.75	nr	16.66
standard; 8–10 cm girth	16.00	0.40	7.90	–	16.00	nr	23.90
selected standard; 10–12 cm girth	22.50	0.58	11.46	–	22.50	nr	33.96
heavy standard; 12–14 cm girth	39.25	0.83	16.39	–	39.25	nr	55.64
extra heavy standard; 14–16 cm girth	47.50	1.00	19.75	–	47.50	nr	67.25

Q31 PLANTING

Item Excluding site overheads and profit	PC £	Labour hours	Labour £	Plant £	Material £	Unit	Total rate £
Prunus avium 'Plena'; including backfillling with excavated material (other operations not included)							
light standard; 6–8 cm girth	9.75	0.36	7.19	–	9.75	nr	**16.94**
standard; 8–10 cm girth	16.00	0.40	7.90	–	16.00	nr	**23.90**
selected standard; 10–12 cm girth	29.50	0.58	11.46	–	29.50	nr	**40.96**
heavy standard; 12–14 cm girth	49.00	0.83	16.46	–	49.00	nr	**65.46**
extra heavy standard; 14–16 cm girth	56.00	1.00	19.75	–	56.00	nr	**75.75**
Quercus robur; including backfillling with excavated material (other operations not included)							
light standard; 6–8 cm girth	18.25	0.35	6.91	–	18.25	nr	**25.16**
standard; 8–10 cm girth	30.75	0.40	7.90	–	30.75	nr	**38.65**
selected standard; 10–12 cm girth	42.00	0.58	11.46	–	42.00	nr	**53.46**
heavy standard; 12–14 cm girth	67.25	0.83	16.46	–	67.25	nr	**83.71**
Robinia pseudoacacia Frisia; including backfillling with excavated material (other operations not included)							
light standard; 6–8 cm girth	21.00	0.35	6.91	–	21.00	nr	**27.91**
standard; 8–10 cm girth	28.00	0.40	7.90	–	28.00	nr	**35.90**
selected standard; 10–12 cm girth	44.75	0.58	11.46	–	44.75	nr	**56.21**
Tree planting; root balled trees; **nursery stock; James Coles & Sons** **(Nurseries) Ltd**							
Acer platanoides; including backfillling with excavated material (other operations not included)							
standard; 8–10 cm girth	24.00	0.48	9.48	–	24.00	nr	**33.48**
selected standard; 10–12 cm girth	29.50	0.56	11.07	–	29.50	nr	**40.57**
heavy standard; 12–14 cm girth	67.00	0.76	15.10	–	67.00	nr	**82.10**
extra heavy standard; 14–16 cm girth	71.00	1.20	23.70	–	71.00	nr	**94.70**
Carpinus betulus; including backfillling with excavated material (other operations not included)							
standard; 8–10 cm girth	29.50	0.48	9.48	–	29.50	nr	**38.98**
selected standard; 10–12 cm girth	38.00	0.56	11.07	–	38.00	nr	**49.07**
heavy standard; 12–14 cm girth	66.00	0.76	15.10	–	66.00	nr	**81.10**
extra heavy standard; 14–16 cm girth	92.00	1.20	23.70	–	92.00	nr	**115.70**
Fraxinus excelsior; including backfillling with excavated material (other operations not included)							
standard; 8–10 cm girth	26.00	0.48	9.48	–	26.00	nr	**35.48**
selected standard; 10–12 cm girth	32.50	0.56	11.07	–	32.50	nr	**43.57**
heavy standard; 12–14 cm girth	49.25	0.76	15.10	–	49.25	nr	**64.35**
extra heavy standard; 14–16 cm girth	62.50	1.20	23.70	–	62.50	nr	**86.20**
Prunus avium 'Plena'; including backfillling with excavated material (other operations not included)							
standard; 8–10 cm girth	33.00	0.40	7.90	–	33.00	nr	**40.90**
selected standard; 10–12 cm girth	33.00	0.56	11.07	–	33.00	nr	**44.07**
heavy standard; 12–14 cm girth	59.00	0.76	15.10	–	59.00	nr	**74.10**
extra heavy standard; 14–16 cm girth	71.00	1.20	23.70	–	71.00	nr	**94.70**

Q31 PLANTING

Item Excluding site overheads and profit	PC £	Labour hours	Labour £	Plant £	Material £	Unit	Total rate £
TREE PLANTING – cont							
Tree planting – cont							
Quercus robur; including backfilling with excavated material (other operations not included)							
standard; 8–10 cm girth	40.75	0.48	9.48	–	40.75	nr	**50.23**
selected standard; 10–12 cm girth	52.00	0.56	11.07	–	52.00	nr	**63.07**
heavy standard; 12–14 cm girth	77.25	0.76	15.10	–	77.25	nr	**92.35**
extra heavy standard; 14–16 cm girth	134.00	1.20	23.70	–	134.00	nr	**157.70**
Robinia pseudoacacia 'Frisia'; including backfilling with excavated material (other operations not included)							
standard; 8–10 cm girth	40.00	0.48	9.48	–	40.00	nr	**49.48**
selected standard; 10–12 cm girth	62.25	0.56	11.07	–	62.25	nr	**73.32**
heavy standard; 12–14 cm girth	98.00	0.76	15.10	–	98.00	nr	**113.10**
Tree planting; Airpot container grown trees; advanced nursery stock and semi-mature; Deepdale Trees Ltd							
Acer platanoides 'Emerald Queen'; including backfilling with excavated material (other operations not included)							
16–18 cm girth	95.00	1.98	39.10	41.88	95.00	nr	**175.98**
18–20 cm girth	130.00	2.18	43.02	45.00	130.00	nr	**218.02**
20–25 cm girth	190.00	2.38	46.93	50.26	190.00	nr	**287.19**
25–30 cm girth	250.00	2.97	58.66	76.93	250.00	nr	**385.59**
30–35 cm girth	450.00	3.96	78.21	83.77	450.00	nr	**611.98**
Prunus avium 'Flora Plena'; including backfilling with excavated material (other operations not included)							
16–18 cm girth	95.00	1.98	39.10	41.88	95.00	nr	**175.98**
18–20 cm girth	130.00	1.60	31.60	45.00	130.00	nr	**206.60**
20–25 cm girth	190.00	2.38	46.93	50.26	190.00	nr	**287.19**
25–30 cm girth	250.00	2.97	58.66	76.93	250.00	nr	**385.59**
30–35 cm girth	350.00	3.96	78.21	83.77	350.00	nr	**511.98**
Quercus palustris 'Pin Oak'; including backfilling with excavated material (other operations not included)							
16–18 cm girth	100.00	1.98	39.10	41.88	100.00	nr	**180.98**
18–20 cm girth	140.00	1.60	31.60	45.00	140.00	nr	**216.60**
20–25 cm girth	210.00	2.38	46.93	50.26	210.00	nr	**307.19**
25–30 cm girth	275.00	2.97	58.66	76.93	275.00	nr	**410.59**
30–35 cm girth	400.00	3.96	78.21	83.77	400.00	nr	**561.98**
Tree planting; Airpot container grown trees; semi-mature and mature trees; Deepdale Trees Ltd; planting and back filling; planted by telehandler or by crane; delivery included; all other operations priced separately							
Semi-mature trees; indicative prices							
40–45 cm girth	550.00	4.00	79.00	52.16	550.00	nr	**681.16**
45–50 cm girth	750.00	4.00	79.00	52.16	750.00	nr	**881.16**

Q31 PLANTING

Item Excluding site overheads and profit	PC £	Labour hours	Labour £	Plant £	Material £	Unit	Total rate £
55–60 cm girth	1350.00	6.00	118.50	52.16	1350.00	nr	**1520.66**
60–70 cm girth	2500.00	7.00	138.25	70.37	2500.00	nr	**2708.62**
70–80 cm girth	3500.00	7.50	148.13	88.58	3500.00	nr	**3736.71**
80–90 cm girth	4500.00	8.00	158.00	104.31	4500.00	nr	**4762.31**
Tree planting; rootballed trees; **advanced nursery stock and** **semi-mature; Lorenz von Ehren** Acer platanoides 'Emerald Queen'; including backfilling with excavated material (other operations not included)							
16–18 cm girth	85.00	1.30	25.60	3.75	85.00	nr	**114.35**
18–20 cm girth	105.00	1.60	31.60	3.75	105.00	nr	**140.35**
20–25 cm girth	130.00	4.50	88.88	16.37	130.00	nr	**235.25**
25–30 cm girth	170.00	6.00	118.50	20.40	170.00	nr	**308.90**
30–35 cm girth	310.00	11.00	217.25	27.38	310.00	nr	**554.63**
Aesculus carnea 'Briotti'; including backfilling with excavated material (other operations not included)							
16–18 cm girth	160.00	1.30	25.60	3.75	160.00	nr	**189.35**
18–20 cm girth	190.00	1.60	31.60	3.75	190.00	nr	**225.35**
20–25 cm girth	230.00	4.50	88.88	16.37	230.00	nr	**335.25**
25–30 cm girth	320.00	6.00	118.50	20.40	320.00	nr	**458.90**
30–35 cm girth	420.00	11.00	217.25	27.38	420.00	nr	**664.63**
Prunus avium 'Plena'; including backfilling with excavated material (other operations not included)							
16–18 cm girth	115.00	1.30	25.60	3.75	115.00	nr	**144.35**
18–20 cm girth	130.00	1.60	31.60	3.75	130.00	nr	**165.35**
20–25 cm girth	150.00	4.50	88.88	16.37	150.00	nr	**255.25**
25–30 cm girth	200.00	6.00	118.50	20.40	200.00	nr	**338.90**
30–35 cm girth	270.00	11.00	217.25	24.15	270.00	nr	**511.40**
Quercus palustris 'Pin Oak'; including backfilling with excavated material (other operations not included)							
16–18 cm girth	140.00	1.30	25.60	3.75	140.00	nr	**169.35**
18–20 cm girth	160.00	1.60	31.60	3.75	160.00	nr	**195.35**
20–25 cm girth	200.00	4.50	88.88	16.37	200.00	nr	**305.25**
25–30 cm girth	240.00	6.00	118.50	20.40	240.00	nr	**378.90**
30–35 cm girth	350.00	11.00	217.25	27.38	350.00	nr	**594.63**
Tilia cordata 'Greenspire'; including backfilling with excavated material (other operations not included)							
16–18 cm girth	95.00	1.30	25.60	3.75	95.00	nr	**124.35**
18–20 cm girth	115.00	1.60	31.60	3.75	115.00	nr	**150.35**
20–25 cm girth	140.00	4.50	88.88	16.37	140.00	nr	**245.25**
25–30 cm girth; 5 × transplanted; 4.0–5.0 m tall	170.00	6.00	118.50	20.40	170.00	nr	**308.90**
30–35 cm girth; 5 × transplanted; 5.0–7.0 m tall	220.00	11.00	217.25	27.38	220.00	nr	**464.63**

Q31 PLANTING

Item Excluding site overheads and profit	PC £	Labour hours	Labour £	Plant £	Material £	Unit	Total rate £
TREE PLANTING – cont							
Tree planting – cont							
Betula pendula (3 stems); including							
backfilling with excavated material (other							
operations not included)							
3.0–3.5 m high	65.00	1.98	39.10	41.88	65.00	nr	**145.98**
3.5–4.0 m high	95.00	1.60	31.60	45.00	95.00	nr	**171.60**
4.0–4.5 m high	135.00	2.38	46.93	50.26	135.00	nr	**232.19**
4.5–5.0 m high	155.00	2.97	58.66	76.93	155.00	nr	**290.59**
5.0–6.0 m high	220.00	3.96	78.21	83.77	220.00	nr	**381.98**
6.0–7.0 m high	360.00	4.50	88.88	98.30	360.00	nr	**547.18**
Pinus sylvestris; including backfilling with							
excavated material (other operations not							
included)							
3.0–3.5 m high	410.00	1.98	39.10	41.88	410.00	nr	**490.98**
3.5–4.0 m high	510.00	1.60	31.60	45.00	510.00	nr	**586.60**
4.0–4.5 m high	700.00	2.38	46.93	50.26	700.00	nr	**797.19**
4.5–5.0 m high	960.00	2.97	58.66	76.93	960.00	nr	**1095.59**
5.0–6.0 m high	1520.00	3.96	78.21	83.77	1520.00	nr	**1681.98**
6.0–7.0 m high	2540.00	4.50	88.88	101.44	2540.00	nr	**2730.32**
Pleached trees; Carpinus betulus;							
specimen trees; Lorenz Von Ehren;							
supply and planting only; excavation,							
support and treepit additives not							
included							
Box shaped trees to provide 'floating							
hedge' or screen effect; planting							
distance 1 tree per m run; trunk and box							
alignment; wire root balls							
4 × transplanted; 20–25 cm; 1.00 m							
centres	320.00	5.00	98.75	16.37	320.00	m	**435.12**
5 × transplanted; 25–30 cm; 1.00 m							
centres	450.00	6.50	128.38	20.40	450.00	m	**598.78**
5 × transplanted; 30–35 cm; 1.20 m							
centres	720.00	11.00	217.25	27.38	720.00	m	**964.63**
Pleached trees; Tilia europaea							
'Pallida'; specimen trees; Lorenz Von							
Ehren; supply and planting only;							
excavation, support and treepit							
additives not included							
Box pleached trees to provide 'floating							
hedge' or screen effect; wire root balls							
4 × transplanted; 20–25 cm; 1.00 m							
centres	320.00	5.00	98.75	16.37	320.00	m	**435.12**
4 × transplanted; 25–30 cm; 1.00 m							
centres	380.00	5.00	98.75	16.37	380.00	m	**495.12**
5 × transplanted; 30–35 cm; 1.20 m							
centres	374.99	5.00	98.75	16.37	374.99	m	**490.11**

Q31 PLANTING

Item Excluding site overheads and profit	PC £	Labour hours	Labour £	Plant £	Material £	Unit	Total rate £
5 × transplanted; 35–40 cm; 1.50 m centres	393.35	5.00	98.75	16.37	393.35	m	**508.47**
6 × transplanted; 40–45 cm; 1.80 m centres	455.59	5.00	98.75	16.37	455.59	m	**570.71**
6 × transplanted; 45–50 cm; 1.80 m centres	583.38	5.00	98.75	16.37	583.38	m	**698.50**
6 × transplanted; 50–60 cm; 2.00 m centres	700.00	5.00	98.75	16.37	700.00	m	**815.12**
Espalier pleached trees; specimen trees; Lorenz Von Ehren; supply and planting only; excavation, support and treepit additives not included Frame pleached trees; to provide floating screen effects; wire root balls							
Carpinus betulus; 20–25 cm; 1.00 m centres	450.00	5.00	98.75	16.37	450.00	m	**565.12**
Carpinus betulus; 25–30 cm; 1.00 m centres	600.00	5.00	98.75	16.37	600.00	m	**715.12**
Tilia europaea 'Pallida'; 20–25 cm; 1.00 m centres	400.00	5.00	98.75	16.37	400.00	m	**515.12**
Tilia europaea 'Pallida'; 25–30 cm; 1.00 m centres	550.00	5.00	98.75	16.37	550.00	m	**665.12**
Tilia europaea 'Pallida'; 30–35 cm; 1.20 m centres	574.98	5.00	98.75	16.37	574.98	m	**690.10**
Umbrella shaped pleaches Umbrella or roof pleached trees to provide umbrella effect; wire root balls							
Tilia euchlora; specimen; 4 × transplanted; 20–25 cm	390.00	5.00	98.75	16.37	390.00	nr	**505.12**
Tilia euchlora; specimen; 5 × transplanted; 25–30 cm	520.00	5.00	98.75	16.37	520.00	nr	**635.12**
Platanus acerifolia; clear stem; 4 × transplanted; 20–25 cm	390.00	5.00	98.75	16.37	390.00	nr	**505.12**
Platanus acerifolia; specimen; 5 × transplanted; 30–35 cm	650.00	5.00	98.75	16.37	650.00	nr	**765.12**
Tilia europaea; specimen; 5 × transplanted; 30–35 cm	690.00	5.00	98.75	16.37	690.00	nr	**805.12**
Platanus acerifolia; specimen; 5 × transplanted; 25–30 cm	520.00	5.00	98.75	16.37	520.00	nr	**635.12**

Q31 PLANTING

Item Excluding site overheads and profit	PC £	Labour hours	Labour £	Plant £	Material £	Unit	Total rate £
TREE PLANTING – cont							
Tree planting; containerized trees;							
Lorenz von Ehren; to the tree prices							
above add for Airpot containerization							
only							
Tree size							
20–25 cm	–	–	–	–	50.00	nr	**50.00**
25–30 cm	–	–	–	–	80.00	nr	**80.00**
30–35 cm	–	–	–	–	90.00	nr	**90.00**
35–40 cm	–	–	–	–	150.00	nr	**150.00**
40–45 cm	–	–	–	–	180.00	nr	**180.00**
45–50 cm	–	–	–	–	260.00	nr	**260.00**
50–60 cm	–	–	–	–	380.00	nr	**380.00**
60–70 cm	–	–	–	–	520.00	nr	**520.00**
70–80 cm	–	–	–	–	660.00	nr	**660.00**
80–90 cm	–	–	–	–	820.00	nr	**820.00**
Tree planting; rootballed trees;							
semi-mature and mature trees; Lorenz							
von Ehren; planting and back filling;							
planted by telehandler or by crane;							
delivery included; all other operations							
priced separately							
Semi-mature trees							
40–45 cm girth	600.00	8.00	158.00	66.25	600.00	nr	**824.25**
45–50 cm girth	850.00	8.00	158.00	87.00	850.00	nr	**1095.00**
50–60 cm girth	1200.00	10.00	197.50	105.63	1200.00	nr	**1503.13**
60–70 cm girth	2200.00	15.00	296.25	222.11	2200.00	nr	**2718.36**
70–80 cm girth	3200.00	18.00	355.50	206.73	3200.00	nr	**3762.23**
80–90 cm girth	4800.00	18.00	355.50	206.73	4800.00	nr	**5362.23**
90–100 cm girth	5900.00	18.00	355.50	206.73	5900.00	nr	**6462.23**
SHRUB PLANTING							
Shrub planting – General							
Preamble: For preparation of planting							
areas see 'Cultivation' at the beginning							
of the section on planting.							
Shrub planting							
Setting out; selecting planting from							
holding area; loading to wheelbarrows;							
planting as plan or as directed; distance							
from holding area maximum 50 m; plants							
2–3 litre containers							
single plants not grouped	–	0.04	0.79	–	–	nr	**0.79**
plants in groups of 3–5 nr	–	0.03	0.49	–	–	nr	**0.49**
plants in groups of 10–100 nr	–	0.02	0.33	–	–	nr	**0.33**
plants in groups of 100 nr minimum	–	0.01	0.23	–	–	nr	**0.23**

Q31 PLANTING

Item Excluding site overheads and profit	PC £	Labour hours	Labour £	Plant £	Material £	Unit	Total rate £
Forming planting holes; in cultivated ground (cultivating not included); by mechanical auger; trimming holes by hand; depositing excavated material alongside holes							
250 mm dia.	–	0.03	0.66	0.05	–	nr	**0.71**
250 × 250 mm	–	0.04	0.79	0.09	–	nr	**0.88**
300 × 300 mm	–	0.08	1.48	0.12	–	nr	**1.60**
Hand excavation; forming planting holes; in cultivated ground (cultivating not included); depositing excavated material alongside holes							
100 × 100 × 100 mm deep; with mattock or hoe	–	0.01	0.13	–	–	nr	**0.13**
250 × 250 × 300 mm deep	–	0.04	0.79	–	–	nr	**0.79**
300 × 300 × 300 mm deep	–	0.06	1.10	–	–	nr	**1.10**
400 × 400 × 400 mm deep	–	0.13	2.47	–	–	nr	**2.47**
500 × 500 × 500 mm deep	–	0.25	4.94	–	–	nr	**4.94**
600 × 600 × 600 mm deep	–	0.43	8.55	–	–	nr	**8.55**
900 × 900 × 600 mm deep	–	1.00	19.75	–	–	nr	**19.75**
1.00 × 1.00x 600 mm deep	–	1.23	24.29	–	–	nr	**24.29**
1.25 × 1.25x 600 mm deep	–	1.93	38.12	–	–	nr	**38.12**
Hand excavation; forming planting holes; in uncultivated ground; depositing excavated material alongside holes							
100 × 100 × 100 mm deep; with mattock or hoe	–	0.03	0.49	–	–	nr	**0.49**
250 × 250 × 300 mm deep	–	0.06	1.10	–	–	nr	**1.10**
300 × 300 × 300 mm deep	–	0.06	1.23	–	–	nr	**1.23**
400 × 400 × 400 mm deep	–	0.25	4.94	–	–	nr	**4.94**
500 × 500 × 500 mm deep	–	0.33	6.43	–	–	nr	**6.43**
600 × 600 × 600 mm deep	–	0.55	10.86	–	–	nr	**10.86**
900 × 900 × 600 mm deep	–	1.25	24.69	–	–	nr	**24.69**
1.00 × 1.00x 600 mm deep	–	1.54	30.37	–	–	nr	**30.37**
1.25 × 1.25x 600 mm deep	–	2.41	47.65	–	–	nr	**47.65**
Bare root planting; to planting holes (forming holes not included); including backfilling with excavated material (bare root plants not included)							
bare root 1+1; 30–90 mm high	–	0.02	0.33	–	–	nr	**0.33**
bare root 1+2; 90–120 mm high	–	0.02	0.33	–	–	nr	**0.33**
Containerized planting; to planting holes (forming holes not included); including backfilling with excavated material (shrub or ground cover not included)							
9 cm pot	–	0.01	0.20	–	–	nr	**0.20**
2 litre container	–	0.02	0.40	–	–	nr	**0.40**
3 litre container	–	0.02	0.44	–	–	nr	**0.44**
5 litre container	–	0.03	0.66	–	–	nr	**0.66**
10 litre container	–	0.05	0.99	–	–	nr	**0.99**
15 litre container	–	0.07	1.32	–	–	nr	**1.32**
20 litre container	–	0.08	1.65	–	–	nr	**1.65**

Q31 PLANTING

Item Excluding site overheads and profit	PC £	Labour hours	Labour £	Plant £	Material £	Unit	Total rate £
SHRUB PLANTING – cont							
Shrub planting – cont							
Shrub planting; 2 litre containerized plants; in cultivated ground (cultivating not included); PC £2.90/nr							
average 2 plants per m²	–	0.06	1.11	–	5.80	m²	6.91
average 3 plants per m²	–	0.08	1.66	–	8.70	m²	10.36
average 4 plants per m²	–	0.11	2.21	–	11.60	m²	13.81
average 6 plants per m²	–	0.17	3.32	–	17.40	m²	20.72
Shrub planting; 3 litre containerized plants; in cultivated ground (cultivating not included); PC £3.40/nr							
average 2 plants per m²	–	0.07	1.38	–	6.80	m²	8.18
average 3 plants per m²	–	0.11	2.07	–	10.20	m²	12.27
average 4 plants per m²	–	0.14	2.77	–	13.60	m²	16.37
average 6 plants per m²	–	0.21	4.15	–	20.40	m²	24.55
Extra over shrubs for stakes	0.66	0.02	0.41	–	0.66	nr	1.07
HEDGE PLANTING							
Hedges							
Excavating trench for hedges; depositing soil alongside trench; by machine							
300 mm deep × 300 mm wide	–	0.03	1.20	1.35	–	m	2.55
300 mm deep × 600 mm wide	–	0.05	1.80	2.05	–	m	3.85
300 mm deep × 450 mm wide	–	0.05	1.78	2.03	–	m	3.81
500 mm deep × 500 mm wide	–	0.06	2.48	2.83	–	m	5.31
500 mm deep × 700 mm wide	–	0.09	3.46	3.96	–	m	7.42
600 mm deep × 750 mm wide	–	0.11	4.45	5.08	–	m	9.53
750 mm deep × 900 mm wide	–	0.17	6.68	7.63	–	m	14.31
Excavating trench for hedges; depositing soil alongside trench; by hand							
300 mm deep × 300 mm wide	–	0.08	4.94	–	–	m	4.94
300 mm deep × 450 mm wide	–	0.13	7.41	–	–	m	7.41
300 mm deep × 600 mm wide	–	0.17	9.88	–	–	m	9.88
500 mm deep × 500 mm wide	–	0.23	13.40	–	–	m	13.40
500 mm deep × 700 mm wide	–	0.32	18.79	–	–	m	18.79
600 mm deep × 750 mm wide	–	0.41	24.15	–	–	m	24.15
750 mm deep × 900 mm wide	–	0.61	36.16	–	–	m	36.16
Setting out; notching out; excavating trench; breaking up subsoil to minimum depth 300 mm							
minimum 400 mm deep	–	0.25	4.94	–	–	m	4.94
Disposal of excavated soil to stockpile							
Up to 25 m distant							
by hand	–	2.40	47.40	–	–	m³	47.40
by machine	–	–	–	2.80	–	m³	2.80

Q31 PLANTING

Item Excluding site overheads and profit	PC £	Labour hours	Labour £	Plant £	Material £	Unit	Total rate £
Instant Hedge planting; Practicality Brown Ltd – Elveden Hedges; continuous strips of instant hedges; delivered in 2.50 m lengths; planting and backfilling with excavated material of hedging plants to trenches (not included); including minor trimming and shaping after planting							
Beech (Fagus) hedges							
height 1.40/1.60 m × 400 mm wide	149.00	0.20	7.90	6.45	149.00	m	**163.35**
height 1.60/1.80 m × 500 mm wide	162.00	0.22	8.78	7.17	162.00	m	**177.95**
Yew (Taxus) hedges							
height 1.40/1.60 m × 400 mm wide	189.00	0.20	7.90	6.45	189.00	m	**203.35**
height 1.60/1.80 m × 500 mm wide	215.00	0.22	8.78	7.17	215.00	m	**230.95**
height 1.80/2.0 m × 500 mm wide	238.00	0.25	9.88	8.06	238.00	m	**255.94**
Prunus Lauruscerasus hedges							
height 1.40/1.60 m × 400 mm wide	147.00	0.20	7.90	6.45	147.00	m	**161.35**
height 1.60/1.80 m × 500 mm wide	160.00	0.22	8.78	7.17	160.00	m	**175.95**
height 1.80/2.00 m × 500 mm wide	173.00	0.25	9.88	8.06	173.00	m	**190.94**
Box (Buxus) hedges							
height 0.80/1.00 m × 300 mm wide	135.00	0.10	3.95	3.23	135.00	m	**142.18**
height 1.00/1.20 m × 300 mm wide	148.00	0.13	4.94	4.03	148.00	m	**156.97**
Instant Hedge planting; Practicality Brown Ltd; individual pre-clipped hedge plants to form mature hedge; planting and backfilling of hedging plants to trenches excavated separately; including minor trimming and shaping after planting							
Beech and Hornbeam hedges							
1.50 m high × 500 mm wide; 500 mm centres	220.00	0.50	29.63	2.74	220.00	m	**252.37**
1.50 m high × 500 mm wide; 750 mm centres	146.30	0.19	11.14	2.74	146.30	m	**160.18**
2.00 m high × 500 mm wide; 500 mm centres	304.00	0.57	33.86	2.74	304.00	m	**340.60**
2.00 m high × 500 mm wide; 750 mm centres	202.16	0.57	33.86	2.74	202.16	m	**238.76**
Yew (Taxus) hedges							
1.50 m high × 500 mm wide; 500 mm centres	300.00	0.50	29.63	2.74	300.00	m	**332.37**
1.50 m high × 500 mm wide; 750 mm centres	199.50	0.19	11.14	2.74	199.50	m	**213.38**
1.75 m high × 500 mm wide; 500 mm centres	340.00	0.57	33.86	2.74	340.00	m	**376.60**

Q31 PLANTING

Item Excluding site overheads and profit	PC £	Labour hours	Labour £	Plant £	Material £	Unit	Total rate £
HEDGE PLANTING – cont							
Instant Hedge planting – cont							
Yew (Taxus) hedges – cont							
1.75 m high × 500 mm wide; 750 mm centres	226.10	0.21	12.47	2.74	226.10	m	**241.31**
2.00 m high × 500 mm wide; 500 mm centres	400.00	0.67	39.50	2.74	400.00	m	**442.24**
2.00 m high × 500 mm wide; 750 mm centres	266.00	0.57	33.86	2.74	266.00	m	**302.60**
2.00 m high × 500 mm wide; 900 mm centres	222.00	0.27	16.02	2.74	222.00	m	**240.76**
Feathered hedges; Beech or Hornbeam							
1.75 m high × 300 mm wide; 400 mm centres	75.00	0.20	11.85	1.37	75.00	m	**88.22**
1.75 m high × 300 mm wide; 500 mm centres	60.00	0.16	9.48	1.37	60.00	m	**70.85**
1.75 m high × 300 mm wide; 750 mm centres	42.60	0.11	6.77	1.37	42.60	m	**50.74**
1.75 m high × 300 mm wide; 900 mm centres	33.30	0.09	5.27	1.37	33.30	m	**39.94**
1.75–2.00 m high × 300 mm wide; 600 mm centres	50.00	0.33	19.75	19.71	50.00	m	**89.46**
1.75–2.00 m high × 300 mm wide; 900 mm centres	33.30	0.29	16.93	17.15	33.30	m	**67.38**
2.00–2.25 m high 300 wide; 600 mm centres	61.13	0.40	23.70	23.74	61.13	m	**108.57**
2.00–2.25 m high × 300 mm wide; 900 mm centres	40.34	0.33	19.75	19.71	40.34	m	**79.80**
2.00–2.25 m high × 300 mm wide; 1.20 m centres	30.56	0.25	14.81	15.24	30.56	m	**60.61**
Hedge planting; Griffin Nurseries Ltd; feathered hedge plants to form mature hedge; planting and backfilling of hedging plants to trenches excavated separately; including minor trimming and shaping after planting							
(Yew) Taxus hedging; root balled							
600–800 mm high × 300 mm wide; 250 mm centres	50.60	0.20	11.85	1.37	50.60	m	**63.82**
600–800 mm high × 300 mm wide; 500 mm centres	25.30	0.11	6.58	1.37	25.30	m	**33.25**
800 mm–1.00 m high × 300 mm wide; 250 mm centres	59.40	0.20	11.85	1.37	59.40	m	**72.62**

Q31 PLANTING

Item Excluding site overheads and profit	PC £	Labour hours	Labour £	Plant £	Material £	Unit	Total rate £
800 mm–1.00 m high × 300 mm wide; 500 mm centres	29.70	0.11	6.58	1.37	29.70	m	**37.65**
1.00–1.20 m high × 300/350 mm wide; 3 nr/m	74.92	0.17	9.88	1.37	74.92	m	**86.17**
1.00–1.20 m high × 300/350 mm wide; 2 nr/m	45.00	0.13	7.41	1.37	45.00	m	**53.78**
1.25–1.50 m high × 300/350 mm wide; 3 nr/m	111.39	0.17	9.88	1.37	111.39	m	**122.64**
1.25–1.50 m high × 300/350 mm wide; 2 nr/m	66.90	0.13	7.90	1.37	66.90	m	**76.17**
1.50–1.75 m high × 400 mm wide; 450 mm centres	124.32	0.33	19.75	14.78	124.32	m	**158.85**
1.50–1.75 m high × 400 mm wide; 600 mm centres	93.34	0.22	13.17	13.29	93.34	m	**119.80**
1.50–1.75 m high × 400 mm wide; 900 mm centres	62.16	0.20	11.85	12.10	62.16	m	**86.11**
1.75–2.00 m high × 500 mm wide; 600 mm centres	123.75	0.33	19.75	19.71	123.75	m	**163.21**
1.75–2.00 m high × 500 mm wide; 900 mm centres	101.01	0.29	16.93	17.15	183.43	m	**217.51**
1.75–2.00 m high × 500 mm wide; 1.20 m centres	61.87	0.25	14.81	15.24	61.87	m	**91.92**
2.00–2.25 m high × 600 mm wide; 600 mm centres	151.67	0.40	23.70	23.74	151.67	m	**199.11**
2.00–2.25 m high × 600 mm wide; 900 mm centres	101.01	0.33	19.75	19.71	101.01	m	**140.47**
2.00–2.25 m high × 600 mm wide; 1.20 m centres	75.83	0.31	18.23	18.33	75.83	m	**112.39**
(Yew) Taxus hedging; wire root balled							
1.50–1.75 m high × 400 mm wide; 450 mm centres	193.14	0.33	19.75	14.78	193.14	m	**227.67**
1.50–1.75 m high × 400 mm wide; 600 mm centres	145.00	0.22	13.17	13.29	145.00	m	**171.46**
1.50–1.75 m high × 400 mm wide; 900 mm centres	95.70	0.20	11.85	12.10	95.70	m	**119.65**
1.75–2.00 m high × 600 mm wide; 600 mm centres	169.62	0.33	19.75	19.71	169.62	m	**209.08**
1.75–2.00 m high × 600 mm wide; 900 mm centres	112.94	0.29	16.93	17.15	112.94	m	**147.02**
1.75–2.00 m high × 600 mm wide; 1.20 m centres	84.79	0.25	14.81	15.24	84.79	m	**114.84**
2.00–2.25 m high × 650 mm wide; 700 mm centres	163.30	0.44	26.33	26.12	163.30	m	**215.75**
2.00–2.25 m high × 600 mm wide; 900 mm centres	127.65	0.36	21.54	21.33	127.65	m	**170.52**
2.00–2.25 m high × 600 mm wide; 1.20 m centres	75.83	0.31	18.23	18.33	75.83	m	**112.39**

Q31 PLANTING

Item Excluding site overheads and profit	PC £	Labour hours	Labour £	Plant £	Material £	Unit	Total rate £
HEDGE PLANTING – cont							
Hedge planting – cont							
Buxus (Box) hedges – cont							
Buxus (Box) hedges; rootballed							
400–500 mm high; 200 mm rootball;							
200 mm centres	33.25	0.09	1.76	–	33.25	m	**35.01**
400–500 mm high; 200 mm rootball;							
300 mm centres	22.14	0.07	1.32	–	22.14	m	**23.46**
400–500 mm high; 200 mm rootball;							
500 mm centres	13.30	0.05	0.99	–	13.30	m	**14.29**
500–600 mm high; 200 mm rootball;							
200 mm centres	52.75	0.09	1.76	–	52.75	m	**54.51**
500–600 mm high; 200 mm rootball;							
300 mm centres	35.13	0.07	1.32	–	35.13	m	**36.45**
500–600 mm high; 200 mm rootball;							
500 mm centres	21.10	0.08	1.52	–	21.10	m	**22.62**
600–800 mm high; 250 mm rootball;							
250 mm centres	–	0.09	1.80	–	53.20	m	**55.00**
600–800 mm high; 250 mm rootball;							
400 mm centres	33.25	0.08	1.65	–	33.25	m	**34.90**
600–800 mm high; 250 mm rootball;							
600 mm centres	17.58	0.08	1.52	–	17.58	m	**19.10**
800 mm–1.00 m high; 300 mm							
rootball; 300 mm centres	71.43	0.10	1.98	–	71.43	m	**73.41**
800 mm–1.00 m high; 300 mm							
rootball; 500 mm centres	42.90	0.11	2.19	–	42.90	m	**45.09**
800 mm–1.00 m high; 300 mm							
rootball; 750 mm centres	28.53	0.13	2.47	–	28.53	m	**31.00**
1.00 mm–1.25 m high; 350 mm							
rootball; 400 mm centres	74.13	0.13	2.47	–	74.13	m	**76.60**
1.00 mm–1.25 m high; 350 mm							
rootball; 600 mm centres	49.43	0.11	2.19	–	49.43	m	**51.62**
1.00 mm–1.25 m high; 350 mm							
rootball; 900 mm centres	32.91	0.10	1.98	–	32.91	m	**34.89**
Native species and bare root hedge planting							
Hedge planting operations only (excavation of trenches priced separately) including backfill with excavated topsoil; PC £0.48/nr							
single row; 200 mm centres	2.40	0.06	1.23	–	2.40	m	**3.63**
single row; 300 mm centres	1.60	0.06	1.10	–	1.60	m	**2.70**
single row; 400 mm centres	1.20	0.04	0.82	–	1.20	m	**2.02**
single row; 500 mm centres	0.96	0.03	0.66	–	0.96	m	**1.62**
double row; 200 mm centres	4.80	0.17	3.29	–	4.80	m	**8.09**
double row; 300 mm centres	3.20	0.13	2.63	–	3.20	m	**5.83**
double row; 400 mm centres	2.40	0.08	1.65	–	2.40	m	**4.05**
double row; 500 mm centres	1.92	0.07	1.32	–	1.92	m	**3.24**
Extra over hedges for incorporating manure; at 1 m³ per 30 m	0.77	0.03	0.49	–	0.77	m	**1.26**

Q31 PLANTING

Item Excluding site overheads and profit	PC £	Labour hours	Labour £	Plant £	Material £	Unit	Total rate £
TOPIARY							
Topiary							
Clipped topiary; Lorenz von Ehren;							
German field grown clipped and							
transplanted as detailed; planted to							
plant pit; including backfilling with							
excavated material and TPMC							
Buxus sempervirens (box); balls							
300 mm dia.; 3 × transplanted;							
container grown or rootballed	16.00	0.25	4.94	–	26.15	nr	**31.09**
500 mm dia.; 4 × transplanted; wire							
rootballed	50.00	1.50	29.63	4.45	125.62	nr	**159.70**
900 mm dia.; 5 × transplanted; wire							
rootballed	210.00	2.75	54.31	4.45	335.15	nr	**393.91**
1300 mm dia.; 6 × transplanted; wire							
rootballed	660.00	3.10	61.23	5.34	782.34	nr	**848.91**
Buxus sempervirens (box); pyramids							
500 mm high; 3 × transplanted;							
container grown or rootballed	35.00	1.50	29.63	4.45	110.62	nr	**144.70**
900 mm high; 4 × transplanted; wire							
rootballed	100.00	2.75	54.31	4.45	225.15	nr	**283.91**
1300 mm high; 5 × transplanted; wire							
rootballed	400.00	3.10	61.23	5.34	522.34	nr	**588.91**
Buxus sempervirens (box); truncated							
pyramids							
500 mm high; 3 × transplanted;							
container grown or rootballed	85.00	1.50	29.63	4.45	160.62	nr	**194.70**
900 mm high; 4 × transplanted; wire							
rootballed	310.00	2.75	54.31	4.45	435.15	nr	**493.91**
Buxus sempervirens (box); truncated							
cone							
1300 mm high; 5 × transplanted; wire							
rootballed	960.00	3.10	61.23	5.34	1082.34	nr	**1148.91**
Buxus sempervirens (box); cubes							
500 mm square; 4 × transplanted;							
rootballed	85.00	1.50	29.63	4.45	160.62	nr	**194.70**
900 mm square; 5 × transplanted;							
wire rootballed	365.00	2.75	54.31	4.45	490.15	nr	**548.91**
Taxus baccata (yew); balls							
500 mm dia.; 4 × transplanted;							
rootballed	60.00	1.50	29.63	4.45	135.62	nr	**169.70**
900 mm dia.; 5 × transplanted; wire							
rootballed	185.00	2.75	54.31	4.45	310.15	nr	**368.91**
1300 mm dia.; 6 × transplanted; wire							
rootballed	570.00	3.10	61.23	5.34	692.34	nr	**758.91**

Q31 PLANTING

Item Excluding site overheads and profit	PC £	Labour hours	Labour £	Plant £	Material £	Unit	Total rate £
TOPIARY – cont							
Clipped topiary							
Taxus baccata (yew); cones							
800 mm high; 4 × transplanted; wire rootballed	60.00	1.50	29.63	4.45	135.62	nr	169.70
1500 mm high; 5 × transplanted; wire rootballed	145.00	2.00	39.50	4.45	238.12	nr	282.07
2500 mm high; 6 × transplanted; wire rootballed	480.00	3.10	61.23	5.34	602.34	nr	668.91
Taxus baccata (yew); cubes							
500 mm square; 4 × transplanted; rootballed	70.00	1.50	29.63	4.45	145.62	nr	179.70
900 mm square; 5 × transplanted; wire rootballed	245.00	2.75	54.31	4.45	370.15	nr	428.91
Taxus baccata (yew); pyramids							
900 mm high; 4 × transplanted; wire rootballed	125.00	1.50	29.63	4.45	200.62	nr	234.70
1500 mm high; 5 × transplanted; wire rootballed	250.00	3.10	61.23	5.34	372.34	nr	438.91
2500 mm high; 6 × transplanted; wire rootballed	700.00	4.00	79.00	8.91	857.01	nr	944.92
Carpinus betulus (common hornbeam); columns; round base							
800 mm wide × 2000 mm high; 4 × transplanted; wire rootballed	155.00	3.10	61.23	5.34	277.34	nr	343.91
800 mm wide × 2750 mm high; 4 × transplanted; wire rootballed	305.00	4.00	79.00	8.91	462.01	nr	549.92
Carpinus betulus (common hornbeam); cones							
3 m high; 5 × transplanted; wire rootballed	275.00	4.00	79.00	8.91	432.01	nr	519.92
4 m high; 5 × transplanted; wire rootballed	570.00	5.00	98.75	8.91	756.23	nr	863.89
Carpinus betulus 'Fastigiata'; pyramids							
4 m high; 6 × transplanted; wire rootballed	970.00	5.00	98.75	8.91	1156.23	nr	1263.89
7 m high; 7 × transplanted; wire rootballed	2650.00	7.50	148.13	13.36	2879.98	nr	3041.47

Q31 PLANTING

Item Excluding site overheads and profit	PC £	Labour hours	Labour £	Plant £	Material £	Unit	Total rate £
LIVING WALLS							
ANS Living Wall; soil-based system; Scotscape Ltd; self contained, irrigated vertical planting system; HDPE Modules, fully-established upon installation; saturated weight 76 kg per m²; access costs excluded ANS System units 250 × 500 × 100 mm overall; fixed to horizontal hanging rails on softwood batten and DPM; planting depth 150 mm at 30 degree angle to vertical surface; inclusive of automatic dripline irrigation system and guttering with running outlets to drain (not included)							
areas above 35 m²	–	–	–	–	–	m²	562.00
areas below 35 m²	–	–	–	–	–	m²	669.50
Fully automatic irrigation systems to the above; comprising breaktank, station controller, solenoid valves, plant feeders and dripline emitters							
areas above 35–60 m²	–	–	–	–	–	nr	3995.00
areas below 35 m²	–	–	–	–	–	nr	1560.00
Biotecture Living Wall; Scotscape Ltd; hydroponic based system; self contained, vertical planting system; for internal or load restricted installations Biotecture; 73 mm thick backless cladding panels on 18 mm Eco sheet backing board fixt to timber battens; 4.5 mm void former of geocomposite root barrier and waterproof backing sheet; all with stainless steel fixings; drainage and guttering							
areas above 35 m²	–	–	–	–	–	m²	457.62
areas below 35 m²	–	–	–	–	–	m²	530.80
Fully automatic irrigation systems to the above; comprising breaktank, submersible pump; filter, regulator station controller, solenoid valves, plant feeders and dripline emitters							
areas above 35–60 m²	–	–	–	–	–	nr	7462.50
areas below 35 m²	–	–	–	–	–	nr	2611.00
Preliminary costs for all living walls general preliminaries	–	–	–	–	–	nr	–
design costs; site visits, consultations and presentations	–	–	–	–	–	nr	345.00

Q31 PLANTING

Item Excluding site overheads and profit	PC £	Labour hours	Labour £	Plant £	Material £	Unit	Total rate £
LIVING WALLS – cont							
Maintenance to living walls							
Annual maintenance to living walls; regular visits to maintain planting and systems; inclusive of feeding; pest control and calibration of irrigation systems							
areas 35–60 m²	–	–	–	–	–	nr	3663.00
areas below 35 m²	–	–	–	–	–	nr	1898.00
Accessories to living walls							
planters to base of living wall in GRP; black; inclusive of drainage and planting	–	–	–	–	–	m	308.75
Green Screen; Mobilane Ltd							
Fully installed vertical green screen of Hedera hibernica; 65 plants per linear m; all on 5 mm galvanized steel weldmesh; attached to existing wall fence or hoarding							
screen 1.80 high	–	–	–	–	95.00	m	95.00
screen 2.20 m high	–	–	–	–	170.00	m	170.00
screen 4.00 m high	–	–	–	–	300.00	m	300.00
HERBACEOUS/GROUNDCOVER/ BULB/BEDDING PLANTING							
Herbaceous and groundcover planting							
Herbaceous plants; PC £1.50/nr; including forming planting holes in cultivated ground (cultivating not included); backfilling with excavated material; 1 litre containers							
average 4 plants per m²; 500 mm centres	6.00	0.09	1.84	–	6.00	m²	7.84
average 6 plants per m²; 408 mm centres	9.00	0.14	2.77	–	9.00	m²	11.77
average 8 plants per m²; 354 mm centres	12.00	0.19	3.69	–	12.00	m²	15.69
Note: For machine incorporation of fertilizers and soil conditioners see 'Cultivation'.							
Bulb planting							
Bulbs; including forming planting holes in cultivated area (cultivating not included); backfilling with excavated material							
small	13.00	0.83	16.46	–	13.00	100 nr	29.46
medium	22.00	0.83	16.46	–	22.00	100 nr	38.46
large	25.00	0.91	17.95	–	25.00	100 nr	42.95

Q31 PLANTING

Item Excluding site overheads and profit	PC £	Labour hours	Labour £	Plant £	Material £	Unit	Total rate £
Bulbs; in grassed area; using bulb planter; including backfilling with screened topsoil or peat and cut turf plug							
small	13.00	1.67	32.92	–	13.00	100 nr	**45.92**
medium	22.00	1.67	32.92	–	22.00	100 nr	**54.92**
large	25.00	2.00	39.50	–	25.00	100 nr	**64.50**
Aquatic planting							
Aquatic plants; in prepared growing medium in pool; plant size 2–3 litre containerized (plants not included)	–	0.04	0.79	–	–	nr	**0.79**
MARKET PRICES OF NATIVE SPECIES							
Oakover Nurseries Ltd; the following prices are typically some of the more popular species used in native plantations; prices vary between species; readers should check the catalogue of the supplier							
1+1; 1 year seedling transplanted and grown for a year							
30–40 cm	–	–	–	–	0.48	nr	**0.48**
40–60 cm	–	–	–	–	0.55	nr	**0.55**
1+2; 1 year seedling transplanted and grown for 2 years							
60–80 cm	–	–	–	–	0.65	nr	**0.65**
Whips							
80–100 cm	–	–	–	–	0.90	nr	**0.90**
100–125 cm	–	–	–	–	1.40	nr	**1.40**
Feathered trees							
125–150 cm	–	–	–	–	2.40	nr	**2.40**
150–180 cm	–	–	–	–	2.70	nr	**2.70**
180–200 cm	–	–	–	–	5.50	nr	**5.50**
200–250 cm	–	–	–	–	8.00	nr	**8.00**
250–300 cm	–	–	–	–	8.93	nr	**8.93**
Native species planting							
Forming planting holes; in cultivated ground (cultivating not included); by mechanical auger; trimming holes by hand; depositing excavated material alongside holes							
250 mm dia.	–	0.03	0.66	0.05	–	nr	**0.71**
250 × 250 mm	–	0.04	0.79	0.09	–	nr	**0.88**
300 × 300 mm	–	0.08	1.48	0.12	–	nr	**1.60**

Q31 PLANTING

Item Excluding site overheads and profit	PC £	Labour hours	Labour £	Plant £	Material £	Unit	Total rate £
MARKET PRICES OF NATIVE SPECIES – cont							
Native species planting – cont							
Hand excavation; forming planting holes; in cultivated ground (cultivating not included); depositing excavated material alongside holes							
100 × 100 × 100 mm deep; with							
mattock or hoe	–	0.01	0.13	–	–	nr	**0.13**
250 × 250 × 300 mm deep	–	0.04	0.79	–	–	nr	**0.79**
300 × 300 × 300 mm deep	–	0.06	1.10	–	–	nr	**1.10**
TREE/SHRUB PROTECTION							
Tree planting; tree protection – General							
Preamble: Care must be taken to ensure that tree grids and guards are removed when trees grow beyond the specified dia. of guard.							
Tree planting; tree protection							
Tree tube; olive green							
1200 mm high × 80 × 80 mm	1.16	0.07	1.32	–	1.16	nr	**2.48**
stakes; 1500 mm high for Crowders							
Tree Tube; driving into ground	0.66	0.05	0.99	–	0.66	nr	**1.65**
Expandable plastic tree guards; including 25 mm softwood stakes							
500 mm high	0.55	0.17	3.29	–	0.55	nr	**3.84**
1.00 m high	0.64	0.17	3.29	–	0.64	nr	**3.93**
English Woodlands; Weldmesh tree guards; nailing to tree stakes (tree stakes not included)							
1800 mm high × 300 mm dia.	18.00	0.25	4.94	–	18.00	nr	**22.94**
J. Toms Ltd; spiral rabbit guards; clear or brown							
450 × 38 mm	0.20	0.03	0.66	–	0.20	nr	**0.86**
610 × 38 mm	0.22	0.03	0.66	–	0.22	nr	**0.88**
English Woodlands; Plastic Mesh Tree Guards; black; supplied in 50 m rolls							
13 × 13 mm small mesh; roll width							
60 cm	1.10	0.04	0.71	–	1.44	nr	**2.15**
13 × 13 mm small mesh; roll width							
120 cm	2.06	0.06	1.16	–	2.40	nr	**3.56**
Tree guards of 3 nr 2.40 m × 100 mm stakes; driving 600 mm into firm ground; bracing with timber braces at top and bottom; including 3 strands barbed wire	0.48	1.00	19.75	–	0.48	nr	**20.23**
English Woodlands; strimmer guard in heavy duty black plastic; 225 mm high	2.42	0.07	1.32	–	2.42	nr	**3.74**

Q31 PLANTING

Item Excluding site overheads and profit	PC £	Labour hours	Labour £	Plant £	Material £	Unit	Total rate £
Tubex Ltd; Standard Treeshelter inclusive of 25 mm stake; prices shown for quantities of 500 nr							
0.6 m high	0.70	0.05	0.99	–	0.92	nr	**1.91**
0.75 m high	0.84	0.05	0.99	–	1.16	nr	**2.15**
1.2 m high	1.16	0.05	0.99	–	1.53	nr	**2.52**
1.5 m high	1.58	0.05	0.99	–	2.03	nr	**3.02**
Tubex Ltd; Shrubshelter inclusive of 25 mm stake; prices shown for quantities of 500 nr							
Ecostart shelter for forestry transplants and seedlings	0.79	0.07	1.32	–	0.79	nr	**2.11**
0.6 m high	1.99	0.07	1.32	–	2.21	nr	**3.53**
0.75 m high	1.99	0.07	1.32	–	2.27	nr	**3.59**
Extra over trees for spraying with antidesiccant spray; Wiltpruf selected standards; standards; light							
standards	2.62	0.20	3.95	–	2.62	nr	**6.57**
standards; heavy standards	4.36	0.25	4.94	–	4.36	nr	**9.30**
FORESTRY PLANTING							
Forestry planting							
Deep ploughing rough ground to form planting ridges at							
2.00 m centres	–	0.63	12.34	13.47	–	100 m²	**25.81**
3.00 m centres	–	0.59	11.62	12.68	–	100 m²	**24.30**
4.00 m centres	–	0.40	7.90	8.62	–	100 m²	**16.52**
Notching plant forestry seedlings; T or L notch	41.00	0.75	14.81	–	41.00	100 nr	**55.81**
Turf planting forestry seedlings	41.00	2.00	39.50	–	41.00	100 nr	**80.50**
Tree tubes; to young trees	182.00	0.30	5.92	–	182.00	100 nr	**187.92**
Cleaning and weeding around seedlings; once	–	0.50	9.88	–	–	100 nr	**9.88**
Treading in and firming ground around seedlings planted; at 2500 per ha after frost or other ground disturbance; once	–	0.33	6.58	–	–	100 nr	**6.58**
Beating up initial planting; once (including supply of replacement seedlings at 10% of original planting)	4.10	0.25	4.94	–	4.10	100 nr	**9.04**

Q31 PLANTING

Item Excluding site overheads and profit	PC £	Labour hours	Labour £	Plant £	Material £	Unit	Total rate £
OPERATIONS AFTER PLANTING							
Operations after planting							
Initial cutting back to shrubs and hedge plants; including disposal of all cuttings	–	1.00	19.75	–	–	100 m^2	**19.75**
Mulch; Melcourt Industries Ltd; Bark Nuggets®; to plant beds; delivered in 80 m^3 loads; maximum distance 25 m							
50 mm thick	2.19	0.04	0.88	–	2.19	m^2	**3.07**
75 mm thick	3.29	0.07	1.32	–	3.29	m^2	**4.61**
100 mm thick	4.38	0.09	1.76	–	4.38	m^2	**6.14**
Mulch; Melcourt Industries Ltd; Bark Nuggets®; to plant beds; delivered in 25 m^3 loads; maximum distance 25 m							
50 mm thick	2.55	0.03	0.58	–	2.55	m^2	**3.13**
75 mm thick	4.31	0.07	1.32	–	4.31	m^2	**5.63**
100 mm thick	5.74	0.09	1.76	–	5.74	m^2	**7.50**
Mulch; Melcourt Industries Ltd; Amenity Bark Mulch FSC; to plant beds; delivered in 80 m^3 loads; maximum distance 25 m							
50 mm thick	1.40	0.04	0.88	–	1.40	m^2	**2.28**
75 mm thick	2.10	0.07	1.32	–	2.10	m^2	**3.42**
100 mm thick	2.80	0.09	1.76	–	2.80	m^2	**4.56**
Mulch; Melcourt Industries Ltd; Amenity Bark Mulch FSC; to plant beds; delivered in 25 m^3 loads; maximum distance 25 m							
50 mm thick	2.08	0.04	0.88	–	2.08	m^2	**2.96**
75 mm thick	3.12	0.07	1.32	–	3.12	m^2	**4.44**
100 mm thick	4.16	0.09	1.76	–	4.16	m^2	**5.92**
Mulch mats; English Woodlands; lay mulch mat to planted area or plant station; mat secured with metal J pins 240 × 3 mm							
Mats to individual plants and plant stations							
Hemcore Biodegradable; 50 × 50 cm; square mat	0.75	0.05	0.99	–	1.11	nr	**2.10**
Woven Polypropylene; 50 × 50 cm; square mat	0.24	0.05	0.99	–	0.60	nr	**1.59**
Woven Polypropylene; 1 × 1 m; square mat	0.68	–	–	–	–	nr	**–**
Hedging mats							
Woven Polypropylene; 1 × 100 m roll; hedge planting	0.39	0.01	0.25	–	0.57	m^2	**0.82**
General planting areas							
Plantex membrane; 1 × 14 m roll; weed control	0.88	0.01	0.20	–	1.06	m^2	**1.26**

Q31 PLANTING

Item Excluding site overheads and profit	PC £	Labour hours	Labour £	Plant £	Material £	Unit	Total rate £
Fertilizers; application during aftercare period							
Fertilizers; in top 150 mm of topsoil at 35 g/m^2							
Mascot Microfine; turf fertilizer; 8+0+6 + 2% Mg + 4% Fe	5.29	0.12	2.42	–	5.29	100 m^2	**7.71**
Enmag; controlled release fertilizer (8–9 months); 11+22+09	8.27	0.12	2.42	–	8.27	100 m^2	**10.69**
Mascot Outfield; turf fertilizer; 8+12+8	4.64	0.12	2.42	–	4.64	100 m^2	**7.06**
Mascot Outfield; turf fertilizer; 9+5+5	3.67	0.12	2.42	–	3.67	100 m^2	**6.09**
Fertilizers; in top 150 mm of topsoil at 70 g/m^2							
Mascot Microfine; turf fertilizer; 8+0+6 + 2% Mg + 4% Fe	10.57	0.12	2.42	–	10.57	100 m^2	**12.99**
Enmag; controlled release fertilizer (8–9 months); 11+22+09	16.55	0.12	2.42	–	16.55	100 m^2	**18.97**
Mascot Outfield; turf fertilizer; 8+12+8	9.28	0.12	2.42	–	9.28	100 m^2	**11.70**
Mascot Outfield; turf fertilizer; 9+5+5	7.35	0.12	2.42	–	7.35	100 m^2	**9.77**
AFTERCARE AS PART OF A LANDSCAPE CONTRACT							
Maintenance operations (Note: the following rates apply to aftercare maintenance executed as part of a landscaping contract only – Please see section Q35 of this publication for further maintenance rates)							
Weeding and hand forking planted areas; including disposing weeds and debris on site; areas maintained weekly	–	–	0.08	–	–	m^2	**0.08**
Weeding and hand forking planted areas; including disposing weeds and debris on site; areas maintained monthly	–	0.01	0.20	–	–	m^2	**0.20**
Extra over weeding and hand forking planted areas for disposing excavated material off site; loaded to maintenance vehicel and transported to the contractors yard							
green waste for composting	–	–	–	–	–	m^3	**2.60**
green waste to contractors yard deposited in skips	–	–	–	36.66	2.60	m^3	**39.26**

Q31 PLANTING

Item Excluding site overheads and profit	PC £	Labour hours	Labour £	Plant £	Material £	Unit	Total rate £
AFTERCARE AS PART OF A **LANDSCAPE CONTRACT – cont**							
Edging to plant beds; maintaing edge **against turfed surround or boundary;** **edging by half moon tool**							
planting beds							
maintaining edges	–	0.04	0.79	–	–	m	**0.79**
reforming edges; removing arisings	–	0.07	1.32	–	–	m	**1.32**
Treepits edges							
maintaining edges	–	0.10	1.98	–	–	nr	**1.98**
reforming edges; removing arisings	–	0.20	3.95	–	–	nr	**3.95**
Sweeping and cleaning pathways and **surfaces**							
Sweep paths or surfaces to high footfall public or commercial areas							
weekly visits	–	–	–	–	–	m^2	–
sweeping snow	–	0.03	0.49	–	–	m^2	**0.49**
gritting and clearing ice	–	0.03	0.49	–	0.03	m^2	**0.52**
Top up mulch at the end of the **12 month maintenance period to the** **full thickness as shown below**							
Mulch; Melcourt Industries Ltd; Bark Nuggets®; to plant beds; delivered in $80\,m^3$ loads; maximum distance 25 m							
50 mm thick	0.44	0.03	0.56	–	0.44	m^2	**1.00**
75 mm thick	0.66	0.01	0.10	–	0.66	m^2	**0.76**
Mulch; Melcourt Industries Ltd; Bark Nuggets®; to plant beds; delivered in $25\,m^3$ loads; maximum distance 25 m							
50 mm thick	0.57	0.03	0.56	–	0.57	m^2	**1.13**
75 mm thick	0.86	0.03	0.66	–	0.86	m^2	**1.52**
Mulch; Melcourt Industries Ltd; Amenity Bark Mulch FSC; to plant beds; delivered in $80\,m^3$ loads; maximum distance 25 m							
50 mm thick	0.70	0.03	0.56	–	0.70	m^2	**1.26**
75 mm thick	0.98	0.03	0.66	–	0.98	m^2	**1.64**
Mulch; Melcourt Industries Ltd; Amenity Bark Mulch FSC; to plant beds; delivered in $25\,m^3$ loads; maximum distance 25 m							
50 mm thick	1.04	0.03	0.49	–	1.04	m^2	**1.53**
75 mm thick	1.35	0.03	0.56	–	1.35	m^2	**1.91**
Fertilizers; at $35\,g/m^2$							
Mascot Microfine; turf fertilizer; 8+0+6 + 2% Mg + 4% Fe	5.29	0.12	2.42	–	5.29	$100\,m^2$	**7.71**
Enmag; controlled release fertilizer (8–9 months); 11+22+09	8.27	0.12	2.42	–	8.27	$100\,m^2$	**10.69**
Mascot Outfield; turf fertilizer; 8+12+8	4.64	0.12	2.42	–	4.64	$100\,m^2$	**7.06**
Mascot Outfield; turf fertilizer; 9+5+5	3.67	0.12	2.42	–	3.67	$100\,m^2$	**6.09**

Q31 PLANTING

Item Excluding site overheads and profit	PC £	Labour hours	Labour £	Plant £	Material £	Unit	Total rate £
Fertilizers; at 70 g/m^2							
Mascot Microfine; turf fertilizer; 8+0+6							
+ 2% Mg + 4% Fe	10.57	0.12	2.42	–	10.57	100 m^2	**12.99**
Enmag; controlled release fertilizer (							
8–9 months); 11+22+09	16.55	0.12	2.42	–	16.55	100 m^2	**18.97**
Mascot Outfield; turf fertilizer; 8+12+8	9.28	0.12	2.42	–	9.28	100 m^2	**11.70**
Mascot Outfield; turf fertilizer; 9+5+5	7.35	0.12	2.42	–	7.35	100 m^2	**9.77**
Watering planting; evenly; at a rate of 5 l/m^2							
using hand-held watering equipment	–	0.25	4.94	–	–	100 m^2	**4.94**
using sprinkler equipment and with sufficient water pressure to run one 15 m radius sprinkler	–	0.14	2.75	–	–	100 m^2	**2.75**
using movable spray lines powering three sprinkler heads with a radius of 15 m and allowing for 60% overlap (irrigation machinery costs not included)	–	0.02	0.31	–	–	100 m^2	**0.31**

Q32 PLANTING IN SPECIAL ENVIRONMENTS – GREEN ROOF SYSTEMS

Item Excluding site overheads and profit	PC £	Labour hours	Labour £	Plant £	Material £	Unit	Total rate £
PREAMBLE							
Preamble: A variety of systems are available which address all the varied requirements for a successful Green Roof. For installation by approved contractors only. The prices shown are for budgeting purposes only, as each installation is site specific and may incorporate some or all of the resources shown. Specifiers should verify that the systems specified include for design liability and inspections by the suppliers. The systems below assume commercial insulation levels are required to the space below the proposed Green Roof. Extensive Green Roofs are those of generally lightweight construction with low maintenance planting and shallow soil designed for aesthetics only; Intensive Green Roofs are designed to allow use for recreation and trafficking. They require more maintenance and allow a greater variety of surfaces and plant types.							
INTENSIVE GREEN ROOFS							
Intensive Green Roof; Bauder Ltd; soil-based systems able to provide a variety of hard and soft landscaping; laid to the surface of an unprepared roof deck							
Vapour barrier laid to prevent intersticial condensation from spaces below the roof applied by torching to the roof deck							
VB4-Expal aluminium lined	–	–	–	–	–	m²	**13.27**
Insulation laid and hot bitumen bonded to vapour barrier							
PIR Insulation 100 mm	–	–	–	–	–	m²	**33.42**
Underlayer to receive root barrier partially bonded to insulation by torching							
G4E	–	–	–	–	–	m²	**13.85**
Root barrier							
Plant E; chemically treated root resistant capping sheet fully bonded to G4E underlayer by torching	–	–	–	–	–	m²	**20.64**
Slip layers to absorb differential movement							
PE Foil; 2 layers laid to root barriers	–	–	–	–	–	m²	**3.34**

Q32 PLANTING IN SPECIAL ENVIRONMENTS – GREEN ROOF SYSTEMS

Item Excluding site overheads and profit	PC £	Labour hours	Labour £	Plant £	Material £	Unit	Total rate £
Optional protection layer to prevent mechanical damage							
Protection Mat; 6 mm thick rubber matting; loose laid	–	–	–	–	–	m²	12.46
Drainage medium laid to root barrier							
Drainage Board; free draining EPS 50 mm thick	–	–	–	–	–	m²	13.38
Reservoir Board; up to 21.5 litre water storage capacity EPS; 50 mm thick	–	–	–	–	–	m²	15.45
Filtration to prevent soil migration to drainage system							
Filter Fleece; 3 mm thick polyester geotextile; loose laid over drainage/ reservoir layer	–	–	–	–	–	m²	2.89
For hard landscaped areas incorporate rigid drainage board laid to the protection mat							
PLT 60 drainage board	–	–	–	–	–	m²	31.08
Intensive Green Roof Systems; BBS Green roofing Ltd; components laid to the inverted build-up; insulation over waterproof rootproof membrane							
Drainage layer; BBS TDC; recycled polypropylene; providing water reservoir, multi-directional drainage and mechanical damage protection							
BBS TDC40; 40 mm deep	–	–	–	–	–	m²	14.70
BBS TDC60; 60 mm deep	–	–	–	–	–	m²	18.20
Optional drainage unfill to BBSTDC layers							
BBS TDC 40; 16 litres/m²	–	–	–	–	–	m²	5.80
BBS TDC60; 26 litres/m²	–	–	–	–	–	m²	7.60
Filter sheet rolled out to drainage layer							
BBSF filter sheet	–	–	–	–	–	m²	2.80
Growing medium and substrates							
BBS INT – Intensive substrate; lightweight growing medium laid to filter sheet 300 mm deep	–	–	–	–	–	m²	38.20
BBS SMi – Semi-intensive substrate; Lightweight growing meduim laid to filter sheet 200 mm deep	–	–	–	–	–	m²	26.76

Q32 PLANTING IN SPECIAL ENVIRONMENTS – GREEN ROOF SYSTEMS

Item Excluding site overheads and profit	PC £	Labour hours	Labour £	Plant £	Material £	Unit	Total rate £
EXTENSIVE GREEN ROOFS							
Extensive Green Roof System; BBS Green Roofing Ltd; low maintenance Sedum system for flat roof 1:60 min falls up to 15°							
Moisture/protection layer; underlay moisture layer membrane protection							
BBSU	–	–	–	–	–	m²	2.60
Drainage layer; BBS TDC recycled polypropylene; providing water reservoir, multi-directional drainage							
BBS TCD20 20 mm deep	–	–	–	–	–	m²	9.56
Filter sheet rolled to drainage layer							
BBSF filter sheet	–	–	–	–	–	m²	2.80
Growing media							
BBS EXT; extensive substrate; lightweight growing medium laid to filter sheet; 80 mm thick	–	–	–	–	–	m²	14.44
BBS SED; pre grown Sedum turf	–	–	–	–	–	m²	23.09
Extensive Lightweight Green Roof System; BBS Green Roofing Ltd; low maintenance Sedum nutrimat system for flat roof 1:60 min falls up to 15°; 60 kg/m² saturated							
Moisture/Protection layer – Underlay moisture layer which protects the membrane							
BBSU	–	–	–	–	–	m²	2.60
Drainage/reservoir / root-zone / filter – recycled PUR foam hydroponic mat							
BBS Nutrimat	–	–	–	–	–	m²	19.88
BBS EXT – Extensive substrate; lightweight growing medium laid to filter sheet 30 mm	–	–	–	–	–	m²	8.20
BBS EXT – Extensive substrate; lightweight growing medium laid to filter sheet 80 mm	–	–	–	–	–	m²	14.44
BBS SED – Pre grown Sedum turf	–	–	–	–	–	m²	23.09

Q32 PLANTING IN SPECIAL ENVIRONMENTS – GREEN ROOF SYSTEMS

Item Excluding site overheads and profit	PC £	Labour hours	Labour £	Plant £	Material £	Unit	Total rate £
EXTENSIVE PITCH ROOF SYSTEMS							
Extensive Lightweight Green Roof System; BBS Green Roofing Ltd; low maintenance Sedum system for Pitch roof 3° to 20°. 60/kg/m² saturated NB// 20° – 50° requires extra restraints							
BBSFG Nutrimat – Fully Grown Sedum Tile – modular assembly to form monolithic green roof layer low maintenance							
Sedum system; 60 kg/m² saturated	–	–	–	–	–	m²	68.50
EXTENSIVE GREEN ROOFS – SOIL FREE							
Extensive Green Roof System; Bauder Ltd; low maintenance soil free system incorporating single layer growing and planting medium							
Vapour barrier laid to prevent intersticial condensation applied by torching to the roof deck							
VB4-Expal aluminium lined	–	–	–	–	–	m²	13.27
Insulation laid and hot bitumen bonded to vapour barrier							
PIR Insulation 100 mm	–	–	–	–	–	m²	33.42
Underlayer to receive root barrier partially bonded to insulation by torching							
G4E	–	–	–	–	–	m²	13.85
Root barrier							
Plant E; chemically treated root resistant capping sheet fully bonded to G4E underlayer by torching	–	–	–	–	–	m²	20.64
LANDSCAPE OPTIONS TO GREEN ROOF SYSTEMS							
Landscape options to the above systems; Bauder Ltd							
Hydroplanting system; to Extensive Green Roof as detailed above							
SDF mat 20 mm thick drainage layer; loose laid	–	–	–	–	–	m²	10.27
Xeroflor vegetation blanket; Bauder Ltd; to Extensive Green Roof as detailed above							
Xeroflor Xf 301 pre-cultivated sedum blanket incorporating 800 gram recycled fibre water retention layer; laid loose	–	–	–	–	–	m²	44.98

Q32 PLANTING IN SPECIAL ENVIRONMENTS – GREEN ROOF SYSTEMS

Item Excluding site overheads and profit	PC £	Labour hours	Labour £	Plant £	Material £	Unit	Total rate £
LANDSCAPE OPTIONS TO GREEN ROOF SYSTEMS – cont							
Reservoir and drainage boards; Bauder; non-specialist installed products for laying under pavings or planting for collection or drainage DSE20 Drainage/Reservoir Board; 20 mm thick drainage board	–	–	–	–	–	m²	10.44
DSE40 Drainage/Reservoir Board; for intensive green roof applications	–	–	–	–	–	m²	19.86
Versicell 20 Drainage Board; specialist installed drainage layer for intensive green roof applications; predominantly hard landscape	–	–	–	–	–	m²	21.88
Bauder Extensive Substrate; 80 mm deep; for use with sedum and hardy perennials and biodiverse vegetation	–	–	–	–	–	m²	18.15
Bauder Intensive Substrate; 200 mm deep; for use in traditional green roof applications e.g. grass and bedding plants	–	–	–	–	–	m²	43.12
Green Roof components; BBS Green roof systems Outlet inspection chambers							
BBS CHM; 150 mm deep	–	–	–	–	–	nr	70.00
BBS CHM; 150 mm deep	–	–	–	–	–	nr	85.00
BBS CHM; 150 mm deep	–	–	–	–	–	nr	95.00
WATERPROOFING AND DRAINAGE OPTIONS							
Warm Roof waterproofing systems; to roof surfaces to receive Green Roof systems Vapour barriers							
Aluminium lined vapour barrier; 2 mm thick; bonded in hot bitumen to the roof deck	–	–	–	–	–	m²	11.16
Korklite insulation bonded to the vapour barrier in hot bitumen; U value dependent; 80 mm thick	–	–	–	–	–	m²	21.13
Universal 2 mm thick underlayer; fully bonded to the insulation	–	–	–	–	–	m²	9.09
Anti-Root cap sheet impregnated with root resisting chemical bonded to the underlayer	–	–	–	–	–	m²	18.70

Q32 PLANTING IN SPECIAL ENVIRONMENTS – GREEN ROOF SYSTEMS

Item Excluding site overheads and profit	PC £	Labour hours	Labour £	Plant £	Material £	Unit	Total rate £
Waterproofing to upstands; Bauder Ltd Bauder Vapour Barrier; Bauder G4E & Bauder Plant E							
up to 200 mm high	–	–	–	–	–	m	**19.11**
up to 400 mm high	–	–	–	–	–	m	**26.34**
up to 600 mm high	–	–	–	–	–	m	**31.23**
Inverted waterproofing systems; to roof surfaces to receive Green Roof systems Monolithic hot melt rubberised bitumen; applied in two 3 mm layers incorporating a polyester reinforcing sheet with 4 mm thick protection sheet and chemically impregnated root barrier; fully bonded; applied to plywood or suitably prepared wood float finish and primed concrete deck or screeds							
10 mm thick	–	–	–	–	–	m²	**35.62**
Roofmate; extruded polystyrene insulation; optional system; thickness to suit required U value; calculated at design stage; indicative thicknesses; laid to Hydrotech 6125							
0.25 U value; average requirement 120 mm	–	–	–	–	–	m²	**24.69**

Q35 LANDSCAPE MAINTENANCE

Item Excluding site overheads and profit	PC £	Labour hours	Labour £	Plant £	Material £	Unit	Total rate £
LONG TERM MAINTENANCE							
Preamble: Long-term landscape maintenance							
Maintenance on long-term contracts differs in cost from that of maintenance as part of a landscape contract. In this section the contract period is generally 3–5 years. Staff are generally allocated to a single project only and therefore productivity is higher whilst overhead costs are lower. Labour costs in this section are lower than the costs used in other parts of the book. Machinery is assumed to be leased over a five year period and written off over the same period. The costs of maintenance and consumables for the various machinery types have been included in the information that follows. Finance costs for the machinery have not been allowed for.							
The rates shown below are for machines working in unconfined contiguous areas. Users should adjust the times and rates if working in smaller spaces or spaces with obstructions.							
MARKET PRICES OF LANDSCAPE CHEMICALS							
The following table provides the areas of spraying in each planting scenario for plants which require 1.00 m dia. weed free circles							
Area of weed free circles required in each planting density							
plants at 500 mm centres;							
400 nr/100 m²	–	–	–	–	400.00	m²	**400.00**
plants at 600 m centres;							
278 nr/100 m²	–	–	–	–	278.00	m²	**278.00**
plants at 750 mm centres;							
178 nr/100 m²	–	–	–	–	178.00	m²	**178.00**
plants at 1.00 m centres;							
100 nr/100 m²	–	–	–	–	100.00	m²	**100.00**
plants at 1.50 m centres; 4444 nr/ha	–	–	–	–	3490.00	m²	**3490.00**
plants at 1.75 m centres; 3265 nr/ha	–	–	–	–	2564.00	m²	**2564.00**
plants at 2.00 m centres; 2500 nr/ha	–	–	–	–	1963.00	m²	**1963.00**

Q35 LANDSCAPE MAINTENANCE

Item Excluding site overheads and profit	PC £	Labour hours	Labour £	Plant £	Material £	Unit	**Total rate £**
MARKET PRICES OF LANDSCAPE CHEMICALS AT SUGGESTED APPLICATION RATES Note: All chemicals are standard knapsack or backpack applied unless specifically stated as CDA or TDA.							
Total herbicides							
Herbicide applications; CDA; chemical application via low pressure specialized wands to landscape planting; application to maintain 1.00 m dia. clear circles (0.79 m^2) around new planting Vanquish Biactive; ALS Ltd; Bayer; enhanced movement glyphosate; application rate 15 litre/ha							
plants at 1.50 m centres; 4444 nr/ha	–	–	–	–	59.46	ha	**59.46**
plants at 1.75 m centres; 3265 nr/ha	–	–	–	–	43.88	ha	**43.88**
plants at 2.00 m centres; 2500 nr/ha	–	–	–	–	33.65	ha	**33.65**
mass spraying	–	–	–	–	170.53	ha	**170.53**
spot spraying	–	–	–	–	1.71	ha	**1.71**
Total herbicides; CDA; low pressure sustainable applications Hilite; Nomix Enviro Ltd; glyphosate oil-emulsion; Total Droplet Control (TDC formerly CDA); low volume sustainable chemical applications; formulation controls annual and perennial grasses and non-woody broad-leaved weeds in amenity and industrial areas, parks, highways, paved areas and factory sites; also effective for clearing vegetation prior to planting of ornamental species, around the bases of trees, shrubs and roses, as well as in forestry areas							
10 l/ha	–	–	–	–	1.63	100 m^2	**1.63**
7.5 l/ ha	–	–	–	–	1.22	100 m^2	**1.22**
Tribute Plus; Nomix Enviro Ltd; TDA; MCPA/dicamba/mecaprop-p; TDC; for selective control of broad-leaved weeds in turf							
6 l/ha; white clover, plantains, creeping buttercup, thistle	–	–	–	–	0.66	100 m^2	**0.66**
8 l/ha; yarrow, ragwort, parsley, piert, cinquefoil, mouse eared chickweed, selfheal, pearlwort, sorrel, black meddick, dandelion and related weeds, bulbous buttercup	–	–	–	–	0.87	100 m^2	**0.87**

Q35 LANDSCAPE MAINTENANCE

Item Excluding site overheads and profit	PC £	Labour hours	Labour £	Plant £	Material £	Unit	Total rate £
MARKET PRICES OF LANDSCAPE CHEMICALS – cont							
Total herbicides – cont							
Stirrup; Rigby Taylor Ltd; glyphosate; CDA; 192 g							
10 l/ha	–	–	–	–	1.63	100 m²	1.63
7.5 l/ha	–	–	–	–	1.63	100 m²	1.63
Gallup Biograde Amenity; Rigby Taylor Ltd; glyphosate 360 g/l formulation							
general use; woody weeds; ash, beech, bracken, bramble; 3 l/ha	–	–	–	–	0.40	100 m²	0.40
annual and perennial grasses; heather (peat soils); 4 l/ha	–	–	–	–	0.53	100 m²	0.53
pre-planting; general clearance; 5 l/ha	–	–	–	–	0.66	100 m²	0.66
heather; mineral soils; 6 l/ha	–	–	–	–	0.79	100 m²	0.79
rhododendron; 10 l/ha	–	–	–	–	1.32	100 m²	1.32
Gallup Hi-aktiv Amenity; Rigby Taylor Ltd; 490 g/l formulation							
general use; woody weeds; ash, beech, bracken, bramble; 2.2 l/ha	–	–	–	–	0.37	100 m²	0.37
annual and perennial grasses; heather (peat soils); 2.9 l/ha	–	–	–	–	0.49	100 m²	0.49
pre-planting; general clearance; 3.7 l/ha	–	–	–	–	0.62	100 m²	0.62
heather; mineral soils; 4.4 l/ha	–	–	–	–	0.74	100 m²	0.74
rhododendron; 7.3 l/ha	–	–	–	–	1.22	100 m²	1.22
Discman CDA Biograde; Rigby Taylor Ltd							
application rate; 7 l/ha; light vegetation and annual weeds	–	–	–	–	1.61	100 m²	1.61
application rate; 8.5 l/ha; forestry/ woodlands; before planting, to control broad-leaved and grass weeds	–	–	–	–	1.96	100 m²	1.96
application rate; 10 l/ha; established deep-rooted annuals, perennial grasses and broad-leaved weeds	–	–	–	–	2.30	100 m²	2.30
Roundup Pro Biactive 360; Everris							
application rate; 5 l/ha; land not intended for cropping	–	–	–	–	0.39	100 m²	0.39
application rate; 10 l/ha; forestry; weed control	–	–	–	–	0.77	100 m²	0.77

Q35 LANDSCAPE MAINTENANCE

Item Excluding site overheads and profit	PC £	Labour hours	Labour £	Plant £	Material £	Unit	Total rate £
Roundup Pro Biactive 450; Everris; **450 g/litre glyphosate** application rate 4 l/ha; natural surfaces not intended to bear vegetation; amenity vegetation control; permeable surfaces overlaying soil; hard surfaces	–	–	–	–	0.35	100 m²	0.35
application rate 8 l/ha; forest; forest nursery weed control	–	–	–	–	0.69	100 m²	0.69
application rate 16 mm/l; stump application	–	–	–	–	0.14	litre	0.14
Roundup Pro Biactive 450; Everris; **enhanced movement glyphosate;** **application rate 4 litre/ha** plants at 1.50 m centres; 4444 nr/ha	50.22	–	–	–	50.22	ha	50.22
plants at 1.75 m centres; 3265 nr/ha	36.94	–	–	–	36.94	ha	36.94
plants at 2.00 m centres; 2500 nr/ha	28.26	–	–	–	28.26	ha	28.26
mass spraying	38.19	–	–	–	38.19	ha	38.19
spot spraying	1.43	–	–	–	1.43	100 m²	1.43
Dual; Nomix Enviro Ltd; glyphosate/ **sulfosulfuron; TDC; low volume** **sustainable chemical applications;** **residual effect lasting up to 6 months;** **shrub beds, gravelled areas, tree** **bases, parks, industrial areas, hard** **areas and around obstacles** application rate 9 l/ha	–	–	–	–	2.03	100 m²	2.03
Glyfos Dakar Pro; Headland Amenity **Ltd; 68% glyphosate as water** **dispersible granules; broad-leaved** **weeds and grasses** application rate 2.5 kg/ha	–	–	–	–	0.32	100 m²	0.32
Total herbicides with residual **pre-emergent action**							
Pistol; ALS; Bayer; barrier prevention **for emergence control and glyphosate** **for control of emerged weeds at a** **single application per season;** **application rate 4.5 litre/ha; 40 g/l** **diflufenican and 250 g/l glyphosate** plants at 1.50 m centres; 4444 nr/ha	–	–	–	–	45.13	ha	45.13
plants at 1.75 m centres; 3265 nr/ha	–	–	–	–	33.05	ha	33.05
plants at 2.00 m centres; 2500 nr/ha	–	–	–	–	32.34	ha	32.34

Q35 LANDSCAPE MAINTENANCE

Item Excluding site overheads and profit	PC £	Labour hours	Labour £	Plant £	Material £	Unit	Total rate £
MARKET PRICES OF LANDSCAPE CHEMICALS – cont							
Residual herbicides							
Chikara; Rigby Taylor Ltd; Belchim Crop Protection; flazasulfuron; systemic pre-emergent and early post-emergent herbicide for the control of annual and perennial weeds on natural surfaces not intended to bear vegetation and permeable surfaces over-lying soil							
150 g/ha	–	–	–	–	2.56	100 m²	**2.56**
Contact herbicides							
Jewel; Everris; carfentrazone and mecoprop-p; selective weed and mosskiller							
application rate; 1.5 kg/ha	–	–	–	–	1.28	100 m²	**1.28**
Natural Weed and Moss Spray; Headland Amenity Ltd; acetic acid							
application rate; 100 l/ha	–	–	–	–	0.45	100 m²	**0.45**
Selective herbicides CDA							
Hilite; Nomix Enviro; glyphosate oil-emulsion; TD; low volume sustainable chemical applications; formulation controls annual and perennial grasses and non-woody broad-leaved weeds in amenity and industrial areas, parks, highways, paved areas and factory sites; also effective for clearing vegetation prior to planting of ornamental species, around the bases of trees, shrubs and roses, as well as in forestry areas							
10 l/ha	–	–	–	–	1.63	100 m²	**1.63**
7.5 l/ha	–	–	–	–	1.22	100 m²	**1.22**
Tribute Plus; Nomix Enviro; MCPA/ dicamba/mecoprop-p; TDC; for selective control of broad-leaved weeds in turf							
6 l/ha; white clover, plantains, creeping buttercup, thistle	–	–	–	–	0.66	100 m²	**0.66**
8 l/ha; yarrow, ragwort, parsley, piert, cinquefoil, mouse eared chickweed, selfheal, pearlwort, sorrel, black meddick, dandelion and related weeds, bulbous buttercup	–	–	–	–	0.87	100 m²	**0.87**

Q35 LANDSCAPE MAINTENANCE

Item Excluding site overheads and profit	PC £	Labour hours	Labour £	Plant £	Material £	Unit	Total rate £
Contact and systemic combined herbicides Jewel; Everris; carfentrazone and mecoprop-p; selective weed and mosskiller							
application rate; 1.5 kg/ha	–	–	–	–	1.28	100 m²	**1.28**
Finale; ALS; Bayer; glufosinate-ammonium; controls grasses and broad-leaved weeds; it can also be used for weed control around non-edible ornamental crops including trees and shrubs and in tree nurseries; may be used to 'burn off' weeds in shrub beds prior to treatment with a residual herbicide; is also ideal for preparing sports turf for line marking							
application rate; 5 l/ha; seedlings of all species; established annual weeds and grasses as specified	–	–	–	–	0.90	100 m²	**0.90**
application rate; 8 l/ha; established annual and perennial weeds and grasses as specified in the product guide	–	–	–	–	1.44	100 m²	**1.44**
Selective herbicides; power or knapsack sprayer Junction; Rigby Taylor Ltd; selective post-emergence turf herbicide containing 6.25 g/litre florasulam plus 452 g/litre 2, 4-D; specially designed for use by contractors for maintenance treatments of a wide range of broad-leaved weeds whilst giving excellent safety to turfgrass							
application rate; 1.2 l/ha	–	–	–	–	0.62	100 m²	**0.62**
Greenor; Rigby Taylor Ltd; fluroxypyr, clopyralid and MCPA; systemic selective treatment of weeds in turfgrass							
application rate; 4 l/ha	–	–	–	–	1.06	100 m²	**1.06**
Super Selective Plus; Rigby Taylor Ltd; combination of three active ingredients (mecoprop-p, MCPA and dicamba) enables this formulation to give good control of most commonly occurring broad-leaved weeds found in sports and amenity turf							
application rate; 3.5 l/ha	–	–	–	–	0.36	100 m²	**0.36**
Crossbar Herbicide; Rigby Taylor Ltd; fluroxypyr, 2,4-D and dicamba for control of weeds in amenity turf including fine turf							
application rate; 2 l/ha	–	–	–	–	0.98	100 m²	**0.98**

Q35 LANDSCAPE MAINTENANCE

Item Excluding site overheads and profit	PC £	Labour hours	Labour £	Plant £	Material £	Unit	Total rate £
MARKET PRICES OF LANDSCAPE CHEMICALS – cont							
Selective herbicides – cont							
Intrepid 2; Everris; for broad-leaved weed control in grass							
application rate; 7700 m²/5 l	–	–	–	–	0.53	100 m²	**0.53**
Re-Act; Everris; MCPA, mecoprop-p and dicamba; selective herbicide; established managed amenity turf and newly seeded grass; controls many annual and perennial weeds; will not vaporise in hot conditions							
application rate; 9090–14,285 m²/5 l	–	–	–	–	0.51	100 m²	**0.51**
Praxys; Everris; fluroxypyr-meptyl, clopyralid and florasulam; selective, systemic post-emergence herbicide for managed amenity turf, lawns and amenity grassland with high selectivity to established and young turf							
application rate; 9090–14,285 m²/5 l	–	–	–	–	1.20	100 m²	**1.20**
Asulox; ALS; Bayer; post-emergence translocated herbicide for the control of bracken and docks							
application rate; docks; 17857 m²/5 l	–	–	–	–	0.32	100 m²	**0.32**
application rate; bracken; 4545 m²/5 l	–	–	–	–	1.27	100 m²	**1.27**
Cabadex; Headland Amenity Ltd; fluroxypyr and florasulam; newly sown and established turf (particularly speedwell, yarrow and yellow suckling clover)							
application rate 2 l/ha	–	–	–	–	0.63	100 m²	**0.63**
Relay Turf; Headland Amenity Ltd; mecoprop-p, MCPA and dicamba; many broad-leaved weeds in managed amenity turf							
application rate; 5 l/ha	–	–	–	–	0.45	100 m²	**0.45**
Blaster; Headland Amenity Ltd; triclopyr clopyralid; nettles, docks, thistles, brambles, woody weeds in amenity grassland							
application rate; 4 l/ha	–	–	–	–	1.66	100 m²	**1.66**

Q35 LANDSCAPE MAINTENANCE

Item Excluding site overheads and profit	PC £	Labour hours	Labour £	Plant £	Material £	Unit	Total rate £
Woody weed herbicides; backpack or power sprayers							
Garlon 4; Nomix Enviro Ltd; 667 g/litre triclopyr; control of brushwood, scrub, woody and perennial herbaceous weeds in amenity, forestry, rail and industrial areas; also suitable for forestry production							
2 l/ha; bramble, briar, broom, gorse, nettle	–	–	–	–	0.20	100 m²	0.20
4 l/ha; alder, birch, blackthorn, dogwood, elder, poplar, rosebay, willow-herb, sycamore	–	–	–	–	0.41	100 m²	0.41
6 l/ha; beech, box, elm, buckthorn, hazel, hornbeam, horse chestnut, lime, maple, privet, rowan, Spanish chestnut, willow, wild pear	–	–	–	–	0.61	100 m²	0.61
8 l/ha; rhododendron, ash, oak	–	–	–	–	0.82	100 m²	0.82
Tordon 22 K; Nomix Enviro; picloram; selective action; control of tough woody weeds, invasive species, scrub and deep-rooted perennials; residual action grass cover is maintained							
2.8 l/ha; white clover, creeping thistle, hogweed, plantains	–	–	–	–	0.24	100 m²	0.24
4.3–5.6 l/ha; common bird's-foot trefoil, field bindweed, creeping buttercup, burdock, cat's ear, common chickweed, colt's foot, creeping cinquefoil, dandelion, docks, hawkweed, Japanese knotweed, mugwort, common nettle, silverweed, tansy, cotton thistle, rosebay willowherb, yarrow	–	–	–	–	0.43	100 m²	0.43
8.4 l/ha; wild mignonette, common ragwort	–	–	–	–	0.73	100 m²	0.73
11.2 l/ha; bracken, brambles	–	–	–	–	1.58	100 m²	1.58
Aquatic herbicides; backpack or power sprayers							
Roundup Pro Biactive 360; Everris							
application rate; 6 l/ha; aquatic weed control	–	–	–	–	0.46	100 m²	0.46
Roundup Pro Biactive 450; Everris; 450 g/litre glyphosate							
application rate 4.8 l/ha; enclosed waters, open waters, land immediately adjacent to aquatic areas	–	–	–	–	0.42	100 m²	0.42

Prices for Measured Works

Q35 LANDSCAPE MAINTENANCE

Item Excluding site overheads and profit	PC £	Labour hours	Labour £	Plant £	Material £	Unit	Total rate £
MARKET PRICES OF LANDSCAPE CHEMICALS – cont							
Aquatic herbicides; TDC							
Conqueror; Nomix Enviro; low volume sustainable chemical applications glyphosate with aquatic approval							
10 l/ha; established annuals, deep-rooted perennials and broad-leaved weeds	–	–	–	–	0.18	100 m²	0.18
7.5 l/ha; light vegetation and small annual weeds, aquatic situations	–	–	–	–	1.32	100 m²	1.32
12.5 l/ha; aquatic situations, emergent weeds e.g. reed, grasses and watercress	–	–	–	–	0.22	100 m²	0.22
15 l/ha; aquatic situations, floating weeds such as water lilies	–	–	–	–	0.26	100 m²	0.26
Fungicides							
Masalon; Rigby Taylor Ltd; systemic							
application rate; 8 l/ha	–	–	–	–	9.11	100 m²	9.11
Defender; Rigby Taylor Ltd; broad spectrum fungicide displaying protective and curative activity against fusarium patch and red thread							
application rate; 700 g/ha	–	–	–	–	7.26	100 m²	7.26
Fusion; Rigby Taylor Ltd; contact and systemic fungicide with curative and preventative activity against major turf diseases such as fusarium, red thread, dollar spot, anthracnose and rust							
application rate; 700 g/ha	–	–	–	–	7.28	100 m²	7.28
Rayzor; Rigby Taylor Ltd; broad spectrum contact fungicide with both protectant and eradicant activity for the control of fusarium patch and red thread in managed amenity turf							
application rate; 5 l/2000 m²	–	–	–	–	9.25	100 m²	9.25
Moss control							
Jewel; Everris; carfentrazone and mecoprop-p; selective weed and moss killer							
application rate; 1.5 kg/ha	–	–	–	–	1.28	100 m²	1.28
Insect control							
Merit Turf; Bayer; insecticide for the control of chafer grubs and leatherjackets							
application rate; 10 kg/3333 m²	–	–	–	–	2.85	100 m²	2.85

Q35 LANDSCAPE MAINTENANCE

Item Excluding site overheads and profit	PC £	Labour hours	Labour £	Plant £	Material £	Unit	Total rate £
Clarification notes on labour costs in this section							
General landscape maintenance team Generally a three man team is used in this section; The column Labour hours reports team hours. The column Labour £ reports the total cost of the team for the unit of work shown							
3 man team	–	1.00	54.75	–	–	hr	**54.75**
2 man team	–	–	–	–	–	hr	**-**
LANDSCAPE CHEMICAL APPLICATION							
Chemical applications labours only							
Controlled droplet application (CDA)							
plants at 1.50 m centres; 4444 nr/ha	–	9.26	344.94	–	–	ha	**344.94**
plants at 1.50 m centres; 44.44 nr/100 m²	–	0.93	34.49	–	–	100 m²	**34.49**
plants at 1.75 m centres; 3265 nr/ha	–	6.80	253.30	–	–	ha	**253.30**
plants at 1.75 m centres; 3265 nr/ha	–	0.68	25.33	–	–	100 m²	**25.33**
plants at 2.00 m centres; 2500 nr/ha	–	5.21	194.07	–	–	ha	**194.07**
plants at 2.00 m centres; 2500 nr/ha	–	0.52	19.41	–	–	100 m²	**19.41**
spot spraying to shrub beds	–	0.20	3.95	–	–	100 m²	**3.95**
spot spraying to hard surfaces	–	0.13	2.47	–	–	100 m²	**2.47**
mass spraying	–	2.50	93.13	–	–	ha	**93.13**
mass spraying	–	0.25	9.31	–	–	100 m²	**9.31**
Power spraying							
mass spraying to areas containing vegetation	–	2.00	39.50	60.02	–	ha	**99.52**
mass spraying to hard surfaces	–	–	0.08	0.12	–	100 m²	**0.20**
spraying to kerbs and channels along roadways 300 mm wide	–	1.25	24.69	37.51	–	1 km	**62.20**
General herbicides; in accordance with manufacturer's instructions; Knapsack spray application; see 'Market rates of chemicals' for costs of specific applications							
knapsack sprayer; selective spraying around bases of plants	–	0.18	3.46	–	–	100 m²	**3.46**
knapsack sprayer; spot spraying	–	0.10	1.98	–	–	100 m²	**1.98**
knapsack sprayer; mass spraying	–	0.10	1.98	–	–	100 m²	**1.98**
granular distribution by hand or hand applicator	–	0.20	3.95	–	–	100 m²	**3.95**
Backpack spraying; keeping weed free at bases of plants							
plants at 1.50 m centres; 4444 nr/ha	–	13.89	517.40	–	–	ha	**517.40**
plants at 1.75 m centres; 3265 nr/ha	–	10.20	379.95	–	–	ha	**379.95**
plants at 2.00 m centres; 2500 nr/ha	–	7.82	291.11	–	–	ha	**291.11**
mass spraying	–	7.50	279.38	–	–	ha	**279.38**

Q35 LANDSCAPE MAINTENANCE

Item Excluding site overheads and profit	PC £	Labour hours	Labour £	Plant £	Material £	Unit	Total rate £
LANDSCAPE CHEMICAL APPLICATION – cont							
Herbicide applications; standard backpack spray applicators; application to maintain 1.00 m dia. clear circles (0.79 m²) around new planting							
Everris; Roundup Pro Biactive 360; glyphosate; enhanced movement glyphosate; application rate 5 litre/ha							
plants at 1.50 m centres; 4444 nr/ha	14.86	12.35	459.85	–	14.86	ha	**474.71**
plants at 1.75 m centres; 3265 nr/ha	10.87	9.07	337.86	–	10.87	ha	**348.73**
plants at 2.00 m centres; 2500 nr/ha	8.38	6.95	258.70	–	8.38	ha	**267.08**
mass spraying	–	–	0.03	–	–	m²	**0.03**
GRASS CUTTING							
Grass cutting only; ride-on or tractor drawn equipment; Norris & Gardiner Ltd; works carried out on one site only such as large fields or amenity areas where labour and machinery is present for a full day; grass cutting only; strimming not included							
Using multiple-gang mower with cylindrical cutters; contiguous areas such as playing fields and the like larger than 3000 m²							
3 gang; 2.13 m cutting width	–	0.01	0.19	0.12	–	100 m²	**0.31**
5 gang; 3.40 m cutting width	–	–	0.18	0.14	–	100 m²	**0.32**
7 gang; 4.65 m cutting width	–	–	0.11	0.11	–	100 m²	**0.22**
Using multiple-gang mower with cylindrical cutters; non-contiguous areas such as verges and general turf areas							
3 gang	–	0.01	0.39	0.12	–	100 m²	**0.51**
5 gang	–	0.01	0.25	0.14	–	100 m²	**0.39**
Using multiple-rotary mower with vertical drive shaft and horizontally rotating bar or disc cutters; contiguous areas larger than 3000 m²							
cutting grass, overgrowth or the like using flail mower or reaper	–	0.01	0.37	0.27	–	100 m²	**0.64**

Q35 LANDSCAPE MAINTENANCE

Item Excluding site overheads and profit	PC £	Labour hours	Labour £	Plant £	Material £	Unit	Total rate £
Grass cutting with strimming; ride-on or tractor drawn equipment; Norris & Gardiner Ltd; works carried out on one site only such as large fields or amenity areas where labour and machinery is present for a full day; includes accompanying strimmer operative							
Using multiple-gang mower with cylindrical cutters; contiguous areas such as playing fields and the like larger than 3000 m²							
3 gang; 2.13 m cutting width	–	0.01	0.39	0.13	–	100 m²	0.52
5 gang; 3.40 m cutting width	–	0.01	0.35	0.15	–	100 m²	0.50
7 gang; 4.65 m cutting width	–	0.01	0.22	0.14	–	100 m²	0.36
Using multiple-gang mower with cylindrical cutters; non-contiguous areas such as verges and general turf areas							
3 gang	–	0.02	0.78	0.18	–	100 m²	0.96
5 gang	–	0.01	0.50	0.16	–	100 m²	0.66
Using multiple-rotary mower with vertical drive shaft and horizontally rotating bar or disc cutters; contiguous areas larger than 3000 m²							
cutting grass, overgrowth or the like using flail mower or reaper	–	0.02	0.75	0.33	–	100 m²	1.08
Grass cutting only; ride-on or tractor drawn equipment; Norris & Gardiner Ltd; works carried out on multiple sites; machinery moved by trailer between sites; strimming not included							
Using multiple-gang mower with cylindrical cutters; contiguous areas such as playing fields and the like larger than 3000 m²							
3 gang; 2.13 m cutting width	–	0.01	0.32	0.12	–	100 m²	0.44
5 gang; 3.40 m cutting width	–	0.01	0.29	0.14	–	100 m²	0.43
7 gang; 4.65 m cutting width	–	–	0.18	0.11	–	100 m²	0.29
Using multiple-gang mower with cylindrical cutters; non-contiguous areas such as verges and general turf areas							
3 gang	–	0.02	0.63	0.12	–	100 m²	0.75
5 gang	–	0.01	0.40	0.14	–	100 m²	0.54
Using multiple-rotary mower with vertical drive shaft and horizontally rotating bar or disc cutters; contiguous areas larger than 3000 m²							
cutting grass, overgrowth or the like using flail mower or reaper	–	0.02	0.61	0.27	–	100 m²	0.88

Q35 LANDSCAPE MAINTENANCE

Item Excluding site overheads and profit	PC £	Labour hours	Labour £	Plant £	Material £	Unit	Total rate £
GRASS CUTTING – cont							
Grass cutting only – cont							
Cutting grass, overgrowth or the like; using tractor-mounted side-arm flail mower; in areas inaccessible to alternative machine; on surface							
not exceeding 30° from horizontal	–	0.01	0.44	0.34	–	100 m²	0.78
30–50° from horizontal	–	0.02	0.87	0.34	–	100 m²	1.21
Grass cutting and strimming; ride-on or tractor drawn equipment; Norris & Gardiner Ltd; works carried out on multiple sites; machinery moved by trailer between sites; includes for accompanying strimmer operative							
Using multiple-gang mower with cylindrical cutters; contiguous areas such as playing fields and the like larger than 3000 m²							
3 gang; 2.13 m cutting width	–	0.02	0.63	0.14	–	100 m²	0.77
5 gang; 3.40 m cutting width	–	0.02	0.57	0.16	–	100 m²	0.73
7 gang; 4.65 m cutting width	–	0.01	0.35	0.12	–	100 m²	0.47
Using multiple-gang mower with cylindrical cutters; non-contiguous areas such as verges and general turf areas							
3 gang	–	0.03	1.26	0.17	–	100 m²	1.43
5 gang	–	0.02	0.81	0.20	–	100 m²	1.01
Using multiple-rotary mower with vertical drive shaft and horizontally rotating bar or disc cutters; contiguous areas larger than 3000 m²							
cutting grass, overgrowth or the like using flail mower or reaper	–	0.03	1.21	0.32	–	100 m²	1.53
Cutting grass, overgrowth or the like; using tractor-mounted side-arm flail mower; in areas inaccessible to alternative machine; on surface							
not exceeding 30° from horizontal	–	0.02	0.89	0.37	–	100 m²	1.26
30° to 50° from horizontal	–	0.05	1.74	0.40	–	100 m²	2.14

Q35 LANDSCAPE MAINTENANCE

Item Excluding site overheads and profit	PC £	Labour hours	Labour £	Plant £	Material £	Unit	Total rate £
Grass cutting; pedestrian operated equipment; Norris & Gardiner Ltd; the following rates are for grass cutting only; there is no allowance for accompanying strimming operations							
Using cylinder lawn mower fitted with not less than five cutting blades, front and rear rollers; on surface not exceeding 30° from horizontal; arisings let fly; width of cut							
51 cm	–	0.03	1.14	0.19	–	100 m	**1.33**
61 cm	–	0.03	0.95	0.17	–	100 m	**1.12**
71 cm	–	0.02	0.82	0.17	–	100 m	**0.99**
91 cm	–	0.02	0.64	0.20	–	100 m	**0.84**
Using rotary self-propelled mower; width of cut							
45 cm	–	0.02	0.74	0.08	–	100 m²	**0.82**
81 cm	–	0.01	0.41	0.08	–	100 m²	**0.49**
91 cm	–	0.01	0.36	0.11	–	100 m²	**0.47**
120 cm	–	0.01	0.28	0.36	–	100 m²	**0.64**
Add for using grass box for collecting and depositing arisings							
removing and depositing arisings	–	0.03	0.93	–	–	100 m²	**0.93**
Grass cutting; pedestrian operated equipment; Norris & Gardiner Ltd; the following rates are for grass cutting with accompanying strimming operations							
Using cylinder lawn mower fitted with not less than five cutting blades, front and rear rollers; on surface not exceeding 30° from horizontal; arisings let fly; width of cut							
51 cm	–	0.06	2.27	0.28	–	100 m	**2.55**
61 cm	–	0.05	1.90	0.24	–	100 m	**2.14**
71 cm	–	0.04	1.63	0.23	–	100 m	**1.86**
91 cm	–	0.03	1.27	0.25	–	100 m	**1.52**
Using rotary self-propelled mower; width of cut							
45 cm	–	0.04	1.48	0.14	–	100 m²	**1.62**
81 cm	–	0.02	0.82	0.11	–	100 m²	**0.93**
91 cm	–	0.02	0.73	0.14	–	100 m²	**0.87**
120 cm	–	0.01	0.55	0.11	–	100 m²	**0.66**
Add for using grass box for collecting and depositing arisings							
removing and depositing arisings	–	0.03	0.93	–	–	100 m²	**0.93**
Add for 30–50° from horizontal	–	–	–	–	–	33%	–
Add for slopes exceeding 50°	–	–	–	–	–	100%	–

Q35 LANDSCAPE MAINTENANCE

Item Excluding site overheads and profit	PC £	Labour hours	Labour £	Plant £	Material £	Unit	Total rate £
GRASS CUTTING – cont							
Grass cutting – cont							
Cutting grass or light woody							
undergrowth; using trimmer with nylon							
cord or metal disc cutter; on surface							
not exceeding 30° from horizontal	–	0.10	3.73	0.28	–	100 m²	4.01
30–50° from horizontal	–	0.20	7.45	0.56	–	100 m²	8.01
exceeding 50° from horizontal	–	0.25	9.31	0.70	–	100 m²	10.01
Grass cutting; collecting arisings;							
Norris & Gardener Ltd							
Extra over for tractor drawn and							
self-propelled machinery using attached							
grass boxes; depositing arisings							
22 cuts per year	–	0.03	0.93	–	–	100 m²	0.93
18 cuts per year	–	0.04	1.40	–	–	100 m²	1.40
12 cuts per year	–	0.05	1.86	–	–	100 m²	1.86
4 cuts per year	–	0.13	4.66	–	–	100 m²	4.66
bedding plant	–	–	–	–	0.29	nr	0.29
Arisings collected by trailed sweepers							
22 cuts per year	–	0.01	0.31	0.26	–	100 m²	0.57
18 cuts per year	–	0.01	0.41	0.26	–	100 m²	0.67
12 cuts per year	–	0.02	0.62	0.26	–	100 m²	0.88
4 cuts per year	–	0.03	1.24	0.26	–	100 m²	1.50
Disposing arisings on site; 100 m							
distance maximum							
22 cuts per year	–	–	0.15	0.02	–	100 m²	0.17
18 cuts per year	–	–	0.18	0.03	–	100 m²	0.21
12 cuts per year	–	0.01	0.27	0.04	–	100 m²	0.31
4 cuts per year	–	0.02	0.81	0.13	–	100 m²	0.94
Disposal of arisings off site							
22 cuts per year	–	–	0.16	0.04	0.07	100 m²	0.27
18 cuts per year	–	0.01	0.19	0.05	0.09	100 m²	0.33
12 cuts per year	–	–	0.12	0.14	0.16	100 m²	0.42
4 cuts per year	–	0.01	0.37	0.82	0.26	100 m²	1.45
GROUNDCARE OPERATIONS							
Harrowing							
Harrowing grassed area with							
drag harrow	–	0.01	0.23	0.15	–	100 m²	0.38
chain or light flexible spiked harrow	–	0.01	0.31	0.15	–	100 m²	0.46

Q35 LANDSCAPE MAINTENANCE

Item Excluding site overheads and profit	PC £	Labour hours	Labour £	Plant £	Material £	Unit	Total rate £
Scarifying							
Mechanical							
Sisis ARP4; including grass collection box; towed by tractor; area scarified annually	–	0.01	0.44	0.10	–	100 m²	**0.54**
Sisis ARP4; including grass collection box; towed by tractor; area scarified 2 years previously	–	0.01	0.52	0.12	–	100 m²	**0.64**
pedestrian operated self-powered equipment	–	0.04	1.30	0.20	–	100 m²	**1.50**
add for disposal of arisings	–	0.01	0.47	1.02	1.30	100 m²	**2.79**
By hand							
hand implement	–	0.25	9.31	–	–	100 m²	**9.31**
add for disposal of arisings	–	0.01	0.47	1.02	1.30	100 m²	**2.79**
Rolling							
Rolling grassed area; equipment towed by tractor; once over; using							
smooth roller	–	0.01	0.27	0.15	–	100 m²	**0.42**
Turf aeration							
By machine							
Vertidrain turf aeration equipment towed by tractor to effect a minimum penetration of 100–250 mm at 100 mm centres	–	0.02	0.73	1.31	–	100 m²	**2.04**
Ryan GA 30; self-propelled turf aerating equipment; to effect a minimum penetration of 100 mm at varying centres	–	0.02	0.73	1.77	–	100 m²	**2.50**
Groundsman; pedestrian operated; self-powered solid or slitting tine turf aerating equipment to effect a minimum penetration of 100 mm	–	0.07	2.61	1.45	–	100 m²	**4.06**
Cushman core harvester; self-propelled; for collection of arisings	–	0.01	0.36	0.76	–	100 m²	**1.12**
By hand							
hand fork; to effect a minimum penetration of 100 mm and spaced 150 mm apart	–	0.67	24.83	–	–	100 m²	**24.83**
hollow tine hand implement; to effect a minimum penetration of 100 mm and spaced 150 mm apart	–	1.00	37.25	–	–	100 m²	**37.25**
collection of arisings by hand	–	1.50	55.88	–	–	100 m²	**55.88**

Prices for Measured Works

Q35 LANDSCAPE MAINTENANCE

Item Excluding site overheads and profit	PC £	Labour hours	Labour £	Plant £	Material £	Unit	Total rate £
GROUNDCARE OPERATIONS – cont							
Turf areas; surface treatments and top dressing;British sugar topsoil							
Apply screened topdressing to grass surfaces; spread using Tru-Lute							
Sand soil mixes 90/10 to 50/50	0.15	–	0.04	0.03	0.15	m²	**0.22**
Apply screened soil 3 mm Kettering loam to goal mouths and worn areas							
20 mm thick	0.88	0.01	0.19	–	0.88	m²	**1.07**
10 mm thick	0.44	0.01	0.19	–	0.44	m²	**0.63**
Leaf clearance; clearing grassed area of leaves and other extraneous debris							
Using equipment towed by tractor							
large grassed areas with perimeters of mature trees such as sports fields and amenity areas	–	0.01	0.23	0.07	–	100 m²	**0.30**
large grassed areas containing ornamental trees and shrub beds	–	–	0.08	0.98	–	100 m²	**1.06**
Using pedestrian operated mechanical equipment and blowers							
grassed areas with perimeters of mature trees such as sports fields and amenity areas	–	0.01	0.31	1.57	–	100 m²	**1.88**
grassed areas containing ornamental trees and shrub beds	–	0.05	1.86	0.67	–	100 m²	**2.53**
verges	–	0.03	1.24	0.09	–	100 m²	**1.33**
By hand							
grassed areas with perimeters of mature trees such as sports fields and amenity areas	–	0.05	1.86	0.20	–	100 m²	**2.06**
grassed areas containing ornamental trees and shrub beds	–	0.08	3.10	0.33	–	100 m²	**3.43**
verges	–	0.17	6.21	0.66	–	100 m²	**6.87**
Removal of arisings							
areas with perimeters of mature trees	–	–	0.13	0.09	0.10	100 m²	**0.32**
areas containing ornamental trees and shrub beds	–	0.01	0.37	0.34	0.26	100 m²	**0.97**
Litter clearance							
Collection and disposal of litter from grassed area	–	0.01	0.19	–	0.07	100 m²	**0.26**
Collection and disposal of litter from isolated grassed area not exceeding 1000 m²	–	0.02	0.75	–	0.07	100 m²	**0.82**
Tree guards, stakes and ties							
Adjusting existing tree tie	–	0.02	0.62	–	–	nr	**0.62**
Taking up single or double tree stake and ties; removing and disposing	–	0.03	0.49	–	–	nr	**0.49**

Q35 LANDSCAPE MAINTENANCE

Item Excluding site overheads and profit	PC £	Labour hours	Labour £	Plant £	Material £	Unit	Total rate £
SHRUB BED MAINTENANCE							
Edge maintenance							
Maintain edges where lawn abuts							
pathway or hard surface using							
strimmer	–	–	0.09	0.01	–	m	0.10
shears	–	0.01	0.30	–	–	m	0.30
Maintain edges where lawn abuts plant							
bed using							
mechanical edging tool	–	–	0.12	0.03	–	m	0.15
shears	–	–	0.20	–	–	m	0.20
half moon edging tool	–	0.01	0.36	–	–	m	0.36
Pruning shrubs							
Trimming ground cover planting							
soft groundcover; vinca ivy and the							
like	–	0.33	18.07	–	–	100 m²	18.07
woody groundcover; cotoneaster and							
the like	–	0.50	27.10	–	–	100 m²	27.10
Pruning massed shrub border (measure							
ground area)							
shrub beds pruned annually	–	–	0.18	–	–	m²	0.18
shrub beds pruned hard every 3 years	–	0.01	0.50	–	–	m²	0.50
Cutting off dead heads							
bush or standard rose	–	0.02	0.90	–	–	nr	0.90
climbing rose	–	0.03	1.51	–	–	nr	1.51
Pruning roses							
bush or standard rose	–	0.02	0.90	–	–	nr	0.90
climbing rose or rambling rose; tying							
in as required	–	0.02	1.20	–	–	nr	1.20
Pruning ornamental shrub; height before							
pruning (increase these rates by 50% if							
pruning work has not been executed							
during the previous 2 years)							
not exceeding 1 m	–	0.01	0.72	–	–	nr	0.72
1 to 2 m	–	0.02	1.00	–	–	nr	1.00
exceeding 2 m	–	0.04	2.26	–	–	nr	2.26
Removing excess growth etc. from face							
of building etc.; height before pruning							
not exceeding 2 m	–	0.01	0.51	–	–	nr	0.51
2 to 4 m	–	0.02	0.90	–	–	nr	0.90
4 to 6 m	–	0.03	1.51	–	–	nr	1.51
6 to 8 m	–	0.04	2.26	–	–	nr	2.26
8 to 10 m	–	0.05	2.58	–	–	nr	2.58
Removing epicormic growth from base of							
shrub or trunk and base of tree (any							
height, any dia.); number of growths							
not exceeding 10	–	0.02	0.90	–	–	nr	0.90
10 to 20	–	0.02	1.20	–	–	nr	1.20

Q35 LANDSCAPE MAINTENANCE

Item Excluding site overheads and profit	PC £	Labour hours	Labour £	Plant £	Material £	Unit	Total rate £
SHRUB BED MAINTENANCE – cont							
Shrub beds, borders and planters							
Lifting							
bulbs	–	0.17	9.03	–	–	100 nr	9.03
tubers or corms	–	0.13	7.23	–	–	100 nr	7.23
established herbaceous plants;							
hoeing and depositing for replanting	–	0.66	36.13	–	–	100 nr	36.13
Temporary staking and tying in							
herbaceous plant	–	0.01	0.60	–	0.13	nr	0.73
Cutting down spent growth of							
herbaceous plant; clearing arisings							
unstaked	–	0.01	0.36	–	–	nr	0.36
staked; not exceeding 4 stakes per							
plant; removing stakes and putting							
into store	–	0.01	0.45	–	–	nr	0.45
Hand weeding; regular visits; costs per							
occasion							
newly planted areas; mulched beds;	–	0.03	1.51	–	–	100 m²	1.51
newly planted areas; un-mulched	–	0.04	2.41	–	–	100 m²	2.41
established areas	–	0.01	0.60	–	–	100 m²	0.60
Removing grasses from groundcover							
areas	–	0.02	1.20	–	–	100 m²	1.20
Hand digging with fork between shrubs;							
not exceeding 150 mm deep; breaking							
down lumps; leaving surface with a							
medium tilth	–	0.44	24.09	–	–	100 m²	24.09
Hand digging with fork or spade to an							
average depth of 230 mm; breaking							
down lumps; leaving surface with a							
medium tilth	–	0.66	36.13	–	–	100 m²	36.13
Hand hoeing; not exceeding 50 mm							
deep; leaving surface with a medium tilth	–	0.13	7.23	–	–	100 m²	7.23
Hand raking to remove stones etc.;							
breaking down lumps: leaving surface							
with a fine tilth prior to planting	–	0.22	12.04	–	–	100 m²	12.04
Hand weeding; planter, window box; not							
exceeding 1.00 m²							
ground level box	–	0.02	0.90	–	–	nr	0.90
box accessed by stepladder	–	0.03	1.51	–	–	nr	1.51
Spreading only compost, mulch or							
processed bark to a depth of 75 mm							
on shrub bed with existing mature							
planting	–	0.03	1.64	–	–	m²	1.64
recently planted areas	–	0.02	1.20	–	–	m²	1.20
groundcover and herbaceous areas	–	0.02	1.36	–	–	m²	1.36
Clearing cultivated area of litter and							
other extraneous debris; using hand							
implement (exludes winter leaf clearance							
weekly maintenance; private areas	–	0.01	0.30	–	–	100 m²	0.30
weekly maintenance; public areas	–	0.01	0.60	–	–	100 m²	0.60
daily maintenance public areas	–	0.01	0.30	–	–	100 m²	0.30

Q35 LANDSCAPE MAINTENANCE

Item Excluding site overheads and profit	PC £	Labour hours	Labour £	Plant £	Material £	Unit	Total rate £
Clearing cultivated area of winter leaf fall, extraneous debris; using hand implement; winter leaf; removal of cleared material to stockpile on site							
per occasion; occasional trees and deciduous shrubs within 100 m²;	–	0.01	0.76	–	–	100 m²	0.76
per occasion; dense tall trees and deciduous shrubs in close proximity	–	0.03	1.51	–	–	100 m²	1.51

Q35 LANDSCAPE MAINTENANCE

Item Excluding site overheads and profit	PC £	Labour hours	Labour £	Plant £	Material £	Unit	Total rate £
BEDDING MAINTENANCE							
Works to established bedding areas							
Lifting							
bedding plants; hoeing and depositing for disposal	–	0.90	49.14	–	–	100 m²	**49.14**
Hand digging with fork; not exceeding 150 mm deep: breaking down lumps; leaving surface with a medium tilth	–	0.24	13.07	–	–	100 m²	**13.07**
Hand weeding							
newly planted areas	–	0.64	34.85	–	–	100 m²	**34.85**
established areas	–	0.16	8.71	–	–	100 m²	**8.71**
Hand digging with fork or spade to an average depth of 230 mm; breaking down lumps; leaving surface with a medium tilth	–	0.16	8.71	–	–	100 m²	**8.71**
Hand hoeing: not exceeding 50 mm deep; leaving surface with a medium tilth	–	0.13	6.97	–	–	100 m²	**6.97**
Hand raking to remove stones etc.; breaking down lumps; leaving surface with a fine tilth prior to planting	–	0.21	11.61	–	–	100 m²	**11.61**
Hand weeding; planter, window box; not exceeding 1.00 m²							
ground level box	–	0.02	0.88	–	–	nr	**0.88**
box accessed by stepladder	–	0.03	1.45	–	–	nr	**1.45**
Spreading only; compost, mulch or processed bark to a depth of 75 mm							
on shrub bed with existing mature planting	–	0.03	1.58	–	–	m²	**1.58**
recently planted areas	–	0.02	1.17	–	–	m²	**1.17**
groundcover and herbaceous areas	–	0.02	1.30	–	–	m²	**1.30**
Collecting bedding from nursery	–	0.95	52.27	13.64	–	100 m²	**65.91**
Setting out							
mass planting single variety	–	0.04	2.18	–	–	m²	**2.18**
pattern	–	0.11	5.80	–	–	m²	**5.80**
Planting only							
massed bedding plants	–	0.06	3.49	–	–	m²	**3.49**
Clearing cultivated area of leaves, litter and other extraneous debris; using hand implement							
weekly maintenance	–	0.04	2.18	–	–	100 m²	**2.18**
daily maintenance	–	–	–	–	–	100 m²	**–**
IRRIGATION AND WATERING							
Irrigation and watering							
Hand held hosepipe; flow rate 25 litres per minute; irrigation requirement							
10 litres/m²	–	0.74	12.90	–	–	100 m²	**12.90**
15 litres/m²	–	1.10	19.25	–	–	100 m²	**19.25**
20 litres/m²	–	1.46	25.60	–	–	100 m²	**25.60**
25 litres/m²	–	1.84	32.15	–	–	100 m²	**32.15**

Q35 LANDSCAPE MAINTENANCE

Item Excluding site overheads and profit	PC £	Labour hours	Labour £	Plant £	Material £	Unit	Total rate £
Hand held hosepipe; flow rate 40 litres per minute; irrigation requirement							
10 litres/m^2	–	0.46	8.09	–	–	100 m^2	**8.09**
15 litres/m^2	–	0.69	12.13	–	–	100 m^2	**12.13**
20 litres/m^2	–	0.91	15.98	–	–	100 m^2	**15.98**
25 litres/m^2	–	1.15	20.06	–	–	100 m^2	**20.06**
HEDGE MAINTENANCE							
Hedge cutting; field hedges cut once or twice annually							
Trimming sides and top using hand tool or hand held mechanical tools							
not exceeding 2 m high	–	0.05	1.86	0.14	–	10 m^2	**2.00**
2 to 4 m high	–	0.17	6.21	0.47	–	10 m^2	**6.68**
Hedge cutting; ornamental							
Trimming sides and top using hand tool or hand held mechanical tools							
not exceeding 2 m high	–	0.06	2.33	0.18	–	10 m^2	**2.51**
2 to 4 m high	–	0.25	9.31	0.70	–	10 m^2	**10.01**
Hedge cutting; reducing width; hand tool or hand held mechanical tools							
Not exceeding 2 m high							
average depth of cut not exceeding 300 mm	–	0.01	0.47	0.04	–	m^2	**0.51**
average depth of cut 300–600 mm	–	0.03	1.04	0.08	–	m^2	**1.12**
average depth of cut 600–900 mm	–	0.03	1.17	0.09	–	m^2	**1.26**
2–4 m high							
average depth of cut not exceeding 300 mm	–	0.02	0.75	0.06	–	m^2	**0.81**
average depth of cut 300–600 mm	–	0.03	1.17	0.09	–	m^2	**1.26**
average depth of cut 600–900 mm	–	0.05	1.86	0.14	–	m^2	**2.00**
4–6 m high							
average depth of cut not exceeding 300 mm	–	0.05	1.86	0.14	–	m^2	**2.00**
average depth of cut 300–600 mm	–	0.08	3.10	0.23	–	m^2	**3.33**
average depth of cut 600–900 mm	–	0.25	9.31	0.70	–	m^2	**10.01**
Hedge cutting; reducing width; tractor mounted hedge cutting equipment							
Not exceeding 2 m high							
average depth of cut not exceeding 300 mm	–	0.02	0.75	0.67	–	10 m^2	**1.42**
average depth of cut 300–600 mm	–	0.03	0.93	0.83	–	10 m^2	**1.76**
average depth of cut 600–900 mm	–	0.10	3.73	3.33	–	10 m^2	**7.06**
2–4 m high							
average depth of cut not exceeding 300 mm	–	0.01	0.23	0.21	–	10 m^2	**0.44**
average depth of cut 300–600 mm	–	0.01	0.47	0.42	–	10 m^2	**0.89**
average depth of cut 600–900 mm	–	0.01	0.37	0.33	–	10 m^2	**0.70**

Q35 LANDSCAPE MAINTENANCE

Item Excluding site overheads and profit	PC £	Labour hours	Labour £	Plant £	Material £	Unit	Total rate £
HEDGE MAINTENANCE – cont							
Hedge cutting; reducing height; hand tool or hand held mechanical tools							
Not exceeding 2 m high							
average depth of cut not exceeding							
300 mm	–	0.03	1.24	0.09	–	10 m²	1.33
average depth of cut 300–600 mm	–	0.07	2.48	0.19	–	10 m²	2.67
average depth of cut 600–900 mm	–	0.20	7.45	0.56	–	10 m²	8.01
2–4 m high							
average depth of cut not exceeding							
300 mm	–	0.02	0.62	0.05	–	m²	0.67
average depth of cut 300–600 mm	–	0.03	1.24	0.09	–	m²	1.33
average depth of cut 600–900 mm	–	0.10	3.73	0.28	–	m²	4.01
4–6 m high							
average depth of cut not exceeding							
300 mm	–	0.03	1.24	0.09	–	m²	1.33
average depth of cut 300–600 mm	–	0.06	2.33	0.18	–	m²	2.51
average depth of cut 600–900 mm	–	0.13	4.66	0.35	–	m²	5.01
Hedge cutting; removal and disposal of arisings							
Sweeping up and depositing arisings							
300 mm cut	–	0.03	0.93	–	–	10 m²	0.93
600 mm cut	–	0.10	3.73	–	–	10 m²	3.73
900 mm cut	–	0.20	7.45	–	–	10 m²	7.45
Chipping arisings							
300 mm cut	–	0.01	0.37	0.11	–	10 m²	0.48
600 mm cut	–	0.04	1.55	0.48	–	10 m²	2.03
900 mm cut	–	0.10	3.73	1.14	–	10 m²	4.87
Disposal of unchipped arisings							
300 mm cut	–	0.01	0.31	0.51	0.17	10 m²	0.99
600 mm cut	–	0.02	0.62	1.02	0.26	10 m²	1.90
900 mm cut	–	0.04	1.55	2.56	0.65	10 m²	4.76
Disposal of chipped arisings							
300 mm cut	–	–	0.06	0.17	0.26	10 m²	0.49
600 mm cut	–	0.01	0.31	0.17	0.52	10 m²	1.00
900 mm cut	–	0.02	0.62	0.34	1.30	10 m²	2.26

Q35 LANDSCAPE MAINTENANCE

Item Excluding site overheads and profit	PC £	Labour hours	Labour £	Plant £	Material £	Unit	Total rate £
Cyclic maintenance operations							
Note: These new items supplement items for long term landscape maintenance already featured in section Q35 of this publication							
Path or hard surface maintenance							
Cyclic maintenance to surfaces based on annual requirements for long term maintenance contractors							
Check or clear daily for litter or dog faeces	–	12.00	210.00	–	–	100 m²	210.00
Sweeping monthly	–	1.00	17.50	–	–	100 m²	17.50
Jet washing bi-annually	–	0.33	5.78	–	–	100 m²	5.78
Surface grit application; 6 occasions per annum	–	3.00	52.50	–	24.00	100 m²	76.50
Snow clearance; per occasion	–	1.25	21.88	–	–	100 m²	21.88
Childrens play area maintenance							
Visit daily to inspect for safety (Play area as part of larger maintained staffed area)	–	0.10	1.75	–	–	nr	1.75
Visit daily to inspect for safety (Play area as isolated play area)	–	0.50	8.75	–	–	nr	8.75
Monthly inspections of play equipment integrity and safety	–	1.00	17.50	–	–	nr	17.50
Annual repairs and maintenance to equipment in play areas; high density populations	–	4.00	70.00	–	–	nr	70.00
Annual repairs and maintenance to equipment in play areas; low density populations	–	1.00	17.50	–	–	nr	17.50
Street furniture maintenance – annual cost							
Clean benches weekly	–	2.60	51.35	–	–	nr	51.35
Sand timber bench bi-annually; remove splinters	–	0.50	9.88	–	–	nr	9.88
Maintain lighting bollard inclusive of cleaning and bulb replacement	–	0.50	9.88	–	2.48	nr	12.36
Maintain lighting standard inclusive of cleaning and bulb replacement	PC	0.75	14.81	–	2.48	nr	17.29
Maintain clean exterior to litter and dog bins	–	3.00	52.50	–	–	nr	52.50
Litter maintenance							
Check and clear litter bins daily (high density) based on 240 days per annum	–	20.00	350.00	256.62	14.40	nr	621.02
Check and clear litter bins weekly	–	4.33	75.83	171.57	3.12	nr	250.52
Clear dog litter bins daily (in conjunction with bin clearance on site)	–	6.00	105.00	240.00	3.60	nr	348.60
Clear dog litter bins weekly (in conjunction with bin clearance on site)	–	1.00	17.50	52.00	14.40	nr	83.90

Q35 LANDSCAPE MAINTENANCE

Item Excluding site overheads and profit	PC £	Labour hours	Labour £	Plant £	Material £	Unit	Total rate £
Cyclic maintenance operations – cont							
Sports maintenance							
Take football posts from store and erect	–	2.00	79.00	–	–	set	**79.00**
Remove football posts and remove to store	–	0.75	29.63	–	–	set	**29.63**
Mark football/rugby field initial mark	19.50	5.00	87.50	–	19.50	nr	**107.00**
Mark football/rugby field remark per occasion	4.50	1.00	17.50	–	4.50	nr	**22.00**
Graffiti maintenance per occasion based on term graffiti removal contract							
Remove graffiti from street furniture;	–	–	–	–	45.00	nr	**45.00**
Remove graffiti from masonry	–	–	–	–	45.00	nr	**45.00**
Japanese Knotweed							
Japanese Knotweed Management Plan; PBA Solutions Ltd Formulate Japanese Knotweed Management Plan (KMP) as recommended by the Environment Agency Code of Practice 2006 (EA CoP)	–	–	–	–	–	nr	**1200.00**
Treatment of Japanese Knotweed; PBA Solutions Ltd; removal and/or treatment in accordance with 'The Knotweed Code of Practice' published by the Environment Agency 2006 (EA CoP); Japanese knotweed must be treated by qualified practitioners Commercial chemical treatments based on areas less than 50 m² knotweed area; allows for the provision of contractual arrangements including method statements and spraying records; also allows for the provision of risk, environment and Coshh assessments							
rate per visit	–	–	–	–	–	nr	**247.50**
rate per year	–	–	–	–	–	nr	**990.00**
rate for complete treatment programme lasting 3 years	–	–	–	–	–	nr	**2970.00**

Q35 LANDSCAPE MAINTENANCE

Item Excluding site overheads and profit	PC £	Labour hours	Labour £	Plant £	Material £	Unit	Total rate £
Commercial chemical treatments based on areas between 50–300 m^2 knotweed area; allows for the provision of contractual arrangements including method statements and spraying records also allows for the provision of risk, environment and Coshh assessments							
rate per visit	–	–	–	–	–	nr	330.00
rate per year	–	–	–	–	–	nr	1320.00
rate for complete treatment programme lasting 3 years	–	–	–	–	–	nr	3960.00
Extra over minimum rate above; for areas 300–900 m^2							
rate per visit	–	–	–	–	–	m^2	0.53
rate per year	–	–	–	–	–	m^2	2.11
rate for complete treatment programme lasting 3 years	–	–	–	–	–	m^2	6.34
Extra over minimum rate above; for areas over 900 m^2							
rate per visit	–	–	–	–	–	m^2	0.44
rate per year	–	–	–	–	–	m^2	170.00
rate for complete treatment programme lasting 3 years	–	–	–	–	–	m^2	5.28
Note: Taking waste off site will require pre-treatment. Landfill operators are required by the Environment Agency to obtain evidence of pre-treatment to show that the waste has been treated to either reduce its volume or hazardousness nature, or facilitate its handling, or enhance its recovery by the waste produce (site of origin) prior being taken to landfill. Generally pre-treatment of Japanese Knotweed can include chemical treatment, cutting or burning. The cost to implement this operation would be similar to item 4/5 with a minimum charge of £300.00. To enable waste to go to landfill soil testing would need to be completed and consultation with a landfill operator or invasive weed specialist is recommended in the first instance.							

Q35 LANDSCAPE MAINTENANCE

Item Excluding site overheads and profit	PC £	Labour hours	Labour £	Plant £	Material £	Unit	Total rate £
Japanese Knotweed – cont							
Japanese Knotweed contaminated							
waste removal; PBA Solutions Ltd							
(see note on page 347)							
Removal off site to a licensed landfill site using 20 tonne gross vehicle weight lorries; price is for haulage and tipping only; the mileage stated below is the distance from point of collection to licensed landfill site; price excludes landfill tax and loading of material into lorries; rates will fluctuate between landfill operators and the figures given are for initial guidance only							
less than 20 miles	–	–	–	–	–	load	**750.00**
20–40 miles	–	–	–	–	–	load	**800.00**
40–60 miles	–	–	–	–	–	load	**995.00**
Root barriers; PBA Solutions Ltd; used in conjunction with on site control methods such as bund treatment, cell burial, vertical boundary protection and capping. The following barriers have been reliably used in connection with Japanese Knotweed control and as such are effective when correctly installed as part of a control programme which would usually include for chemical control.							
Linear root deflection barriers; installed to trench measured separately							
Flexiroot non permeable; reinforced LDPE coated polypropylene geomembrane	5.89	0.05	0.99	–	5.89	m	**6.88**
Bioroot-barrier X permeable root barrier; mechanically bonded geocomposite consisting of a cooper-foil bonded between 2 layers of geotextile	10.80	0.05	0.99	–	10.80	m²	**11.79**

Q35 LANDSCAPE MAINTENANCE

Item Excluding site overheads and profit	PC £	Labour hours	Labour £	Plant £	Material £	Unit	**Total rate £**
Landfill tax rates; based on waste being within landfill operator thresholds for controlled waste and with Japanese Knotweed being the only form of contamination; the rates below are based on visual inspections of the load at the gate of the tip							
Rates to April 2013							
standard rate from April 2014 (where knotweed is visible at more than 5% on inspection at landfill)	–	–	–	–	–	tonne	–
lower rate (where knotweed is visible at less than 5% on inspection at landfill)	–	–	–	–	2.50	tonne	**2.50**

Q40 FENCING

Item Excluding site overheads and profit	PC £	Labour hours	Labour £	Plant £	Material £	Unit	Total rate £
Clarification notes on labour costs in this section							
Fencing team Generally a two man fencing team is used in this section; The column 'Labour hours' reports team hours. The column 'Labour £' reports the total cost of the team for the unit of fencing shown							
2 man team	–	1.00	39.50	–	–	hr	**39.50**
TEMPORARY FENCING							
Temporary security fence; HSS Hire; mesh framed unclimbable fencing; including precast concrete supports and couplings Weekly hire; 2.85 × 2.00high							
weekly hire rate	–	–	–	2.49	–	m	**2.49**
erection of fencing; labour only	–	0.07	2.63	–	–	m	**2.63**
removal of fencing loading to collection vehicle	–	0.03	1.32	–	–	m	**1.32**
delivery charge	–	–	–	0.80	–	m	**0.80**
return haulage charge	–	–	–	0.60	–	m	**0.60**
Protective fencing; AVS Fencing Supplies Ltd Cleft chestnut rolled fencing; fixing to 100 mm dia. chestnut posts; driving into firm ground at 3 m centres							
900 mm high	4.06	0.05	2.11	–	4.80	m	**6.91**
1200 mm high	5.20	0.08	3.16	–	5.94	m	**9.10**
1500 mm high; 3 strand	6.40	0.11	4.22	–	7.14	m	**11.36**
Enclosures; Earth Anchors 　Rootfast anchored galvanized steel enclosures post ADP 20–1000; 1000 mm high × 20 mm dia. with							
AA25–750 socket and padlocking ring	35.00	0.05	1.98	–	37.27	nr	**39.25**
steel cable, orange plastic coated	1.25	–	0.04	–	1.25	m	**1.29**
TIMBER FENCING							
Timber fencing; AVS Fencing Supplies Ltd; tanalized softwood fencing; all timber posts are kiln dried redwood with 15 year guarantee Timber lap panels; pressure treated; fixed to timber posts 75 × 75 mm in 1:3:6 concrete; at 1.90 m centres							
900 mm high	9.08	0.33	13.17	–	16.29	m	**29.46**
1200 mm high	9.45	0.38	14.81	–	16.32	m	**31.13**
1500 mm high	9.45	0.40	15.80	–	16.68	m	**32.48**
1800 mm high	9.53	0.45	17.77	–	17.85	m	**35.62**

Q40 FENCING

Item Excluding site overheads and profit	PC £	Labour hours	Labour £	Plant £	Material £	Unit	Total rate £
Timber lap panels; fixed to slotted concrete posts 100 × 100 mm 1:3:6 concrete at 1.88 m centres							
900 mm high	9.08	0.38	14.81	–	19.39	m	**34.20**
1200 mm high	9.28	0.40	15.80	–	19.60	m	**35.40**
1500 mm high	10.60	0.42	16.79	–	22.60	m	**39.39**
1800 mm high	10.68	0.50	19.75	–	22.73	m	**42.48**
extra for corner posts	21.92	0.45	17.77	–	28.66	nr	**46.43**
extra over panel fencing for 300 mm high trellis tops; slats at 100 mm centres; including additional length of posts	4.49	0.50	19.75	–	4.49	m	**24.24**
Closeboarded fencing; to concrete posts 100 × 100 mm; 2 nr softwood arris rails; 100 × 22 mm softwood pales lapped 13 mm; including excavating and backfilling into firm ground at 3.00 m centres; setting in concrete 1:3:6; concrete gravel board 150 × 150 mm							
900 mm high	10.03	0.50	19.75	–	12.56	m	**32.31**
1050 mm high	10.51	0.55	21.73	–	13.04	m	**34.77**
1500 mm high	11.83	0.57	22.71	–	14.36	m	**37.07**
Closeboarded fencing; to concrete posts 100 × 100 mm; 3 nr softwood arris rails; 100 × 22 mm softwood pales lapped 13 mm; including excavating and backfilling into firm ground at 3.00 m centres; setting in concrete 1:3:6							
1650 mm high	13.52	0.63	24.69	–	16.05	m	**40.74**
1800 mm high	14.00	0.65	25.68	–	16.53	m	**42.21**
Closeboarded fencing; to timber redwood posts 100 × 100 mm; 2 nr softwood arris rails; 100 × 22 mm softwood pales lapped 13 mm; including excavating and backfilling into firm ground at 3.00 m centres; setting in concrete 1:3:6							
1350 mm high	11.59	0.50	19.75	–	14.12	m	**33.87**
1650 mm high	12.91	0.63	24.69	–	15.44	m	**40.13**
1800 mm high	13.79	0.65	25.68	–	16.32	m	**42.00**
Closeboarded fencing; to timber posts 100 × 100 mm; 3 nr softwood arris rails; 100 × 22 mm softwood pales lapped 13 mm; including excavating and backfilling into firm ground at 3.00 m centres; setting in concrete 1:3:6							
1350 mm high	12.64	0.48	18.81	–	15.17	m	**33.98**
1650 mm high	13.96	0.65	25.68	–	16.49	m	**42.17**
1800 mm high	14.84	0.68	26.66	–	17.37	m	**44.03**
extra over for post 125 × 100 mm; for 1800 mm high fencing	13.47	–	–	–	13.47	m	**13.47**
extra over to the above for counter rail	0.52	–	–	–	0.52	m	**0.52**
extra over to the above for capping rail	0.52	0.05	1.98	–	0.52	m	**2.50**
extra over to the above for concrete gravel board 150 × 50 mm	5.20	–	–	–	5.20	m	**5.20**

Q40 FENCING

Item Excluding site overheads and profit	PC £	Labour hours	Labour £	Plant £	Material £	Unit	Total rate £
TIMBER FENCING – cont							
Timber fencing – cont							
Feather edge board fencing; to timber posts 100 × 100 mm; 2 nr softwood cant rails; 125 × 22 mm feather edge boards lapped 13 mm; including excavating and backfilling into firm ground at 3.00 m centres; setting in concrete 1:3:6							
1350 mm high	9.90	0.28	11.06	–	11.03	m	**22.09**
1650 mm high	11.39	0.28	11.06	–	12.53	m	**23.59**
1800 mm high	12.47	0.28	11.06	–	13.61	m	**24.67**
Feather edge board fencing; to timber posts 100 × 100 mm; 3 nr softwood cant rails; 125 × 22 mm feather edge boards lapped 13 mm; including excavating and backfilling into firm ground at 3.00 m centres; setting in concrete 1:3:6							
1350 mm high	10.81	0.29	11.29	–	11.94	m	**23.23**
1650 mm high	12.30	0.29	11.29	–	13.44	m	**24.73**
1800 mm high	12.43	0.29	11.29	–	14.52	m	**25.81**
extra over for post 125 × 100 mm; for 1800 mm high fencing	13.47	–	–	–	13.47	nr	**13.47**
extra over to the above for counter rail and capping	1.66	0.05	1.98	–	1.66	m	**3.64**
Palisade fencing; 22 × 75 mm softwood vertical palings with pointed tops; nailing to two 50 × 100 mm horizontal softwood arris rails; morticed to 100 × 100 mm softwood posts with weathered tops at 3.00 m centres; setting in concrete							
900 mm high	9.51	0.45	17.77	–	12.04	m	**29.81**
1050 mm high	8.08	0.45	17.77	–	10.61	m	**28.38**
1200 mm high	8.28	0.47	18.76	–	10.81	m	**29.57**
extra over for rounded tops	0.66	–	–	–	0.66	m	**0.66**
Post-and-rail fencing; 90 × 38 mm softwood horizontal rails; fixing with galvanized nails to 150 × 75 mm softwood posts; including excavating and backfilling into firm ground at 1.80 m centres; all treated timber							
1200 mm high; 3 horizontal rails	15.26	0.18	6.92	–	15.48	m	**22.40**
1200 mm high; 4 horizontal rails	18.87	0.18	6.92	–	18.87	m	**25.79**
Cleft rail fencing; chestnut adze tapered rails 2.80 m long; morticed into joints; to 125 × 100 mm softwood posts 1.95 m long; including excavating and backfilling into firm ground at 2.80 m centres							
two rails	9.50	0.13	4.94	–	9.50	m	**14.44**
three rails	12.03	0.14	5.53	–	12.03	m	**17.56**
four rails	14.56	0.18	6.92	–	14.56	m	**21.48**

Q40 FENCING

Item Excluding site overheads and profit	PC £	Labour hours	Labour £	Plant £	Material £	Unit	Total rate £
Hit and Miss horizontal rail fencing; 87 × 38 mm top and bottom rails; 100 × 22 mm vertical boards arranged alternately on opposite side of rails; to 100 × 100 mm posts; including excavating and backfilling into firm ground; setting in concrete at 1.8 m centres							
treated softwood; 1600 mm high	22.93	0.60	23.70	–	25.20	m	**48.90**
treated softwood; 1800 mm high	25.08	0.67	26.33	–	27.58	m	**53.91**
treated softwood; 2000 mm high	27.25	0.70	27.65	–	29.75	m	**57.40**
Trellis tops to fencing Extra over screen fencing for 300 mm high trellis tops; slats at 100 mm centres; including additional length of posts	4.49	0.05	1.98	–	4.49	m	**6.47**
TRELLIS							
Traditional trellis panels; The Garden Trellis Company; bespoke trellis panels for decorative, screening or security applications; timber planed all round; height of trellis 1800 mm Free-standing panels; posts 70 × 70 mm set in concrete; timber frames mitred and grooved 45 × 34 mm; heights and widths to suit; slats 32 × 10 mm joinery quality tanalized timber at 100 mm ccs; elements fixed by galvanized staples; capping rail 70 × 34 mm							
HV68 horizontal and vertical slats; softwood	97.20	1.00	19.75	–	103.46	m	**123.21**
D68 diagonal slats; softwood	111.60	1.00	19.75	–	117.86	m	**137.61**
HV68 horizontal and vertical slats; hardwood iroko	234.00	1.00	19.75	–	240.26	m	**260.01**
D68 diagonal slats; hardwood iroko or Western Red Cedar	255.60	1.00	19.75	–	261.86	m	**281.61**
Trellis panels fixed to face of existing wall or railings; timber frames mitred and grooved 45 × 34 mm; heights and widths to suit; slats 32 × 10 mm joinery quality tanalized timber at 100 mm ccs; elements fixed by galvanized staples; capping rail 70 × 34 mm							
HV68 horizontal and vertical slats; softwood	95.40	0.50	9.88	–	97.20	m	**107.08**
D68 diagonal slats; softwood	104.40	0.50	9.88	–	106.20	m	**116.08**
HV68 horizontal and vertical slats; iroko or Western Red Cedar	234.00	0.50	9.88	–	235.80	m	**245.68**
D68 diagonal slats; iroko or Western Red Cedar	237.60	0.50	9.88	–	239.40	m	**249.28**

Q40 FENCING

Item Excluding site overheads and profit	PC £	Labour hours	Labour £	Plant £	Material £	Unit	Total rate £
TRELLIS – cont							
Contemporary style trellis panels; The Garden Trellis Company; bespoke trellis panels for decorative, screening or security applications; timber planed all round							
Free-standing panels 30/15; posts 70 × 70 mm set in concrete with 90 × 30 mm top capping; slats 30 × 14 mm with 15 mm gaps; vertical support at 450 mm centres							
joinery treated softwood	147.60	1.00	19.75	–	153.86	m	**173.61**
hardwood iroko or Western Red Cedar	252.00	1.00	19.75	–	258.26	m	**278.01**
Panels 30/15 face fixed to existing wall or fence; posts 70 × 70 mm set in concrete with 90 × 30 mm top capping; slats 30 × 14 mm with 15 mm gaps; vertical support at 450 mm centres							
joinery treated softwood	144.00	0.50	9.88	–	145.80	m	**155.68**
hardwood iroko or Western Red Cedar	230.40	0.50	9.88	–	232.20	m	**242.08**
Integral arches to ornamental trellis panels; The Garden Trellis Company							
Arches to trellis panels in 45 × 34 mm grooved timbers to match framing; fixed to posts							
R450 1/4 circle; joinery treated softwood	45.00	1.50	29.63	–	51.26	nr	**80.89**
R450 1/4 circle; hardwood iroko or Western Red Cedar	60.00	1.50	29.63	–	66.26	nr	**95.89**
Arches to span 1800 mm wide							
joinery treated softwood	60.00	1.50	29.63	–	66.26	nr	**95.89**
hardwood iroko or Western Red Cedar	80.00	1.50	29.63	–	86.26	nr	**115.89**
Painting or staining of trellis panels; high quality coatings							
microporous opaque paint or spirit based stain	–	–	–	–	28.00	m²	**28.00**
STRAINED WIRE FENCING/FIELD FENCING							
Posts for strained wire and wire mesh; AVS Fencing Supplies Ltd							
Strained wire fencing; concrete inter posts only at 2750 mm centres; 610 mm below ground; excavating holes; filling with concrete; replacing topsoil; disposing surplus soil off site							
900 mm high	0.03	0.28	10.97	0.73	1.33	m	**13.03**
1200 mm high	3.81	0.28	10.97	0.73	5.11	m	**16.81**
1800 mm high	5.20	0.39	15.36	1.82	6.49	m	**23.67**

Q40 FENCING

Item Excluding site overheads and profit	PC £	Labour hours	Labour £	Plant £	Material £	Unit	Total rate £
Extra over strained wire fencing for concrete straining posts with one strut; posts and struts 610 mm below ground; struts, cleats, stretchers, winders, bolts and eye bolts; excavating holes; filling to within 150 mm of ground level with concrete (1:12) – 40 mm aggregate; replacing topsoil; disposing surplus soil off site							
900 mm high	26.76	0.34	13.23	1.64	51.21	nr	**66.08**
1200 mm high	29.84	0.33	13.17	1.64	55.06	nr	**69.87**
1800 mm high	42.18	0.33	13.17	1.64	69.09	nr	**83.90**
Extra over strained wire fencing for concrete straining posts with two struts; posts and struts 610 mm below ground; excavating holes; filling to within 150 mm of ground level with concrete (1:12) – 40 mm aggregate; replacing topsoil; disposing surplus soil off site							
900 mm high	38.58	0.46	17.97	2.10	75.05	nr	**95.12**
1200 mm high	43.14	0.42	16.79	1.64	79.80	nr	**98.23**
1800 mm high	61.10	0.42	16.79	1.64	104.55	nr	**122.98**
Strained wire fencing; galvanized steel angle posts only at 2750 mm centres; 610 mm below ground; driving in							
900 mm high; 40 × 40 × 5 mm	3.89	0.02	0.60	–	3.89	m	**4.49**
1200 mm high; 40 × 40 × 5 mm	3.49	0.02	0.66	–	3.49	m	**4.15**
1500 mm high; 40 × 40 × 5 mm	5.89	0.03	1.20	–	5.89	m	**7.09**
1800 mm high; 40 × 40 × 5 mm	6.34	0.04	1.44	–	6.34	m	**7.78**
Galvanized steel straining posts with two struts for strained wire fencing; setting in concrete							
900 mm high; 50 × 50 × 6 mm	68.20	0.50	19.75	–	70.84	nr	**90.59**
1200 mm high; 50 × 50 × 6 mm	82.36	0.50	19.75	–	85.00	nr	**104.75**
1500 mm high; 50 × 50 × 6 mm	102.90	0.50	19.75	–	105.54	nr	**125.29**
1800 mm high; 50 × 50 × 6 mm	106.63	0.50	19.75	–	109.27	nr	**129.02**
Fixing of fencing to posts							
Strained wire; to posts (posts not included); 3 mm galvanized wire; fixing with galvanized stirrups							
900 mm high; 2 wire	0.20	0.02	0.66	–	0.29	m	**0.95**
1200 mm high; 3 wire	0.29	0.02	0.92	–	0.38	m	**1.30**
1400 mm high; 3 wire	0.29	0.02	0.92	–	0.38	m	**1.30**
1800 mm high; 3 wire	0.29	0.02	0.92	–	0.38	m	**1.30**
Barbed wire; to posts (posts not included); 3 mm galvanized wire; fixing with galvanized stirrups							
900 mm high; 2 wire	0.30	0.03	1.32	–	0.39	m	**1.71**
1200 mm high; 3 wire	0.45	0.05	1.84	–	0.54	m	**2.38**
1400 mm high; 3 wire	0.45	0.05	1.84	–	0.54	m	**2.38**
1800 mm high; 3 wire	0.45	0.05	1.84	–	0.54	m	**2.38**

Q40 FENCING

Item Excluding site overheads and profit	PC £	Labour hours	Labour £	Plant £	Material £	Unit	Total rate £
STRAINED WIRE FENCING/FIELD FENCING – cont							
Fencing only to posts and strained wire Chain link fencing; AVS Fencing Supplies Ltd; to strained wire and posts priced separately; 3 mm galvanized wire; 51 mm mesh; galvanized steel components; fixing to line wires threaded through posts and strained with eye-bolts; posts (not included)							
900 mm high	5.03	0.03	1.32	–	5.12	m	6.44
1200 mm high	6.11	0.03	1.32	–	6.20	m	7.52
1800 mm high	7.19	0.05	1.98	–	7.28	m	9.26
Chain link fencing; to strained wire and posts priced separately; 3.15 mm plastic coated galvanized wire (wire only 2.5 mm); 51 mm mesh; galvanized steel components; fencing with line wires threaded through posts and strained with eye-bolts; posts (not included) (Note: plastic coated fencing can be cheaper than galvanized finish as wire of a smaller cross-sectional area can be used)							
900 mm high	2.51	0.03	1.32	–	2.69	m	4.01
1200 mm high	3.39	0.03	1.32	–	3.57	m	4.89
1800 mm high	4.92	0.06	2.47	–	5.64	m	8.11
Extra over strained wire fencing for cranked arms and galvanized barbed wire							
1 row	1.07	0.01	0.33	–	1.07	m	1.40
2 row	1.22	0.03	0.99	–	1.22	m	2.21
3 row	1.37	0.03	0.99	–	1.37	m	2.36
Field fencing; AVS Fencing Supplies Ltd; woven wire mesh; fixed to posts and straining wires measured separately							
cattle fence; 1200 m high; 114 × 300 mm at bottom to 230 × 300 mm at top	0.92	0.05	1.98	–	1.12	m	3.10
sheep fence; 800 mm high; 140 × 300 mm at bottom to 230 × 300 mm at top	0.74	0.05	1.98	–	0.94	m	2.92
deer fence; 2.0 m high; 89 × 150 mm at bottom to 267 × 300 mm at top	1.88	0.06	2.47	–	2.08	m	4.55
Extra for concreting in posts	–	0.25	9.88	–	4.04	nr	13.92
Extra for straining post	12.35	0.38	14.81	–	12.35	nr	27.16

Q40 FENCING

Item Excluding site overheads and profit	PC £	Labour hours	Labour £	Plant £	Material £	Unit	Total rate £
Boundary fencing; strained wire and wire mesh; Jacksons Fencing Tubular chain link fencing; galvanized; plastic coated; 60.3 mm dia. posts at 3.0 m centres; setting 700 mm into ground; choice of 10 mesh colours; including excavating holes, backfilling and removing surplus soil; with top rail only							
900 mm high	–	–	–	–	–	m	37.04
1200 mm high	–	–	–	–	–	m	39.35
1800 mm high	–	–	–	–	–	m	43.85
2000 mm high	–	–	–	–	–	m	46.31
Tubular chain link fencing; galvanized; plastic coated; 60.3 mm dia. posts at 3.0 m centres; cranked arms and three lines barbed wire; setting 700 mm into ground; including excavating holes; backfilling and removing surplus soil; with top rail only							
2000 mm high	–	–	–	–	–	m	46.95
1800 mm high	–	–	–	–	–	m	45.62
RABBIT NETTING							
Rabbit netting; AVS Fencing Supplies Ltd Timber stakes; peeled kiln dried pressure treated; pointed; 1.8 m posts driven 750 mm into ground at 3 m centres (line wires and netting priced separately)							
75–100 mm stakes	3.65	0.13	4.94	–	3.65	m	8.59
Corner posts or straining posts 150 mm dia. × 2.4 m high set in concrete; centres to suit local conditions or changes of direction							
1 strut	14.85	0.50	19.75	4.68	20.98	nr	45.41
2 strut	17.85	0.50	19.75	4.68	23.98	nr	48.41
Strained wire; to posts (posts not included); 3 mm galvanized wire; fixing with galvanized stirrups							
900 mm high; 2 wire	0.20	0.02	0.66	–	0.29	m	0.95
1200 mm high; 3 wire	0.29	0.02	0.92	–	0.38	m	1.30
Rabbit netting; 31 mm 19 gauge 1050 mm high netting fixed to posts line wires and straining posts or corner posts all priced separately							
900 mm high; turned in	0.89	0.02	0.79	–	0.89	m	1.68
900 mm high; buried 150 mm in trench	0.89	0.04	1.65	–	0.89	m	2.54

Q40 FENCING

Item Excluding site overheads and profit	PC £	Labour hours	Labour £	Plant £	Material £	Unit	Total rate £
TUBULAR FENCING							
Boundary fencing;							
Steelway-Fensecure Ltd							
Classic two rail tubular fencing; top and							
bottom with stretcher bars and straining							
wires in between; comprising 60.3 mm							
tubular posts at 3.00 m centres; setting in							
concrete; 35 mm top rail tied with							
aluminium and steel fittings;							
50 × 50 × 355/2.5 mm PVC coated chain							
link; all components galvanized and							
coated in green nylon							
964 mm high	–	–	–	–	–	m	39.53
1269 mm high	–	–	–	–	–	m	43.52
1574 mm high	–	–	–	–	–	m	49.48
1878 mm high	–	–	–	–	–	m	51.95
2188 mm high	–	–	–	–	–	m	55.37
2458 mm high	–	–	–	–	–	m	61.93
2948 mm high	–	–	–	–	–	m	78.48
3562 mm high	–	–	–	–	–	m	78.28
End posts; Classic range; 60.3 mm dia.;							
setting in concrete							
964 mm high	–	–	–	–	–	nr	58.34
1269 mm high	–	–	–	–	–	nr	58.34
1574 mm high	–	–	–	–	–	nr	64.97
1878 mm high	–	–	–	–	–	nr	70.97
2188 mm high	–	–	–	–	–	nr	77.19
2458 mm high	–	–	–	–	–	nr	88.69
2948 mm high	–	–	–	–	–	nr	101.42
3562 mm high	–	–	–	–	–	nr	115.96
Corner posts; 60.3 mm dia.; setting in							
concrete							
964 mm high	–	–	–	–	–	nr	44.36
1269 mm high	–	–	–	–	–	nr	26.59
1574 mm high	–	–	–	–	–	nr	51.90
1878 mm high	–	–	–	–	–	nr	54.37
2188 mm high	–	–	–	–	–	nr	63.24
2458 mm high	–	–	–	–	–	nr	67.37
2948 mm high	–	–	–	–	–	nr	80.90
3562 mm high	–	–	–	–	–	nr	86.14

Q40 FENCING

Item Excluding site overheads and profit	PC £	Labour hours	Labour £	Plant £	Material £	Unit	Total rate £
WELDED MESH FENCING							
Boundary fencing; McArthur Group							
Paladin welded mesh colour coated							
green fencing; fixing to metal posts at							
2.975 m centres with manufacturer's							
fixings; setting 600 mm deep in firm							
ground; including excavating holes;							
backfilling and removing surplus							
excavated material; includes 10 year							
product guarantee							
1800 mm high	39.65	0.33	13.04	–	39.65	m	**52.69**
2000 mm high	37.58	0.33	13.04	–	45.63	m	**58.67**
2400 mm high	47.57	0.33	13.04	–	56.94	m	**69.98**
Extra over welded galvanized plastic							
coated mesh fencing for concreting in							
posts	8.09	0.06	2.20	–	8.09	m	**10.29**
Boundary fencing; Lang & Fulton							
Orsogril rectangular steel bar mesh							
fence panels; pleione pattern; bolting to							
60 × 8 mm uprights at 2 m centres; mesh							
62 × 66 mm; setting in concrete							
930 mm high panels	49.00	0.08	4.94	–	61.76	m	**66.70**
1326 mm high panels	67.00	0.10	5.92	–	79.76	m	**85.68**
1722 mm high panels	83.00	0.17	9.87	–	95.76	m	**105.63**
SECURITY FENCING							
Security fencing; Jacksons Fencing							
Barbican galvanized steel paling fencing;							
on 60 × 60 mm posts at 3 m centres;							
setting in concrete							
1250 mm high	–	–	–	–	–	m	**70.63**
1500 mm high	–	–	–	–	–	m	**80.34**
2000 mm high	–	–	–	–	–	m	**99.33**
2500 mm high	–	–	–	–	–	m	**120.27**
Gates; to match Barbican galvanized							
steel paling fencing							
width 1 m	–	–	–	–	–	nr	**1400.00**
width 2 m	–	–	–	–	–	nr	**1480.00**
width 3 m	–	–	–	–	–	nr	**1566.00**
width 4 m	–	–	–	–	–	nr	**1606.00**
width 8 m	–	–	–	–	–	pair	**2244.00**
width 9 m	–	–	–	–	–	pair	**2810.00**
width 10 m	–	–	–	–	–	pair	**3344.00**

Q40 FENCING

Item Excluding site overheads and profit	PC £	Labour hours	Labour £	Plant £	Material £	Unit	Total rate £
SECURITY FENCING – cont							
Security fencing; Jacksons Fencing; intruder guards Viper Spike Intruder Guards; to existing structures; including fixing bolts							
Viper 1; 40 × 5 mm × 1.1 m long; with base plate	34.50	1.00	39.50	–	34.50	nr	**74.00**
Viper 3; 160 × 190 mm wide; U shape; to prevent intruders climbing pipes	38.95	0.50	19.75	–	38.95	nr	**58.70**
Security fencing; Jacksons Fencing Razor Barb Concertina; spiral wire security barriers; fixing to 600 mm steel ground stakes							
Ref 3275; 450 mm dia. roll; medium barb; galvanized	2.85	0.05	1.98	–	3.05	m	**5.03**
Ref 3276; 730 mm dia. roll; medium barb; galvanized	4.28	0.05	1.98	–	4.48	m	**6.46**
Ref 3277; 950 mm dia. roll; medium barb; galvanized	5.22	0.05	1.98	–	5.42	m	**7.40**
Three lines Barbed Tape; medium barb; on 50 × 50 mm mild steel angle posts; setting in concrete							
Ref 3283; barbed tape; medium barb; galvanized	1.14	0.21	8.32	–	12.72	m	**21.04**
Five lines Barbed Tape; medium barb; on 50 × 50 mm mild steel angle posts; setting in concrete							
Ref 3283; barbed tape; medium barb; galvanized	1.90	0.17	6.58	–	13.48	m	**20.06**
TRIP RAILS							
Trip rails; Birdsmouth; Jacksons Fencing Diamond rail fencing; Birdsmouth; posts 100 × 100 mm softwood planed at 1.35 m centres set in 1:3:6 concrete; rail 75 × 75 mm nominal secured with galvanized straps nailed to posts							
posts 900 mm (600 mm above ground); at 1.35 m centres	9.94	0.25	9.88	0.31	11.63	m	**21.82**
posts 1.20 m (900 mm above ground); at 1.35 m centres	11.50	0.25	9.88	0.31	13.18	m	**23.37**
posts 900 mm (600 mm above ground); at 1.80 m centres	8.36	0.19	7.43	0.23	9.64	m	**17.30**
posts 1.20 m (900 mm above ground); at 1.80 m centres	9.52	0.19	7.43	0.23	10.79	m	**18.45**

Q40 FENCING

Item Excluding site overheads and profit	PC £	Labour hours	Labour £	Plant £	Material £	Unit	Total rate £
Trip rails; Townscape Products Ltd							
Hollow steel section knee rails; galvanized; 500 mm high; setting in concrete							
1000 mm bays	121.72	1.00	39.50	–	128.96	m	168.46
1200 mm bays	25.65	1.00	39.50	–	112.06	m	151.56
WINDBREAK FENCING							
Windbreak fencing							
Fencing; English Woodlands; Shade and Shelter Netting windbreak fencing; green; to 100 mm dia. treated softwood posts; setting 450 mm into ground; fixing with 50 × 25 mm treated softwood battens nailed to posts; including excavating and backfilling into firm ground; setting in concrete at 3 m centres							
1200 mm high	1.06	0.08	3.07	–	4.91	m	7.98
1800 mm high	1.87	0.08	3.07	–	5.72	m	8.79
BALLSTOP FENCING							
Ball stop fencing; **Steelway-Fensecure Ltd**							
Ball stop net; 30 × 30 mm netting fixed to 60.3 mm dia. 12 mm solid bar lattice; galvanized; dual posts; top, middle and bottom rails							
4.5 m high	–	–	–	–	–	m	109.84
5 .0 m high	–	–	–	–	–	m	138.67
6.0 m high	–	–	–	–	–	m	154.50
7.0 m high	–	–	–	–	–	m	170.36
8.0 m high	–	–	–	–	–	m	186.16
9.0 m high	–	–	–	–	–	m	202.02
10.0 m high	–	–	–	–	–	m	184.58
Ball stop net; corner posts							
4.5 m high	–	–	–	–	–	m	75.37
5.0 m high	–	–	–	–	–	m	82.49
6.0 m high	–	–	–	–	–	m	128.15
7.0 m high	–	–	–	–	–	m	114.16
8.0 m high	–	–	–	–	–	m	130.00
9.0 m high	–	–	–	–	–	m	135.12
10 m high	–	–	–	–	–	m	161.67
Ball stop net; end posts							
4.5 m high	–	–	–	–	–	m	62.69
5.0 m high	–	–	–	–	–	m	69.82
6.0 m high	–	–	–	–	–	m	85.66
7.0 m high	–	–	–	–	–	m	101.52
8.0 m high	–	–	–	–	–	m	117.32
9.0 m high	–	–	–	–	–	m	133.19
10 m high	–	–	–	–	–	m	135.12

Q40 FENCING

Item Excluding site overheads and profit	PC £	Labour hours	Labour £	Plant £	Material £	Unit	Total rate £
RAILINGS							
Railings; Steelway-Fensecure Ltd							
Mild steel bar railings of balusters at 115 mm centres welded to flat rail top and bottom; bays 2.00 m long; bolting to 51 × 51 mm hollow square section posts; setting in concrete							
galvanized; 900 mm high	–	–	–	–	–	m	82.28
galvanized; 1200 mm high	–	–	–	–	–	m	99.66
galvanized; 1500 mm high	–	–	–	–	–	m	110.76
galvanized; 1800 mm high	–	–	–	–	–	m	132.87
primed; 900 mm high	–	–	–	–	–	m	77.55
primed; 1200 mm high	–	–	–	–	–	m	94.16
primed; 1500 mm high	–	–	–	–	–	m	104.09
primed; 1800 mm high	–	–	–	–	–	m	124.06
Mild steel blunt top railings of 19 mm balusters at 130 mm centres welded to bottom rail; passing through holes in top rail and welded; top and bottom rails 40 × 10 mm; bolting to 51 × 51 mm hollow square section posts; setting in concrete							
galvanized; 800 mm high	–	–	–	–	–	m	88.59
galvanized; 1000 mm high	–	–	–	–	–	m	93.04
galvanized; 1300 mm high	–	–	–	–	–	m	110.76
galvanized; 1500 mm high	–	–	–	–	–	m	119.61
primed; 800 mm high	–	–	–	–	–	m	85.28
primed; 1000 mm high	–	–	–	–	–	m	88.59
primed; 1300 mm high	–	–	–	–	–	m	105.22
primed; 1500 mm high	–	–	–	–	–	m	112.99
Railings; traditional pattern; 16 mm dia. verticals at 127 mm intervals with horizontal bars near top and bottom; balusters with spiked tops; 51 × 20 mm standards; including setting 520 mm into concrete at 2.75 m centres							
primed; 1200 mm high	–	–	–	–	–	m	99.66
primed; 1500 mm high	–	–	–	–	–	m	112.99
primed; 1800 mm high	–	–	–	–	–	m	121.83
Interlaced bow-top mild steel railings; traditional park type; 16 mm dia. verticals at 80 mm intervals; welded at bottom to 50 × 10 mm flat and slotted through 38 × 8 mm top rail to form hooped top profile; 50 × 10 mm standards; setting 560 mm into concrete at 2.75 m centres							
galvanized; 900 mm high	–	–	–	–	–	m	150.63
galvanized; 1200 mm high	–	–	–	–	–	m	183.87
galvanized; 1500 mm high	–	–	–	–	–	m	219.30

Q40 FENCING

Item Excluding site overheads and profit	PC £	Labour hours	Labour £	Plant £	Material £	Unit	Total rate £
galvanized; 1800 mm high	–	–	–	–	–	m	254.74
primed; 900 mm high	–	–	–	–	–	m	146.19
primed; 1200 mm high	–	–	–	–	–	m	177.21
primed; 1500 mm high	–	–	–	–	–	m	210.45
primed; 1800 mm high	–	–	–	–	–	m	243.65
ESTATE FENCING/HORIZONTAL BAR FENCING							
Metal estate and parkland fencing; Stonebank Ironcraft Ltd; continously welded, mild steel, 40 × 10 mm uprights manually driven into normal soil to a minimum depth of 600 mm at 1.00 m centres; 40 × 10 mm flat steel supports welded to uprights beneath topsoil at typically every third upright; horizontal rails threaded through holes within uprights; 20 mm solid steel top rail; 4 nr 25 × 8 mm × 16 mm dia. solid steel lower rails; painted fencing shot-blasted, primed and finished with two coats of paint							
Bare metal; 1200 mm high							
10–50 m	–	–	–	–	–	m	65.33
51–100 m	–	–	–	–	–	m	55.17
101–500 m	–	–	–	–	–	m	51.50
above 500 m	–	–	–	–	–	m	49.50
Primed and painted							
10–50 m	–	–	–	–	–	m	78.04
51–100 m	–	–	–	–	–	m	66.08
101–500 m	–	–	–	–	–	m	61.77
above 500 m	–	–	–	–	–	m	59.41
Galvanized and painted							
10–50 m	–	–	–	–	–	m	98.56
51–100 m	–	–	–	–	–	m	83.60
101–500 m	–	–	–	–	–	m	78.21
above 500 m	–	–	–	–	–	m	75.26
End posts and corner posts; 40 × 40 mm; standard hollow section; end or corner posts with capped tops set into concrete							
bare metal	–	–	–	–	–	nr	65.00
primed and painted	–	–	–	–	–	nr	85.00
galvanized and painted	–	–	–	–	–	nr	100.00

Q40 FENCING

Item Excluding site overheads and profit	PC £	Labour hours	Labour £	Plant £	Material £	Unit	Total rate £
ESTATE FENCING/HORIZONTAL BAR FENCING – cont							
Gates for metal estate fencing; Stonebank Ironcraft Ltd; fully welded gates to match fence; scroll at hinge end; turnover at latch end; posts topped with finials and set into concrete							
Pedestrian gates; 1.0–1.5 m wide							
bare metal	–	–	–	–	–	nr	450.00
galvanized and painted	–	–	–	–	–	nr	600.00
Field gates; 1.50–4.0 m wide							
bare metal	–	–	–	–	–	nr	600.00
galvanized and painted	–	–	–	–	–	nr	800.00
Kissing gates							
bare metal	–	–	–	–	–	nr	550.00
galvanized and painted	–	–	–	–	–	nr	745.00
Quadrant; 4 nr uprights with rails welded to fence							
bare metal	–	–	–	–	–	nr	230.00
primed and painted	–	–	–	–	–	nr	300.00
Step stile; 2 treads							
bare metal	–	–	–	–	–	nr	140.00
primed and painted	–	–	–	–	–	nr	200.00
GATES							
Gates – General							
Preamble: Gates in fences; see specification for fencing as gates in traditional or proprietary fencing systems are usually constructed of the same materials and finished as the fencing itself.							
Gates; hardwood; AVS Fencing Supplies Ltd							
Hardwood entrance gate; five bar diamond braced; curved hanging stile; planed iroko; fixed to 150 × 150 mm softwood posts; inclusive of hinges and furniture							
Ref 1100 040; 0.9 m wide	205.00	2.50	98.75	–	264.72	nr	363.47
Ref 1100 041; 1.2 m wide	210.00	2.50	98.75	–	269.72	nr	368.47
Ref 1100 042; 1.5 m wide	215.00	2.50	98.75	–	274.72	nr	373.47
Ref 1100 043; 1.8 m wide	225.00	2.50	98.75	–	284.72	nr	383.47
Ref 1100 044; 2.1 m wide	230.00	2.50	98.75	–	289.72	nr	388.47
Ref 1100 045; 2.4 m wide	245.00	2.50	98.75	–	304.72	nr	403.47
Ref 1100 047; 3.0 m wide	255.00	2.50	98.75	–	314.72	nr	413.47
Ref 1100 048; 3.3 m wide	270.00	2.50	98.75	–	329.72	nr	428.47
Ref 1100 049; 3.6 m wide	280.00	2.50	98.75	–	339.72	nr	438.47

Q40 FENCING

Item Excluding site overheads and profit	PC £	Labour hours	Labour £	Plant £	Material £	Unit	Total rate £
Hardwood field gate; five bar diamond braced; planed iroko; fixed to 150 × 150 mm softwood posts; inclusive of hinges and furniture							
Ref 1100 100; 0.9 m wide	125.00	2.00	79.00	–	184.72	nr	**263.72**
Ref 1100 101; 1.2 m wide	140.00	2.00	79.00	–	199.72	nr	**278.72**
Ref 1100 102; 1.5 m wide	155.00	2.00	79.00	–	214.72	nr	**293.72**
Ref 1100 103; 1.8 m wide	170.00	2.00	79.00	–	229.72	nr	**308.72**
Ref 1100 104; 2.1 m wide	180.00	2.00	79.00	–	239.72	nr	**318.72**
Ref 1100 105; 2.4 m wide	190.00	2.00	79.00	–	249.72	nr	**328.72**
Ref 1100 106; 2.7 m wide	215.00	2.00	79.00	–	274.72	nr	**353.72**
Ref 1100 108; 3.3 m wide	225.00	2.00	79.00	–	284.72	nr	**363.72**
Gates; treated softwood; Jacksons Fencing							
Timber field gates; including wrought iron ironmongery; five bar type; diamond braced; 1.80 m high; to 200 × 200 mm posts; setting 750 mm into firm ground							
width 2400 mm	111.80	3.00	118.50	–	253.14	nr	**371.64**
width 2700 mm	120.55	3.00	118.50	–	261.89	nr	**380.39**
width 3000 mm	129.25	3.50	138.25	–	270.59	nr	**408.84**
width 3300 mm	138.55	3.75	148.13	–	279.89	nr	**428.02**
Featherboard garden gates; including ironmongery; to 100 × 120 mm posts; one diagonal brace							
1.0 × 1.2 m high	30.90	1.50	59.25	–	135.96	nr	**195.21**
1.0 × 1.5 m high	40.60	1.50	59.25	–	154.71	nr	**213.96**
1.0 × 1.8 m high	115.45	1.50	59.25	–	168.71	nr	**227.96**
Picket garden gates; including ironmongery; to match picket fence; width 1000 mm; to 100 × 120 mm posts; one diagonal brace							
950 mm high	75.79	1.50	59.25	–	114.45	nr	**173.70**
1200 mm high	79.89	1.50	59.25	–	118.55	nr	**177.80**
1800 mm high	93.74	1.50	59.25	–	147.00	nr	**206.25**
Gates; tubular mild steel; Jacksons Fencing							
Field gates; galvanized; including ironmongery; diamond braced; 1.80 m high; to tubular steel posts; setting in concrete							
width 3000 mm	126.27	2.50	98.75	–	233.91	nr	**332.66**
width 3300 mm	133.88	2.50	98.75	–	241.52	nr	**340.27**
width 3600 mm	141.66	2.50	98.75	–	249.30	nr	**348.05**
width 4200 mm	157.23	2.50	98.75	–	264.87	nr	**363.62**

Q40 FENCING

Item Excluding site overheads and profit	PC £	Labour hours	Labour £	Plant £	Material £	Unit	Total rate £
GATES – cont							
Gates; sliding; Jacksons Fencing							
Sliding Gate; including all galvanized							
rails and vertical rail infill panels; special							
guide and shutting frame posts (Note:							
foundations installed by suppliers)							
access width 4.00 m; 1.5 m high gates	–	–	–	–	–	nr	**5734.00**
access width 4.00 m; 2.0 m high gates	–	–	–	–	–	nr	**5854.00**
access width 4.00 m; 2.5 m high gates	–	–	–	–	–	nr	**5971.00**
access width 6.00 m; 1.5 m high gates	–	–	–	–	–	nr	**6507.00**
access width 6.00 m; 2.0 m high gates	–	–	–	–	–	nr	**6662.00**
access width 6.00 m; 2.5 m high gates	–	–	–	–	–	nr	**6818.00**
access width 8.00 m; 1.5 m high gates	–	–	–	–	–	nr	**7228.00**
access width 8.00 m; 2.0 m high gates	–	–	–	–	–	nr	**7421.00**
access width 8.00 m; 2.5 m high gates	–	–	–	–	–	nr	**7613.00**
access width 10.00 m; 2.0 m high gates	–	–	–	–	–	nr	**8764.00**
access width 10.00 m; 2.5 m high gates	–	–	–	–	–	nr	**9002.00**
KISSING GATES/STILES							
Kissing gates							
Kissing gates; Jacksons Fencing; in							
galvanized metal bar; fixing to fencing							
posts (posts not included);							
1.65 × 1.30 × 1.00high	270.00	2.50	98.75	–	270.00	nr	**368.75**
Stiles							
Stiles; two posts; setting into firm							
ground; three rails; two treads	80.00	1.50	59.25	–	90.78	nr	**150.03**
GUARD RAILS							
Pedestrian guard rails and barriers							
Mild steel pedestrian guard rails; Broxap							
Street Furniture; 1.00 m high with							
150 mm toe space; to posts at 2.00 m							
centres; galvanized finish							
vertical in-line bar infill panel	44.00	1.00	39.50	–	50.13	m	**89.63**
vertical staggered bar infill with							
230 mm visibility gap at top	46.00	1.00	39.50	–	52.13	m	**91.63**

Q50 SITE/STREET FURNITURE/EQUIPMENT

Item Excluding site overheads and profit	PC £	Labour hours	Labour £	Plant £	Material £	Unit	Total rate £
GENERALLY							
Note: The costs of delivery from the supplier are generally not included in these items. Readers should check with the supplier to verify the delivery costs.							
Furniture/equipment – General							
Preamble: The following items include fixing to manufacturer's instructions; holding down bolts or other fittings and making good (excavating, backfilling and tarmac, concrete or paving bases not included).							
BASES FOR STREET FURNITURE							
Bases for street furniture							
Excavating; filling with concrete 1:3:6; bases for street furniture							
300 × 450 × 500 mm deep	–	0.75	14.81	–	7.15	nr	**21.96**
300 × 600 × 500 mm deep	–	1.11	21.92	–	9.54	nr	**31.46**
300 × 900 × 500 mm deep	–	1.67	32.98	–	14.31	nr	**47.29**
1750 × 900 × 300 mm deep	–	2.33	46.02	–	50.07	nr	**96.09**
2000 × 900 × 300 mm deep	–	2.67	52.73	–	57.23	nr	**109.96**
2400 × 900 × 300 mm deep	–	3.20	63.20	–	68.68	nr	**131.88**
2400 × 1000 × 300 mm deep	–	3.56	70.31	–	76.31	nr	**146.62**
Precast concrete flags; to concrete bases (not included); bedding and jointing in cement: mortar (1:4)							
450 × 600 × 50 mm	12.33	1.17	23.04	–	18.78	m²	**41.82**
Precast concrete paving blocks; to concrete bases (not included); bedding in sharp sand; butt joints							
200 × 100 × 65 mm	9.21	0.50	9.88	–	11.16	m²	**21.04**
200 × 100 × 80 mm	10.00	0.50	9.88	–	12.01	m²	**21.89**
Engineering paving bricks; to concrete bases (not included); bedding and jointing in sulphate-resisting cement: lime: sand mortar (1:1:6)							
over 300 mm wide	11.43	0.56	11.10	–	25.44	m²	**36.54**
Edge restraints to pavings; haunching in concrete (1:3:6)							
200 × 300 mm	5.05	0.10	1.98	–	5.05	m	**7.03**

Q50 SITE/STREET FURNITURE/EQUIPMENT

Item Excluding site overheads and profit	PC £	Labour hours	Labour £	Plant £	Material £	Unit	Total rate £
BASES FOR STREET FURNITURE – cont							
Bases for street furniture – cont							
Bases; Earth Anchors Ltd							
Rootfast ancillary anchors; A1; 500 mm long × 25 mm dia. and top strap F2; including bolting to site furniture (site furniture not included)	9.50	0.50	9.88	–	9.50	set	**19.38**
installation tool for above	23.50	–	–	–	23.50	nr	**23.50**
Rootfast ancillary anchors; A4; heavy duty 40 mm square fixed head anchors; including bolting to site furniture (site furniture not included)	29.00	0.33	6.58	–	29.00	set	**35.58**
Rootfast ancillary anchors; F4; vertical socket; including bolting to site furniture (site furniture not included)	12.75	0.05	0.99	–	12.75	set	**13.74**
Rootfast ancillary anchors; F3; horizontal socket; including bolting to site furniture (site furniture not included)	14.00	0.05	0.99	–	14.00	set	**14.99**
installation tools for the above	68.00	–	–	–	68.00	nr	**68.00**
Rootfast ancillary anchors; A3; anchored bases; including bolting to site furniture (site furniture not included)	47.00	0.33	6.58	–	47.00	set	**53.58**
installation tools for the above	68.00	–	–	–	68.00	nr	**68.00**

Q50 SITE/STREET FURNITURE/EQUIPMENT

Item Excluding site overheads and profit	PC £	Labour hours	Labour £	Plant £	Material £	Unit	Total rate £
BARRIERS							
Barriers – General Preamble: The provision of car and truck control barriers may form part of the landscape contract. Barriers range from simple manual counterweighted poles to fully automated remote-control security gates, and the exact degree of control required must be specified. Complex barriers may need special maintenance and repair.							
Barriers; Barriers Direct Manually operated pole barriers; counterbalance; to tubular steel supports; bolting to concrete foundation (foundation not included); aluminium boom; various finishes, including arm rest.							
clear opening up to 3.00 m	893.20	6.00	118.50	–	913.89	nr	**1032.39**
clear opening up to 4.00 m	893.20	6.00	118.50	–	913.89	nr	**1032.39**
clear opening 5.00 m	893.20	6.00	118.50	–	913.89	nr	**1032.39**
clear opening 6.00 m	1124.20	6.00	118.50	–	1144.89	nr	**1263.39**
clear opening 7.00 m	1201.20	6.00	118.50	–	1221.89	nr	**1340.39**
Electrically operated pole barriers; enclosed fan-cooled motor; double worm reduction gear; overload clutch coupling with remote controls; aluminium boom; various finishes, including catch pole (exclusive of electrical connections by electrician)							
Barriers Direct 2.65 m boom with catchpost	1528.00	6.00	118.50	–	1548.69	nr	**1667.19**
Barriers Direct 4.5 m boom with catchpost	1443.00	6.00	118.50	–	1463.69	nr	**1582.19**
Barriers Direct 6.0 m boom with catchpost	1783.32	6.00	118.50	–	1804.01	nr	**1922.51**
BOLLARDS							
Excavating; for bollards and barriers; by hand Holes for bollards							
400 × 400 × 400 mm; disposing off site	–	0.42	8.23	–	1.39	nr	**9.62**
600 × 600 × 600 mm; disposing off site	–	0.97	19.16	–	3.92	nr	**23.08**

Q50 SITE/STREET FURNITURE/EQUIPMENT

Item Excluding site overheads and profit	PC £	Labour hours	Labour £	Plant £	Material £	Unit	Total rate £
BOLLARDS – cont							
Concrete bollards – General							
Preamble: Precast concrete bollards are available in a very wide range of shapes and sizes. The bollards listed here are the most commonly used sizes and shapes; manufacturer's catalogues should be consulted for the full range. Most manufacturers produce bollards to match their suites of street furniture, which may include planters, benches, litter bins and cycle stands. Most parallel sided bollards can be supplied in removable form, with a reduced shank, precast concrete socket and lifting hole to permit removal with a bar.							
Concrete bollards							
Marshalls Plc; cylinder; straight or tapered; 200–400 mm dia.; plain grey concrete; setting into firm ground (excavating and backfilling not included)							
Bridgford; 915 mm high above ground	74.00	2.00	39.50	–	84.53	nr	**124.03**
Marshalls Plc; cylinder; straight or tapered; 200–400 mm dia.; Beadalite reflective finish; setting into firm ground (excavating and backfilling not included)							
Wexham concrete bollard; exposed silver grey	165.00	2.00	39.50	–	170.66	nr	**210.16**
Wexham Major concrete bollard; exposed silver grey	265.00	2.00	39.50	–	275.53	nr	**315.03**
Precast concrete verge markers; various shapes; 450 mm high							
plain grey concrete	55.78	1.00	19.75	–	59.82	nr	**79.57**
white concrete	58.49	1.00	19.75	–	62.53	nr	**82.28**
exposed aggregate	61.92	1.00	19.75	–	65.96	nr	**85.71**
Bollards							
Recycled plastic bollards; Furnitubes International Ltd							
Aberdeen Circular ABR150; 1000 × 150 mm dia.	65.00	2.00	39.50	–	70.66	nr	**110.16**
Aberdeen Square ABR140; 1000 × 140 × 140 mm	47.00	2.00	39.50	–	52.66	nr	**92.16**
Service bollards; Furnitubes International Ltd; stainless steel; for housing services connection points to electricity, water etc. (services connections not included)							
Kenton KEN717; 900 × 250 mm dia.	647.00	2.00	39.50	–	652.66	nr	**692.16**
Zenith ZEN707; 900 × 250 mm dia.	579.00	2.00	39.50	–	584.66	nr	**624.16**

Q50 SITE/STREET FURNITURE/EQUIPMENT

Item Excluding site overheads and profit	PC £	Labour hours	Labour £	Plant £	Material £	Unit	Total rate £
Other bollards							
Removable parking posts; Marshalls Plc							
RT\RD4; domestic telescopic bollard	163.00	2.00	39.50	–	168.66	nr	**208.16**
RT\R8; heavy duty telescopic bollard	216.00	2.00	39.50	–	221.66	nr	**261.16**
Plastic bollards; Marshalls Plc							
Lismore; three ring recycled plastic							
bollard	78.00	2.00	39.50	–	83.66	nr	**123.16**
Cast iron bollards – General							
Preamble: The following bollards are particularly suitable for conservation areas. Logos for civic crests can be incorporated to order.							
Cast iron bollards							
Bollards; Furnitubes International Ltd (excavating and backfilling not included)							
Doric Round; 920 mm high × 170 mm							
dia.	96.00	2.00	39.50	–	101.39	nr	**140.89**
Manchester Round; 975 mm high;							
225 mm square base	104.00	2.00	39.50	–	109.39	nr	**148.89**
Kenton; heavy duty galvanized steel;							
900 mm high; 350 mm dia.	197.00	2.00	39.50	–	202.39	nr	**241.89**
Bollards; Marshalls Plc (excavating and backfilling not included)							
MSF103; cast iron bollard; Small							
Manchester	120.00	2.00	39.50	–	125.66	nr	**165.16**
MSF102; cast iron bollard;							
Manchester	135.00	2.00	39.50	–	140.66	nr	**180.16**
Cast iron bollards with rails – General							
Preamble: The following cast iron bollards are suitable for conservation areas.							
Cast iron bollards with rails							
Cast iron posts with steel tubular rails; Broxap Street Furniture; setting into firm ground (excavating not included)							
Sheffield Short; 420 mm high; one rail,							
type A	65.00	1.50	29.63	–	67.59	nr	**97.22**
Type C mild steel tubular rail;							
including connector	8.50	0.03	0.66	–	10.05	m	**10.71**
Steel bollards							
Steel bollards; Marshalls Plc (excavating and backfilling not included)							
SSB01; stainless steel bollard;							
101 × 1250 mm	103.00	2.00	39.50	–	108.66	nr	**148.16**
RS001; stainless steel bollard;							
114 × 1500 mm	118.00	2.00	39.50	–	123.66	nr	**163.16**
RB119 Brunel; steel bollard;							
168 × 1500 mm	115.00	2.00	39.50	–	125.53	nr	**165.03**

Q50 SITE/STREET FURNITURE/EQUIPMENT

Item Excluding site overheads and profit	PC £	Labour hours	Labour £	Plant £	Material £	Unit	Total rate £
BOLLARDS – cont							
Timber bollards							
Woodscape Ltd; durable hardwood							
RP 250/1500; 250 mm dia. ×							
1500 mm long	246.70	1.00	19.75	–	246.94	nr	266.69
SP 250/1500; 250 mm square ×							
1500 mm long	246.70	1.00	19.75	–	246.94	nr	266.69
SP 150/1200; 150 mm square ×							
1200 mm long	88.94	1.00	19.75	–	89.18	nr	108.93
SP 125/750; 125 mm square ×							
750 mm long	49.09	1.00	19.75	–	49.33	nr	69.08
RP 125/750; 125 mm dia. × 750 mm							
long	49.09	1.00	19.75	–	49.33	nr	69.08
Deterrent bollards							
Semi-mountable vehicle deterrent and							
kerb protection bollards; Furnitubes							
International Ltd (excavating and							
backfilling not included)							
bell decorative	500.00	2.00	39.50	–	507.67	nr	547.17
half bell	555.00	2.00	39.50	–	562.67	nr	602.17
full bell ®	708.00	2.00	39.50	–	715.67	nr	755.17
three quarter bell	611.00	2.00	39.50	–	618.67	nr	658.17
Security bollards							
Security bollards; Furnitubes							
International Ltd (excavating and							
backfilling not included)							
Burr Bloc Type 6; removable steel							
security bollard; 750 mm high above							
ground; 410 × 285 mm	538.00	2.00	39.50	–	556.19	nr	595.69
SEATING							
Outdoor seats							
CED Ltd; stone bench; Sinuous							
bench; stone type bench to organic S							
pattern; laid to concrete base; 500 mm							
high × 400 mm wide (not included)							
2.00 m long	1441.80	4.00	79.00	–	1441.80	nr	1520.80
5.00 m long	3539.16	8.00	158.00	–	3539.16	nr	3697.16
10.00 m long	7076.16	11.00	217.25	–	7076.16	nr	7293.41

Q50 SITE/STREET FURNITURE/EQUIPMENT

Item Excluding site overheads and profit	PC £	Labour hours	Labour £	Plant £	Material £	Unit	**Total rate £**
Outdoor seats; concrete framed – **General** Preamble: Prices for the following concrete framed seats and benches with hardwood slats include for fixing (where necessary) by bolting into existing paving or concrete bases (bases not included) or building into walls or concrete foundations (walls and foundations not included). Delivery generally not included.							
Outdoor seats; concrete framed Outdoor seats; Townscape Products Ltd							
Oxford benches; 1800 × 430 × 440 mm	398.64	2.00	39.50	–	410.90	nr	**450.40**
Maidstone seats; 1800 × 610 × 785 mm	596.00	2.00	39.50	–	608.26	nr	**647.76**
Outdoor seats; concrete Outdoor seats; Marshalls Ltd							
Boulevard 2000 concrete seat	730.00	2.00	39.50	–	745.46	nr	**784.96**
Outdoor seats; Furnitubes International Ltd							
Amesbury concrete seat, 2.2 m long	1260.00	2.00	39.50	–	1275.46	nr	**1314.96**
Marlborough concrete seat, 1.8 m long	1260.00	2.00	39.50	–	1275.46	nr	**1314.96**
Outdoor seats; metal framed – **General** Preamble: Metal framed seats with hardwood backs to various designs can be bolted to ground anchors.							
Outdoor seats; metal framed Outdoor seats; Furnitubes International Ltd							
NS6 Newstead; steel standards with iroko slats; 1.80 m long	308.00	2.00	39.50	–	323.46	nr	**362.96**
NEB6 New Forest Single Bench; cast iron standards with iroko slats; 1.83 m long	258.00	2.00	39.50	–	273.46	nr	**312.96**
EA6 Eastgate; cast iron standards with iroko slats; 1.86 m long	324.00	2.00	39.50	–	339.46	nr	**378.96**
NE6 New Forest Seat; cast iron standards with iroko slats; 1.83 m long	345.00	2.00	39.50	–	360.46	nr	**399.96**
Zenith long bench, satin stainless steel, iroko slats, 1.8 m long	1200.00	2.00	39.50	–	1215.46	nr	**1254.96**
Ashburton; steel with recycled plastic slats; 1.8 m long	302.00	–	–	–	–	nr	**–**

Q50 SITE/STREET FURNITURE/EQUIPMENT

Item Excluding site overheads and profit	PC £	Labour hours	Labour £	Plant £	Material £	Unit	Total rate £
SEATING – cont							
Outdoor seats – cont							
Outdoor seats; Orchard Street Furniture Ltd; Bramley; broad iroko slats to steel frame							
1.20 m long	197.53	2.00	39.50	–	205.90	nr	**245.40**
1.80 m long	240.89	2.00	39.50	–	249.26	nr	**288.76**
2.40 m long	276.57	2.00	39.50	–	284.94	nr	**324.44**
Outdoor seats; Orchard Street Furniture Ltd; Laxton; narrow iroko slats to steel frame							
1.20 m long	211.03	2.00	39.50	–	219.40	nr	**258.90**
1.80 m long	259.31	2.00	39.50	–	267.68	nr	**307.18**
2.40 m long	294.69	2.00	39.50	–	303.06	nr	**342.56**
Outdoor seats; Orchard Street Furniture Ltd; Lambourne; iroko slats to cast iron frame							
1.80 m long	548.18	2.00	39.50	–	556.55	nr	**596.05**
2.40 m long	645.08	2.00	39.50	–	653.45	nr	**692.95**
Outdoor seats; Broxap Street Furniture; metal and timber							
Eastgate	379.00	2.00	39.50	–	387.37	nr	**426.87**
Outdoor seats; Earth Anchors Ltd; Forest-Saver; steel frame and recycled slats							
bench; 1.8 m	199.00	1.00	19.75	–	199.00	nr	**218.75**
seat; 1.8 m	329.00	1.00	19.75	–	329.00	nr	**348.75**
Outdoor seats; Earth Anchors Ltd; Evergreen; cast iron frame and recycled slats							
bench; 1.8 m	428.00	1.00	19.75	–	428.00	nr	**447.75**
seat; 1.8 m	552.00	1.00	19.75	–	552.00	nr	**571.75**
Outdoor seats; all steel							
Outdoor seats; Marshalls Plc							
MSF Central; stainless steel seat	1198.00	0.50	9.88	–	1198.00	nr	**1207.88**
MSF Central; steel seat	598.00	2.00	39.50	–	598.00	nr	**637.50**
MSF 502; cast iron Heritage seat	510.00	2.00	39.50	–	510.00	nr	**549.50**
Outdoor seats; Earth Anchors Ltd; Ranger							
bench; 1.8 m	275.00	1.00	19.75	–	275.00	nr	**294.75**
seat; 1.8 m	431.00	1.00	19.75	–	431.00	nr	**450.75**
Outdoor seats; all timber							
Outdoor seats; Lister Lutyens Co Ltd; Mendip; teak							
1.524 m long	536.00	1.00	19.75	–	555.30	nr	**575.05**
1.829 m long	578.00	1.00	19.75	–	597.30	nr	**617.05**
2.438 m long (including centre leg)	764.00	1.00	19.75	–	783.30	nr	**803.05**

Q50 SITE/STREET FURNITURE/EQUIPMENT

Item Excluding site overheads and profit	PC £	Labour hours	Labour £	Plant £	Material £	Unit	Total rate £
Outdoor seats; Lister Lutyens Co Ltd; Sussex; hardwood							
1.5 m	266.00	2.00	39.50	–	285.30	nr	**324.80**
Outdoor seats; Woodscape Ltd; solid hardwood							
seat type 3 with back; 2.00 m long; free-standing	870.98	2.00	39.50	–	890.28	nr	**929.78**
seat type 3 with back; 2.00 m long; building in	915.08	4.00	79.00	–	934.38	nr	**1013.38**
seat type 4; 2.00 m long; fixing to wall	402.41	2.00	39.50	–	421.71	nr	**461.21**
seat type 4 with back; 2.00 m long; fixing to wall	534.71	3.00	59.25	–	539.51	nr	**598.76**
seat type 4 with back; 2.50 m long; fixing to wall	622.91	3.00	59.25	–	627.71	nr	**686.96**
seat type 5 with back; 2.00 m long; free-standing	887.51	2.00	39.50	–	906.81	nr	**946.31**
seat type 5 with back; 2.50 m long; free-standing	975.71	2.00	39.50	–	995.01	nr	**1034.51**
seat type 5 with back; 2.00 m long; building in	931.61	2.00	39.50	–	950.91	nr	**990.41**
seat type 5 with back; 2.50 m long; building in	1019.81	3.00	59.25	–	1039.11	nr	**1098.36**
bench type 1; 2.00 m long; free- standing	689.06	2.00	39.50	–	708.36	nr	**747.86**
bench type 1; 2.00 m long; building in	733.16	4.00	79.00	–	752.46	nr	**831.46**
bench type 2; 2.00 m long; free- standing	661.50	2.00	39.50	–	680.80	nr	**720.30**
bench type 2; 2.50 m long; free- standing	749.70	2.00	39.50	–	769.00	nr	**808.50**
bench type 2; 2.00 m long; building in	705.60	4.00	79.00	–	724.90	nr	**803.90**
bench type 2; 2.50 m long; building in	793.80	4.00	79.00	–	813.10	nr	**892.10**
bench type 2; 2.00 m long overall; curved to 5 m radius; building in	904.05	4.00	79.00	–	923.35	nr	**1002.35**
Outdoor seats; Orchard Street Furniture Ltd; Allington; all iroko							
1.20 m long	294.08	2.00	39.50	–	313.38	nr	**352.88**
1.80 m long	338.07	2.00	39.50	–	357.37	nr	**396.87**
2.40 m long	411.28	2.00	39.50	–	430.58	nr	**470.08**
Outdoor seats; tree benches/seats							
Tree bench; Neptune Outdoor Furniture Ltd; Beaufort; hexagonal; timber							
SF34–15 A; 1500 mm dia.	1213.00	0.50	9.88	–	1213.00	nr	**1222.88**
Tree seat; Neptune Outdoor Furniture Ltd; Beaufort; hexagonal; timber; with back							
SF32–10 A; 720 mm dia.	1395.00	0.50	9.88	–	1395.00	nr	**1404.88**
SF32–20 A; 1720 mm dia.	1737.00	0.50	9.88	–	1737.00	nr	**1746.88**

Q50 SITE/STREET FURNITURE/EQUIPMENT

Item Excluding site overheads and profit	PC £	Labour hours	Labour £	Plant £	Material £	Unit	Total rate £
SEATING – cont							
Outdoor seats; recycled plastic							
Furnitubes International Ltd							
Aberdeen seat, brown recycled plastic; 1 slat; 2 m long	389.00	2.00	39.50	–	404.46	nr	443.96
Aberdeen seat, brown recycled plastic; 2 slats; 2 m long	474.00	2.00	39.50	–	489.46	nr	528.96
Anti skate board devices							
Stainless steel; Furnitubes International Ltd							
for fitting to seats	20.00	0.20	3.95	–	20.00	nr	23.95
for fitting to benches	40.00	0.20	3.95	–	40.00	nr	43.95
SIGNAGE							
Directional signage; cast aluminium							
Signage; Furnitubes International Ltd							
FFL1 Lancer; cast aluminium finials	64.50	0.07	1.32	–	64.50	nr	65.82
FAAIS; arrow end type cast aluminium directional arms; single line; 90 mm wide	120.00	2.00	39.50	–	120.00	nr	159.50
FAAID; arrow end type cast aluminium directional arms; double line; 145 mm wide	153.00	0.13	2.63	–	153.00	nr	155.63
FAAIS; arrow end type cast aluminium directional arms; treble line; 200 mm wide	180.00	0.20	3.95	–	180.00	nr	183.95
FCK1211G Kingston; composite standard root columns	385.00	2.00	39.50	–	396.25	nr	435.75
Park signage							
Entrance signs and map boards; vitreous enamel							
entrance map board; 1250 × 1000 mm high with two support posts; all associated works to post bases	–	–	–	–	–	nr	2159.00
information board with two locking cabinets; 1250 × 1000 mm high with two support posts; all associated works to post bases	–	–	–	–	–	nr	2061.20
Miscellaneous park signage; vitreous enamel							
'No dogs' sign; 200 × 150 mm; fixing to fencing or gates	–	–	–	–	–	nr	319.58
'Dog exercise area' sign; 300 × 400 mm; fixing to fencing or gates	–	–	–	–	–	nr	397.92
'Nature conservation area' sign; 900 × 400 mm high with two support posts; all associated works to post bases	–	–	–	–	–	nr	1140.00

Q50 SITE/STREET FURNITURE/EQUIPMENT

Item Excluding site overheads and profit	PC £	Labour hours	Labour £	Plant £	Material £	Unit	Total rate £
Monolith signage boards							
Furnitubes International Ltd; Fulham monolith signage board; stainless steel frame; vinyl graphics; 2600 mm above ground; 600 mm below ground; 350 mm width × 120 mm depth							
without baseplate	5080.00	2.00	39.50	–	5085.17	nr	**5124.67**
with baseplate	5230.00	2.00	39.50	–	5235.17	nr	**5274.67**
TREE GRILLES/TREE PROTECTION							
Tree grilles; cast iron							
Cast iron tree grilles; Furnitubes International Ltd							
GS 1070 Greenwich; two part; 1000 mm square × 700 mm dia. tree hole	124.00	2.00	39.50	–	124.00	nr	**163.50**
GC 1270 Greenwich; two part, 1200 mm dia. × 700 mm dia. tree hole	75.00	2.00	39.50	–	75.00	nr	**114.50**
Cast iron tree grilles; Marshalls Plc							
Heritage; cast iron grille plus frame; 1 × 1 m	415.00	3.00	59.25	–	427.06	nr	**486.31**
Cast iron tree grilles; Townscape Products Ltd							
Baltimore; 1200 mm square × 460 mm dia. tree hole	557.74	2.00	39.50	–	557.74	nr	**597.24**
Baltimore; hexagonal; maximum width 1440 mm nominal × 600 mm dia. tree hole	897.43	2.00	39.50	–	897.43	nr	**936.93**
Tree grilles; steel							
Steel tree grilles; Furnitubes International Ltd							
GSF 102G Greenwich; steel tree grille frame for GS 1045; one part	123.00	2.00	39.50	–	123.00	nr	**162.50**
GSF 122G Greenwich; steel tree grille frame for GS 1270 and GC 1245; one part	131.00	2.00	39.50	–	131.00	nr	**170.50**
Note: Care must be taken to ensure that tree grids and guards are removed when trees grow beyond the specified dia. of guard.							

Q50 SITE/STREET FURNITURE/EQUIPMENT

Item Excluding site overheads and profit	PC £	Labour hours	Labour £	Plant £	Material £	Unit	Total rate £
BINS – LITTER/WASTE/DOGWASTE/ SALT AND GRIT BINS							
Dog waste bins							
Earth Anchors Ltd							
HG45 A; steel; 45 l; earth anchored; post mounted	179.00	0.33	6.58	–	179.00	nr	**185.58**
HG45 A; steel; 45 l; as above with pedal operation	209.00	0.33	6.58	–	209.00	nr	**215.58**
Furnitubes International Ltd							
PED 701; Pedigree; post mounted cast iron dog waste bins; 1250 mm total height above ground; 400 mm square bin	457.00	0.75	14.81	–	463.13	nr	**477.94**
LUK745 P; Lucky; steel dog bin; post mounted; 47 l	298.00	1.50	29.63	–	304.13	nr	**333.76**
LUK745 W; Lucky; steel dog bin; wall mounted; 47 l	251.00	0.75	14.81	–	257.13	nr	**271.94**
TER801 P; Terrier; polythene dog bin; post mounted; 40 l	122.00	1.50	29.63	–	128.13	nr	**157.76**
TER801 W; Terrier; polythene dog bin; wall mounted; 40 l	97.00	0.75	14.81	–	103.13	nr	**117.94**
Litter bins; precast concrete in textured white or exposed aggregate finish; with wire baskets and drainage holes							
Bins; Marshalls Plc							
Boulevard 700; concrete circular litter bin	440.00	0.50	9.88	–	440.00	nr	**449.88**
Bins; Neptune Outdoor Furniture Ltd							
SF16; 42 l	232.00	0.50	9.88	–	232.00	nr	**241.88**
SF14; 100 l	325.00	0.50	9.88	–	325.00	nr	**334.88**
Bins; Townscape Products Ltd							
Sutton; 750 mm high × 500 mm dia.; 70 l capacity; including GRP canopy	273.22	0.33	6.58	–	468.46	nr	**475.04**
Braunton; 750 mm high × 500 mm dia.; 70 l capacity; including GRP canopy	289.68	1.50	29.63	–	484.92	nr	**514.55**
Litter bins; metal; stove-enamelled perforated metal for holder and container							
Bins; Townscape Products Ltd							
Metro; 440 × 420 × 800 mm high; 62 l capacity	805.00	0.33	6.58	–	805.00	nr	**811.58**
Voltan; large round; 460 mm dia. × 780 mm high; 56 l capacity	720.74	0.33	6.58	–	720.74	nr	**727.32**
Voltan; small round with pedestal; 410 mm dia. × 760 mm high; 31 l capacity	608.38	0.33	6.58	–	608.38	nr	**614.96**

Q50 SITE/STREET FURNITURE/EQUIPMENT

Item Excluding site overheads and profit	PC £	Labour hours	Labour £	Plant £	Material £	Unit	Total rate £
Litter bins; all-steel							
Bins; Marshall Plc							
MSF Central; steel litter bin	310.00	1.00	19.75	–	310.00	nr	**329.75**
MSF Central; stainless steel litter bin	602.00	0.67	13.17	–	602.00	nr	**615.17**
Bins; Furnitubes International Ltd							
Wave Bin; WVB 440; free-standing 55 l liners; 440 mm dia. × 850 mm high	432.00	0.50	9.88	–	433.60	nr	**443.48**
Wave Bin WVB 520; free-standing 85 l liners; 520 mm dia. × 850 mm high; cast iron plinth	459.00	0.50	9.88	–	460.60	nr	**470.48**
Wave Bin; WVB 440 S304; free-standing; steel with satin finish; cast bronze plinth; open top; 55 l	800.00	0.50	9.88	–	800.00	nr	**809.88**
Wave Bin; WVB 520 S304; free-standing; steel with satin finish; cast bronze plinth; open top; 80 l	945.00	0.50	9.88	–	945.00	nr	**954.88**
Liverpool Bin; LVR520 S304; free-standing; stainless steel; side opening; 965 mm high; 125 l	1151.00	0.50	9.88	–	1151.00	nr	**1160.88**
Bins; Earth Anchors Ltd							
Ranger; 100 l; pedestal mounted	445.00	0.50	9.88	–	445.00	nr	**454.88**
Big Ben; 82 l; steel frame and liner; colour coated finish; earth anchored	298.00	1.00	19.75	–	298.00	nr	**317.75**
Beau; 42 l; steel frame and liner; colour coated finish; earth anchored	243.00	1.00	19.75	–	243.00	nr	**262.75**
Bins; Townscape Products Ltd							
Baltimore Major with GRP canopy; 560 mm dia. × 960 mm high; 140 l capacity	1163.18	1.00	19.75	–	1163.18	nr	**1182.93**
Litter bins; cast iron							
Bins; Marshalls Plc							
MSF5501 Heritage; cast iron litter bin	625.00	1.00	19.75	3.13	625.00	nr	**647.88**
Bins; Furnitubes International Ltd							
Covent Garden COV 702; side opening, 500 mm square × 1050 mm high; 105 l capacity	269.50	0.50	9.88	–	273.69	nr	**283.57**
Covent Garden COV 803; side opening, 500 mm dia. × 1025 mm high; 85 l capacity	248.50	0.50	9.88	–	252.69	nr	**262.57**
Covent Garden COV 912; open top; 500 mm A/F octagonal × 820 mm high; 85 l capacity	380.00	0.50	9.88	–	384.19	nr	**394.07**
Albert ALB 800; open top; 400 mm dia. × 845 mm high; 55 l capacity	306.00	0.50	9.88	–	310.19	nr	**320.07**
Bins; Broxap Street Furniture							
Derby Hercules; post mounted; steel; 40 l	161.60	1.00	19.75	–	165.79	nr	**185.54**
Bins; Townscape Products Ltd							
York Major; 650 mm dia. × 1060 mm high; 140 l capacity	1392.67	1.00	19.75	–	1393.07	nr	**1412.82**

Q50 SITE/STREET FURNITURE/EQUIPMENT

Item Excluding site overheads and profit	PC £	Labour hours	Labour £	Plant £	Material £	Unit	Total rate £
BINS – LITTER/WASTE/DOGWASTE/ SALT AND GRIT BINS – cont							
Litter bins; timber faced; hardwood slatted casings with removable metal litter containers; ground or wall fixing							
Bins; Lister Lutyens Co Ltd							
Monmouth; 675 mm high × 450 mm							
wide; free-standing	180.00	0.33	6.58	–	180.00	nr	**186.58**
Monmouth; 675 mm high × 450 mm							
wide; bolting to ground (without legs)	160.00	1.00	19.75	–	164.99	nr	**184.74**
Bins; Woodscape Ltd							
square; 580 × 580 × 950 mm high; with							
lockable lid	727.65	0.50	9.88	–	727.65	nr	**737.53**
round; 580 mm dia. × 950 mm high;							
with lockable lid	727.65	0.50	9.88	–	727.65	nr	**737.53**
Plastic litter and grit bins; glassfibre reinforced polyester grit bins; yellow body; hinged lids							
Bins; Wybone Ltd; Victoriana glass fibre; cast iron effect litter bins; including lockable liner							
LBV/2; 521 × 521 × 673 mm high; open							
top; square shape; 0.078 m^3 capacity	204.90	1.00	19.75	–	209.09	nr	**228.84**
LVC/3; 457 mm dia. × 648 mm high;							
open top; drum shape; 0.084 m^3							
capacity; with lockable liner	223.52	1.00	19.75	–	227.71	nr	**247.46**
Grit bins; Furnitubes International Ltd							
Grit and salt bin; yellow glass fibre; hinged lid; 170 l	241.00	–	–	–	–	nr	–
Cigarette bins							
Furnitubes International Ltd							
ZEN 275 Zenith cigarette bin;							
stainless steel; wall or post mounted;							
1.7 L	39.00	0.75	14.81	–	45.13	nr	**59.94**
LVR250 PC Liverpool cigarette bin;							
steel; wall mounted; 1.9 L	32.99	0.50	9.88	–	32.99	nr	**42.87**
SMK500F Smoke King cigarette bin;							
cast aluminium; bolt down; 3.5 L	219.99	0.75	14.81	–	226.12	nr	**240.93**
CYCLE HOLDERS/CYCLE SHELTERS							
Cycle holders							
Cycle stands; Marshalls Plc							
Sheffield; steel cycle stand; RCS1	42.00	0.50	9.88	–	42.00	nr	**51.88**
Sheffield; stainless steel cycle stand;							
RSCS1	115.00	0.50	9.88	–	115.00	nr	**124.88**

Q50 SITE/STREET FURNITURE/EQUIPMENT

Item Excluding site overheads and profit	PC £	Labour hours	Labour £	Plant £	Material £	Unit	**Total rate £**
Cycle holders; Autopa; VELOPA; galvanized steel							
R; fixing to wall or post; making good	34.00	1.00	19.75	–	42.37	nr	**62.12**
SR(V); fixing in ground; making good	43.00	1.00	19.75	–	51.37	nr	**71.12**
Sheffield cycle stands; ragged steel	54.00	1.00	19.75	–	62.37	nr	**82.12**
Cycle holders; Townscape Products Ltd							
Guardian cycle holders; tubular steel frame; setting in concrete; 1250 × 550 × 775 mm high; making good	390.00	1.00	19.75	–	398.37	nr	**418.12**
Penny cycle stands; 600 mm dia.; exposed aggregate bollards with 8 nr cycle holders; in galvanized steel; setting in concrete; making good	877.00	1.00	19.75	–	885.37	nr	**905.12**
Cycle holders; Broxap Street Furniture							
Neath cycle rack; BX/MW/AG; for 6 nr cycles; semi-vertical; galvanized and polyester powder coated; 1.32 m wide × 2.542 m long × 1.80 m high	379.00	10.00	197.50	–	395.75	nr	**593.25**
Premier Senior economy combined shelter and rack; BX/MW/AW; for 10 nr cycles; horizontal; galvanized only; 2.13 m wide × 3.05 m long × 2.15 m high	815.00	10.00	197.50	–	831.75	nr	**1029.25**
Toast Rack double sided free- standing cycle rack; BX/MW/GH; for 10 nr cycles; galvanized and polyester coated; 3.25 m long	240.00	4.00	79.00	–	250.35	nr	**329.35**
Cycle shelters							
Furnitubes International Ltd; Academy free-standing bolt down shelter, galvanized steel frame and roof; 3450 mm overall width; 2150 mm maximum height; 2670 mm depth							
corrugated roof	1873.00	6.00	118.50	–	1885.26	nr	**2003.76**
clear polycarbonate roof	1957.00	6.00	118.50	–	1969.26	nr	**2087.76**
PICNIC TABLES/ SUNDRY FURNITURE							
Street furniture ranges							
Townscape Products Ltd; Belgrave; natural grey concrete							
bollards; 250 mm dia. × 500 mm high	126.15	1.00	19.75	–	129.37	nr	**149.12**
seats; 1800 × 600 × 736 mm high	491.83	2.00	39.50	–	491.83	nr	**531.33**
Picnic benches – General							
Preamble: The following items include for fixing to ground to manufacturer's instructions or concreting in.							

Q50 SITE/STREET FURNITURE/EQUIPMENT

Item Excluding site overheads and profit	PC £	Labour hours	Labour £	Plant £	Material £	Unit	Total rate £
PICNIC TABLES/ SUNDRY FURNITURE – cont							
Picnic tables and benches							
Picnic tables and benches; Broxap Street Furniture							
Eastgate picnic unit	629.00	2.00	39.50	–	637.37	nr	676.87
Picnic tables and benches; Woodscape Ltd							
table and benches built in; 2 m long	2251.58	2.00	39.50	–	2259.95	nr	2299.45
Picnic table; recycled plastic							
Furnitubes International Ltd							
Dundee picnic table; brown recycled plastic; 1.8 m long	621.00	2.00	39.50	–	636.46	nr	675.96
Lifebuoy stations							
Lifebuoy stations; Earth Anchors Ltd							
Rootfast lifebuoy station complete with post AP44; SOLAS approved lifebouys 590 mm and lifeline	165.00	2.00	39.50	–	174.58	nr	214.08
installation tool for above	–	–	–	–	86.00	nr	86.00
SMOKING SHELTERS							
Smoking shelters							
Furnitubes International Ltd; Ashby free-standing bolt down; aluminium frame; PET glazing; 2050 mm length × 2330 mm high							
3290 mm depth; maximum 3 people	2046.33	4.00	79.00	–	2058.59	nr	2137.59
6050 mm depth; maximum 6 people	2571.03	5.00	98.75	–	2583.29	nr	2682.04
FLAGPOLES							
Flagpoles							
Flagpoles; ground mounted; Harrison External Display Systems; in glass fibre; smooth white finish; including hinged baseplate, nylon halyard system and revolving gold onion finial; setting in concrete; to manufacturer's recommendations (excavating not included)							
6 m high; external halyard system	177.00	4.00	79.00	–	197.69	nr	276.69
6 m high; internal halyard system	279.99	4.00	79.00	–	300.68	nr	379.68
10 m high; external halyard system	322.00	5.00	98.75	–	342.69	nr	441.44
10 m high; internal halyard system	440.99	5.00	98.75	–	461.68	nr	560.43
12 m high; external halyard system	390.00	6.00	118.50	–	410.69	nr	529.19
12 m high; internal halyard system	502.99	6.00	118.50	–	523.68	nr	642.18

Q50 SITE/STREET FURNITURE/EQUIPMENT

Item Excluding site overheads and profit	PC £	Labour hours	Labour £	Plant £	Material £	Unit	**Total rate £**
Flagpoles; wall mounted; Harrison External Display Systems; vertical or angled poles; in glass fibre; smooth white finish; including base, top bracket, external nylon halyard rope, cleat and revolving gold onion finial							
3 m pole	187.00	2.00	39.50	–	187.00	nr	**226.50**
Banner flagpoles; Harrison External Display Systems; 90 mm dia. aluminium pole; suitable for 1 × 2 m banner							
6 m high	458.62	4.00	79.00	–	479.31	nr	**558.31**
PLANT CONTAINERS							
Market prices of containers							
Plant containers; terracotta							
Capital Garden Products Ltd							
Large Pot LP63; weathered terracotta; 1170 × 1600 mm dia.	–	–	–	–	793.00	nr	**793.00**
Large Pot LP38; weathered terracotta; 610 × 970 mm dia.	–	–	–	–	340.00	nr	**340.00**
Large Pot LP23; weathered terracotta; 480 × 580 mm dia.	–	–	–	–	183.00	nr	**183.00**
Indian style Shimmer Pot 2322; 585 × 560 mm dia.	–	–	–	–	167.00	nr	**167.00**
Indian style Shimmer Pot 1717; 430 × 430 mm dia.	–	–	–	–	108.00	nr	**108.00**
Indian style Shimmer Pot 1314; 330 × 355 mm dia.	–	–	–	–	98.00	nr	**98.00**
Plant containers; faux lead							
Capital Garden Products Ltd							
Trough 2508 Tudor Rose; 620 × 220 × 230 mm high	–	–	–	–	62.00	nr	**62.00**
Tub 2004 Elizabethan; 510 mm square	–	–	–	–	98.00	nr	**98.00**
Tub 1513 Elizabethan; 380 mm square	–	–	–	–	66.00	nr	**66.00**
Tub 1601 Tudor Rose; 420 × 400 mm dia.	–	–	–	–	69.00	nr	**69.00**
Plant containers; window boxes							
Capital Garden Products Ltd							
Adam 5401; faux lead; 1370 × 270 × 210 mm high	–	–	–	–	108.00	nr	**108.00**
Oakleaf OAK24; terracotta; 610 × 230 × 240 mm high	–	–	–	–	87.00	nr	**87.00**
Swag 2402; faux lead; 610 × 200 × 210 mm high	–	–	–	–	54.00	nr	**54.00**

Q50 SITE/STREET FURNITURE/EQUIPMENT

Item Excluding site overheads and profit	PC £	Labour hours	Labour £	Plant £	Material £	Unit	Total rate £
PLANT CONTAINERS – cont							
Plant containers; timber							
Plant containers; hardwood; Neptune Outdoor Furniture Ltd							
Beaufort T38–4D;							
1500 × 1500 × 900 mm high	–	–	–	–	1397.00	nr	**1397.00**
Beaufort T38–3C;							
1000 × 1500 × 700 mm high	–	–	–	–	1023.00	nr	**1023.00**
Beaufort T38–2 A;							
1000 × 500 × 500 mm high	–	–	–	–	539.00	nr	**539.00**
Kara T42–4D; 1500 × 1500 × 900 mm high	–	–	–	–	1496.00	nr	**1496.00**
Kara T42–3C; 1000 × 1500 × 700 mm high	–	–	–	–	1100.00	nr	**1100.00**
Kara T42–2 A; 1000 × 500 × 500 mm high	–	–	–	–	583.00	nr	**583.00**
Plant containers; hardwood; Woodscape Ltd							
square; 900 × 900 × 420 mm high	–	–	–	–	377.74	nr	**377.74**
Measured works							
Plant containers; precast concrete							
Plant containers; Marshalls Plc							
Boulevard 700; circular base and ring	605.00	2.00	39.50	20.63	605.00	nr	**665.13**
Boulevard 1200; circular base and ring	765.00	1.00	19.75	20.63	765.00	nr	**805.38**
PERGOLAS							
Pergolas; AVS Fencing Supplies Ltd; construct timber pergola; posts 150 × 150 mm × 2.40 m finished height in 600 mm deep minimum concrete pits; beams of 200 × 50 mm × 2.00 m wide; fixed to posts with dowels; rafters 200 × 38 mm notched to beams; inclusive of all mechanical excavation and disposal off site							
Pergola 2.00 m wide in green oak; posts at 2.00 m centres; dowel fixed							
rafters at 600 mm centres	297.55	8.40	165.90	3.75	357.38	m	**527.03**
rafters at 400 mm centres	356.00	10.80	213.30	3.75	420.84	m	**637.89**
Pergola 3.00 m wide in green oak; posts at 1.50 m centres							
rafters at 600 mm centres	393.29	7.20	142.20	4.50	462.77	m	**609.47**
rafters at 400 mm centres	480.79	8.00	158.00	4.50	557.77	m	**720.27**

Q50 SITE/STREET FURNITURE/EQUIPMENT

Item Excluding site overheads and profit	PC £	Labour hours	Labour £	Plant £	Material £	Unit	Total rate £
Pergola 2.00 wide in prepared softwood; **beams fixed with bolts; posts at 2.00 m** **centres**							
rafters at 600 mm centres	423.75	5.00	98.75	3.75	484.10	m	**586.60**
rafters at 400 mm centres	546.50	6.00	118.50	3.75	611.86	m	**734.11**
Pergola 3.00 m wide in prepared softwood; posts at 1.50 m centres							
rafters at 600 mm centres	353.59	7.20	142.20	4.50	423.07	m	**569.77**
rafters at 400 mm centres	440.70	8.00	158.00	4.50	517.68	m	**680.18**
STONES AND BOULDERS							
Standing stones; CED Ltd; erect **standing stones; vertical height above** **ground; in concrete base; including** **excavation setting in concrete to 1/3** **depth and crane offload into position** Purple schist							
1.00 m high	97.37	1.50	29.63	15.72	114.26	nr	**159.61**
1.25 m high	123.35	1.50	29.63	15.72	135.66	nr	**181.01**
1.50 m high	197.34	2.00	39.50	18.87	211.50	nr	**269.87**
2.00 m high	324.58	3.00	59.25	31.45	365.60	nr	**456.30**
2.50 m high	493.35	1.50	29.63	62.90	572.62	nr	**665.15**
Rockery stone – General Preamble: Rockery stone prices vary considerably with source, carriage, distance and load. Typical PC prices are in the range of £60–80 per tonne collected.							
Rockery stone; CED Ltd Boulders; maximum distance 25 m; by machine							
750 mm dia.	100.29	0.90	17.77	5.59	100.29	nr	**123.65**
1 m dia.	237.74	2.00	39.50	11.17	237.74	nr	**288.41**
1.5 m dia.	803.37	2.00	39.50	49.18	803.37	nr	**892.05**
2 m dia.	1901.94	2.00	39.50	49.18	1901.94	nr	**1990.62**
Boulders; maximum distance 25 m; by hand							
750 mm dia.	100.29	0.75	14.81	–	100.29	nr	**115.10**
1 m dia.	237.74	1.89	37.24	–	237.74	nr	**274.98**

Q50 SITE/STREET FURNITURE/EQUIPMENT

Item Excluding site overheads and profit	PC £	Labour hours	Labour £	Plant £	Material £	Unit	Total rate £
PLAYGROUND EQUIPMENT							
Playground equipment – General							
Preamble: The range of equipment manufactured or available in the UK is so great that comprehensive coverage would be impossible, especially as designs, specifications and prices change fairly frequently. The following information should be sufficient to give guidance to anyone designing or equipping a playground. In comparing prices note that only outline specification details are given here and that other refinements which are not mentioned may be the reason for some difference in price between two apparently identical elements. The fact that a particular manufacturer does not appear under one item heading does not necessarily imply that they do not make it. Landscape designers are advised to check that equipment complies with ever more stringent safety standards before specifying.							
Playground equipment – Installation							
The rates below include for installation of the specified equipment by the manufacturers. Most manufacturers will offer an option to install the equipment they have supplied.							
Play systems; Kompan Ltd; Galaxy; multiple play activity systems for non-prescribed play; for children 6–14 years; galvanized steel and high density polyethylene							
GXY906 Adara; 14 different play activities	–	–	–	–	–	nr	15971.70
GXY8011 Sirius; 10 different play activities	–	–	–	–	–	nr	9977.25

Q50 SITE/STREET FURNITURE/EQUIPMENT

Item Excluding site overheads and profit	PC £	Labour hours	Labour £	Plant £	Material £	Unit	Total rate £
Sports and social areas; Freegame multi-use games areas; Kompan Ltd; enclosed sports areas; complete with surfacing boundary and goals and targets; galvanized steel framework with high density polyethylene panels, galvanized steel goals and equipment; surfacing priced separately							
FRE1211 Classic Multigoal; 7 m Pitch; complete; suitable for multiple ball sports; suitable for use with natural artificial or hard landscape surfaces; fully enclosed including two end sports walls	–	–	–	–	–	nr	5343.70
FRE2110 Cosmos; 12 × 20 m	–	–	–	–	–	nr	31702.40
FRE2116 Cosmos; 19 × 36 m	–	–	–	–	–	nr	46897.00
FRE3000; Meeting Point; shelter or social area	–	–	–	–	–	nr	4365.60
Swings – General							
Preamble: Prices for the following vary considerably. Those given represent the middle of the range and include multiple swings and timber ,rubber or tyre seats; ground fixing and priming only.							
Swings							
Swings; Wicksteed Ltd							
Flexi Swing; 1 bay 1 flat seat	–	–	–	–	–	nr	2112.00
Flexi Swing; 1 bay 2 flat seats	–	–	–	–	–	nr	2594.00
Flexi Swing; 1 bay 2 cradle seats	–	–	–	–	–	nr	2751.00
Flexi Swing; 1 bay 1 flat seat & 1 cradle seat	–	–	–	–	–	nr	2672.00
Hurricane Swing; 2 button seats	–	–	–	–	–	nr	6307.00
Pendulum Swing; 1 basket seat	–	–	–	–	–	nr	8484.00
1.8 m Timber Swing; 2 cradle seats	–	–	–	–	–	nr	1592.00
2.4 m Timber Swing; 2 cradle seats	–	–	–	–	–	nr	1878.00
2.4 m Timber Swing; 2 flat seats	–	–	–	–	–	nr	2990.00
2.4 m Birds Nest Timber Swing; 1 basket seat	–	–	–	–	–	nr	1612.00
Swings; Lappset UK Ltd							
020414M; swing frame with two flat seats	–	–	–	–	–	nr	1198.60
Swings; Kompan Ltd							
M951P Sunflower swing; 1–3 years	–	–	–	–	–	nr	1330.25
M947P double swings; 1–6 years	–	–	–	–	–	nr	2116.20
M961P double swings; 6–12 years	–	–	–	–	–	nr	1782.00

Q50 SITE/STREET FURNITURE/EQUIPMENT

Item Excluding site overheads and profit	PC £	Labour hours	Labour £	Plant £	Material £	Unit	Total rate £
PLAYGROUND EQUIPMENT – cont							
Slides							
Slides; Wicksteed Ltd							
Pedestal slides; 3.40 m	–	–	–	–	–	nr	3755.00
Pedestal slides; 4.40 m	–	–	–	–	–	nr	4889.00
Pedestal slides; 5.80 m	–	–	–	–	–	nr	5689.00
Embankment slides; 3.40 m	–	–	–	–	–	nr	2876.00
Embankment slides; 4.40 m	–	–	–	–	–	nr	3740.00
Embankment slides; 5.80 m	–	–	–	–	–	nr	4760.00
Embankment slides; 7.30 m	–	–	–	–	–	nr	6048.00
Embankment slides; 9.10 m	–	–	–	–	–	nr	7576.00
Embankment slides; 11.00 m	–	–	–	–	–	nr	9061.00
Slides; Lappset UK Ltd							
142015M slide	–	–	–	–	–	nr	3266.90
141115M Jumbo slide	–	–	–	–	–	nr	5088.20
Slides; Kompan Ltd							
M351P slide	–	–	–	–	–	nr	2972.65
M326P Aladdin's Cave slide	–	–	–	–	–	nr	2754.75
Moving equipment – General							
Preamble: The following standard items							
of playground equipment vary							
considerably in quality and price; the							
following prices are middle of the range.							
Moving equipment							
Roundabouts; Wicksteed Ltd							
Dual axis roundabout; Dizzy	–	–	–	–	–	nr	3655.00
Dual axis roundabout; Whizzy	–	–	–	–	–	nr	2489.00
Dual axis roundabout; Solar Spinner	–	–	–	–	–	nr	3088.00
Roundabouts; Kompan Ltd							
Supernova GXY 916; multifunctional							
spinning and balancing disc; capacity							
approximately 15 children	–	–	–	–	–	nr	4535.75
Seesaws							
Seesaws; Lappset UK Ltd							
010300; seesaws	–	–	–	–	–	nr	590.20
010237; seesaws	–	–	–	–	–	nr	2028.00
Seesaws; Wicksteed Ltd							
Buddy Board	–	–	–	–	–	nr	2260.00
Glow Worm	–	–	–	–	–	nr	1933.00
Cobra	–	–	–	–	–	nr	2255.00
Sidewinder	–	–	–	–	–	nr	2026.00

Q50 SITE/STREET FURNITURE/EQUIPMENT

Item Excluding site overheads and profit	PC £	Labour hours	Labour £	Plant £	Material £	Unit	Total rate £
Play sculptures – General Preamble: Many variants on the shapes of playground equipment are available, simulating spacecraft, trains, cars, houses etc., and these designs are frequently changed. The basic principles remain constant but manufacturer's catalogues should be checked for the latest styles.							
Climbing equipment and play structures – General Preamble: Climbing equipment generally consists of individually designed modules. Play structures generally consist of interlinked and modular pieces of equipment and sculptures. These may consist of climbing, play and skill based modules, nets and various other activities. Both are set into either safety surfacing or defined sand pit areas.The equipment below outlines a range from various manufacturers. Individual catalogues should be consulted in each instance. Safety areas should be allowed round all equipment.							
Climbing equipment and play structures Climbing equipment; Wicksteed Ltd							
Funrun Fitness Trail; Under Starter's Orders; set of 12 units	–	–	–	–	–	nr	19309.00
Climbing equipment; Lappset UK Ltd							
138401M Storks Nest	–	–	–	–	–	nr	3738.80
122457M Playhouse	–	–	–	–	–	nr	2919.80
120100M Activity Tower	–	–	–	–	–	nr	12330.50
120124M Tower and Climbing Frame	–	–	–	–	–	nr	11804.00
SMP Playgrounds Ltd							
Nexus – The Core; multi-play structure	–	–	–	–	–	nr	11240.25
Spring equipment Spring based equipment for 1–8 year olds; Kompan Ltd							
M101P Crazy Hen	–	–	–	–	–	nr	594.05
M128P Crazy Daisy	–	–	–	–	–	nr	868.10
M141P Spring Seesaw	–	–	–	–	–	nr	2072.15
M155P Quartet Seesaw	–	–	–	–	–	nr	1422.25
Spring based equipment for under 12's; Lappset UK Ltd							
010501 Horse Springer	–	–	–	–	–	nr	764.40

Q50 SITE/STREET FURNITURE/EQUIPMENT

Item Excluding site overheads and profit	PC £	Labour hours	Labour £	Plant £	Material £	Unit	Total rate £
PLAYGROUND EQUIPMENT – cont							
Sandpits							
Market prices							
play pit sand; Boughton Loam Ltd	122.40	–	–	–	–	m³	-
Kompan Ltd							
Basic550; 276 × 154 × 31 cm deep	–	–	–	–	–	nr	1115.40
SPORTS EQUIPMENT							
Sports equipment							
Tennis posts; steel; suitable for hard or							
grass tennis courts; including winder,							
sockets and dust cap							
round	216.00	1.00	19.75	–	216.00	set	235.75
square	225.00	1.00	19.75	–	225.00	set	244.75
Tennis nets; not including posts or							
fixings							
Tournament	146.00	3.00	59.25	–	152.90	set	212.15
Match	129.00	3.00	59.25	–	135.90	set	195.15
Club	114.00	3.00	59.25	–	120.90	set	180.15
Football goals; full size; socketed;							
including international net supports and							
nets							
aluminium	1482.00	4.00	79.00	–	1488.90	set	1567.90
steel; heavyweight	1155.00	0.14	2.84	–	1346.61	set	1349.45
Football goals; full size; free-standing;							
including nets							
aluminium	1518.00	4.00	79.00	–	1518.00	set	1597.00
steel	1362.00	4.00	79.00	–	1362.00	set	1441.00
Mini-soccer goals; free-standing;							
including nets							
aluminium	701.00	4.00	79.00	–	701.00	set	780.00
steel	478.00	4.00	79.00	–	478.00	set	557.00
Rugby posts; socketed							
aluminium; 10 m high	1459.00	6.00	118.50	–	1477.39	set	1595.89
aluminium; 12 m high	1841.00	6.00	118.50	–	1859.39	set	1977.89
steel; 12 m high	1643.00	6.00	118.50	–	1661.39	set	1779.89
Hockey goals; steel; including							
backboards and nets							
socketed	956.00	1.00	19.75	–	956.00	set	975.75
free-standing	1036.00	1.00	19.75	–	1036.00	set	1055.75
Cricket cages; steel; including netting							
free-standing	968.00	12.00	237.00	–	968.00	set	1205.00
wheelaway	1091.00	12.00	237.00	–	1091.00	set	1328.00

R12 DRAINAGE

Item Excluding site overheads and profit	PC £	Labour hours	Labour £	Plant £	Material £	Unit	Total rate £
Clarification notes on labour costs in this section							
General groundworks team Generally a three man team is used in this section; The column Labour hours reports team hours. The column Labour £ reports the total cost of the team for the unit of work shown							
3 man team	–	1.00	59.25	–	–	hr	**59.25**
Excavation for Drainage							
Machine excavation Excavating pits; starting from ground level; works exclude for earthwork retention							
maximum depth not exceeding 1.00 m	–	0.50	9.88	22.55	–	m³	**32.43**
maximum depth not exceeding 2.00 m	–	0.50	9.88	40.23	–	m³	**50.11**
maximum depth not exceeding 4.00 m	–	0.50	9.88	61.87	–	m³	**71.75**
Disposal of excavated material; depositing on site in permanent spoil heaps; average 50 m	–	0.04	0.82	1.92	–	m³	**2.74**
Filling to excavations; obtained from on site spoil heaps; average thickness not exceeding 0.25 m	–	0.13	2.63	5.50	–	m³	**8.13**
Hand excavation Excavating pits; starting from ground level; works exclude for earthwork retention							
maximum depth not exceeding 1.00 m	–	1.20	47.40	–	–	m³	**47.40**
maximum depth not exceeding 2.00 m	–	1.50	59.25	–	–	m³	**59.25**
Works to pits or drainage excavations Surface treatments; compacting; bottoms of excavations	–	0.05	0.99	–	–	m²	**0.99**
Earthwork support; distance between opposing faces not exceeding 2.00 m							
maximum depth not exceeding 1.00 m	–	0.20	3.95	–	4.44	m³	**8.39**
maximum depth not exceeding 2.00 m	–	0.30	5.92	–	4.44	m³	**10.36**
maximum depth not exceeding 4.00 m	–	0.67	13.17	–	1.96	m³	**15.13**
Inspection Chambers and Manholes							
Inspection chambers; in situ concrete Beds; plain in situ concrete; 11.50 N/mm²–40 mm aggregate							
thickness not exceeding 150 mm	–	2.00	39.50	–	95.81	m³	**135.31**
thickness 150–450 mm	–	1.75	34.56	–	95.81	m³	**130.37**

R12 DRAINAGE

Item Excluding site overheads and profit	PC £	Labour hours	Labour £	Plant £	Material £	Unit	Total rate £
Inspection Chambers and Manholes – cont							
Inspection chambers – cont							
Benchings in bottoms; plain in situ concrete; 25.50 N/mm^2–20 mm aggregate							
thickness 150–450 mm	–	2.00	39.50	–	95.81	m^3	**135.31**
Isolated cover slabs; reinforced In situ concrete; 21.00 N/mm^2–20 mm aggregate							
thickness not exceeding 150 mm	95.81	4.00	79.00	–	95.81	m^3	**174.81**
Fabric reinforcement; A193 (3.02 kg/m^2) in cover slabs	1.89	0.06	1.24	–	1.89	m^2	**3.13**
Formwork to reinforced in situ concrete; isolated cover slabs							
soffits; horizontal	–	3.28	64.78	–	4.21	m^2	**68.99**
height not exceeding 250 mm	–	0.97	19.16	–	2.24	m	**21.40**
Inspection chambers; precast concrete units							
Precast concrete inspection chamber units; FP McCann Ltd; bedding, jointing and pointing in cement mortar (1:3); 600 × 450 mm internally							
600 mm deep	54.94	6.00	118.50	–	59.31	nr	**177.81**
900 mm deep	68.67	7.00	138.25	–	73.04	nr	**211.29**
Drainage chambers; FP McCann Ltd; 1200 × 750 mm reducing to 600 × 600 mm; no base unit; depth of invert							
1050 mm deep	416.13	9.00	177.75	–	423.41	nr	**601.16**
1650 mm deep	616.05	11.00	217.25	–	627.69	nr	**844.94**
2250 mm deep	815.96	12.50	246.88	–	830.51	nr	**1077.39**
Cover slabs for chambers or shaft sections; FP McCann Ltd; heavy duty							
900 mm dia. internally	65.72	0.67	13.15	–	65.72	nr	**78.87**
1050 mm dia. internally	72.08	2.00	39.50	13.75	72.08	nr	**125.33**
1200 mm dia. internally	86.92	1.00	19.75	13.75	86.92	nr	**120.42**
1500 mm dia. internally	147.38	1.00	19.75	13.75	147.38	nr	**180.88**
1800 mm dia. internally	220.98	2.00	39.50	30.94	220.98	nr	**291.42**
Brickwork							
Walls to manholes; bricks; PC £300.00/ 1000; in cement mortar (1:3)							
one brick thick	37.80	3.00	59.25	–	46.53	m^2	**105.78**
one and a half brick thick	56.70	4.00	79.00	–	69.80	m^2	**148.80**
two brick thick projection of footing or the like	75.60	4.80	94.80	–	93.06	m^2	**187.86**

R12 DRAINAGE

Item Excluding site overheads and profit	PC £	Labour hours	Labour £	Plant £	Material £	Unit	Total rate £
Walls to manholes; engineering bricks; in cement mortar (1:3)							
one brick thick	34.78	3.00	59.25	–	43.51	m²	102.76
one and a half brick thick	52.16	4.00	79.00	–	65.26	m²	144.26
two brick thick projection of footing or the like	69.55	4.80	94.80	–	87.02	m²	181.82
Extra over common or engineering bricks in any mortar for fair face; flush pointing as work proceeds; English bond walls or the like	–	0.13	2.63	–	–	m²	2.63
In situ finishings; cement: sand mortar (1:3); steel trowelled; 13 mm one coat work to manhole walls; to brickwork or blockwork base; over 300 mm wide	–	0.80	15.80	–	2.91	m²	18.71
Building into brickwork; ends of pipes; making good facings or renderings							
small	–	0.20	3.95	–	–	nr	3.95
large	–	0.30	5.92	–	–	nr	5.92
extra large	–	0.40	7.90	–	–	nr	7.90
extra large; including forming ring arch cover	–	0.50	9.88	–	–	nr	9.88
Inspection chambers; polypropylene; Hepworth Plc							
Mini access chamber; up to 600 mm deep; including cover and frame							
300 mm dia. × 600 mm deep; three 100/110 mm inlets	160.40	3.00	59.25	–	162.92	nr	222.17
Up to 1200 mm deep; including polymer cover and frame with screw down lid							
475 mm dia. × 940 mm deep; five 100/110 mm inlets; supplied with four stoppers in inlets	237.13	4.00	79.00	–	316.33	nr	395.33
Extra for square ductile iron cover and frame to 1 tonne load; screw down lid	110.00	–	–	–	33.34	nr	33.34
Step irons; drainage systems; malleable cast iron; galvanized; building into joints							
General purpose pattern; for one brick walls	4.14	0.17	3.36	–	4.14	nr	7.50
Accessories in PVC-u							
110 mm screwed access cover	15.78	–	–	–	15.78	nr	15.78
110 mm rodding eye	30.52	0.50	9.88	–	34.74	nr	44.62
gully with P traps; 110 mm; 154 × 154 mm grating	74.47	1.00	19.75	–	76.16	nr	95.91
Kerbs; to gullies; in one course Class B engineering bricks; to four sides; rendering in cement: mortar (1:3); dished to gully gratings	2.32	1.00	19.75	–	3.19	nr	22.94

R12 DRAINAGE

Item Excluding site overheads and profit	PC £	Labour hours	Labour £	Plant £	Material £	Unit	Total rate £
Channels							
Best quality vitrified clay half section channels; Hepworth Plc; bedding and jointing in cement: mortar (1:2)							
Channels; straight							
100 mm	6.76	0.80	15.80	–	9.59	m	**25.39**
150 mm	11.24	1.00	19.75	–	14.07	m	**33.82**
225 mm	25.24	1.35	26.66	–	28.07	m	**54.73**
300 mm	51.80	1.80	35.55	–	54.64	m	**90.19**
Bends; 15, 30, 45 or 90°							
100 mm bends	6.08	0.75	14.81	–	7.50	nr	**22.31**
150 mm bends	10.51	0.90	17.77	–	12.64	nr	**30.41**
225 mm bends	40.72	1.20	23.70	–	43.56	nr	**67.26**
300 mm bends	83.02	1.10	21.73	–	86.57	nr	**108.30**
Best quality vitrified clay channels; Hepworth Plc; bedding and jointing in cement: mortar (1:2)							
Branch bends; 15, 30, 45 or 90°; left or right hand							
100 mm	6.08	0.75	14.81	–	7.50	nr	**22.31**
150 mm	10.51	0.90	17.77	–	12.64	nr	**30.41**
Pipe Laying							
Excavating trenches; to receive pipes; grading bottoms to falls; backfilling with excavated material and compacting; disposal of surplus material off site; volumes allow for bedding materials which are priced separately below							
Trenches 300 mm wide							
depth of pipe 750 mm	–	–	–	10.81	3.09	m³	**13.90**
depth of pipe 900 mm	–	–	–	12.35	3.91	m³	**16.26**
depth of pipe 1.20 m	–	–	–	12.97	5.54	m³	**18.51**
depth of pipe 1.50 m	–	–	–	14.41	7.17	m³	**21.58**
depth of pipe 2.00 m	–	–	–	11.92	9.89	m³	**21.81**
Earthwork support; providing support to opposing faces of excavation; moving along as work proceeds							
Maximum depth not exceeding 2.00 m							
distance between opposing faces not exceeding 2.00 m	–	0.80	15.80	21.00	–	m	**36.80**

R12 DRAINAGE

Item Excluding site overheads and profit	PC £	Labour hours	Labour £	Plant £	Material £	Unit	Total rate £
Excavating trenches; using 3 tonne tracked excavator; to receive pipes; grading bottoms; earthwork support; filling with excavated material to within 150 mm of finished surfaces and compacting; completing fill with topsoil; disposal of surplus soil							
Services not exceeding 200 mm nominal size							
average depth of run not exceeding 0.50 m	1.51	0.12	2.37	1.06	1.51	m	**4.94**
average depth of run not exceeding 0.75 m	1.51	0.16	3.22	1.44	1.51	m	**6.17**
average depth of run not exceeding 1.00 m	1.51	0.28	5.60	2.51	1.51	m	**9.62**
average depth of run not exceeding 1.25 m	1.26	0.38	7.57	3.40	1.26	m	**12.23**
Granular beds to trenches; lay granular material to trenches excavated separately; to receive pipes (not included)							
300 mm wide × 100 mm thick							
reject sand	–	0.05	0.99	0.23	3.55	m	**4.77**
reject gravel	–	0.05	0.99	0.23	3.14	m	**4.36**
shingle 40 mm aggregate	–	0.05	0.99	0.23	3.72	m	**4.94**
sharp sand	3.74	0.05	0.99	0.23	3.74	m	**4.96**
300 mm wide × 150 mm thick							
reject sand	–	0.08	1.48	0.34	5.33	m	**7.15**
reject gravel	–	0.08	1.48	0.34	4.71	m	**6.53**
shingle 40 mm aggregate	–	0.08	1.48	0.34	5.58	m	**7.40**
sharp sand	5.61	0.08	1.48	0.34	5.61	m	**7.43**
Excavating trenches; using 3 tonne tracked excavator; to receive pipes; grading bottoms; earthwork support; filling with imported granular material type 2 and compacting; disposal of surplus soil							
Services not exceeding 200 mm nominal size							
average depth of run not exceeding 0.50 m	1.50	0.09	1.71	0.76	4.77	m	**7.24**
average depth of run not exceeding 0.75 m	2.25	0.11	2.14	0.95	7.15	m	**10.24**
average depth of run not exceeding 1.00 m	3.00	0.14	2.76	1.23	9.53	m	**13.52**
average depth of run not exceeding 1.25 m	3.74	0.23	4.51	2.03	11.90	m	**18.44**

R12 DRAINAGE

Item Excluding site overheads and profit	PC £	Labour hours	Labour £	Plant £	Material £	Unit	Total rate £
Pipe Laying – cont							
Excavating trenches; using 3 tonne **tracked excavator; to receive pipes;** **grading bottoms; earthwork support;** **filling with concrete, ready mixed** **ST2; disposal of surplus soil**							
Services not exceeding 200 mm nominal size							
average depth of run not exceeding 0.50 m	11.54	0.11	2.11	0.36	14.81	m	**17.28**
average depth of run not exceeding 0.75 m	17.30	0.13	2.57	0.45	22.20	m	**25.22**
average depth of run not exceeding 1.00 m	23.06	0.17	3.29	0.60	29.59	m	**33.48**
average depth of run not exceeding 1.25 m	28.83	0.23	4.44	0.90	36.99	m	**42.33**
Earthwork support; providing support **to opposing faces of excavation;** **moving along as work proceeds**							
Maximum depth not exceeding 2.00 m							
trenchbox; distance between opposing faces not exceeding 2.00 m	–	0.80	15.80	21.00	–	m	**36.80**
timber; distance between opposing faces not exceeding 500 mm	–	0.20	7.90	–	0.02	m	**7.92**
Clay pipes and fittings; Hepworth Plc; **Supersleve**							
100 mm clay pipes; polypropylene slip coupling; in trenches (trenches not included)							
laid straight	4.88	0.25	4.94	–	6.52	m	**11.46**
short runs under 3.00 m	4.88	0.31	6.17	–	6.52	m	**12.69**
Extra over 100 mm clay pipes for							
bends; 15–90 degree; single socket	7.41	0.25	4.94	–	7.41	nr	**12.35**
junction; 45 or 90 degree; double socket	15.61	0.25	4.94	–	15.61	nr	**20.55**
slip couplings; polypropylene	2.62	0.08	1.65	–	2.62	nr	**4.27**
gully with P trap; 100 mm; 154 × 154 mm plastic grating	24.18	1.00	19.75	–	36.24	nr	**55.99**
150 mm clay pipes; polypropylene slip coupling; in trenches (trenches not included)							
laid straight	16.30	0.30	5.92	–	16.30	m	**22.22**
short runs under 3.00 m	16.30	0.60	11.85	–	16.30	m	**28.15**
Extra over 150 mm clay pipes for							
bends; 15–90 degree	9.89	0.28	5.53	–	14.66	nr	**20.19**
junction; 45 or 90 degree; 100 × 150 mm	14.53	0.40	7.90	–	24.07	nr	**31.97**
junction; 45 or 90 degree; 150 × 150 mm	13.23	0.40	7.90	–	22.78	nr	**30.68**

R12 DRAINAGE

Item Excluding site overheads and profit	PC £	Labour hours	Labour £	Plant £	Material £	Unit	Total rate £
slip couplings; polypropylene	4.77	0.05	0.99	–	4.77	nr	**5.76**
tapered pipe; 100–150 mm	14.87	0.50	9.88	–	14.87	nr	**24.75**
tapered pipe; 150–225 mm	38.19	0.50	9.88	–	38.19	nr	**48.07**
socket adaptor; connection to							
traditional pipes and fittings	10.03	0.33	6.52	–	14.80	nr	**21.32**
Accessories in clay							
access pipe; 150 mm	37.29	–	–	–	46.83	nr	**46.83**
rodding eye; 150 mm	35.67	0.50	9.88	–	39.88	nr	**49.76**
gully with P traps; 150 mm;							
154 × 154 mm plastic grating	48.44	0.80	15.80	–	54.90	nr	**70.70**
PVC-u pipes and fittings; Wavin							
Plastics Ltd; OsmaDrain							
110 mm PVC-u pipes; in trenches							
(trenches not included)							
laid straight	6.66	0.08	1.58	–	6.66	m	**8.24**
short runs under 3.00 m	7.62	0.12	2.37	–	7.62	m	**9.99**
Extra over 110 mm PVC-u pipes for							
bends; short radius	14.94	0.25	4.94	–	14.94	nr	**19.88**
bends; long radius	27.99	0.25	4.94	–	27.99	nr	**32.93**
junctions; equal; double socket	17.82	0.25	4.94	–	17.82	nr	**22.76**
slip couplings	5.62	0.25	4.94	–	5.62	nr	**10.56**
adaptors to clay	16.84	0.50	9.88	–	16.84	nr	**26.72**
160 mm PVC-u pipes; in trenches							
(trenches not included)							
laid straight	18.00	0.08	1.58	–	18.00	m	**19.58**
short runs under 3.00 m	34.19	0.12	2.37	–	34.19	m	**36.56**
Extra over 160 mm PVC-u pipes for							
socket bend; double; 90 or 45°	58.43	0.20	3.95	–	58.43	nr	**62.38**
socket bend; double; 15 or 30°	54.55	0.20	3.95	–	54.55	nr	**58.50**
socket bend; single; 87.5 or 45°	33.94	0.20	3.95	–	33.94	nr	**37.89**
socket bend; single; 15 or 30°	30.10	0.20	3.95	–	30.10	nr	**34.05**
bends; short radius	41.97	0.25	4.94	–	56.90	nr	**61.84**
bends; long radius	113.10	0.25	4.94	–	113.10	nr	**118.04**
junctions; single	106.23	0.33	6.58	–	106.23	nr	**112.81**
pipe coupler	20.98	0.05	0.99	–	20.98	nr	**21.97**
slip couplings PVC-u	10.46	0.05	0.99	–	10.46	nr	**11.45**
adaptors to clay	42.74	0.50	9.88	–	42.74	nr	**52.62**
level invert reducer	17.22	0.50	9.88	–	17.22	nr	**27.10**
spiggot	38.21	0.20	3.95	–	38.21	nr	**42.16**

R12 DRAINAGE

Item Excluding site overheads and profit	PC £	Labour hours	Labour £	Plant £	Material £	Unit	Total rate £
Gullies and Interception Traps							
Intercepting traps; Hepworth Plc Vitrified clay; inspection arms; brass stoppers; iron levers; chains and staples; galvanized; staples cut and pinned to brickwork; cement: mortar (1:2) joints to vitrified clay pipes and channels; bedding and surrounding in concrete; 11.50 N/mm^2–40 mm aggregate; cutting and fitting brickwork; making good facings							
100 mm inlet; 100 mm outlet	88.33	3.00	59.25	–	95.26	nr	**154.51**
150 mm inlet; 150 mm outlet	127.37	2.00	39.50	–	138.60	nr	**178.10**
Gullies; concrete; FP McCann Concrete road gullies; trapped; cement: mortar (1:2) joints to concrete pipes; bedding and surrounding in concrete; 11.50 N/mm^2–40 mm aggregate; 450 mm dia. × 1.07 m deep; rodding eye; stoppers	48.76	6.00	118.50	–	69.52	nr	**188.02**
Gullies; vitrified clay; Hepworth Plc; **bedding in concrete; 11.50 N/mm^2–** **40 mm aggregate** Yard gullies (mud); trapped; domestic duty (up to 1 tonne)							
100 mm outlet; 100 mm dia.; 225 mm internal width; 585 mm internal depth	114.87	3.50	69.13	–	115.46	nr	**184.59**
150 mm outlet; 100 mm dia.; 225 mm internal width; 585 mm internal depth	114.87	3.50	69.13	–	115.46	nr	**184.59**
Yard gullies (mud); trapped; medium duty (up to 5 tonnes)							
100 mm outlet; 100 mm dia.; 225 mm internal width; 585 mm internal depth	162.38	3.50	69.13	–	162.98	nr	**232.11**
150 mm outlet; 100 mm dia.; 225 mm internal width; 585 mm internal depth	177.92	3.50	69.13	–	178.52	nr	**247.65**
Combined filter and silt bucket for yard gullies							
225 mm dia.	41.54	–	–	–	41.54	nr	**41.54**
Road gullies; trapped with rodding eye							
100 mm outlet; 300 mm internal dia.; 600 mm internal depth	–	3.50	69.13	–	111.07	nr	**180.20**
150 mm outlet; 300 mm internal dia.; 600 mm internal depth	113.14	3.50	69.13	–	113.73	nr	**182.86**
150 mm outlet; 400 mm internal dia.; 750 mm internal depth	131.21	3.50	69.13	–	131.80	nr	**200.93**
150 mm outlet; 450 mm internal dia.; 900 mm internal depth	177.52	3.50	69.13	–	178.12	nr	**247.25**

R12 DRAINAGE

Item Excluding site overheads and profit	PC £	Labour hours	Labour £	Plant £	Material £	Unit	Total rate £
Hinged gratings and frames for gullies; alloy							
193 mm for 150 mm dia. gully	–	–	–	–	27.53	nr	27.53
120 × 120 mm	–	–	–	–	9.79	nr	9.79
150 × 150 mm	–	–	–	–	17.71	nr	17.71
230 × 230 mm	–	–	–	–	32.38	nr	32.38
316 × 316 mm	–	–	–	–	85.84	nr	85.84
Hinged gratings and frames for gullies; cast iron							
265 mm for 225 mm dia. gully	–	–	–	–	55.00	nr	55.00
150 × 150 mm	–	–	–	–	17.71	nr	17.71
230 × 230 mm	–	–	–	–	32.38	nr	32.38
316 × 316 mm	–	–	–	–	85.84	nr	85.84
Universal gully trap PVC-u; Wavin Plastics Ltd; OsmaDrain system; bedding in concrete; 11.50 N/mm²– 40 mm aggregate							
Universal gully fitting; comprising gully trap only							
110 mm outlet; 110 mm dia.; 205 mm internal depth	12.67	3.50	69.13	–	13.26	nr	82.39
Vertical inlet hopper; c\w plastic grate							
272 × 183 mm	17.02	0.25	4.94	–	17.02	nr	21.96
Sealed access hopper							
110 × 110 mm	40.02	0.25	4.94	–	40.02	nr	44.96
Universal gully PVC-u; Wavin Plastics Ltd; OsmaDrain system; accessories to universal gully trap							
Hoppers; backfilling with clean granular material; tamping; surrounding in lean mix concrete							
plain hopper; with 110 mm spigot; 150 mm long	13.92	0.40	7.90	–	14.27	nr	22.17
vertical inlet hopper; with 110 mm spigot; 150 mm long	17.02	0.40	7.90	–	17.02	nr	24.92
sealed access hopper; with 110 mm spigot; 150 mm long	40.02	0.40	7.90	–	40.02	nr	47.92
plain hopper; solvent weld to trap	9.50	0.40	7.90	–	9.50	nr	17.40
vertical inlet hopper; solvent wed to trap	16.33	0.40	7.90	–	16.33	nr	24.23
sealed access cover; PVC-u	20.38	0.10	1.98	–	20.38	nr	22.36

R12 DRAINAGE

Item Excluding site overheads and profit	PC £	Labour hours	Labour £	Plant £	Material £	Unit	Total rate £
Gullies and Interception Traps – cont							
Gullies PVC-u; Wavin Plastics Ltd; OsmaDrain system; bedding in concrete; 11.50 N/mm²–40 mm aggregate							
Bottle gully; providing access to the drainage system for cleaning							
bottle gully; 228 × 228 × 317 mm deep	33.79	0.50	9.88	–	34.30	nr	**44.18**
sealed access cover; PVC-u;							
217 × 217 mm	26.27	0.10	1.98	–	26.27	nr	**28.25**
grating; ductile iron; 215 × 215 mm	20.25	0.10	1.98	–	20.25	nr	**22.23**
bottle gully riser; 325 mm	4.36	0.50	9.88	–	5.45	nr	**15.33**
Yard gully; trapped; 300 mm dia. × 600 mm deep; including catchment bucket and ductile iron cover and frame; medium duty loading							
305 mm dia. × 600 mm deep	204.90	2.50	49.38	–	208.11	nr	**257.49**
Kerbs to gullies							
One course Class B engineering bricks to four sides; rendering in cement: mortar (1:3); dished to gully gratings							
150 × 150 mm	1.38	0.33	6.58	–	1.74	nr	**8.32**
Linear Drainage							
Linear drainage							
Marshalls Plc; Mini Beany combined kerb and channel drainage system; to trenches (not included)							
Precast concrete drainage channel base; 185–385 mm deep; bedding, jointing and pointing in cement mortar (1:3); on 150 mm deep concrete (ready mixed) foundation; including haunching with in situ concrete; 11.50 N/mm²– 40 mm aggregate one side; channels 250 mm wide × 1.00 m long							
straight; 1.00 m long	25.81	1.00	19.75	–	29.35	m	**49.10**
straight; 500 mm long	28.93	1.05	20.74	–	32.47	m	**53.21**
radial; 30/10 or 9/6 internal or external	26.73	1.33	26.33	–	30.27	m	**56.60**
angles 45 or 90 degree	26.73	1.00	19.75	–	30.27	nr	**50.02**
Mini Beany Top Block; perforated kerb unit to drainage channel above; natural grey							
straight	30.67	0.33	6.58	–	31.93	m	**38.51**
radial; 30/10 or 9/6 internal or external	19.05	0.50	9.88	–	20.31	m	**30.19**
angles 45 or 90 degree	19.05	0.50	9.88	–	20.31	nr	**30.19**

R12 DRAINAGE

Item Excluding site overheads and profit	PC £	Labour hours	Labour £	Plant £	Material £	Unit	Total rate £
Mini Beany; outfalls; two section concrete trapped outfall with Mini Beany cast iron access cover and frame; to concrete foundation							
high capacity outfalls; silt box 150/ 225 mm outlet; two section trapped outfall silt box and cast iron access cover	552.44	1.00	19.75	–	843.43	nr	**863.18**
inline side or end outlet; outfall 150 mm; 2 section concrete trapped outfall; cast iron Mini Beany access cover and frame	529.95	1.00	19.75	–	530.68	nr	**550.43**
Ancillaries to Mini Beany							
end cap	15.29	0.25	4.94	–	15.29	nr	**20.23**
end cap outlets	15.29	0.25	4.94	–	15.29	nr	**20.23**
Precast concrete channels; Charcon Hard Landscaping; on 150 mm deep concrete foundations; including haunching with in situ concrete; 21.00 N/mm²–20 mm aggregate; both sides							
Charcon Safeticurb; slotted safety channels; for pedestrians and light vehicles							
DBJ; 305 × 305 mm	82.03	0.67	13.17	–	98.32	m	**111.49**
DBA; 250 × 250 mm	42.23	0.67	13.17	–	58.52	m	**71.69**
Charcon Safeticurb; slotted safety channels; for heavy vehicles							
DBM; 248 × 248 mm	75.19	0.80	15.80	–	91.48	m	**107.28**
Clearway; 324 × 257 mm	86.24	0.80	15.80	–	102.53	m	**118.33**
Charcon Safeticurb; inspection units; ductile iron lids; including jointing to drainage channels							
248 × 248 × 914 mm	102.34	1.50	29.63	–	103.30	nr	**132.93**
Silt box tops; concrete frame; cast iron grid lids; type 1; set over gully							
457 × 610 mm	458.67	2.00	39.50	–	459.63	nr	**499.13**
Manhole covers; type K; cast iron; providing inspection to blocks and back gullies	582.75	2.00	39.50	–	585.15	nr	**624.65**
Slot and Channel drains							
Loadings for slot drains							
A15; 1.5 tonne; pedestrian							
B125; 12.5 tonne; domestic use							
C250; 25 tonne; car parks, supermarkets, industrial units							
D400; 40 tonne; highways							
E600; 60 tonne; forklifts							

R12 DRAINAGE

Item Excluding site overheads and profit	PC £	Labour hours	Labour £	Plant £	Material £	Unit	Total rate £
Linear Drainage – cont							
Slot drains; ACO Technologies; laid to concrete bed C25 on compacted granular base on 200 mm deep concrete bed; haunched with 200 mm concrete surround; all in 750 × 430 mm wide trench with compacted 200 mm granular base surround (excavation and subbase not included)							
ACO MultiDrain M100PPD; recycled polypropylene drainage channel; range of gratings to complement installations which require discreet slot drainage							
142 mm wide × 150 mm deep	74.70	1.20	23.70	–	85.12	m	**108.82**
Accessories for M100PPD							
connectors; vertical outlet 110 mm	17.25	–	–	–	17.25	nr	**17.25**
connectors; vertical outlet 160 mm	17.25	–	–	–	17.25	nr	**17.25**
sump units; 110 mm	132.85	1.50	29.63	–	152.01	nr	**181.64**
universal gully and bucket;							
440 × 440 × 1315 mm deep	793.05	3.00	59.25	–	828.98	nr	**888.23**
Channel drains; ACO Technologies; ACO MultiDrain MD polymer concrete channel drainage system; traditional channel and grate drainage solution							
ACO MultiDrain MD Brickslot; offset galvanized slot drain grating for M100PPD; load class C250							
brickslot; galvanized steel; 1.00 mm	85.00	–	–	–	85.00	m	**85.00**
ACO MultiDrain MD Brickslot; offset slot drain grating; load class C250–400							
brickslot; galvanized steel; 1000 mm	158.80	1.20	23.70	–	158.80	m	**182.50**
brickslot; stainless steel; 1000 mm	173.80	1.20	23.70	–	347.60	m	**371.30**
Accessories for MultiDrain MD							
sump unit; complete with sediment							
bucket and access unit	132.85	1.00	19.75	–	270.20	nr	**289.95**
end cap; closing piece	7.30	–	–	–	7.30	nr	**7.30**
end cap; inlet/outlet	7.30	–	–	–	7.30	nr	**7.30**
ACO Drainlock Gratings for M100PPD and M100D system							
A15 loading							
slotted galvanized steel	23.05	0.15	2.96	–	23.05	m	**26.01**
perforated galvanized steel	42.10	0.15	2.96	–	42.10	m	**45.06**
C250 loading							
Heelguard composite black; 500 mm							
long with security locking	55.10	–	–	–	55.10	m	**55.10**
Intercept; ductile iron; 500 mm long	126.50	–	–	–	126.50	m	**126.50**
slotted galvanized steel; 1.00 mm long	59.60	–	–	–	59.60	m	**59.60**

R12 DRAINAGE

Item Excluding site overheads and profit	PC £	Labour hours	Labour £	Plant £	Material £	Unit	Total rate £
perforated galvanized steel; 1.00 mm long	65.10	–	–	–	65.10	m	**65.10**
mesh galvanized steel; 1.00 mm long	44.60	–	–	–	44.60	m	**44.60**
Wade Ltd; stainless steel channel drains; specialized applications; bespoke manufacture							
Drain in stainless steel; c/w tie in lugs and inbuilt falls to 100 mm spigot outlet; stainless steel gratings							
NE channel; Ref 12430; secured gratings	420.00	1.20	23.70	–	430.42	m	**454.12**
Fin drains							
Fin drains; Cooper Clarke Civils and Lintels; Geofin Shallow Drain; drainage of sports fields and grassed landscaped areas							
Geofin Shallow Drain; to trenches; excavation by trenching machine; backfilled with single size aggregate 20 mm and covered with sharp sand rootzone 200 mm thick							
Geofin; 25 mm thick × 150 mm deep	5.50	0.05	0.99	0.79	10.32	m	**12.10**
Geofin; 25 mm thick × 450 mm deep	4.20	0.07	1.32	1.05	14.03	m	**16.40**
Geofin; 25 mm thick × 900 mm deep	8.70	0.10	1.98	1.05	27.02	m	**30.05**
Fin drains; Cooper Clarke Civils and Lintels; Geofin Geocomposite Finn Drain; laid to slabs							
Geofin fin drain laid horizontally to slab or blinded ground; covered with 20 mm shingle 200 mm thick							
Geofin; 25 mm thick × 900 mm wide; laid flat to falls	9.66	0.02	0.40	0.16	16.75	m²	**17.31**
Drainage to roof decks and planters							
Maxit Ltd; Leca (light expanded clay aggregate); drainage aggregate to roofdecks and planters							
Placed mechanically to planters; average 100 mm thick; by mechanical plant tipped into planters							
aggregate size 10–20 mm; delivered in 30 m³ loads	47.00	0.40	7.90	6.74	47.00	m³	**61.64**
aggregate size 10–20 mm; delivered in 70 m³ loads	44.00	0.20	3.95	6.05	44.00	m³	**54.00**

Prices for Measured Works

R12 DRAINAGE

Item Excluding site overheads and profit	PC £	Labour hours	Labour £	Plant £	Material £	Unit	Total rate £
Drainage to roof decks and planters – cont							
Maxit Ltd – cont							
Placed by light aggregate blower (maximum 40 m)							
aggregate size 10–20 mm; delivered in 30 m³ loads	47.00	0.14	2.82	–	51.59	m³	**54.41**
aggregate size 10–20 mm; delivered in 55 m³ loads on blower vehicle	67.00	0.14	2.82	–	67.00	m³	**69.82**
aggregate size 10–20 mm; delivered in 70 m³ loads	44.00	0.14	2.82	–	48.59	m³	**51.41**
By hand							
aggregate size 10–20 mm; delivered in 30 m³ loads	47.00	1.33	26.33	–	47.00	m³	**73.33**
aggregate size 10–20 mm; delivered in 70 m³ loads	44.00	1.33	26.33	–	44.00	m³	**70.33**
Drainage boards laid to insulated slabs on roof decks; boards laid below growing medium and granulated drainage layer and geofabric (all not included) to collect and channel water to drainage outlets (not included)							
Alumasc Floradrain; polyethylene irrigation/drainage layer; inclusive of geofabric laid over the surface of the drainage board							
Floradrain FD40; 0.96 × 2.08panels	12.05	0.05	0.99	–	13.20	m²	**14.19**
Floradrain FD60; 1.00 × 2.00panel	21.42	0.07	1.32	–	22.57	m²	**23.89**
Access covers and Frames							
Load Classes for Access Covers							
FACTA (Fabricated Access Cover Trade Association) class:							
A – 0.5 tonne maximum slow moving wheel load							
AA – 1.5 tonne maximum slow moving wheel load							
AAA – 2.5 tonne maximum slow moving wheel load							
B – 5 tonne maximum slow moving wheel load							
C – 6.5 tonne maximum slow moving wheel load							
D – 11 tonne maximum slow moving wheel load							

R12 DRAINAGE

Item Excluding site overheads and profit	PC £	Labour hours	Labour £	Plant £	Material £	Unit	Total rate £
Access covers and frames; solid top;							
galvanized; Steelway Brickhouse;							
Bristeel; bedding frame in cement							
mortar (1:3); cover in grease and							
sand; clear opening sizes; base size							
shown in brackets (50 mm depth)							
FACTA AA; single seal							
450 × 450 mm (520 × 520 mm)	64.26	1.50	29.63	–	74.70	nr	**104.33**
600 × 450 mm (670 × 520 mm)	71.32	1.80	35.55	–	82.92	nr	**118.47**
600 × 600 mm (670 × 670 mm)	76.26	2.00	39.50	–	90.18	nr	**129.68**
FACTA AA; double seal							
450 × 450 mm (560 × 560 mm)	104.39	1.50	29.63	–	114.84	nr	**144.47**
600 × 450 mm (710 × 560 mm)	114.70	1.80	35.55	–	126.31	nr	**161.86**
600 × 600 mm (710 × 710 mm)	139.05	2.00	39.50	–	152.98	nr	**192.48**
FACTA AAA; single seal							
450 × 450 mm (520 × 520 mm)	98.97	1.50	29.63	–	109.42	nr	**139.05**
600 × 450 mm (670 × 520 mm)	106.81	1.80	35.55	–	118.42	nr	**153.97**
600 × 600 mm (670 × 670 mm)	113.81	2.00	39.50	–	127.73	nr	**167.23**
FACTA AAA; double seal							
450 mm × 450 mm (560 × 560)	140.00	1.50	29.63	–	150.44	nr	**180.07**
600 mm × 450 mm (710 × 560)	152.95	1.80	35.55	–	164.55	nr	**200.10**
600 mm × 600 mm (710 × 710)	170.42	2.00	39.50	–	184.35	nr	**223.85**
FACTA B; single seal							
450 × 450 mm (520 × 520 mm)	83.07	1.50	29.63	–	93.51	nr	**123.14**
600 × 450 mm (670 × 520 mm)	103.71	1.80	35.55	–	115.31	nr	**150.86**
600 × 600 mm (670 × 670 mm)	120.56	2.00	39.50	–	134.48	nr	**173.98**
FACTA B; double seal							
450 × 450 mm (560 × 560 mm)	133.19	1.50	29.63	–	143.63	nr	**173.26**
600 × 450 mm (710 × 560 mm)	148.28	1.80	35.55	–	159.88	nr	**195.43**
600 × 600 mm (710 × 710 mm)	175.40	2.00	39.50	–	189.33	nr	**228.83**
Access covers and frames; recessed;							
galvanized; Steelway Brickhouse;							
Bripave; bedding frame in cement							
mortar (1:3); cover in grease and							
sand; clear opening sizes; base size							
shown in brackets (for block depths							
50 mm, 65 mm, 80 mm or 100 mm)							
FACTA AA							
450 × 450 mm (562 × 562 mm)	206.95	1.50	29.63	–	217.39	nr	**247.02**
600 × 450 mm (712 × 562 mm)	227.00	1.80	35.55	–	238.60	nr	**274.15**
600 × 600 mm (712 × 712 mm)	235.30	2.00	39.50	–	248.67	nr	**288.17**
750 × 600 mm (862 × 712 mm)	262.56	2.40	47.40	–	277.65	nr	**325.05**
FACTA B							
450 × 450 mm (562 × 562 mm)	207.31	1.50	29.63	–	217.75	nr	**247.38**
600 × 450 mm (712 × 562 mm)	241.00	1.80	35.55	–	252.60	nr	**288.15**
600 × 600 mm (712 × 712 mm)	244.00	2.00	39.50	–	257.93	nr	**297.43**
750 × 600 mm (862 × 712 mm)	244.79	2.40	47.40	–	259.88	nr	**307.28**
FACTA D							
450 × 450 mm (562 × 562 mm)	232.35	1.50	29.63	–	242.79	nr	**272.42**
600 × 450 mm (712 × 562 mm)	246.86	1.80	35.55	–	258.46	nr	**294.01**
600 × 600 mm (712 × 712 mm)	255.05	2.00	39.50	–	268.98	nr	**308.48**
750 × 600 mm (862 × 712 mm)	310.40	2.40	47.40	–	325.49	nr	**372.89**

R12 DRAINAGE

Item Excluding site overheads and profit	PC £	Labour hours	Labour £	Plant £	Material £	Unit	Total rate £
Access covers and Frames – cont							
Access covers and frames; Jones of Oswestry; bedding frame in cement mortar (1:3); cover in grease and sand; clear opening sizes							
Access covers and frames; Suprabloc; to paved areas; filling with blocks cut and fitted to match surrounding paving							
pedestrian weight; 300 × 300 mm	194.35	2.50	49.38	–	201.34	nr	250.72
pedestrian weight; 450 × 450 mm	331.63	2.50	49.38	–	331.63	nr	381.01
pedestrian weight; 450 × 600 mm	369.35	2.50	49.38	–	376.63	nr	426.01
light vehicular weight; 300 × 300 mm	249.00	2.40	47.40	–	255.99	nr	303.39
light vehicular weight; 450 × 450 mm	191.97	3.00	59.25	–	199.97	nr	259.22
light vehicular weight; 450 × 600 mm	381.39	4.00	79.00	–	388.38	nr	467.38
heavy vehicular weight; 300 × 300 mm	249.55	3.00	59.25	–	256.83	nr	316.08
heavy vehicular weight; 450 × 600 mm	395.38	3.00	59.25	–	403.38	nr	462.63
heavy vehicular weight; 600 × 600 mm	464.00	4.00	79.00	–	475.64	nr	554.64
Extra over manhole frames and covers for							
filling recessed manhole covers with brick paviors; PC £305.00/1000	13.42	1.00	19.75	–	14.02	m^2	33.77
filling recessed manhole covers with vehicular paving blocks; PC £8.99/m^2	8.99	0.75	14.81	–	9.59	m^2	24.40
filling recessed manhole covers with concrete paving flags; PC £24.49/m^2	14.76	0.35	6.91	–	15.12	m^2	22.03

R13 LAND DRAINAGE

Item Excluding site overheads and profit	PC £	Labour hours	Labour £	Plant £	Material £	Unit	Total rate £
MARKET PRICES OF BACKFILLING MATERIALS							
Market prices of commonly used materials used for backfilling							
The prices below show the market price with standard settlement factors allowed for							
Sharp sand	–	–	–	–	35.72	m³	**35.72**
Gravel (washed river or pit)	–	–	–	–	33.67	m³	**33.67**
Shingle	–	–	–	–	36.35	m³	**36.35**
Topsoil	–	–	–	–	33.60	m³	**33.60**
Gravel rejects	–	–	–	–	30.68	m³	**30.68**
Selected granular material 6F2	–	–	–	–	13.80	m³	**13.80**
Clarification notes on labour costs in this section							
General groundworks team							
Generally a three man team is used in this section; The column Labour hours reports team hours. The column Labour £ reports the total cost of the team for the unit of work shown							
3 man team	–	1.00	59.25	–	–	hr	**59.25**

SOAKAWAYS
Preamble: Flat rate hourly rainfall = 50 mm/hr and assumes 100 impermeability of the run-off area. A storage capacity of the soakaway should be 1/3 of the hourly rainfall. Formulae for calculating soakaway depths are provided in the publications mentioned below and in the Tables and Memoranda section of this publication. The design of soakaways is dependent on, amongst other factors, soil conditions, permeability, groundwater level and runoff. The definitive documents for design of soakaways are CIRIA 156 and BRE Digest 365 dated September 1991. The suppliers of the systems below will assist through their technical divisions. Excavation earthwork support of pits and disposal not included.

R13 LAND DRAINAGE

Item Excluding site overheads and profit	PC £	Labour hours	Labour £	Plant £	Material £	Unit	Total rate £
SOAKAWAYS – cont							
Soakaway excavation							
Excavating mechanical; to reduce levels							
maximum depth not exceeding							
1.00 m; JCB sitemaster	–	0.05	0.99	2.06	–	m³	3.05
maximum depth not exceeding							
1.00 m; 360 tracked excavator	–	0.04	0.79	2.00	–	m³	2.79
maximum depth not exceeding							
2.00 m; 360 tracked excavator	–	0.06	1.19	3.00	–	m³	4.19
Disposal							
Excavated material; off site; to tip;							
mechanically loaded (JCB)							
inert	–	–	–	–	–	m³	18.14
In situ concrete ring beam							
foundations to base of soakaway;							
300 mm wide × 250 mm deep; poured							
on or against earth or unblinded							
hardcore							
Internal dia.s of rings							
900 mm	19.65	4.00	79.00	–	19.65	nr	98.65
1200 mm	26.19	4.50	88.88	–	26.19	nr	115.07
1500 mm	32.74	5.00	98.75	–	32.74	nr	131.49
2400 mm	52.39	5.50	108.63	–	52.39	nr	161.02
Concrete soakaway rings; Milton							
Pipes Ltd; perforations and step irons							
to concrete rings at manufacturers							
recommended centres; placing of							
concrete ring soakaways to in situ							
concrete ring beams (1:3:6) (not							
included); filling and surrounding							
base with gravel 225 mm deep (not							
included)							
Ring dia. 900 mm							
1.00 m deep; volume 636 litres	76.40	1.25	24.69	21.88	195.84	nr	242.41
1.50 m deep; volume 954 litres	114.60	1.65	32.59	57.75	290.94	nr	381.28
2.00 m deep; volume 1272 litres	152.80	1.65	32.59	57.75	386.04	nr	476.38
Ring dia. 1200 mm							
1.00 m deep; volume 1131 litres	99.75	3.00	59.25	52.50	237.27	nr	349.02
1.50 m deep; volume 1696 litres	149.63	4.50	88.88	78.75	350.89	nr	518.52
2.00 m deep; volume 2261 litres	199.50	4.50	88.88	78.75	464.52	nr	632.15
2.50 m deep; volume 2827 litres	249.38	6.00	118.50	105.00	530.19	nr	753.69
Ring dia. 1500 mm							
1.00 m deep; volume 1767 litres	161.00	4.50	88.88	52.50	331.57	nr	472.95
1.50 m deep; volume 2651 litres	241.50	6.00	118.50	73.50	489.52	nr	681.52
2.00 m deep; volume 3534 litres	322.00	6.00	118.50	73.50	647.47	nr	839.47
2.50 m deep; volume 4418 litres	249.38	7.50	148.13	131.25	652.29	nr	931.67

R13 LAND DRAINAGE

Item Excluding site overheads and profit	PC £	Labour hours	Labour £	Plant £	Material £	Unit	Total rate £
Ring dia. 2400 mm							
1.00 m deep; volume 4524 litres	620.25	6.00	118.50	52.50	852.47	nr	1023.47
1.50 m deep; volume 6786 litres	930.38	7.50	148.13	73.50	1267.44	nr	1489.07
2.00 m deep; volume 9048 litres	1240.50	7.50	148.13	73.50	1689.27	nr	1910.90
2.50 m deep; volume 11310 litres	1550.63	9.00	177.75	131.25	2104.24	nr	2413.24
Extra over for							
250 mm depth chamber ring	–	–	–	–	–	100%	–
500 mm depth chamber ring	–	–	–	–	–	50%	–
Cover slabs to soakaways							
Heavy duty precast concrete							
900 mm dia.	128.50	1.00	19.75	26.25	128.50	nr	174.50
1200 mm dia.	147.25	1.00	19.75	26.25	147.25	nr	193.25
1500 mm dia.	233.75	1.00	19.75	26.25	233.75	nr	279.75
2400 mm dia.	875.95	1.00	19.75	26.25	875.95	nr	921.95
Step irons to concrete chamber rings	31.60	–	–	–	31.60	m	31.60
Extra over soakaways for filter wrapping with a proprietary filter membrane							
900 mm dia. × 1.00 m deep	1.44	1.00	19.75	–	1.44	nr	21.19
900 mm dia. × 2.00 m deep	2.88	1.50	29.63	–	2.88	nr	32.51
1050 mm dia. × 1.00 m deep	1.68	1.50	29.63	–	1.68	nr	31.31
1050 mm dia. × 2.00 m deep	3.36	2.00	39.50	–	3.36	nr	42.86
1200 mm dia. × 1.00 m deep	1.92	2.00	39.50	–	1.92	nr	41.42
1200 mm dia. × 2.00 m deep	3.83	2.50	49.38	–	3.83	nr	53.21
1500 mm dia. × 1.00 m deep	2.39	2.50	49.38	–	2.39	nr	51.77
1500 mm dia. × 2.00 m deep	4.78	2.50	49.38	–	4.78	nr	54.16
1800 mm dia. × 1.00 m deep	2.87	3.00	59.25	–	2.87	nr	62.12
1800 mm dia. × 2.00 m deep	5.74	3.25	64.19	–	5.74	nr	69.93
Gravel surrounding to concrete ring soakaway							
40 mm aggregate backfilled to vertical face of soakaway wrapped with geofabric (not included)							
250 mm thick	35.46	0.20	3.95	10.50	35.46	m³	49.91
Backfilling to face of soakaway; carefully compacting as work proceeds							
Arising from the excavations							
average thickness exceeding 0.25 m; depositing in layers 150 mm maximum thickness	–	0.03	0.66	5.25	–	m³	5.91

R13 LAND DRAINAGE

Item Excluding site overheads and profit	PC £	Labour hours	Labour £	Plant £	Material £	Unit	Total rate £
SOAKAWAYS – cont							
Aquacell soakaway; Wavin Plastics Ltd; preformed polypropylene soakaway infiltration crate units; to trenches; surrounded by geotextile and 40 mm aggregate laid 100 mm thick in trenches (excavation, disposal and backfilling not included)							
1.00 m × 500 × 400 mm; internal volume 190 litres							
4 crates; 2.00 m × 1.00 m × 400 mm; 760 litres	181.22	1.60	31.60	17.32	229.92	nr	278.84
8 crates; 2.00 m × 1.00 m × 800 mm; 1520 litres	362.43	3.20	63.20	22.27	422.52	nr	507.99
12 crates; 6.00 m × 500 mm × 800 mm; 2280 litres	543.65	4.80	94.80	46.20	658.76	nr	799.76
16 crates; 4.00 m × 1.00 m × 800 mm; 3040 litres	724.86	6.40	126.40	41.25	828.59	nr	996.24
20 crates; 5.00 m × 1.00 m × 800 mm; 3800 litres	906.08	8.00	158.00	40.84	1017.72	nr	1216.56
30 crates; 15.00 m × 1.00 m × 400 mm; 5700 litres	1359.12	12.00	237.00	111.38	1632.98	nr	1981.36
60 crates; 15.00 m × 1.00 m × 800 mm; 11400 litres	2718.24	20.00	395.00	145.61	3088.57	nr	3629.18
Geofabric surround to Aquacell units; Terram Ltd							
Terram synthetic fibre filter fabric; to face of concrete rings (not included); anchoring whilst backfilling (not included)							
Terram 1000; 0.70 mm thick; mean water flow 50 l/m²/s	0.49	0.05	0.99	–	0.49	m²	1.48
SAND SLITTING/GROOVING							
Sand slitting; Agripower Ltd							
Drainage slits; at 1.00 m centres; using spinning disc trenching machine; backfilling to 100 mm of surface with pea gravel; finished surface with medium grade sand 100; arisings to be loaded, hauled and tipped onsite; minimum pitch size 6000 m²							
250 mm depth	–	–	–	–	–	m	1.80
300 mm depth	–	–	–	–	–	m	1.97
400 mm depth	–	–	–	–	–	m	2.98

R13 LAND DRAINAGE

Item Excluding site overheads and profit	PC £	Labour hours	Labour £	Plant £	Material £	Unit	Total rate £
Sand grooving or banding; Agripower Ltd Drainage bands at 260 mm m centres; using Blec Sandmaster; to existing grass at 260 mm centres; minimum pitch size 4000 m²							
sand banding or sand grooving	–	–	–	–	–	m²	1.02
LAND DRAINS							
Market prices of backfilling materials							
Sand	38.76	–	–	–	38.76	m³	38.76
Gravel rejects	32.92	–	–	–	32.92	m³	32.92
Imported topsoil (allowing for 20% settlement)	–	–	–	–	33.60	m³	33.60
Land drainage; calculation table For calculation of drainage per hectare the following table can be used; rates show the lengths of drains per unit and not the value							
Lateral drains							
10.00 m centres	–	–	–	–	–	m/ha	1000.00
15.00 m centres	–	–	–	–	–	m/ha	650.00
25.00 m centres	–	–	–	–	–	m/ha	400.00
30.00 m centres	–	–	–	–	–	m/ha	330.00
Main drains							
1 nr (at 100 m centres)	–	–	–	–	–	m/ha	100.00
2 nr (at 50 m centres)	–	–	–	–	–	m/ha	200.00
3 nr (at 33.3 m centres)	–	–	–	–	–	m/ha	-
4 nr (at 25 m centres)	–	–	–	–	–	m/ha	400.00
Land drainage; excavating Removing 150 mm depth of topsoil; 300 mm wide; depositing beside trench; by machine	–	1.50	29.63	13.50	–	100 m	43.13
Removing 150 mm depth of topsoil; 300 mm wide; depositing beside trench; by hand	–	8.00	158.00	–	–	100 m	158.00
Excavated material; disposal on site							
In spoil heaps							
average 25 m distance	–	–	–	2.80	–	m³	2.80
average 50 m distance	–	–	–	3.23	–	m³	3.23
average 100 m distance (1 dumper)	–	–	–	4.05	–	m³	4.05
average 100 m distance (2 dumpers)	–	–	–	4.18	–	m³	4.18
average 200 m distance	–	–	–	4.89	–	m³	4.89
average 200 m distance (2 dumpers)	–	–	–	5.02	–	m³	5.02
Removing excavated material from site to tip; mechanically loaded							
excavated material and clean hardcore rubble	–	–	–	1.56	16.83	m³	18.39

R13 LAND DRAINAGE

Item Excluding site overheads and profit	PC £	Labour hours	Labour £	Plant £	Material £	Unit	Total rate £
LAND DRAINS – cont							
Land drainage; Agripower Ltd;							
excavating trenches (with minimum							
project size of 1000 m) by trenching							
machine; spreading arisings on site;							
laying perforated pipe; as shown							
below; backfilling with shingle to							
350 mm from surface							
Width 150 mm; 80 mm perforated pipe							
depth 750 mm	–	–	–	–	–	m	**5.50**
Width 175 mm; 100 mm perforated pipe							
depth 800 mm	–	–	–	–	–	m	**6.65**
Width 250 mm; 150 mm perforated pipe							
depth 900 mm	–	–	–	–	–	m	**9.70**
Land drainage; excavating for drains;							
by backacter excavator JCB C3X							
Sitemaster; including disposing spoil							
to spoil heaps not exceeding 100 m;							
boning to levels by laser							
Width 150–225 mm							
depth 450 mm	–	9.67	190.92	310.00	–	100 m	**500.92**
depth 600 mm	–	13.00	256.75	465.00	–	100 m	**721.75**
depth 700 mm	–	15.50	306.13	581.25	–	100 m	**887.38**
depth 900 mm	–	23.00	454.25	930.00	–	100 m	**1384.25**
Width 300 mm							
depth 450 mm	–	9.67	190.92	310.00	–	100 m	**500.92**
depth 700 mm	–	14.00	276.50	511.50	–	100 m	**788.00**
depth 900 mm	–	23.00	454.25	930.00	–	100 m	**1384.25**
depth 1000 mm	–	25.00	493.75	1023.00	–	100 m	**1516.75**
depth 1200 mm	–	25.00	493.75	1162.50	–	100 m	**1656.25**
depth 1500 mm	–	31.00	612.25	1302.00	–	100 m	**1914.25**
Land drainage; excavating for drains;							
by 7 tonne tracked excavator;							
including disposing spoil to spoil							
heaps not exceeding 100 m							
Width 225 mm							
depth 450 mm	–	7.00	138.25	65.69	–	100 m	**203.94**
depth 600 mm	–	7.17	141.54	68.43	–	100 m	**209.97**
depth 700 mm	–	7.35	145.12	71.40	–	100 m	**216.52**
depth 900 mm	–	7.76	153.30	78.20	–	100 m	**231.50**
depth 1000 mm	–	8.00	158.00	82.11	–	100 m	**240.11**
depth 1200 mm	–	8.26	163.20	86.43	–	100 m	**249.63**
Width 300 mm							
depth 450 mm	–	7.35	145.12	71.40	–	100 m	**216.52**
depth 600 mm	–	7.76	153.30	78.20	–	100 m	**231.50**
depth 700 mm	–	8.26	163.20	86.43	–	100 m	**249.63**
depth 900 mm	–	9.25	182.69	102.64	–	100 m	**285.33**
depth 1000 mm	–	10.14	200.32	117.30	–	100 m	**317.62**
depth 1200 mm	–	11.00	217.25	131.38	–	100 m	**348.63**

R13 LAND DRAINAGE

Item Excluding site overheads and profit	PC £	Labour hours	Labour £	Plant £	Material £	Unit	Total rate £
Width 600 mm							
depth 450 mm	–	13.00	256.75	164.22	–	100 m	**420.97**
depth 600 mm	–	14.11	278.69	182.47	–	100 m	**461.16**
depth 700 mm	–	16.33	322.58	218.96	–	100 m	**541.54**
depth 900 mm	–	17.29	341.39	234.60	–	100 m	**575.99**
depth 1000 mm	–	18.38	363.10	252.65	–	100 m	**615.75**
depth 1200 mm	–	21.18	418.34	298.58	–	100 m	**716.92**
Land drainage; excavating for drains;							
by hand, including disposing spoil to							
spoil heaps not exceeding 100 m							
Width 150 mm							
depth 450 mm	–	22.03	435.09	–	–	100 m	**435.09**
depth 600 mm	–	29.38	580.25	–	–	100 m	**580.25**
depth 700 mm	–	34.27	676.83	–	–	100 m	**676.83**
depth 900 mm	–	44.06	870.18	–	–	100 m	**870.18**
Width 225 mm							
depth 450 mm	–	33.05	652.74	–	–	100 m	**652.74**
depth 600 mm	–	44.06	870.18	–	–	100 m	**870.18**
depth 700 mm	–	51.41	1015.35	–	–	100 m	**1015.35**
depth 900 mm	–	66.10	1305.47	–	–	100 m	**1305.47**
depth 1000 mm	–	73.44	1450.44	–	–	100 m	**1450.44**
Width 300 mm							
depth 450 mm	–	44.06	870.18	–	–	100 m	**870.18**
depth 600 mm	–	58.75	1160.31	–	–	100 m	**1160.31**
depth 700 mm	–	68.54	1353.66	–	–	100 m	**1353.66**
depth 900 mm	–	88.13	1740.57	–	–	100 m	**1740.57**
depth 1000 mm	–	97.92	1933.92	–	–	100 m	**1933.92**
Width 375 mm							
depth 450 mm	–	55.08	1087.83	–	–	100 m	**1087.83**
depth 600 mm	–	73.44	1450.44	–	–	100 m	**1450.44**
depth 700 mm	–	85.68	1692.18	–	–	100 m	**1692.18**
depth 900 mm	–	110.16	2175.66	–	–	100 m	**2175.66**
depth 1000 mm	–	122.40	2417.40	–	–	100 m	**2417.40**
Width 450 mm							
depth 450 mm	–	66.10	1305.47	–	–	100 m	**1305.47**
depth 600 mm	–	88.13	1740.57	–	–	100 m	**1740.57**
depth 700 mm	–	102.82	2030.69	–	–	100 m	**2030.69**
depth 900 mm	–	132.19	2610.75	–	–	100 m	**2610.75**
depth 1000 mm	–	146.88	2900.88	–	–	100 m	**2900.88**
Width 600 mm							
depth 450 mm	–	88.13	1740.57	–	–	100 m	**1740.57**
depth 600 mm	–	117.50	2320.63	–	–	100 m	**2320.63**
depth 700 mm	–	137.09	2707.53	–	–	100 m	**2707.53**
depth 900 mm	–	176.26	3481.14	–	–	100 m	**3481.14**
depth 1000 mm	–	195.84	3867.84	–	–	100 m	**3867.84**
Width 900 mm							
depth 450 mm	–	132.19	2610.75	–	–	100 m	**2610.75**
depth 600 mm	–	176.26	3481.14	–	–	100 m	**3481.14**
depth 700 mm	–	205.63	4061.19	–	–	100 m	**4061.19**
depth 900 mm	–	264.38	5221.51	–	–	100 m	**5221.51**
depth 1000 mm	–	293.76	5801.76	–	–	100 m	**5801.76**

R13 LAND DRAINAGE

Item Excluding site overheads and profit	PC £	Labour hours	Labour £	Plant £	Material £	Unit	Total rate £
LAND DRAINS – cont							
Earthwork support; moving along as work proceeds							
Maximum depth not exceeding 2.00 m							
distance between opposing faces not exceeding 2.00 m	–	0.80	15.80	21.00	–	m	36.80
Land drainage; pipe laying							
Hepworth; agricultural clay drain pipes; 300 mm length; butt joints; in straight runs							
75 mm dia.	518.47	8.00	158.00	–	518.47	100 m	676.47
100 mm dia.	886.42	9.00	177.75	–	886.42	100 m	1064.17
150 mm dia.	1815.85	10.00	197.50	–	1815.85	100 m	2013.35
Extra over clay drain pipes for filter-wrapping pipes with Terram or similar filter fabric							
Terram 700	0.23	0.04	0.79	–	0.23	m²	1.02
Terram 1000	0.20	0.04	0.79	–	0.20	m²	0.99
Junctions between drains in clay pipes							
75 × 75 mm	17.17	0.25	4.94	–	17.52	nr	22.46
100 × 100 mm	22.79	0.25	4.94	–	23.36	nr	28.30
150 × 150 mm	28.06	0.25	4.94	–	28.91	nr	33.85
Wavin Plastics Ltd; flexible plastic perforated pipes in trenches (not included); to a minimum depth of 450 mm (couplings not included)							
OsmaDrain; flexible plastic perforated pipes in trenches (not included); to a minimum depth of 450 mm (couplings not included)							
80 mm dia.; available in 100 m coil	72.75	2.00	39.50	–	72.75	100 m	112.25
100 mm dia.; available in 100 m coil	117.81	2.00	39.50	–	117.81	100 m	157.31
160 mm dia.; available in 35 m coil	284.91	2.00	39.50	–	284.91	100 m	324.41
WavinCoil; plastic pipe junctions							
80 × 80 mm	3.20	0.05	0.99	–	3.20	nr	4.19
100 × 100 mm	3.58	0.05	0.99	–	3.58	nr	4.57
100 × 60 mm	3.41	0.05	0.99	–	3.41	nr	4.40
100 × 80 mm	3.41	0.05	0.99	–	3.41	nr	4.40
160 × 160 mm	8.52	0.05	0.99	–	8.52	nr	9.51
WavinCoil; couplings for flexible pipes							
80 mm dia.	1.13	0.03	0.66	–	1.13	nr	1.79
100 mm dia.	1.25	0.03	0.66	–	1.25	nr	1.91
160 mm dia.	1.69	0.03	0.66	–	1.69	nr	2.35

R13 LAND DRAINAGE

Item Excluding site overheads and profit	PC £	Labour hours	Labour £	Plant £	Material £	Unit	Total rate £
Land drainage; backfilling trench after laying pipes with gravel rejects or similar; blind filling with ash or sand; topping with 150 mm imported topsoil from dumps not exceeding 100 m; by machine							
Width 150 mm							
depth 450 mm	–	3.30	65.17	66.53	205.54	100 m	**337.24**
depth 600 mm	–	4.30	84.92	87.15	278.69	100 m	**450.76**
depth 750 mm	–	4.96	97.96	100.76	317.77	100 m	**516.49**
depth 900 mm	–	6.30	124.42	128.40	407.57	100 m	**660.39**
Width 225 mm							
depth 450 mm	–	4.95	97.76	99.79	317.34	100 m	**514.89**
depth 600 mm	–	6.45	127.39	130.72	418.20	100 m	**676.31**
depth 750 mm	–	7.95	157.01	161.66	519.37	100 m	**838.04**
depth 900 mm	–	9.45	186.64	192.60	620.23	100 m	**999.47**
Width 375 mm							
depth 450 mm	–	8.25	162.94	166.31	528.69	100 m	**857.94**
depth 600 mm	–	10.75	212.31	217.88	696.90	100 m	**1127.09**
depth 750 mm	–	13.25	261.69	269.44	865.40	100 m	**1396.53**
depth 900 mm	–	15.75	311.06	321.00	1033.61	100 m	**1665.67**
Land drainage; backfilling trench after laying pipes with gravel rejects or similar, blind filling with ash or sand; topping with 150 mm topsoil from dumps not exceeding 100 m; by hand							
Width 150 mm							
depth 450 mm	–	18.63	367.94	–	192.94	100 m	**560.88**
depth 600 mm	–	24.84	490.59	–	278.69	100 m	**769.28**
depth 750 mm	–	31.05	613.24	–	254.78	100 m	**868.02**
depth 900 mm	–	37.26	735.88	–	407.57	100 m	**1143.45**
Width 225 mm							
depth 450 mm	–	27.94	551.82	–	317.34	100 m	**869.16**
depth 600 mm	–	37.26	735.88	–	418.20	100 m	**1154.08**
depth 750 mm	–	46.57	919.76	–	519.37	100 m	**1439.13**
depth 900 mm	–	55.89	1103.83	–	620.23	100 m	**1724.06**
Width 375 mm							
depth 450 mm	–	46.57	919.76	–	528.69	100 m	**1448.45**
depth 600 mm	–	61.10	1206.72	–	696.90	100 m	**1903.62**
depth 750 mm	–	77.63	1533.19	–	865.40	100 m	**2398.59**
depth 900 mm	–	93.15	1839.71	–	1033.61	100 m	**2873.32**

R13 LAND DRAINAGE

Item Excluding site overheads and profit	PC £	Labour hours	Labour £	Plant £	Material £	Unit	Total rate £
SPORTS AND AMENITY DRAINAGE							
Sports or amenity drainage;							
Agripower Ltd Ltd; excavating							
trenches (with minimum project size							
of 1000 m) by trenching machine;							
arisings to spoil heap on site							
maximum 100 m; backfill with shingle							
to within 125 mm from surface and							
with rootzone to ground level							
Width 145 mm; 80 mm perforated pipe							
depth 550 mm	–	–	–	–	–	m	7.94
Width 145 mm; 100 mm perforated pipe							
depth 650 mm	–	–	–	–	–	m	9.12
Width 145 mm; 150 mm perforated pipe							
depth 700 mm	–	–	–	–	–	100 m	11.75
CATCHWATER OR FRENCH DRAINS							
Catchwater or french drains; laying							
pipes to drain excavated separately;							
100 mm dia. non-coilable perforated							
plastic pipes; including straight							
jointing; pipes laid with perforations							
uppermost; lining trench; wrapping							
pipes with filter fabric; backfilling with							
shingle							
Width 300 mm							
depth 450 mm	8386.26	9.10	179.72	37.50	8831.28	100 m	9048.50
depth 600 mm	8390.38	9.84	194.34	58.58	8994.97	100 m	9247.89
depth 750 mm	8394.68	10.60	209.35	74.25	9158.84	100 m	9442.44
depth 900 mm	8398.89	11.34	223.97	89.51	9322.62	100 m	9636.10
depth 1000 mm	8401.69	11.84	233.84	99.83	9431.81	100 m	9765.48
depth 1200 mm	8407.31	12.84	253.59	120.45	9650.18	100 m	10024.22
Width 450 mm							
depth 450 mm	8392.57	10.44	206.19	108.08	9076.95	100 m	9391.22
depth 600 mm	8398.89	11.34	223.97	89.51	9322.62	100 m	9636.10
depth 750 mm	8405.20	12.46	246.09	112.61	9568.29	100 m	9926.99
depth 900 mm	8411.52	13.60	268.60	136.13	9813.96	100 m	10218.69
depth 1000 mm	8415.72	14.34	283.21	151.39	9977.74	100 m	10412.34
depth 1200 mm	8424.14	15.84	312.84	182.32	10305.30	100 m	10800.46
depth 1500 mm	8436.77	18.10	357.48	228.94	10796.63	100 m	11383.05
depth 2000 mm	8457.82	21.84	431.34	306.07	11615.53	100 m	12352.94
Width 600 mm							
depth 450 mm	8398.89	11.24	221.99	89.51	9322.62	100 m	9634.12
depth 600 mm	8407.31	12.84	253.59	120.45	9650.18	100 m	10024.22
depth 750 mm	8415.72	14.34	283.21	151.39	9977.74	100 m	10412.34
depth 900 mm	8424.14	15.84	312.84	182.32	10305.30	100 m	10800.46
depth 1000 mm	8421.36	16.84	332.59	202.95	10515.28	100 m	11050.82
depth 1200 mm	8440.98	18.84	372.09	244.20	10960.41	100 m	11576.70
depth 1500 mm	8457.82	21.84	431.34	576.08	11615.53	100 m	12622.95
depth 2000 mm	8485.88	19.84	391.84	553.58	12707.39	100 m	13652.81

R13 LAND DRAINAGE

Item Excluding site overheads and profit	PC £	Labour hours	Labour £	Plant £	Material £	Unit	Total rate £
Width 900 mm							
depth 450 mm	8411.52	13.60	268.60	136.13	9813.96	100 m	**10218.69**
depth 600 mm	8424.14	7.92	156.42	182.32	10305.30	100 m	**10644.04**
depth 750 mm	8436.77	9.05	178.74	228.94	10796.63	100 m	**11204.31**
depth 900 mm	8449.40	10.17	200.86	275.14	11287.97	100 m	**11763.97**
depth 1000 mm	8457.82	10.92	215.67	306.07	11615.53	100 m	**12137.27**
depth 1200 mm	8474.66	12.42	245.29	367.95	12270.65	100 m	**12883.89**
depth 1500 mm	8499.91	14.67	289.73	460.76	13253.32	100 m	**14003.81**
depth 2000 mm	8542.01	18.92	373.67	705.45	14891.12	100 m	**15970.24**
depth 2500 mm	8584.10	22.17	437.86	860.14	16528.91	100 m	**17826.91**
depth 3000 mm	8626.19	25.92	511.92	924.83	18166.71	100 m	**19603.46**
Catchwater or french drains; laying pipes to drains excavated separately; 160 mm dia. non-coilable perforated plastic pipes; including straight jointing; pipes laid with perforations uppermost; lining trench; wrapping pipes with filter fabric; backfilling with shingle							
Width 600 mm							
depth 450 mm	16024.95	11.20	221.20	86.63	16919.61	100 m	**17227.44**
depth 600 mm	16033.37	12.70	250.82	117.56	17247.17	100 m	**17615.55**
depth 750 mm	16041.79	14.20	280.45	148.50	17574.73	100 m	**18003.68**
depth 900 mm	16050.21	15.70	310.07	179.44	17902.28	100 m	**18391.79**
depth 1000 mm	16055.82	16.70	329.82	200.06	18120.66	100 m	**18650.54**
depth 1200 mm	16067.05	18.70	369.32	241.43	18557.40	100 m	**19168.15**
depth 1500 mm	16083.88	21.70	428.57	303.19	19212.52	100 m	**19944.28**
depth 2000 mm	16111.95	26.70	527.33	406.31	20304.38	100 m	**21238.02**
depth 2500 mm	16140.01	31.70	626.08	509.44	21396.24	100 m	**22531.76**
depth 3000 mm	16168.07	36.70	724.83	612.56	22488.11	100 m	**23825.50**
Width 900 mm							
depth 450 mm	8411.52	13.60	268.60	136.13	9813.96	100 m	**10218.69**
depth 600 mm	8424.14	15.84	312.84	182.32	10305.30	100 m	**10800.46**
depth 750 mm	8436.77	18.10	357.48	228.94	10796.63	100 m	**11383.05**
depth 900 mm	8449.40	20.34	401.71	275.14	11287.97	100 m	**11964.82**
depth 1000 mm	8457.82	21.84	431.34	306.07	11615.53	100 m	**12352.94**
depth 1200 mm	8474.66	24.84	490.59	367.95	12270.65	100 m	**13129.19**
depth 1500 mm	8499.91	29.34	579.47	460.76	13253.32	100 m	**14293.55**
depth 2000 mm	8542.01	36.84	727.59	615.45	14891.12	100 m	**16234.16**
depth 2500 mm	8584.10	44.34	875.72	770.14	16528.91	100 m	**18174.77**
depth 3000 mm	8626.19	51.84	1023.84	924.83	18166.71	100 m	**20115.38**

R13 LAND DRAINAGE

Item Excluding site overheads and profit	PC £	Labour hours	Labour £	Plant £	Material £	Unit	Total rate £
CATCHWATER OR FRENCH DRAINS – cont							
Catchwater or french drains; Exxon Chemical Geopolymers Ltd; Filtram filter drain; in trenches (trenches not included); comprising filter fabric, liquid conducting core and 110 mm uPVC slitpipes; all in accordance with manufacturer's instructions; backfilling with shingle							
Width 600 mm							
depth 1000 mm	8429.76	16.84	332.59	202.95	10523.67	100 m	**11059.21**
depth 1200 mm	8440.98	18.84	372.09	244.20	10960.41	100 m	**11576.70**
depth 1500 mm	8457.82	21.84	431.34	576.08	11615.53	100 m	**12622.95**
depth 2000 mm	8485.88	19.84	391.84	553.58	12707.39	100 m	**13652.81**
Width 900 mm							
depth 450 mm	8411.52	13.60	268.60	136.13	9813.96	100 m	**10218.69**
depth 600 mm	8424.14	15.84	312.84	182.32	10305.30	100 m	**10800.46**
depth 750 mm	8436.77	18.10	357.48	228.94	10796.63	100 m	**11383.05**
depth 900 mm	8449.40	20.34	401.71	275.14	11287.97	100 m	**11964.82**
depth 1000 mm	8457.82	21.84	431.34	306.07	11615.53	100 m	**12352.94**
depth 1200 mm	8474.66	24.84	490.59	367.95	12270.65	100 m	**13129.19**
depth 1500 mm	8499.91	29.34	579.47	460.76	13253.32	100 m	**14293.55**
depth 2000 mm	8542.01	37.84	747.34	705.45	14891.12	100 m	**16343.91**
depth 2500 mm	8584.10	44.34	875.72	860.14	16528.91	100 m	**18264.77**
depth 3000 mm	8626.19	51.84	1023.84	1014.83	18166.71	100 m	**20205.38**
DITCHING							
Ditching; clear silt and bottom ditch not exceeding 1.50 m deep; strim back vegetation; disposing to spoil heaps; by machine							
Up to 1.50 m wide at top	–	6.00	118.50	27.00	–	100 m	**145.50**
1.50–2.50 m wide at top	–	8.00	158.00	36.00	–	100 m	**194.00**
2.50–4.00 m wide at top	–	9.00	177.75	40.50	–	100 m	**218.25**
Ditching; clear only vegetation from ditch not exceeding 1.50 m deep; disposing to spoil heaps; by strimmer							
Up to 1.50 m wide at top	–	5.00	98.75	7.01	–	100 m	**105.76**
1.50–2.50 m wide at top	–	6.00	118.50	12.61	–	100 m	**131.11**
2.50–4.00 m wide at top	–	7.00	138.25	19.62	–	100 m	**157.87**
Ditching; clear silt from ditch not exceeding 1.50 m deep; trimming back vegetation; disposing to spoil heaps; by hand							
Up to 1.50 m wide at top	–	15.00	296.25	7.01	–	100 m	**303.26**
1.50–2.50 m wide at top	–	27.00	533.25	12.61	–	100 m	**545.86**
2.50–4.00 m wide at top	–	42.00	829.50	19.62	–	100 m	**849.12**

R13 LAND DRAINAGE

Item Excluding site overheads and profit	PC £	Labour hours	Labour £	Plant £	Material £	Unit	Total rate £
Ditching; excavating and forming ditch and bank to given profile (normally 45°); in loam or sandy loam; by machine							
Width 300 mm							
depth 600 mm	–	3.70	73.08	33.30	–	100 m	106.38
depth 900 mm	–	5.20	102.70	46.80	–	100 m	149.50
depth 1200 mm	–	7.20	142.20	64.80	–	100 m	207.00
depth 1500 mm	–	9.40	185.65	74.03	–	100 m	259.68
Width 600 mm							
depth 600 mm	–	–	–	69.27	–	100 m	69.27
depth 900 mm	–	–	–	102.78	–	100 m	102.78
depth 1200 mm	–	–	–	139.65	–	100 m	139.65
depth 1500 mm	–	–	–	201.09	–	100 m	201.09
Width 900 mm							
depth 600 mm	–	–	–	393.75	–	100 m	393.75
depth 900 mm	–	–	–	693.75	–	100 m	693.75
depth 1200 mm	–	–	–	932.25	–	100 m	932.25
depth 1500 mm	–	–	–	1155.00	–	100 m	1155.00
Width 1200 mm							
depth 600 mm	–	–	–	618.75	–	100 m	618.75
depth 900 mm	–	–	–	918.75	–	100 m	918.75
depth 1200 mm	–	–	–	1216.88	–	100 m	1216.88
depth 1500 mm	–	–	–	1559.25	–	100 m	1559.25
Width 1500 mm							
depth 600 mm	–	–	–	763.13	–	100 m	763.13
depth 900 mm	–	–	–	1163.25	–	100 m	1163.25
depth 1200 mm	–	–	–	1546.88	–	100 m	1546.88
depth 1500 mm	–	–	–	1938.75	–	100 m	1938.75
Extra for ditching in clay	–	–	–	–	–	20%	–
Ditching; excavating and forming ditch and bank to given profile (normal 45°); in loam or sandy loam; by hand							
Width 300 mm							
depth 600 mm	–	36.00	711.00	–	–	100 m	711.00
depth 900 mm	–	42.00	829.50	–	–	100 m	829.50
depth 1200 mm	–	56.00	1106.00	–	–	100 m	1106.00
depth 1500 mm	–	70.00	1382.50	–	–	100 m	1382.50
Width 600 mm							
depth 600 mm	–	56.00	1106.00	–	–	100 m	1106.00
depth 900 mm	–	84.00	1659.00	–	–	100 m	1659.00
depth 1200 mm	–	112.00	2212.00	–	–	100 m	2212.00
depth 1500 mm	–	140.00	2765.00	–	–	100 m	2765.00
Width 900 mm							
depth 600 mm	–	84.00	1659.00	–	–	100 m	1659.00
depth 900 mm	–	126.00	2488.50	–	–	100 m	2488.50
depth 1200 mm	–	168.00	3318.00	–	–	100 m	3318.00
depth 1500 mm	–	210.00	4147.50	–	–	100 m	4147.50

R13 LAND DRAINAGE

Item Excluding site overheads and profit	PC £	Labour hours	Labour £	Plant £	Material £	Unit	Total rate £
DITCHING – cont							
Ditching – cont							
Width 1200 mm							
depth 600 mm	–	112.00	2212.00	–	–	100 m	**2212.00**
depth 900 mm	–	168.00	3318.00	–	–	100 m	**3318.00**
depth 1200 mm	–	224.00	4424.00	–	–	100 m	**4424.00**
depth 1500 mm	–	280.00	5530.00	–	–	100 m	**5530.00**
Extra for ditching in clay	–	–	–	–	–	50%	–
Extra for ditching in compacted soil	–	–	–	–	–	90%	–
Piped ditching							
Jointed concrete pipes; FP McCann Ltd; including bedding, haunching and topping with 150 mm concrete; 11.50 N/ mm²–40 mm aggregate; to existing ditch							
300 mm dia.	17.10	0.67	13.17	8.25	34.74	m	**56.16**
450 mm dia.	25.41	0.67	13.17	8.25	51.88	m	**73.30**
600 mm dia.	40.98	0.67	13.17	8.25	78.56	m	**99.98**
900 mm dia.	108.17	1.00	19.75	12.38	129.32	m	**161.45**
Jointed concrete pipes; FP McCann Ltd; including bedding, haunching and topping with 150 mm concrete; 11.50 N/ mm²–40 mm aggregate; to existing ditch							
1200 mm dia.	185.91	1.00	19.75	12.38	213.59	m	**245.72**
extra over jointed concrete pipes for bends to 45°	170.98	0.67	13.17	10.31	179.42	nr	**202.90**
extra over jointed concrete pipes for single junctions 300 mm dia.	116.83	0.67	13.17	10.31	125.27	nr	**148.75**
extra over jointed concrete pipes for single junctions 450 mm dia.	173.19	0.67	13.17	9.38	181.63	nr	**204.18**
extra over jointed concrete pipes for single junctions 600 mm dia.	280.55	0.67	13.17	9.38	288.99	nr	**311.54**
extra over jointed concrete pipes for single junctions 900 mm dia.	474.87	0.67	13.17	9.38	483.31	nr	**505.86**
extra over jointed concrete pipes for single junctions 1200 mm dia.	725.37	0.67	13.17	9.38	733.81	nr	**756.36**
Outfalls							
Reinforced concrete outfalls to water course; flank walls; for 150 mm drain outlets; overall dimensions							
900 × 1050 × 900 mm high	–	–	–	–	–	m	**487.60**
GRC outfall headwalls	–	–	–	–	–	nr	**105.00**
JKH Unit; standard small	–	–	–	–	–	nr	**85.00**

S PIPED WATER SUPPLY SYSTEMS

Item Excluding site overheads and profit	PC £	Labour hours	Labour £	Plant £	Material £	Unit	Total rate £
BOREHOLES							
Borehole drilling; Agripower Ltd							
Drilling operations to average 100 m							
depth; lining with 150 mm dia. sieve							
drilling to 100 m	–	–	–	–	–	nr	12000.00
rate per m over 100 m	–	–	–	–	–	m	120.00
pump, rising main, cable and kiosk	–	–	–	–	–	nr	5000.00
COLD WATER							
Blue MDPE polythene pipes; type 50;							
for cold water services; with							
compression fittings; bedding on							
100 mm DOT type 1 granular fill							
material							
Pipes							
20 mm dia.	0.47	0.08	1.58	–	4.02	m	5.60
25 mm dia.	1.08	0.08	1.58	–	4.64	m	6.22
32 mm dia.	1.32	0.08	1.58	–	4.87	m	6.45
50 mm dia.	2.17	0.10	1.98	–	5.73	m	7.71
60 mm dia.	3.48	0.10	1.98	–	7.04	m	9.02
Hose union bib taps; including fixing to							
wall; making good surfaces							
15 mm	13.50	0.75	14.81	–	13.90	nr	28.71
22 mm	18.08	0.75	14.81	–	18.68	nr	33.49
Stopcocks; including fixing to wall;							
making good surfaces							
15 mm	7.90	0.75	14.81	–	8.30	nr	23.11
22 mm	11.85	0.75	14.81	–	12.25	nr	27.06
Standpipes; to existing 25 mm water							
mains							
1.00 m high	–	–	–	–	–	nr	225.00
Hose junction bib taps; to standpipes							
19 mm	–	–	–	–	–	nr	80.00
RAINWATER HARVESTING							

Rainwater harvesting tanks
Notes: Rainwater harvesting tanks must be installed on granular or concrete bases. If installed below ground the tanks may require construction of drained underground chambers to support them. Please see the appropriate sections in this book for excavation, disposal, bases, retaining walls and drainage.

S PIPED WATER SUPPLY SYSTEMS

Item Excluding site overheads and profit	PC £	Labour hours	Labour £	Plant £	Material £	Unit	Total rate £
RAINWATER HARVESTING – cont							
Excavation for underground tanks;							
excavation inclusive of earthwork							
retention for self supporting tanks							
7 tonne tracked excavator (bucket							
volume 0.28 m^3)							
maximum depth not exceeding 1.00 m	–	0.06	1.23	4.10	–	m^3	**5.33**
maximum depth not exceeding 2.00 m	–	0.07	1.41	4.68	–	m^3	**6.09**
maximum depth not exceeding 3.00 m	–	0.09	1.80	5.96	–	m^3	**7.76**
Disposal							
Excavated material; off site; to tip;							
mechanically loaded (JCB)							
inert	–	–	–	–	–	m^3	**18.14**
Type 1 granular fill base; PC £17.95/							
tonne (£39.49/m^3 compacted)							
By machine							
100 mm thick	3.95	0.03	0.55	0.41	3.95	m^2	**4.91**
150 mm thick	5.92	0.03	0.49	0.62	5.92	m^2	**7.03**
Backfilling to surround of rainwater							
tank; carefully compacting as work							
proceeds							
Arising from the excavations							
average thickness exceeding 0.25 m;							
depositing in layers 150 mm							
maximum thickness	–	0.03	0.66	5.25	–	m^3	**5.91**
Rainwater harvesting tanks;							
Combined Harvesters Ltd; self							
supporting underground or above							
ground tanks							
Columbus rainwater tank with max							
1.00 m cover in pedestrian areas;							
complete with tank dome and pedestrian							
lid, submersible automatic pump system,							
supra filtration system, 25 m black and							
green rainwater pipe, rainwater labelling							
kit, 125 mm fine filter (excavation,							
backfilling base, trenching and inlet							
pipework all not included)							
3700 litre; 2.44 × 1.65 × 1.95	1374.00	3.00	118.50	–	1514.00	nr	**1632.50**
4500 litre; 2.44 × 1.84 × 1.84	1508.00	3.50	138.25	–	1648.00	nr	**1786.25**
6500 litre; 2.68 × 2.02 × 2.29	1768.00	4.00	158.00	–	1908.00	nr	**2066.00**
9000 litre; 2.44 × 1.84 × 1.88	2494.00	4.50	177.75	–	2634.00	nr	**2811.75**
13000 litre – 2 × 6500 l tanks;							
2.68 × 4.20 × 2.29	3014.00	7.00	276.50	–	3154.00	nr	**3430.50**

S PIPED WATER SUPPLY SYSTEMS

Item Excluding site overheads and profit	PC £	Labour hours	Labour £	Plant £	Material £	Unit	Total rate £
Cristall rainwater tank; lighter weight tank; complete with tank dome and pedestrian lid, submersible automatic pump system, supra filtration system, 25 m black and green rainwater pipe, rainwater labelling kit, 125 mm fine filter (excavation, backfilling base, trenching and inlet pipework all not included)							
1650 litre; 2.10 × 1.05 × 1.22	1069.00	5.00	197.50	–	1209.00	nr	1406.50
2650 litre; 2.10 × 1.30 × 1.50	1182.00	6.00	237.00	–	1322.00	nr	1559.00
Hercules rainwater tank; above or below ground (pedestrian areas only) installation; extendable to multiple tanks							
1600 litre; 1.35 m dia. × 1.60 m high	1383.00	4.50	177.75	–	1523.00	nr	1700.75
Lilo low profile tanks for reduced excavation; complete with rainwater filters and pressure pump							
1500 litre; 2.10 × 1.25 × 1.02	1124.00	–	–	–	1124.00	nr	1124.00
3000 litre; 2.46 × 2.1 × 1.05 m	1529.00	6.00	237.00	–	1669.00	nr	1906.00
5000 litre; 2.89 × 2.30 × 1.26	1906.00	6.50	256.75	–	2046.00	nr	2302.75
10000 litre 2.89 × 4,60 × 1.26 m	3289.00	7.00	276.50	–	3429.00	nr	3705.50
Combined Harvesters Ltd; accessories for rainwater harvesting tanks							
Downpipe filter for connection to roof downpipes.							
Quattro; for roof areas up to 50 m²	7.88	0.25	4.94	–	7.88	nr	12.82
Regendieb; for roof areas up to 80 m²	27.84	0.25	4.94	–	27.84	nr	32.78
Regendieb Deluxe; for roof areas up to 100 m²	31.84	–	–	–	31.84	nr	31.84
Pumps; inclusive of electrical installation and connections							
Raincatcher direct pump package	318.90	0.25	4.94	–	458.90	nr	463.84
Garden Comfort pump package	292.90	0.75	14.81	–	292.90	nr	307.71
Filters for rainwater tanks; complete with overflow siphons and rodent guard	153.16	–	–	–	153.16	nr	153.16
Extra over to cover rainwater tanks in geofabric prior to backfilling; 10 m² of terram per tank; average cost	–	0.25	9.88	–	4.07	nr	13.95

S PIPED WATER SUPPLY SYSTEMS

Item Excluding site overheads and profit	PC £	Labour hours	Labour £	Plant £	Material £	Unit	Total rate £
IRRIGATION							
Irrigation Infrastructure and pipework; main or ring main supply							
Excavate and lay mains supply pipe to supply irrigated area							
PE 80 20 mm	–	–	–	–	–	m	4.87
PE 80 25 mm	–	–	–	–	–	m	5.06
Pro-Flow PEHD 63 mm PE	–	–	–	–	–	m	8.68
Pro-Flow PEHD 50 mm PE	–	–	–	–	–	m	7.09
Pro-Flow PEHD 40 mm PE	–	–	–	–	–	m	6.11
Pro-Flow PEHD 32 mm PE	–	–	–	–	–	m	5.49
Irrigation systems; head control to sprinkler stations							
Valves installed at supply point on irrigation supply manifold; includes for pressure control and filtration							
2 valve unit	–	–	–	–	–	nr	132.03
4 valve unit	–	–	–	–	–	nr	206.11
6 valve unit	–	–	–	–	–	nr	291.49
12 valve unit	–	–	–	–	–	nr	558.92
Solenoid valve; 25 mm with chamber; extra over for each active station	–	–	–	–	–	nr	106.16
Cable to solenoid valves (alternative to valves on manifold as above)							
4 core	–	–	–	–	–	m	1.57
6 core	–	–	–	–	–	m	7.63
12 core	–	–	–	–	–	m	3.19
Irrigation; water supply and control							
Header tank and submersible pump and pressure stat							
1000 litre (500 gallon 25 mm/10 days to 1500 m²) 2 m³/hr pump	–	–	–	–	–	nr	489.67
4540 litre (1000 gallon 25 mm/10 days to 3500 m²)	–	–	–	–	–	nr	991.89
9080 litre (2000 gallon 25 mm/10 days to 7000 m²)	–	–	–	–	–	nr	1642.27
Electric multistation controllers 240 V							
WeatherMatic Smartline; 4 station; standard	–	–	–	–	–	nr	149.21
WeatherMatic Smartline; 6 station; standard	–	–	–	–	–	nr	167.62
WeatherMatic Smartline; 12 station; standard	–	–	–	–	–	nr	284.18
WeatherMatic Smartline; 12 station; radio controlled	–	–	–	–	–	nr	583.80
WeatherMatic Smartline; 12 station; with moisture sensor	–	–	–	–	–	nr	981.43
WeatherMatic Smartline; 12 station; radio controlled; with moisture sensor	–	–	–	–	–	nr	1178.61

S PIPED WATER SUPPLY SYSTEMS

Item Excluding site overheads and profit	PC £	Labour hours	Labour £	Plant £	Material £	Unit	Total rate £
Irrigation sprinklers; station consisting of multiple sprinklers; inclusive of all trenching, wiring and connections to ring main or main supply							
Sprayheads; WeatherMatic LX; including nozzle; 21 m² coverage							
100 mm (4')	–	–	–	–	–	nr	15.52
150 mm (6')	–	–	–	–	–	nr	21.92
300 mm (12')	–	–	–	–	–	nr	25.28
Matched precipitation sprinklers; WeatherMatic LX							
MP 1000; 16 m² coverage	–	–	–	–	–	nr	35.57
MP 2000; 30 m² coverage	–	–	–	–	–	nr	35.57
MP 3000; 81 m² coverage	–	–	–	–	–	nr	35.57
Gear drive sprinklers; WeatherMatic; placed to provide head to head (100%) overlap; average							
T3; 121 m² coverage	–	–	–	–	–	nr	46.72
CT70; 255 m² coverage	–	–	–	–	–	nr	82.36
CT70SS; 361 m² coverage	–	–	–	–	–	nr	91.46
Drip irrigation; driplines; inclusive of connections and draindown systems							
Metzerplas TechLand 30 cm drip emitter							
fixed on soil surface	–	–	–	–	–	m	1.07
sub-surface	–	–	–	–	–	m	1.57
Metzerplas TechLand 50 cm drip emitter							
fixed on soil surface	–	–	–	–	–	m	0.90
sub-surface	–	–	–	–	–	m	1.41
Commissioning and testing of irrigation system							
per station	–	–	–	–	–	nr	39.56
Annual maintenance costs of irrigation system							
Call out charge per visit	–	–	–	–	–	nr	329.59
Extra over per station	–	–	–	–	–	nr	13.19
Moisture Control moisture sensing control system							
Large garden comprising 7000 m² and 24 stations with four moisture sensors							
turf only	–	–	–	–	–	nr	13333.33
turf/shrub beds; 70/30	–	–	–	–	–	nr	7777.78
Medium garden comprising 3500 m² and 12 stations with two moisture sensors							
turf only	–	–	–	–	–	nr	8888.89
turf/shrub beds; 70/30	–	–	–	–	–	nr	11555.56

S PIPED WATER SUPPLY SYSTEMS

Item Excluding site overheads and profit	PC £	Labour hours	Labour £	Plant £	Material £	Unit	Total rate £
IRRIGATION – cont							
Moisture Control moisture sensing control system – cont							
Smaller gardens garden comprising 1000 m² and 6 stations with one moisture sensor							
turf only	–	–	–	–	–	nr	**5444.44**
turf/shrub beds; 70/30	–	–	–	–	–	nr	**6000.01**
turf/shrub beds; 50/50	–	–	–	–	–	nr	**6444.44**
Leaky Pipe Systems Ltd; Leaky Pipe; moisture leaking pipe irrigation system							
Main supply pipe inclusive of machine excavation; exclusive of connectors							
20 mm LDPE polytubing	1.12	0.05	0.99	0.61	1.12	m	**2.72**
16 mm LDPE polytubing	0.66	0.05	0.99	0.61	0.66	m	**2.26**
Water filters and cartridges							
No 10; 20 mm	–	–	–	–	48.00	nr	**48.00**
Big Blue and RR30 cartridge; 25 mm	–	–	–	–	139.35	nr	**139.35**
Water filters and pressure regulator sets; complete assemblies							
No 10; flow rate 3.1–82 litres per minute	–	–	–	–	59.00	nr	**59.00**
Leaky pipe hose; placed 150 mm sub surface for turf irrigation; distance between laterals 350 mm; excavation and backfilling priced separately							
LP12L low leak	4.28	0.04	0.79	–	4.28	m²	**5.07**
LP12H high leak	3.71	0.04	0.79	–	3.71	m²	**4.50**
LP12UH ultra high leak	4.84	0.04	0.79	–	4.84	m²	**5.63**
Leaky pipe hose; laid to surface for landscape irrigation; distance between laterals 600 mm							
LP12L low leak	2.55	0.03	0.49	–	2.55	m²	**3.04**
LP12H high leak	2.21	0.03	0.49	–	2.21	m²	**2.70**
LP12UH ultra high leak	2.89	0.03	0.49	–	2.89	m²	**3.38**
Leaky pipe hose; laid to surface for landscape irrigation; distance between laterals 900 mm							
LP12L low leak	1.71	0.02	0.33	–	1.71	m²	**2.04**
LP12H high leak	1.48	0.02	0.33	–	1.48	m²	**1.81**
LP12UH ultra high leak	1.93	0.02	0.33	–	1.93	m²	**2.26**
Leaky pipe hose; laid to surface for tree irrigation							
laid around circumference of tree pit							
LP12L low leak	2.52	0.13	2.47	–	2.52	nr	**4.99**
LP12H high leak	2.19	0.13	2.47	–	2.19	nr	**4.66**
LP12UH ultra high leak	2.86	0.13	2.47	–	2.86	nr	**5.33**

S PIPED WATER SUPPLY SYSTEMS

Item Excluding site overheads and profit	PC £	Labour hours	Labour £	Plant £	Material £	Unit	Total rate £
Accessories							
automatic multi-station controller							
stations; inclusive of connections	375.70	2.00	39.50	–	375.70	nr	**415.20**
solenoid valves; inclusive of wiring							
and connections to a multi-station							
controller; nominal distance from							
controller 25 m	62.40	0.50	9.88	–	517.40	nr	**527.28**

Prices for Measured Works

S PIPED WATER SUPPLY SYSTEMS

Item Excluding site overheads and profit	PC £	Labour hours	Labour £	Plant £	Material £	Unit	Total rate £
LAKES AND PONDS							
Lakes and ponds – General Preamble: The pressure of water against a retaining wall or dam is considerable, and where water retaining structures form part of the design of water features, the landscape architect is advised to consult a civil engineer. Artificially contained areas of water in raised reservoirs over 25,000 m³ have to be registered with the local authority and their dams will have to be covered by a civil engineer's certificate of safety.							
Typical linings – General Preamble: In addition to the traditional methods of forming the linings of lakes and ponds in puddled clay or concrete, there are a number of lining materials available. They are mainly used for reservoirs but can also help to form comparatively economic water features especially in soil which is not naturally water retentive. Information on the construction of traditional clay puddle ponds can be obtained from the British Trust for Conservation Volunteers, 36 St. Mary's Street, Wallingford, Oxfordshire, OX10 0EU. Tel: (01491) 39766. The cost of puddled clay ponds depends on the availability of suitable clay, the type of hand or machine labour that can be used and the use to which the pond is to be put.							
Lake liners; Fairwater Water Garden Design Consultants Ltd; to evenly graded surface of excavations (excavating not included); all stones over 75 mm; removing debris; including all welding and jointing of liner sheets							
Geotextile underlay; inclusive of spot welding to prevent dragging							
to water features	–	–	–	–	–	m²	2.27
to lakes or large features	–	–	–	–	–	1000 m²	2265.50
Butyl rubber liners; Varnamo; inclusive of site vulcanising laid to geotextile above							
0.75 mm thick	–	–	–	–	–	m²	7.57
0.75 mm thick	–	–	–	–	–	1000 m²	5570.00
1.00 mm thick	–	–	–	–	–	m²	8.62
1.00 mm thick	–	–	–	–	–	1000 m²	8620.00

S PIPED WATER SUPPLY SYSTEMS

Item Excluding site overheads and profit	PC £	Labour hours	Labour £	Plant £	Material £	Unit	Total rate £
Lake liners; Landline Ltd; Landflex or Alkorplan geomembranes; to prepared surfaces (surfaces not included); all joints fully welded; installation by Landline employees Landflex HC polyethylene geomembranes							
0.50 mm thick	–	–	–	–	–	1000 m²	3410.00
0.75 mm thick	–	–	–	–	–	1000 m²	3940.00
1.00 mm thick	–	–	–	–	–	1000 m²	4460.00
1.50 mm thick	–	–	–	–	–	1000 m²	5250.00
2.00 mm thick	–	–	–	–	–	1000 m²	6040.00
Alkorplan PVC geomembranes							
0.80 mm thick	–	–	–	–	–	1000 m	6250.00
1.20 mm thick	–	–	–	–	–	1000 m	9080.00
Lake liners; Monarflex Low density polyethylene (LDPE) lake and reservoir lining system; welding on site by Monarflex technicians (surface preparation and backfilling not included)							
Blackline 500	–	–	–	–	–	100 m²	931.70
Blackline 750	–	–	–	–	–	100 m²	1023.80
Blackline 1000	–	–	–	–	–	100 m²	1264.95
Operations over surfaces of lake liners Dug ballast; evenly spread over excavation already brought to grade							
150 mm thick	533.12	2.00	39.50	22.34	533.12	100 m²	594.96
200 mm thick	710.82	3.00	59.25	33.52	710.82	100 m²	803.59
300 mm thick	1066.23	3.50	69.13	39.10	1066.23	100 m²	1174.46
Imported topsoil; evenly spread over excavation							
100 mm thick	336.00	1.50	29.63	16.76	336.00	100 m²	382.39
150 mm thick	504.00	2.00	39.50	22.34	504.00	100 m²	565.84
200 mm thick	672.00	3.00	59.25	33.52	672.00	100 m²	764.77
Blinding existing subsoil with 50 mm sand	187.11	1.00	19.75	22.34	187.11	100 m²	229.20
Topsoil from excavation; evenly spread over excavation							
100 mm thick	–	–	–	16.76	–	100 m²	16.76
200 mm thick	–	–	–	22.34	–	100 m²	22.34
300 mm thick	–	–	–	33.52	–	100 m²	33.52
Extra over for screening topsoil using a Powergrid screener; removing debris	–	–	–	4.98	0.91	m³	5.89

S PIPED WATER SUPPLY SYSTEMS

Item Excluding site overheads and profit	PC £	Labour hours	Labour £	Plant £	Material £	Unit	Total rate £
LAKES AND PONDS – cont							
Lake construction; Fairwater Water							
Garden Design Consultants Ltd; lake							
construction at ground level;							
excavation; forming of lake;							
commissioning; excavated material							
spread on site							
Lined with existing site clay; natural							
edging with vegetation meeting the water							
500 m²	–	–	–	–	–	nr	9261.00
1000 m²	–	–	–	–	–	nr	16096.00
1500 m²	–	–	–	–	–	nr	22050.00
2000 m²	–	–	–	–	–	nr	27562.00
Lined with 0.75 mm butyl on geotextile							
underlay; natural edging with vegetation							
meeting the water							
500 m²	–	–	–	–	–	nr	14883.00
1000 m²	–	–	–	–	–	nr	25798.00
1500 m²	–	–	–	–	–	nr	35280.00
2000 m²	–	–	–	–	–	nr	44100.00
Extra over to the above for hard block							
edging to secure and protect liner;							
measured at lake perimeter	–	–	–	–	–	m	46.31
Lined with imported puddling clay;							
natural edging with vegetation meeting							
the water							
500 m²	–	–	–	–	–	nr	24255.00
1000 m²	–	–	–	–	–	nr	41895.00
1500 m²	–	–	–	–	–	nr	57330.00
2000 m²	–	–	–	–	–	nr	71662.00

Ornamental pools – General
Preamble: Small pools may be lined with one of the materials mentioned under lakes and ponds, or may be in rendered brickwork, puddled clay or, for the smaller sizes, fibreglass. Most of these tend to be cheaper than waterproof concrete. Basic prices for various sizes of concrete pools are given in the Approximate Estimates section. Prices for excavation, grading, mass concrete, and precast concrete retaining walls are given in the relevant sections. The manufacturers should be consulted before specifying the type and thickness of pool liner, as this depends on the size, shape and proposed use of the pool. The manufacturer's recommendation on foundations and construction should be followed.

S PIPED WATER SUPPLY SYSTEMS

Item Excluding site overheads and profit	PC £	Labour hours	Labour £	Plant £	Material £	Unit	Total rate £
Ornamental pools							
Pool liners; to 50 mm sand blinding to excavation (excavating not included); all stones over 50 mm; removing debris from surfaces of excavation; including underlay, all welding and jointing of liner sheets							
black polythene; 1000 gauge	–	–	–	–	–	m²	5.88
blue polythene; 1000 gauge	–	–	–	–	–	m²	7.77
coloured PVC; 1500 gauge	–	–	–	–	–	m²	8.90
black PVC; 1500 gauge	–	–	–	–	–	m²	6.72
black butyl; 0.75 mm thick	–	–	–	–	–	m²	8.85
black butyl; 1.00 mm thick	–	–	–	–	–	m²	9.28
black butyl; 1.50 mm thick	–	–	–	–	–	m²	32.44
Fine gravel; 100 mm; evenly spread over area of pool; by hand	3.29	0.13	2.47	–	3.29	m²	5.76
Selected topsoil from excavation; 100 mm; evenly spread over area of pool; by hand	–	0.13	2.47	–	–	m²	2.47
Extra over selected topsoil for spreading imported topsoil over area of pool; by hand	33.60	–	–	–	33.60	m³	33.60
Pool surrounds and ornament; Haddonstone Ltd; Portland Bath or terracotta cast stone							
Pool surrounds; installed to pools or water feature construction priced separately; surrounds and copings to 112.5 mm internal brickwork							
C4HSKVP half small pool surround; internal dia. 1780 mm; kerb features continuous moulding enriched with ovolvo and palmette designs; inclusive of plinth and integral conch shell vases flanked by dolphins	1308.00	16.00	316.00	–	1321.28	nr	1637.28
C4SKVP small pool surround as above but with full circular construction; internal dia. 1780 mm	2624.00	48.00	948.00	–	2641.05	nr	3589.05
C4MKVP medium pool surround; internal dia. 2705 mm; inclusive of plinth and integral vases	3936.00	48.00	948.00	–	3971.84	nr	4919.84
C4XLKVP extra large pool surround; internal dia. 5450 mm	5248.00	140.00	2765.00	–	5291.05	nr	8056.05

S PIPED WATER SUPPLY SYSTEMS

Item Excluding site overheads and profit	PC £	Labour hours	Labour £	Plant £	Material £	Unit	Total rate £
LAKES AND PONDS – cont							
Pool surrounds and ornament – cont							
Pool centre pieces and fountains;							
inclusive of plumbing and pumps							
HC350 Lotus Bowl; 1830 mm wide							
with C1700 triple Dolphin Fountain;							
HD2900 Doric Pedestal	3843.00	8.00	158.00	–	3846.22	nr	**4004.22**
C251 Gothic Fountain and Gothic							
Upper Base A350; free-standing							
fountain	1079.00	4.00	79.00	–	1082.22	nr	**1161.22**
HC521 Romanesque Fountain; free-							
standing bowl with self-circulating							
fountain; filled with cobbles; 815 mm							
dia. × 348 mm high	475.00	2.00	39.50	–	560.82	nr	**600.32**
C300 Lion Fountain; 610 mm high on							
fountain base C305; 280 mm high	157.00	2.00	39.50	–	160.22	nr	**199.72**
Wall fountains, watertanks and							
fountains; inclusive of installation							
drainage, automatic top up, pump and							
balancing tank							
Capital Garden Products Ltd							
Lion Wall Fountain F010;							
970 × 940 × 520 mm	–	4.00	79.00	–	1710.00	nr	**1789.00**
Dolphin Wall Fountain F001;							
740 × 510 mm	–	4.00	79.00	–	1542.00	nr	**1621.00**
Dutch Master Wall Fountain F012;							
740 × 410 mm	–	4.00	79.00	–	1569.00	nr	**1648.00**
James II Watertank 2801;							
730 × 730 × 760 mm high; 405 litres	–	4.00	79.00	–	1722.00	nr	**1801.00**
James II Fountain 4901 bp;							
710 × 1300 × 1440 mm high; 655 litres	–	4.00	79.00	–	1565.00	nr	**1644.00**
Fountain surrounds; Architectural							
Heritage Ltd; to pools constructe							
dseparately inclusive of installation,							
drainage, automatic top up, pump and							
balancing tank;							
Parterre Pool Surround; age patinated							
artificial stone; reproduction; Overall dia.							
3.65 m, overall height 280 mm	3800.00	4.00	79.00	–	5250.00	nr	**5329.00**
Great Westwood pool surround;							
age-patinated artificial stone pool;							
reproduction; Overall dia. 6 m; height							
350 mm	–	4.00	79.00	–	7450.00	nr	**7529.00**

S PIPED WATER SUPPLY SYSTEMS

Item Excluding site overheads and profit	PC £	Labour hours	Labour £	Plant £	Material £	Unit	Total rate £
Fountain kits; typical prices of submersible units comprising fountain pumps, fountain nozzles, underwater spotlights, nozzle extension armatures, underwater terminal boxes and electrical control panels							
Single aerated white foamy water columns; ascending jet 70 mm dia.; descending water up to four times larger; jet height adjustable between 1.00–1.70 m	4095.00	–	–	–	4694.00	nr	4694.00
Single aerated white foamy water columns; ascending jet 110 mm dia.; descending water up to four times larger; jet height adjustable between 1.50–3.00 m	6825.00	–	–	–	7749.80	nr	7749.80
FOUNTAINS AND WATER FEATURES							
Waterfall construction; Fairwater Water Garden Design Consultants Ltd							
Stone placed on top of butyl liner; securing with concrete and dressing to form natural rock pools and edgings							
Portland stone; m³ rate	–	–	–	–	–	m³	697.31
Portland stone; tonne rate	–	–	–	–	–	tonne	386.12
Balancing tank; blockwork construction; inclusive of recirculation pump and pond level control; 110 mm balancing pipe to pond; waterproofed with preformed polypropylene membrane; pipework mains water top-up and overflow							
450 × 600 × 1000 mm	–	–	–	–	–	nr	1450.00
Extra for pump							
2000 gallons per hour; submersible	–	–	–	–	–	nr	215.00
Water features; Fairwater Water Garden Design Consultants Ltd							
Natural stream; butyl lined level changes of 1.00 m; water pumped from lower pond (not included) via balancing tank; level changes via Purbeck stone water falls							
20 m long stream	–	–	–	–	–	nr	20000.00
Bespoke free-standing water wall; steel or glass panel; self contained recirculation system and reservoir							
water wall; 2.00 m high × 850 mm wide	–	–	–	–	–	nr	9371.00

S PIPED WATER SUPPLY SYSTEMS

Item Excluding site overheads and profit	PC £	Labour hours	Labour £	Plant £	Material £	Unit	Total rate £
SWIMMING POOLS							
Natural swimming pools; Fairwater Water Garden Design Consultants Ltd Natural swimming pool of 50 m² area; shingle regeneration zone 50 m² planted with marginal and aquatic plants	–	–	–	–	–	nr	60637.00
Guncast Swimming Pools Ltd; excavation and blinding of swimming pools; works by machine inclusive of allowances for working space and wall thicknesses; excavation of all trenches for pool pipeworks and electrical services Regular shaped pools							
100 m² pool	–	–	–	–	–	nr	8750.00
72 m² pool	–	–	–	–	–	nr	6750.00
60 m² pool	–	–	–	–	–	nr	6750.00
Irregular shaped pools							
100 m² pool	–	–	–	–	–	nr	9187.50
72 m² pool	–	–	–	–	–	nr	7087.50
60 m² pool	–	–	–	–	–	nr	7087.50
Guncast Swimming Pools Ltd; construction of a rectangular swimming pool in situ concrete corner steps, graduated shell depths of 1.1–2.5 m, water depth of 1.0–2.4 m as standard skimmer pool, tiled corner steps; excludes pump house construction Inclusive of drainage layer, formwork reinforcement and all Gunite construction, rendered internally and bespoke mosaic surface finish; bullnosed indian sandstone copings							
pool size 17.0 × 6.0 m; 172,500 litre	–	–	–	–	–	nr	13000.00
pool size 12.0 × 6.0 m; 125,750 litre	–	–	–	–	–	nr	102000.00
pool size 12.0 × 5.0 m; 102,150 litre	–	–	–	–	–	nr	76000.00
Guncast Swimming Pools Ltd; options and accessories for swimming pools Underwater lighting							
8 nr underwater floodlights 300w 12V; stainless steel	–	–	–	–	–	nr	5240.00
2 nr underwater floodlights 300w 12 V; stainless steel	–	–	–	–	–	nr	1310.00
Automatic pool cleaner	–	–	–	–	–	nr	1475.00

S PIPED WATER SUPPLY SYSTEMS

Item Excluding site overheads and profit	PC £	Labour hours	Labour £	Plant £	Material £	Unit	Total rate £
Automatic pool covers							
insulated automatic slatted cover; 12 × 6 m inclusive of additional surface skimmer and balance pipe in pit; housed in decking (not included)	–	–	–	–	–	nr	19930.00
solar cover; 17 × 6 m Ocea; housed in wall cave pit; logic motor; concealed with stainless steel frame and removable pvc panels; tiled or rendered to match pool	–	–	–	–	–	nr	33850.00
Driglide pool cover; 17 × 6 m selected colours on trackway	–	–	–	–	–	nr	35100.00
Ozone disinfection system	–	–	–	–	–	nr	5112.00
Guncast Swimming Pools Ltd; refurbishment of an existing swimming pool; with Roman bay end steps; no leaks present in existing pool; existing pipework and existing filtration systems and pump							
Empty pool, retain existing blockwork structure; retile pool; replace copings as indian sandstone bullnose; prepare shell; fix mosaic; pressure test							
pool size 8.50 × 4.25 59,000 litre	–	–	–	–	–	nr	25000.00
Guncast Swimming Pools Ltd; options and accessories for swimming pools							
Pool heating options							
propane boiler 250,000 BTU	–	–	–	–	–	nr	3870.00
pool heating via propane boiler 400000 BTU	–	–	–	–	–	nr	4895.00
heat pump 100 × 1000 mm concrete base	–	–	–	–	–	nr	4500.00
pool heating via heat pump	–	–	–	–	–	nr	4500.00
Diving board	–	–	–	–	–	nr	2200.00
Pool maintenance kit							
telescopic pole, wall/floor brush; skimmer net; flexible vacuum hose; underwater vacuum cleaner; pool water test kit; thermometer; pool treatment chemicals	–	–	–	–	–	nr	875.00

S PIPED WATER SUPPLY SYSTEMS

Item Excluding site overheads and profit	PC £	Labour hours	Labour £	Plant £	Material £	Unit	Total rate £
INTERNAL WATER FEATURES							
Interanal Water features; Fairwater **Water Garden Design Consultants** **Ltd;** Stainless steel rill; powder coated black 3000 mm × 250 mm; 2 nr letterbox weirs to maintain water level in rill; WRAS compliant header tank; modified balancing tank with chemical dosing unit; circulation pump and pipework;							
3.00 m × 250 mm wide	–	–	–	–	–	nr	3842.00
Brimming bowl with catchment tray; **Fairwater Water Garden Design** **Consultants Ltd;** Marble brimming bowl. recessed Polypro catchment tray 1.50 × 1.50x 120 mm ; pipework; WRAS compliant header tank balancing tank with Chemical dosing unit; circulation pump							
1.00 m dia.	–	–	–	–	–	nr	3190.00
Design mockup and commissioning on the above	–	–	–	–	–	nr	2640.00
Maintenance 12 months; monthly	–	–	–	–	–	nr	1104.00

V ELECTRICAL INSTALLATION

Item Excluding site overheads and profit	PC £	Labour hours	Labour £	Plant £	Material £	Unit	Total rate £
Clarification notes on labour costs in this section							
General groundworks team							
Generally a three man team is used in this section; The column Labour hours reports team hours. The column Labour £ reports the total cost of the team for the unit of work shown							
3 man team	–	1.00	59.25	–	–	hr	**59.25**
CABLES AND SWITCHING							
Trenching for electrical services							
3 tonne excavator (bucket volume 0.13 m^3); arisings laid alongside							
600 mm deep	–	–	–	2.15	–	m	**2.15**
800 mm deep	–	–	–	2.51	–	m	**2.51**
1.00 m deep	–	–	–	3.01	–	m	**3.01**
Extra over any types of excavating irrespective of depth for breaking out existing materials; heavy duty 110 volt breaker tool							
hard rock	–	5.00	98.75	29.00	–	m^3	**127.75**
concrete	–	3.00	59.25	17.40	–	m^3	**76.65**
reinforced concrete	–	4.00	79.00	28.21	–	m^3	**107.21**
brickwork, blockwork or stonework	–	1.50	29.63	8.70	–	m^3	**38.33**
By hand							
600 mm deep	–	0.24	4.75	–	–	m	**4.75**
800 mm deep	–	0.43	8.44	–	–	m	**8.44**
1.00 m deep	–	0.67	13.17	–	–	m	**13.17**
Backfilling of trenches; including laying of electrical marker tape; compacting lightly as work proceeds							
To trenches containing armoured cable							
by machine	0.13	–	–	2.59	0.13	m	**2.72**
by hand	0.13	0.20	3.95	–	0.13	m	**4.08**
To trenches containing ducted cable; including 150 mm sharp sand over the duct							
by machine	2.17	–	–	3.24	2.17	m	**5.41**
by hand	2.17	0.25	4.94	–	2.17	m	**7.11**

V ELECTRICAL INSTALLATION

Item Excluding site overheads and profit	PC £	Labour hours	Labour £	Plant £	Material £	Unit	Total rate £
CABLES AND SWITCHING – cont							
Cable to trenches (trenching operations not included); laying only cable							
Twin core steel wire armoured cable; 50 m drums							
1.5 mm core	0.92	–	–	–	0.92	m	**0.92**
2.5 mm core	1.16	–	–	–	1.16	m	**1.16**
Twin core steel wire armoured cable; lengths less than 50 m runs							
1.5 mm core	0.99	–	–	–	0.99	m	**0.99**
2.5 mm core	1.24	–	–	–	1.24	m	**1.24**
Three core steel wire armoured cable; 50 m drums							
1.5 mm core	1.01	–	–	–	1.01	m	**1.01**
2.5 mm core	1.46	–	–	–	1.46	m	**1.46**
4.0 mm core	1.73	–	–	–	1.73	m	**1.73**
6.0 mm core	2.35	–	–	–	2.35	m	**2.35**
10.0 mm core	4.22	–	–	–	4.22	m	**4.22**
Three core steel wire armoured cable; lengths less than 50 m runs							
1.5 mm core	1.07	–	–	–	1.07	m	**1.07**
2.5 mm core	1.58	–	–	–	1.58	m	**1.58**
Ducts to trenches; twin wall flexible cable ducts with drawstrings; laid on 150 mm clean sharp sand							
Twin wall duct; laying to trenches							
63 mm × 50 m coils	1.00	0.02	0.40	–	1.00	m	**1.40**
110 mm × 50 m coils	1.50	0.02	0.40	–	1.50	m	**1.90**
Cable drawing through ducts							
straight runs up to 50 m lengths	–	1.00	19.75	–	–	nr	**19.75**
Terminations to armoured cables; cutting coiling and taping length of cable to receive connection to light fitting, transformer and the like; fixing to temporary post							
Armoured cable	–	0.33	6.58	–	–	m	**6.58**
Plain ducted cable	–	0.13	2.47	–	–	m	**2.47**
Junction boxes; fixing to cables to receive connections to light fittings or transformers; to IP68							
Underground jointing box							
2 way	32.80	0.33	6.58	–	32.80	m	**39.38**
3 way	30.00	0.57	11.29	–	30.00	m	**41.29**
Underground jointing box							
2 way	11.15	0.33	6.58	–	11.15	m	**17.73**
3 way	14.95	0.40	7.90	–	14.95	m	**22.85**

V ELECTRICAL INSTALLATION

Item Excluding site overheads and profit	PC £	Labour hours	Labour £	Plant £	Material £	Unit	Total rate £
Electrical connections							
Connections to existing distribution boards							
per switching circuit located within 1 m of the distribution board; chasing of cables to walls not included	–	–	–	–	–	nr	**46.68**
Connections to switches; external power cable circuits; connections to internal wall switches; drilling through walls and making good; chasing of cables to walls not included							
price for the first circuit	4.30	1.00	19.75	–	50.98	nr	**70.73**
additional circuits	4.30	–	–	–	15.50	nr	**15.50**
Low voltage							
Low voltage lights connecting to transformers (transformers shown separately); fixing as described							
Hunza Spike adjustable spotlight 63.5 mm dia. × 75 mm long; c/w 20/35/50 w lamp							
stainless steel	324.32	-	-	-	368.77	nr	**368.77**
copper	357.12	-	-	-	368.77	nr	**368.77**
black	239.39	-	-	-	251.04	nr	**251.04**
Hunza Spike spotlight; 63.5 mm dia. × 75 mm long; c/w 20/35/50 w lamp							
stainless steel	148.14	-	-	-	159.79	nr	**159.79**
copper	106.80	-	-	-	118.45	nr	**118.45**
black	76.30	-	-	-	87.95	nr	**87.95**
Hunza pole lights single; fixing to concrete base 150 × 150 × 150 mm							
stainless steel	439.32	0.75	14.81	-	451.34	nr	**466.15**
copper	439.32	0.75	14.81	-	451.33	nr	**466.14**
black	282.71	0.75	14.81	-	294.72	nr	**309.53**
Hunza pole lights double; fixing to concrete base 150 × 150 × 150 mm							
stainless steel	733.66	0.75	14.81	-	745.68	nr	**760.49**
copper	733.66	0.75	14.81	-	745.68	nr	**760.49**
black	733.66	0.75	14.81	-	745.68	nr	**760.49**
Recessed wall lights eyelid steplights; fixing to brick or stone walls; inclusive of core drilling							
stainless steel	32.80	0.75	14.81	-	192.84	nr	**207.65**
copper	32.80	0.75	14.81	-	187.81	nr	**202.62**
black	32.80	0.75	14.81	-	187.81	nr	**202.62**

Prices for Measured Works

V ELECTRICAL INSTALLATION

Item Excluding site overheads and profit	PC £	Labour hours	Labour £	Plant £	Material £	Unit	Total rate £
Low voltage – cont							
Low voltage lights – cont							
Hunza recessed deck path or lawn lights; inclusive of core drilling excavation and all making good							
stainless steel light installed to deck	229.17	0.50	9.88	-	240.82	nr	**250.70**
stainless steel light installed to path	229.17	0.75	14.81	-	240.82	nr	**255.63**
stainless steel light installed to lawn	229.17	0.40	7.90	-	240.82	nr	**248.72**
stainless steel driveway light	229.17	0.50	9.88	-	240.82	nr	**250.70**
Exterior Lighting Transformers							
Outdoor transformers for 12 volt exterior lighting connecting to junction box (not included); IP67							
Single light transformers							
100 × 68 × 72 mm; 50 vA							
fused	24.50	-	-	-	59.50	nr	**59.50**
unfused	22.82	-	-	-	57.82	nr	**57.82**
Two light transformer 130 × 85 × 85 mm; 100vA							
fused	38.90	-	-	-	73.90	nr	**73.90**
unfused	35.50	-	-	-	70.50	nr	**70.50**
Three light transformer 130 × 85 × 85 mm; 150 vA							
fused	40.38	-	-	-	75.38	nr	**75.38**
unfused	-	-	-	-	74.90	nr	**74.90**
Four light transformer 130 × 85 × 100 mm; 200 vA							
fused	51.13	-	-	-	86.13	nr	**86.13**
unfused	-	-	-	-	80.20	nr	**80.20**

V41 STREET/AREA FLOODLIGHTING

Item Excluding site overheads and profit	PC £	Labour hours	Labour £	Plant £	Material £	Unit	Total rate £
GENERAL							
Street area floodlighting – General Preamble: There are an enormous number of luminaires available which are designed for small scale urban and garden projects. The designs are continually changing and the landscape designer is advised to consult the manufacturer's latest catalogue. Most manufacturers supply light fittings suitable for column, bracket, bulkhead, wall or soffit mounting. Highway lamps and columns for trafficked roads are not included in this section as the design of highway lighting is a very specialized subject outside the scope of most landscape contracts. The IP reference number refers to the waterproof properties of the fitting; the higher the number the more waterproof the fitting. Most items can be fitted with time clocks or PIR controls.							
MARKET PRICES OF LAMPS							
Market prices of lamps Lamps							
70 w HQIT-S	–	–	–	–	7.95	nr	7.95
70 w HQIT	–	–	–	–	17.95	nr	17.95
70 w SON	–	–	–	–	9.18	nr	9.18
70 w SONT	–	–	–	–	4.95	nr	4.95
100 w SONT	–	–	–	–	11.46	nr	11.46
150 w SONT	–	–	–	–	12.80	nr	12.80
28 w 2D	–	–	–	–	5.39	nr	5.39
100 w GLS\E27	–	–	–	–	2.90	nr	2.90
Clarification notes on labour costs in this section							
General groundworks team Generally a two man team is used in this section; The column Labour hours reports team hours. The column Labour £ reports the total cost of the team for the unit of work shown							
2 man expert team	–	1.00	39.50	–	–	hr	39.50
craftsman	–	1.00	19.75	–	–	hr	19.75
subcontract electrician	–	–	–	–	–	hr	35.00

V41 STREET/AREA FLOODLIGHTING

Item Excluding site overheads and profit	PC £	Labour hours	Labour £	Plant £	Material £	Unit	Total rate £
BULKHEAD LIGHTING							
Bulkhead and canopy fittings; **including fixing to wall and light** **fitting (lamp, final painting, electric** **wiring, connections or related fixtures** **such as switch gear and time clock** **mechanisms not included unless** **otherwise indicated)**							
Bulkhead and canopy fittings; Targetti Poulsen							
Nyhavn Wall small domed top conical shade with rings; copper wall lantern; finished untreated copper to achieve verdigris finish; also available in white aluminium; 310 mm dia. shade; with wall mounting arm; to IP 44	627.00	0.25	9.88	–	627.60	nr	**637.48**
Bulkhead and canopy fittings; Sugg Lighting							
Princess IP54 backlamp; 375 × 229 mm; copper frame; stove painted in black; with chimney and lampholder	380.00	0.25	9.88	–	380.60	nr	**390.48**
Victoria IP54 backlamp; 502 × 323 mm; copper frame; polished copper finish; with chimney, door and lampholder	478.00	0.25	9.88	–	478.60	nr	**488.48**
Palace IP54 backlamp; 457 × 321 mm; copper frame; stove painted black finish; with chimney, door and lampholder	425.00	0.25	9.88	–	425.60	nr	**435.48**
Windsor IP54 backlamp; 650 × 306 mm; copper frame; polished and lacquered finish; with door and lampholder	468.00	0.25	9.88	–	468.60	nr	**478.48**
Windsor IP54 gas backlamp; 650 × 306 mm; copper frame; polished and lacquered finish; hinged door; double inverted cluster mantle with permanent pilot and mains solenoid	836.00	0.25	9.88	–	836.60	nr	**846.48**

V41 STREET/AREA FLOODLIGHTING

Item Excluding site overheads and profit	PC £	Labour hours	Labour £	Plant £	Material £	Unit	Total rate £
FLOODLIGHTS							
Floodlighting; ground, wall or pole							
mounted; including fixing (lamp, final							
painting, electric wiring, connections							
or related fixtures such as switch							
gear and time clock mechanisms not							
included unless otherwise indicated)							
Floodlighting							
SPR−12; multi purpose ground/spike							
mounted wide angle floodlight; 70 w							
HIT or SON; to IP65	319.50	0.50	19.75	–	322.82	nr	**342.57**
Floodlight accessories							
earth spike for SPR−12	35.50	–	–	–	35.50	nr	**35.50**
louvre for SPR−12	48.00	–	–	–	48.00	nr	**48.00**
cowl for SPR−12	48.00	–	–	–	48.00	nr	**48.00**
barn doors for SPR−12	112.00	–	–	–	112.00	nr	**112.00**
Large area/pitch floodlighting; CU							
Phosco							
FL444 1000 w SON-T; floodlights with							
lamp and loose gear; narrow							
asymmetric beam	488.50	–	–	–	523.50	nr	**523.50**
FL444 2.0 kw MBIOS; floodlight with							
lamp and loose gear; projector beam	795.50	–	–	–	830.50	nr	**830.50**
Large area floodlighting; CU Phosco							
FL345/G/250S; floodlight with lamp							
and integral gear	379.66	–	–	–	414.66	nr	**414.66**
FL345/G/400 MBI; floodlight with							
lamp and integral gear	399.04	–	–	–	434.04	nr	**434.04**
Small area floodlighting; Sugg Lighting;							
floodlight for feature lighting; clear or							
toughened glass							
Scenario lamp; 150 w HPS-T	698.75	–	–	–	733.75	nr	**733.75**
Scenario lamp; 150 w HQI-T	698.75	–	–	–	733.75	nr	**733.75**

V41 STREET/AREA FLOODLIGHTING

Item Excluding site overheads and profit	PC £	Labour hours	Labour £	Plant £	Material £	Unit	Total rate £
SPOTLIGHTS							
Spotlights for uplighting and for illuminating signs and notice boards, statuary and other features) ground, wall or pole mounted; including fixing, light fitting and priming (lamp, final painting, electric wiring, connections or related fixtures such as switch gear and time clock mechanisms not included); all mains voltage (240 v) unless otherwise stated							
Spotlights							
WeeBee Spot SP−05 for low voltage halogen reflector; 20/35/50 w; c/w integral transformer and wall mounting box; to IP65	160.00	0.50	9.88	–	198.32	nr	**208.20**
Spotlighters, uplighters and cowl lighting; Havells Sylvania; TECHNO − SHORT ARM; head adjustable 130°; rotation 350°; projection 215 mm on 210 mm base plate; 355 mm high; integral gear; PG16 cable gland; black or aluminium							
Ref S.3517.09/14; 70 w CDMT/HQIT	747.41	–	–	–	782.41	nr	**782.41**
Ref S.3518.09/14; 150 w CDMT/HQIT	785.65	–	–	–	820.65	nr	**820.65**
RECESSED LIGHTING							
Recessed uplighting; including walk/ drive over fully recessed uplighting; excavating, ground fixing, concreting in and making good surfaces (electric wiring, connections or related fixtures such as switch gear and time-clock mechanisms not included unless otherwise stated) (Note: transformers will power multiple lights dependent on the distance between the light units); all mains voltage (240 v) unless otherwise stated							
Recessed uplighting; 266 mm dia.; diecast aluminium with stainless steel top plate 10 mm toughened safety glass; 2000 kg drive over; integral control gear; to IP67							
IPR−14 HIT metal halide; white light; spot or flood or wall wash distribution	275.00	0.50	9.88	–	313.32	nr	**323.20**
IRR−14 HIT compact fluorescent; white light; low power consumption; flood distribution	275.00	0.50	9.88	–	313.32	nr	**323.20**

V41 STREET/AREA FLOODLIGHTING

Item Excluding site overheads and profit	PC £	Labour hours	Labour £	Plant £	Material £	Unit	Total rate £
Accessories for IPR−14 uplighters;							
rockguard	154.50	–	–	–	154.50	nr	**154.50**
stainless steel installation sleeve	49.50	–	–	–	49.50	nr	**49.50**
anti glare louvre (internal tilt)	98.50	–	–	–	98.50	nr	**98.50**
Accessories for Nimbus uplighters;							
IS installation sleeve	57.00	–	–	–	57.00	nr	**57.00**
Wall recessed; Havells Sylvania; EOS range; integral gear; asymmetric reflector; toughened reeded glass							
MINI EOS; 145 × 90 mm; black or aluminium; 20 W QT9 lamp	143.91	–	–	–	187.66	nr	**187.66**
RECTANGULAR EOS; 270 × 145 mm; black or aluminium; 20 W QT9 lamp	208.41	–	–	–	279.38	nr	**279.38**

LIGHTED BOLLARDS

Lighted bollards; including excavating, ground fixing, concreting in and making good surfaces (lamp, final painting, electric wiring, connections or related fixtures such as switch gear and time-clock mechanisms not included) (Note: all illuminated bollards must be earthed); heights given are from ground level to top of bollards

Item	PC £	Labour hours	Labour £	Plant £	Material £	Unit	Total rate £
Lighted bollards							
Orbiter; vandal-resistant; head of cast aluminium; domed top; anti-glare rings; pole extruded aluminium; diffuser clear UV stabilized polycarbonate; powder coated; 1040 mm high × 255 mm dia.; with root or base plate; IP44	654.00	2.50	98.75	–	686.38	nr	**785.13**
Waterfront; solidly proportioned; head of cast silumin; domed top; symmetrical distribution; pole extruded aluminium sandblasted or painted white; internal diffuser clear UV stabilized polycarbonate; 865 mm high × dia. 260 mm; IP55	736.00	2.50	98.75	–	768.38	nr	**867.13**
Bysted; concentric louvred bollard; head cast iron; post COR-TEN steel; externally untreated to provide natural aging effect of uniform oxidized red surface finish; internal painted white; lamp diffuser rings of clear polycarbonate; 1130 mm high × 280 mm dia.; to IP44	1133.00	2.50	98.75	–	1165.38	nr	**1264.13**

V41 STREET/AREA FLOODLIGHTING

Item Excluding site overheads and profit	PC £	Labour hours	Labour £	Plant £	Material £	Unit	Total rate £
LIGHTED BOLLARDS – cont							
Lighted bollards – cont							
Lighted bollards; Woodscape Ltd							
Illuminated bollard; in 'very durable							
hardwood'; integral die cast							
aluminium lighting unit;							
165 × 165 mm × 1.00 m high	271.58	2.50	98.75	–	309.81	nr	408.56
STREET AND PRECINCT LIGHTING							
Precinct lighting lanterns; ground,							
wall or pole mounted; including fixing							
and light fitting (lamps, poles,							
brackets, final painting, electric							
wiring, connections or related fixtures							
such as switch gear and time clock							
mechanisms not included)							
Precinct lighting lanterns; Targetti							
Poulsen							
Nyhavn Park side entry 90 degree							
curved arm mounted; steel rings over							
domed top; housing cast silumin							
sandblasted with integral gear;							
protected by UV stabilized clear							
polycarbonate diffuser; IP55	650.00	0.50	19.75	–	650.00	nr	669.75
Kipp; bottom entry pole-top; shot							
blasted or lacquered Hanover design							
award; hinged diffuser for simple							
maintenance; indirect lighting							
technique ensures low glare; IP55	638.00	0.50	19.75	–	638.00	nr	657.75
Precinct lighting lanterns; Sugg Lighting							
Juno Dome; IP66; molded GRP body;							
graphite finish	560.00	0.50	19.75	–	560.00	nr	579.75
Juno Cone; IP66; molded GRP body;							
graphite finish	560.00	0.50	19.75	–	560.00	nr	579.75
Sharkon; IP65; cast aluminium body;							
silver and blue finish	560.00	0.50	19.75	–	560.00	nr	579.75

V41 STREET/AREA FLOODLIGHTING

Item Excluding site overheads and profit	PC £	Labour hours	Labour £	Plant £	Material £	Unit	Total rate £
REPRODUCTION STREET LIGHTING							
Reproduction street lanterns and columns; including excavating, concreting in, backfilling and disposing of spoil, or fixing to ground or wall, making good surfaces, light fitting and priming (final painting, electric wiring, connections or related fixtures such as switch gear and time-clock mechanisms not included) (Note: lanterns up to 14 inches and columns up to 7 ft are suitable for residential lighting)							
Reproduction street lanterns; Sugg Lighting; reproduction lanterns handmade to original designs; all with ES lamp holder							
Westminster IP54 hexagonal hinged top lantern with door; two piece folded polycarbonate glazing; lamp 100 w HQI T; integral photo electric cell							
small; 1016 mm high × 356 mm wide	876.13	0.50	9.88	–	876.13	nr	**886.01**
large; 1124 mm high × 760 mm wide	1376.00	0.50	9.88	–	1376.00	nr	**1385.88**
Guildhall IP54; handcrafted copper frame; clear polycarbonate glazing circular tapered lantern with hemispherical top; integral photo electric cell; with door							
small; 711 mm high × 330 mm wide	749.28	0.50	9.88	–	749.88	nr	**759.76**
medium; 1150 mm high × 432 mm wide	790.13	0.50	9.88	–	790.73	nr	**800.61**
large; 1370 mm high × 550 mm wide	1096.50	0.50	9.88	–	1097.10	nr	**1106.98**
Grosvenor circular lantern with door; copper frame; polished copper finish; polycarbonate glazing							
small; 790 × 330 mm; IP54	763.25	0.50	9.88	–	763.85	nr	**773.73**
medium; 1080 × 435 mm; IP65	790.13	0.50	9.88	–	790.73	nr	**800.61**
Classic Globe IP54 lantern with hinged outer frame; cast aluminium frame; black polyester powder coating							
medium; 965 × 483 mm	698.75	0.50	9.88	–	699.35	nr	**709.23**
Windsor lantern; copper frame; polished copper finish							
small; 905 × 356 mm; IP54; with door	568.00	0.50	9.88	–	568.60	nr	**578.48**
small; 905 × 356 mm; IP65; without door	530.00	0.50	9.88	–	530.60	nr	**540.48**
medium; 1124 × 420 mm; IP54; with door	620.00	0.50	9.88	–	620.60	nr	**630.48**
large; 1124 × 470 mm; IP54; with door	720.00	0.50	9.88	–	720.60	nr	**730.48**
medium; gas lantern; 1124 × 420 mm; IP54; with door; integral solenoid and pilot	1101.88	0.50	9.88	–	1102.48	nr	**1112.36**

V41 STREET/AREA FLOODLIGHTING

Item Excluding site overheads and profit	PC £	Labour hours	Labour £	Plant £	Material £	Unit	Total rate £
REPRODUCTION STREET LIGHTING – cont							
Reproduction brackets and suspensions; Sugg Lighting							
Iron brackets							
Bow bracket; 6'0"	1071.57	2.00	39.50	–	1072.17	nr	**1111.67**
Ornate iron bracket; large	478.38	2.00	39.50	–	478.98	nr	**518.48**
Ornate iron bracket; medium	462.25	2.00	39.50	–	462.85	nr	**502.35**
Swan neck iron bracket; large	451.50	2.00	39.50	–	452.10	nr	**491.60**
Swan neck iron bracket; medium	255.85	2.00	39.50	–	256.45	nr	**295.95**
Cast brackets							
Universal cast bracket	341.85	2.00	39.50	–	342.45	nr	**381.95**
Abbey bracket	202.10	2.00	39.50	–	202.70	nr	**242.20**
Short Abbey bracket	199.95	2.00	39.50	–	200.55	nr	**240.05**
Plaza cast bracket	167.16	2.00	39.50	–	167.76	nr	**207.26**
Base mountings							
Universal pedestal	339.70	2.00	39.50	–	340.30	nr	**379.80**
Universal plinth	193.51	2.00	39.50	–	194.11	nr	**233.61**
Reproduction lighting columns							
Reproduction lighting columns; Sugg Lighting							
Harborne C11; fabricated iron/steel heavy duty post with integral cast root; 3–5 m	1483.51	8.00	158.00	–	1491.17	nr	**1649.17**
Aylesbury C12; base fabricated heavy gauge 89 mm aluminium post; 3–5 m	2115.61	8.00	158.00	–	2123.27	nr	**2281.27**
Cannonbury C13; fabricated heavy gauge 89 mm aluminium post; 3–5 m	1558.75	8.00	158.00	–	1566.41	nr	**1724.41**
standard column C14; rooted British Steel 168/89 mm embellished post; 5–8 m	702.10	8.00	158.00	–	709.76	nr	**867.76**
Large Constitution Hill C22X; cast aluminium, steel cored post with extended spigot; 4.3 m	5622.25	6.00	118.50	–	5629.91	nr	**5748.41**
Cardiff C29; cast aluminium post; 3.5 m	2759.54	6.00	118.50	–	2767.20	nr	**2885.70**
Seven Dials C36; cast aluminium rooted post; 3.8 m	2038.20	8.00	158.00	–	2045.86	nr	**2203.86**
Royal Exchange C42; traditional cast aluminium post; welded multi arm construction; 2.1 m	3113.20	6.00	118.50	–	3120.86	nr	**3239.36**

PART 5

Tables and Memoranda

This part of the book contains the following sections:

Environmental Noise Barriers

A Guide To Their Acoustic and Visual Design, Second Edition

Benz Kotzen & Colin English

Environmental Noise Barriers: A Guide to Their Acoustic and Visual Design, Second Edition is a unique one-stop reference for practitioners in transport acoustics, whether engineers, landscape architects, or manufacturers, and for highways departments in local and central authorities. This extensively revised second edition has been updated in line with UK and EU legislation and international provision of barriers. Written for an international audience by a landscape architect and acoustical engineer, this accessible reference includes some 200 full-color illustrations.

April 2009 246x189: 284pp
Hb: 978-0-415-43708-0: £121.00

To Order: Tel: +44 (0) 1235 400524 Fax: +44 (0) 1235 400525
or Post: Taylor and Francis Customer Services,
Bookpoint Ltd, Unit T1, 200 Milton Park, Abingdon, Oxon, OX14 4TA UK
Email: book.orders@tandf.co.uk

For a complete listing of all our titles visit:
www.tandf.co.uk

CONVERSION TABLES

CONVERSION TABLES

Length	Unit	Conversion factors			
Millimetre	mm	1 in	= 25.4 mm	1 mm	= 0.0394 in
Centimetre	cm	1 in	= 2.54 cm	1 cm	= 0.3937 in
Metre	m	1 ft	= 0.3048 m	1 m	= 3.2808 ft
		1 yd	= 0.9144 m		= 1.0936 yd
Kilometre	km	1 mile	= 1.6093 km	1 km	= 0.6214 mile

Note:
1 cm	= 10 mm	1 ft	= 12 in
1 m	= 1 000 mm	1 yd	= 3 ft
1 km	= 1 000 m	1 mile	= 1 760 yd

Area	Unit	Conversion factors			
Square Millimetre	mm^2	$1\ in^2$	$= 645.2\ mm^2$	$1\ mm^2$	$= 0.0016\ in^2$
Square Centimetre	cm^2	$1\ in^2$	$= 6.4516\ cm^2$	$1\ cm^2$	$= 1.1550\ in^2$
Square Metre	m^2	$1\ ft^2$	$= 0.0929\ m^2$	$1\ m^2$	$= 10.764\ ft^2$
		$1\ yd^2$	$= 0.8361\ m^2$	$1\ m^2$	$= 1.1960\ yd^2$
Square Kilometre	km^2	$1\ mile^2$	$= 2.590\ km^2$	$1\ km^2$	$= 0.3861\ mile^2$

Note:
$1\ cm^2$	$= 100\ mm^2$	$1\ ft^2$	$= 144\ in^2$
$1\ m^2$	$= 10\ 000\ cm^2$	$1\ yd^2$	$= 9\ ft^2$
$1\ km^2$	$= 100$ hectares	1 acre	$= 4\ 840\ yd^2$
		$1\ mile^2$	$= 640$ acres

Volume	Unit	Conversion factors			
Cubic Centimetre	cm^3	$1\ cm^3$	$= 0.0610\ in^3$	$1\ in^3$	$= 16.387\ cm^3$
Cubic Decimetre	dm^3	$1\ dm^3$	$= 0.0353\ ft^3$	$1\ ft^3$	$= 28.329\ dm^3$
Cubic Metre	m^3	$1\ m^3$	$= 35.3147\ ft^3$	$1\ ft^3$	$= 0.0283\ m^3$
		$1\ m^3$	$= 1.3080\ yd^3$	$1\ yd^3$	$= 0.7646\ m^3$
Litre	l	1 l	= 1.76 pint	1 pint	= 0.5683 l
			= 2.113 US pt		= 0.4733 US l

Note:
$1\ dm^3$	$= 1\ 000\ cm^3$	$1\ ft^3$	$= 1\ 728\ in^3$	1 pint	= 20 fl oz
$1\ m^3$	$= 1\ 000\ dm^3$	$1\ yd^3$	$= 27\ ft^3$	1 gal	= 8 pints
1 l	$= 1\ dm^3$				

Neither the Centimetre nor Decimetre are SI units, and as such their use, particularly that of the Decimetre, is not widespread outside educational circles.

Mass	Unit	Conversion factors			
Milligram	mg	1 mg	= 0.0154 grain	1 grain	= 64.935 mg
Gram	g	1 g	= 0.0353 oz	1 oz	= 28.35 g
Kilogram	kg	1 kg	= 2.2046 lb	1 lb	= 0.4536 kg
Tonne	t	1 t	= 0.9842 ton	1 ton	= 1.016 t

Note:
1 g	= 1000 mg	1 oz	= 437.5 grains	1 cwt	= 112 lb
1 kg	= 1000 g	1 lb	= 16 oz	1 ton	= 20 cwt
1 t	= 1000 kg	1 stone	= 14 lb		

Force	Unit	Conversion factors			
Newton	N	1 lbf	= 4.448 N	1 kgf	= 9.807 N
Kilonewton	kN	1 lbf	= 0.004448 kN	1 ton f	= 9.964 kN
Meganewton	MN	100 tonf	= 0.9964 MN		

CONVERSION TABLES

Pressure and stress	Unit	Conversion factors	
Kilonewton per square metre	kN/m^2	1 lbf/in^2	= 6.895 kN/m^2
		1 bar	= 100 kN/m^2
Meganewton per square metre	MN/m^2	1 tonf/ft^2	= 107.3 kN/m^2 = 0.1073 MN/m^2
		1 kgf/cm^2	= 98.07 kN/m^2
		1 lbf/ft^2	= 0.04788 kN/m^2

Coefficient of consolidation (Cv) or swelling	Unit	Conversion factors	
Square metre per year	m^2/year	1 cm^2/s	= 3 154 m^2/year
		1 ft^2/year	= 0.0929 m^2/year

Coefficient of permeability	Unit	Conversion factors	
Metre per second	m/s	1 cm/s	= 0.01 m/s
Metre per year	m/year	1 ft/year	= 0.3048 m/year
			= 0.9651 × (10)^8 m/s

Temperature	Unit	Conversion factors	
Degree Celsius	°C	°C = 5/9 × (°F − 32)	°F = (9 × °C)/5 + 32

CONVERSION TABLES

SPEED CONVERSION

km/h	m/min	mph	fpm
1	16.7	0.6	54.7
2	33.3	1.2	109.4
3	50.0	1.9	164.0
4	66.7	2.5	218.7
5	83.3	3.1	273.4
6	100.0	3.7	328.1
7	116.7	4.3	382.8
8	133.3	5.0	437.4
9	150.0	5.6	492.1
10	166.7	6.2	546.8
11	183.3	6.8	601.5
12	200.0	7.5	656.2
13	216.7	8.1	710.8
14	233.3	8.7	765.5
15	250.0	9.3	820.2
16	266.7	9.9	874.9
17	283.3	10.6	929.6
18	300.0	11.2	984.3
19	316.7	11.8	1038.9
20	333.3	12.4	1093.6
21	350.0	13.0	1148.3
22	366.7	13.7	1203.0
23	383.3	14.3	1257.7
24	400.0	14.9	1312.3
25	416.7	15.5	1367.0
26	433.3	16.2	1421.7
27	450.0	16.8	1476.4
28	466.7	17.4	1531.1
29	483.3	18.0	1585.7
30	500.0	18.6	1640.4
31	516.7	19.3	1695.1
32	533.3	19.9	1749.8
33	550.0	20.5	1804.5
34	566.7	21.1	1859.1
35	583.3	21.7	1913.8
36	600.0	22.4	1968.5
37	616.7	23.0	2023.2
38	633.3	23.6	2077.9
39	650.0	24.2	2132.5
40	666.7	24.9	2187.2

CONVERSION TABLES

km/h	m/min	mph	fpm
41	683.3	25.5	2241.9
42	700.0	26.1	2296.6
43	716.7	26.7	2351.3
44	733.3	27.3	2405.9
45	750.0	28.0	2460.6
46	766.7	28.6	2515.3
47	783.3	29.2	2570.0
48	800.0	29.8	2624.7
49	816.7	30.4	2679.4
50	833.3	31.1	2734.0

GEOMETRY

GEOMETRY

Two dimensional figures

Figure	Diagram of figure	Surface area	Perimeter
Square		a^2	$4a$
Rectangle		ab	$2(a+b)$
Triangle		$\frac{1}{2}ch$	$a+b+c$
Circle		πr^2 $\frac{1}{4}\pi d^2$ where $2r = d$	$2\pi r$ πd
Parallelogram		ah	$2(a+b)$
Trapezium		$\frac{1}{2}h(a+b)$	$a+b+c+d$
Ellipse		Approximately πab	$\pi(a+b)$
Hexagon		$2.6 \times a^2$	

GEOMETRY

Figure	Diagram of figure	Surface area	Perimeter
Octagon		$4.83 \times a^2$	$6a$
Sector of a circle		$\frac{1}{2}rb$ or $\frac{q}{360}\pi r^2$ note b = angle $\frac{q}{360} \times \pi 2r$	
Segment of a circle		$S - T$ where S = area of sector, T = area of triangle	
Bellmouth		$\frac{3}{14} \times r^2$	

GEOMETRY

Three dimensional figures

Figure	Diagram of figure	Surface area	Volume
Cube		$6a^2$	a^3
Cuboid/ rectangular block		$2(ab + ac + bc)$	abc
Prism/ triangular block		$bd + hc + dc + ad$	$\frac{1}{2}hcd$
Cylinder		$2\pi r^2 + 2\pi h$	$\pi r^2 h$ $\frac{1}{4}\pi d^2 h$
Sphere		$4\pi r^2$	$\frac{4}{3}\pi r^3$
Segment of sphere		$2\pi Rh$	$\frac{1}{6}\pi h(3r^2 + h^2)$ $\frac{1}{3}\pi h^2(3R - H)$
Pyramid		$(a + b)l + ab$	$\frac{1}{3}abh$

GEOMETRY

Figure	Diagram of figure	Surface area	Volume
Frustum of a pyramid		$l(a + b + c + d) + \sqrt{(ab + cd)}$ [rectangular figure only]	$\frac{h}{3}(ab + cd + \sqrt{abcd})$
Cone		πrl (excluding base) $\pi rl + \pi r^2$ (including base)	$\frac{1}{3}\pi r^2 h$ $\frac{1}{12}\pi d^2 h$
Frustum of a cone		$\pi r^2 + \pi R^2 + \pi l(R + r)$	$\frac{1}{3}\pi(R^2 + Rr + r^2)$

FORMULAE

Formulae

Formula	Description
Pythagoras' Theorem	$A^2 = B^2 + C^2$ where A is the hypotenuse of a right-angled triangle and B and C are the two adjacent sides
Simpson's Rule	The Area is divided into an even number of strips of equal width, and therefore has an odd number of ordinates at the division points $$\text{area} = \frac{S(A + 2B + 4C)}{3}$$ where S = common interval (strip width) A = sum of first and last ordinates B = sum of remaining odd ordinates C = sum of the even ordinates The Volume can be calculated by the same formula, but by substituting the area of each coordinate rather than its length
Trapezoidal Rule	A given trench is divided into two equal sections, giving three ordinates, the first, the middle and the last $$\text{volume} = \frac{S \times (A + B + 2C)}{2}$$ where S = width of the strips A = area of the first section B = area of the last section C = area of the rest of the sections
Prismoidal Rule	A given trench is divided into two equal sections, giving three ordinates, the first, the middle and the last $$\text{volume} = \frac{L \times (A + 4B + C)}{6}$$ where L = total length of trench A = area of the first section B = area of the middle section C = area of the last section

TYPICAL THERMAL CONDUCTIVITY OF BUILDING MATERIALS

TYPICAL THERMAL CONDUCTIVITY OF BUILDING MATERIALS

(Always check manufacturer's details – variation will occur depending on product and nature of materials)

	Thermal conductivity (W/mK)		Thermal conductivity (W/mK)
Acoustic plasterboard	0.25	Oriented strand board	0.13
Aerated concrete slab (500 kg/m³)	0.16	Outer leaf brick	0.77
Aluminium	237	Plasterboard	0.22
Asphalt (1700 kg/m³)	0.5	Plaster dense (1300 kg/m³)	0.5
Bitumen-impregnated fibreboard	0.05	Plaster lightweight (600 kg/m³)	0.16
Blocks (standard grade 600 kg/m³)	0.15	Plywood (950 kg/m³)	0.16
Blocks (solar grade 460 kg/m³)	0.11	Prefabricated timber wall panels (check manufacturer)	0.12
Brickwork (outer leaf 1700 kg/m³)	0.84	Screed (1200 kg/m³)	0.41
Brickwork (inner leaf 1700 kg/m³)	0.62	Stone chippings (1800 kg/m³)	0.96
Dense aggregate concrete block 1800 kg/m³ (exposed)	1.21	Tile hanging (1900 kg/m³)	0.84
Dense aggregate concrete block 1800 kg/m³ (protected)	1.13	Timber (650 kg/m³)	0.14
Calcium silicate board (600 kg/m³)	0.17	Timber flooring (650 kg/m³)	0.14
Concrete general	1.28	Timber rafters	0.13
Concrete (heavyweight 2300 kg/m³)	1.63	Timber roof or floor joists	0.13
Concrete (dense 2100 kg/m³ typical floor)	1.4	Roof tile (1900 kg/m³)	0.84
Concrete (dense 2000 kg/m³ typical floor)	1.13	Timber blocks (650 kg/m³)	0.14
Concrete (medium 1400 kg/m³)	0.51	Cellular glass	0.045
Concrete (lightweight 1200 kg/m³)	0.38	Expanded polystyrene	0.034
Concrete (lightweight 600 kg/m³)	0.19	Expanded polystyrene slab (25 kg/m³)	0.035
Concrete slab (aerated 500 kg/m³)	0.16	Extruded polystyrene	0.035
Copper	390	Glass mineral wool	0.04
External render sand/cement finish	1	Mineral quilt (12 kg/m³)	0.04
External render (1300 kg/m³)	0.5	Mineral wool slab (25 kg/m³)	0.035
Felt – Bitumen layers (1700 kg/m³)	0.5	Phenolic foam	0.022
Fibreboard (300 kg/m³)	0.06	Polyisocyanurate	0.025
Glass	0.93	Polyurethane	0.025
Marble	3	Rigid polyurethane	0.025
Metal tray used in wriggly tin concrete floors (7800 kg/m³)	50	Rock mineral wool	0.038
Mortar (1750 kg/m³)	0.8		

EARTHWORK

EARTHWORK

Weights of Typical Materials Handled by Excavators

The weight of the material is that of the state in its natural bed and includes moisture
Adjustments should be made to allow for loose or compacted states

Material	Mass (kg/m³)	Mass (lb/cu yd)
Ashes, dry	610	1028
Ashes, wet	810	1365
Basalt, broken	1954	3293
Basalt, solid	2933	4943
Bauxite, crushed	1281	2159
Borax, fine	849	1431
Caliche	1440	2427
Cement, clinker	1415	2385
Chalk, fine	1221	2058
Chalk, solid	2406	4055
Cinders, coal, ash	641	1080
Cinders, furnace	913	1538
Clay, compacted	1746	2942
Clay, dry	1073	1808
Clay, wet	1602	2700
Coal, anthracite, solid	1506	2538
Coal, bituminous	1351	2277
Coke	610	1028
Dolomite, lumpy	1522	2565
Dolomite, solid	2886	4864
Earth, dense	2002	3374
Earth, dry, loam	1249	2105
Earth, Fullers, raw	673	1134
Earth, moist	1442	2430
Earth, wet	1602	2700
Felsite	2495	4205
Fieldspar, solid	2613	4404
Fluorite	3093	5213
Gabbro	3093	5213
Gneiss	2696	4544
Granite	2690	4534
Gravel, dry ¼ to 2 inch	1682	2835
Gravel, dry, loose	1522	2565
Gravel, wet ¼ to 2 inch	2002	3374
Gypsum, broken	1450	2444
Gypsum, solid	2787	4697
Hardcore (consolidated)	1928	3249
Lignite, dry	801	1350
Limestone, broken	1554	2619
Limestone, solid	2596	4375
Magnesite, magnesium ore	2993	5044
Marble	2679	4515
Marl, wet	2216	3735
Mica, broken	1602	2700
Mica, solid	2883	4859
Peat, dry	400	674
Peat, moist	700	1179
Peat, wet	1121	1889

EARTHWORK

Material	Mass (kg/m^3)	Mass (lb/cu yd)
Potash	1281	2159
Pumice, stone	640	1078
Quarry waste	1438	2423
Quartz sand	1201	2024
Quartz, solid	2584	4355
Rhyolite	2400	4045
Sand and gravel, dry	1650	2781
Sand and gravel, wet	2020	3404
Sand, dry	1602	2700
Sand, wet	1831	3086
Sandstone, solid	2412	4065
Shale, solid	2637	4444
Slag, broken	2114	3563
Slag, furnace granulated	961	1619
Slate, broken	1370	2309
Slate, solid	2667	4495
Snow, compacted	481	810
Snow, freshly fallen	160	269
Taconite	2803	4724
Trachyte	2400	4045
Trap rock, solid	2791	4704
Turf	400	674
Water	1000	1685

Transport Capacities

Type of vehicle	Capacity of vehicle	
	Payload	Heaped capacity
Wheelbarrow	150	0.10
1 tonne dumper	1250	1.00
2.5 tonne dumper	4000	2.50
Articulated dump truck (Volvo A20 6 × 4)	18500	11.00
Articulated dump truck (Volvo A35 6 × 6)	32000	19.00
Large capacity rear dumper (Euclid R35)	35000	22.00
Large capacity rear dumper (Euclid R85)	85000	50.00

EARTHWORK

Machine Volumes for Excavating and Filling

Machine type	Cycles per minute	Volume per minute (m³)
1.5 tonne excavator	1	0.04
	2	0.08
	3	0.12
3 tonne excavator	1	0.13
	2	0.26
	3	0.39
5 tonne excavator	1	0.28
	2	0.56
	3	0.84
7 tonne excavator	1	0.28
	2	0.56
	3	0.84
21 tonne excavator	1	1.21
	2	2.42
	3	3.63
Backhoe loader JCB3CX excavator Rear bucket capacity 0.28 m³	1	0.28
	2	0.56
	3	0.84
Backhoe loader JCB3CX loading Front bucket capacity 1.00 m³	1	1.00
	2	2.00

Machine Volumes for Excavating and Filling

Machine type	Loads per hour	Volume per hour (m³)
1 tonne high tip skip loader Volume 0.485 m³	5	2.43
	7	3.40
	10	4.85
3 tonne dumper Max volume 2.40 m³ Available volume 1.9 m³	4	7.60
	5	9.50
	7	13.30
	10	19.00
6 tonne dumper Max volume 3.40 m³ Available volume 3.77 m³	4	15.08
	5	18.85
	7	26.39
	10	37.70

EARTHWORK

Bulkage of Soils (after excavation)

Type of soil	Approximate bulking of 1 m³ after excavation
Vegetable soil and loam	25–30%
Soft clay	30–40%
Stiff clay	10–20%
Gravel	20–25%
Sand	40–50%
Chalk	40–50%
Rock, weathered	30–40%
Rock, unweathered	50–60%

Shrinkage of Materials (on being deposited)

Type of soil	Approximate bulking of 1 m³ after excavation
Clay	10%
Gravel	8%
Gravel and sand	9%
Loam and light sandy soils	12%
Loose vegetable soils	15%

Voids in Material Used as Subbases or Beddings

Material	m³ of voids/m³
Alluvium	0.37
River grit	0.29
Quarry sand	0.24
Shingle	0.37
Gravel	0.39
Broken stone	0.45
Broken bricks	0.42

Angles of Repose

Type of soil		Degrees
Clay	– dry	30
	– damp, well drained	45
	– wet	15–20
Earth	– dry	30
	– damp	45
Gravel	– moist	48
Sand	– dry or moist	35
	– wet	25
Loam		40

EARTHWORK

Slopes and Angles

Ratio of base to height	Angle in degrees
5:1	11
4:1	14
3:1	18
2:1	27
1½:1	34
1:1	45
1:1½	56
1:2	63
1:3	72
1:4	76
1:5	79

Grades (in Degrees and Percents)

Degrees	Percent	Degrees	Percent
1	1.8	24	44.5
2	3.5	25	46.6
3	5.2	26	48.8
4	7.0	27	51.0
5	8.8	28	53.2
6	10.5	29	55.4
7	12.3	30	57.7
8	14.0	31	60.0
9	15.8	32	62.5
10	17.6	33	64.9
11	19.4	34	67.4
12	21.3	35	70.0
13	23.1	36	72.7
14	24.9	37	75.4
15	26.8	38	78.1
16	28.7	39	81.0
17	30.6	40	83.9
18	32.5	41	86.9
19	34.4	42	90.0
20	36.4	43	93.3
21	38.4	44	96.6
22	40.4	45	100.0
23	42.4		

EARTHWORK

Bearing Powers

Ground conditions		Bearing power		
		kg/m^2	lb/in^2	Metric t/m^2
Rock,	broken	483	70	50
	solid	2415	350	240
Clay,	dry or hard	380	55	40
	medium dry	190	27	20
	soft or wet	100	14	10
Gravel,	cemented	760	110	80
Sand,	compacted	380	55	40
	clean dry	190	27	20
Swamp and alluvial soils		48	7	5

Earthwork Support

Maximum depth of excavation in various soils without the use of earthwork support

Ground conditions	Feet (ft)	Metres (m)
Compact soil	12	3.66
Drained loam	6	1.83
Dry sand	1	0.3
Gravelly earth	2	0.61
Ordinary earth	3	0.91
Stiff clay	10	3.05

It is important to note that the above table should only be used as a guide. Each case must be taken on its merits and, as the limited distances given above are approached, careful watch must be kept for the slightest signs of caving in

CONCRETE WORK

CONCRETE WORK

Weights of Concrete and Concrete Elements

Type of material		kg/m³	lb/cu ft
Ordinary concrete (dense aggregates)			
Non-reinforced plain or mass concrete			
Nominal weight		2305	144
Aggregate	– limestone	2162 to 2407	135 to 150
	– gravel	2244 to 2407	140 to 150
	– broken brick	2000 (av)	125 (av)
	– other crushed stone	2326 to 2489	145 to 155
Reinforced concrete			
Nominal weight		2407	150
Reinforcement	– 1%	2305 to 2468	144 to 154
	– 2%	2356 to 2519	147 to 157
	– 4%	2448 to 2703	153 to 163
Special concretes			
Heavy concrete			
Aggregates	– barytes, magnetite	3210 (min)	200 (min)
	– steel shot, punchings	5280	330
Lean mixes			
Dry-lean (gravel aggregate)		2244	140
Soil-cement (normal mix)		1601	100

CONCRETE WORK

Type of material		kg/m² per mm thick	lb/sq ft per inch thick
Ordinary concrete (dense aggregates)			
Solid slabs (floors, walls etc.)			
Thickness:	75 mm or 3 in	184	37.5
	100 mm or 4 in	245	50
	150 mm or 6 in	378	75
	250 mm or 10 in	612	125
	300 mm or 12 in	734	150
Ribbed slabs			
Thickness:	125 mm or 5 in	204	42
	150 mm or 6 in	219	45
	225 mm or 9 in	281	57
	300 mm or 12 in	342	70
Special concretes			
Finishes etc.			
	Rendering, screed etc.		
	Granolithic, terrazzo	1928 to 2401	10 to 12.5
	Glass-block (hollow) concrete	1734 (approx)	9 (approx)
Prestressed concrete		Weights as for reinforced concrete (upper limits)	
Air-entrained concrete		Weights as for plain or reinforced concrete	

CONCRETE WORK

Average Weight of Aggregates

Materials	Voids %	Weight kg/m^3
Sand	39	1660
Gravel 10–20 mm	45	1440
Gravel 35–75 mm	42	1555
Crushed stone	50	1330
Crushed granite (over 15 mm)	50	1345
(n.e. 15 mm)	47	1440
'All-in' ballast	32	1800–2000

Material	kg/m^3	lb/cu yd
Vermiculite (aggregate)	64–80	108–135
All-in aggregate	1999	125

Applications and Mix Design

Site mixed concrete

Recommended mix	Class of work suitable for	Cement (kg)	Sand (kg)	Coarse aggregate (kg)	Nr 25 kg bags cement per m^3 of combined aggregate
1:3:6	Roughest type of mass concrete such as footings, road haunching over 300 mm thick	208	905	1509	8.30
1:2.5:5	Mass concrete of better class than 1:3:6 such as bases for machinery, walls below ground etc.	249	881	1474	10.00
1:2:4	Most ordinary uses of concrete, such as mass walls above ground, road slabs etc. and general reinforced concrete work	304	889	1431	12.20
1:1.5:3	Watertight floors, pavements and walls, tanks, pits, steps, paths, surface of 2 course roads, reinforced concrete where extra strength is required	371	801	1336	14.90
1:1:2	Works of thin section such as fence posts and small precast work	511	720	1206	20.40

CONCRETE WORK

Ready mixed concrete

Application	Designated concrete	Standardized prescribed concrete	Recommended consistence (nominal slump class)
Foundations			
Mass concrete fill or blinding	GEN 1	ST2	S3
Strip footings	GEN 1	ST2	S3
Mass concrete foundations			
Single storey buildings	GEN 1	ST2	S3
Double storey buildings	GEN 3	ST4	S3
Trench fill foundations			
Single storey buildings	GEN 1	ST2	S4
Double storey buildings	GEN 3	ST4	S4
General applications			
Kerb bedding and haunching	GEN 0	ST1	S1
Drainage works – immediate support	GEN 1	ST2	S1
Other drainage works	GEN 1	ST2	S3
Oversite below suspended slabs	GEN 1	ST2	S3
Floors			
Garage and house floors with no embedded steel	GEN 3	ST4	S2
Wearing surface: Light foot and trolley traffic	RC30	ST4	S2
Wearing surface: General industrial	RC40	N/A	S2
Wearing surface: Heavy industrial	RC50	N/A	S2
Paving			
House drives, domestic parking and external parking	PAV 1	N/A	S2
Heavy-duty external paving	PAV 2	N/A	S2

CONCRETE WORK

Prescribed Mixes for Ordinary Structural Concrete

Weights of cement and total dry aggregates in kg to produce approximately one cubic metre of fully compacted concrete together with the percentages by weight of fine aggregate in total dry aggregates

Conc. grade	Nominal max size of aggregate (mm)	40		20		14		10	
	Workability	**Med.**	**High**	**Med.**	**High**	**Med.**	**High**	**Med.**	**High**
	Limits to slump that may be expected (mm)	**50–100**	**100–150**	**25–75**	**75–125**	**10–50**	**50–100**	**10–25**	**25–50**
7	Cement (kg)	180	200	210	230	–	–	–	–
	Total aggregate (kg)	1950	1850	1900	1800	–	–	–	–
	Fine aggregate (%)	30–45	30–45	35–50	35–50	–	–	–	–
10	Cement (kg)	210	230	240	260	–	–	–	–
	Total aggregate (kg)	1900	1850	1850	1800	–	–	–	–
	Fine aggregate (%)	30–45	30–45	35–50	35–50	–	–	–	–
15	Cement (kg)	250	270	280	310	–	–	–	–
	Total aggregate (kg)	1850	1800	1800	1750	–	–	–	–
	Fine aggregate (%)	30–45	30–45	35–50	35–50	–	–	–	–
20	Cement (kg)	300	320	320	350	340	380	360	410
	Total aggregate (kg)	1850	1750	1800	1750	1750	1700	1750	1650
	Sand								
	Zone 1 (%)	35	40	40	45	45	50	50	55
	Zone 2 (%)	30	35	35	40	40	45	45	50
	Zone 3 (%)	30	30	30	35	35	40	40	45
25	Cement (kg)	340	360	360	390	380	420	400	450
	Total aggregate (kg)	1800	1750	1750	1700	1700	1650	1700	1600
	Sand								
	Zone 1 (%)	35	40	40	45	45	50	50	55
	Zone 2 (%)	30	35	35	40	40	45	45	50
	Zone 3 (%)	30	30	30	35	35	40	40	45
30	Cement (kg)	370	390	400	430	430	470	460	510
	Total aggregate (kg)	1750	1700	1700	1650	1700	1600	1650	1550
	Sand								
	Zone 1 (%)	35	40	40	45	45	50	50	55
	Zone 2 (%)	30	35	35	40	40	45	45	50
	Zone 3 (%)	30	30	30	35	35	40	40	45

REINFORCEMENT

REINFORCEMENT

Weights of Bar Reinforcement

Nominal sizes (mm)	Cross-sectional area (mm²)	Mass (kg/m)	Length of bar (m/tonne)
6	28.27	0.222	4505
8	50.27	0.395	2534
10	78.54	0.617	1622
12	113.10	0.888	1126
16	201.06	1.578	634
20	314.16	2.466	405
25	490.87	3.853	260
32	804.25	6.313	158
40	1265.64	9.865	101
50	1963.50	15.413	65

Weights of Bars (at specific spacings)

Weights of metric bars in kilogrammes per square metre

Size (mm)	Spacing of bars in millimetres									
	75	100	125	150	175	200	225	250	275	300
6	2.96	2.220	1.776	1.480	1.27	1.110	0.99	0.89	0.81	0.74
8	5.26	3.95	3.16	2.63	2.26	1.97	1.75	1.58	1.44	1.32
10	8.22	6.17	4.93	4.11	3.52	3.08	2.74	2.47	2.24	2.06
12	11.84	8.88	7.10	5.92	5.07	4.44	3.95	3.55	3.23	2.96
16	21.04	15.78	12.63	10.52	9.02	7.89	7.02	6.31	5.74	5.26
20	32.88	24.66	19.73	16.44	14.09	12.33	10.96	9.87	8.97	8.22
25	51.38	38.53	30.83	25.69	22.02	19.27	17.13	15.41	14.01	12.84
32	84.18	63.13	50.51	42.09	36.08	31.57	28.06	25.25	22.96	21.04
40	131.53	98.65	78.92	65.76	56.37	49.32	43.84	39.46	35.87	32.88
50	205.51	154.13	123.31	102.76	88.08	77.07	68.50	61.65	56.05	51.38

Basic weight of steelwork taken as 7850 kg/m³
Basic weight of bar reinforcement per metre run = 0.00785 kg/mm²
The value of π has been taken as 3.141592654

REINFORCEMENT

Fabric Reinforcement

Preferred range of designated fabric types and stock sheet sizes

Fabric reference	Longitudinal wires			Cross wires			
	Nominal wire size (mm)	Pitch (mm)	Area (mm²/m)	Nominal wire size (mm)	Pitch (mm)	Area (mm²/m)	Mass (kg/m²)
Square mesh							
A393	10	200	393	10	200	393	6.16
A252	8	200	252	8	200	252	3.95
A193	7	200	193	7	200	193	3.02
A142	6	200	142	6	200	142	2.22
A98	5	200	98	5	200	98	1.54
Structural mesh							
B1131	12	100	1131	8	200	252	10.90
B785	10	100	785	8	200	252	8.14
B503	8	100	503	8	200	252	5.93
B385	7	100	385	7	200	193	4.53
B283	6	100	283	7	200	193	3.73
B196	5	100	196	7	200	193	3.05
Long mesh							
C785	10	100	785	6	400	70.8	6.72
C636	9	100	636	6	400	70.8	5.55
C503	8	100	503	5	400	49.0	4.34
C385	7	100	385	5	400	49.0	3.41
C283	6	100	283	5	400	49.0	2.61
Wrapping mesh							
D98	5	200	98	5	200	98	1.54
D49	2.5	100	49	2.5	100	49	0.77

Stock sheet size 4.8 m × 2.4 m, Area 11.52 m²

Average weight kg/m³ of steelwork reinforcement in concrete for various building elements

Substructure	kg/m³ concrete	Substructure	kg/m³ concrete
Pile caps	110–150	Plate slab	150–220
Tie beams	130–170	Cant slab	145–210
Ground beams	230–330	Ribbed floors	130–200
Bases	125–180	Topping to block floor	30–40
Footings	100–150	Columns	210–310
Retaining walls	150–210	Beams	250–350
Raft	60–70	Stairs	130–170
Slabs – one way	120–200	Walls – normal	40–100
Slabs – two way	110–220	Walls – wind	70–125

Note: For exposed elements add the following %:
Walls 50%, Beams 100%, Columns 15%

FORMWORK

Formwork Stripping Times – Normal Curing Periods

Conditions under which concrete is maturing	Minimum periods of protection for different types of cement					
	Number of days (where the average surface temperature of the concrete exceeds 10°C during the whole period)			Equivalent maturity (degree hours) calculated as the age of the concrete in hours multiplied by the number of degrees Celsius by which the average surface temperature of the concrete exceeds 10°C		
	Other	SRPC	OPC or RHPC	Other	SRPC	OPC or RHPC
1. Hot weather or drying winds	7	4	3	3500	2000	1500
2. Conditions not covered by 1	4	3	2	2000	1500	1000

KEY
OPC – Ordinary Portland Cement
RHPC – Rapid-hardening Portland Cement
SRPC – Sulphate-resisting Portland Cement

Minimum Period before Striking Formwork

	Minimum period before striking		
	Surface temperature of concrete		
	16°C	17°C	t°C (0–25)
Vertical formwork to columns, walls and large beams	12 hours	18 hours	300 hours t+10
Soffit formwork to slabs	4 days	6 days	100 days t+10
Props to slabs	10 days	15 days	250 days t+10
Soffit formwork to beams	9 days	14 days	230 days t+10
Props to beams	14 days	21 days	360 days t+10

MASONRY

Number of Bricks Required for Various Types of Work per m² of Walling

Description	Brick size	
	215 × 102.5 × 50 mm	215 × 102.5 × 65 mm
Half brick thick		
Stretcher bond	74	59
English bond	108	86
English garden wall bond	90	72
Flemish bond	96	79
Flemish garden wall bond	83	66
One brick thick and cavity wall of two half brick skins		
Stretcher bond	148	119

Quantities of Bricks and Mortar required per m² of Walling

	Unit	No of bricks required	Mortar required (cubic metres)		
Standard bricks			No frogs	Single frogs	Double frogs
Brick size 215 × 102.5 × 50 mm					
half brick wall (103 mm)	m²	72	0.022	0.027	0.032
2 × half brick cavity wall (270 mm)	m²	144	0.044	0.054	0.064
one brick wall (215 mm)	m²	144	0.052	0.064	0.076
one and a half brick wall (322 mm)	m²	216	0.073	0.091	0.108
Mass brickwork	m³	576	0.347	0.413	0.480
Brick size 215 × 102.5 × 65 mm					
half brick wall (103 mm)	m²	58	0.019	0.022	0.026
2 × half brick cavity wall (270 mm)	m²	116	0.038	0.045	0.055
one brick wall (215 mm)	m²	116	0.046	0.055	0.064
one and a half brick wall (322 mm)	m²	174	0.063	0.074	0.088
Mass brickwork	m³	464	0.307	0.360	0.413
Metric modular bricks			Perforated		
Brick size 200 × 100 × 75 mm					
90 mm thick	m²	67	0.016	0.019	
190 mm thick	m²	133	0.042	0.048	
290 mm thick	m²	200	0.068	0.078	
Brick size 200 × 100 × 100 mm					
90 mm thick	m²	50	0.013	0.016	
190 mm thick	m²	100	0.036	0.041	
290 mm thick	m²	150	0.059	0.067	
Brick size 300 × 100 × 75 mm					
90 mm thick	m²	33	–	0.015	
Brick size 300 × 100 × 100 mm					
90 mm thick	m²	44	0.015	0.018	

Note: Assuming 10 mm thick joints

MASONRY

Mortar Required per m² Blockwork (9.88 blocks/m²)

Wall thickness	75	90	100	125	140	190	215
Mortar m³/m²	0.005	0.006	0.007	0.008	0.009	0.013	0.014

Mortar group	Cement: lime: sand	Masonry cement: sand	Cement: sand with plasticizer
1	1:0–0.25:3		
2	1:0.5:4–4.5	1:2.5-3.5	1:3–4
3	1:1:5–6	1:4–5	1:5–6
4	1:2:8–9	1:5.5–6.5	1:7–8
5	1:3:10–12	1:6.5–7	1:8

Group 1: strong inflexible mortar
Group 5: weak but flexible

All mixes within a group are of approximately similar strength
Frost resistance increases with the use of plasticizers
Cement: lime: sand mixes give the strongest bond and greatest resistance to rain penetration
Masonry cement equals ordinary Portland cement plus a fine neutral mineral filler and an air entraining agent

Calcium Silicate Bricks

Type	Strength	Location
Class 2 crushing strength	14.0 N/mm²	not suitable for walls
Class 3	20.5 N/mm²	walls above DPC
Class 4	27.5 N/mm²	cappings and copings
Class 5	34.5 N/mm²	retaining walls
Class 6	41.5 N/mm²	walls below ground
Class 7	48.5 N/mm²	walls below ground

The Class 7 calcium silicate bricks are therefore equal in strength to Class B bricks
Calcium silicate bricks are not suitable for DPCs

Durability of bricks

FL	Frost resistant with low salt content
FN	Frost resistant with normal salt content
ML	Moderately frost resistant with low salt content
MN	Moderately frost resistant with normal salt content

MASONRY

Brickwork Dimensions

No. of horizontal bricks	Dimensions (mm)	No. of vertical courses	Height of vertical courses (mm)
½	112.5	1	75
1	225.0	2	150
1½	337.5	3	225
2	450.0	4	300
2½	562.5	5	375
3	675.0	6	450
3½	787.5	7	525
4	900.0	8	600
4½	1012.5	9	675
5	1125.0	10	750
5½	1237.5	11	825
6	1350.0	12	900
6½	1462.5	13	975
7	1575.0	14	1050
7½	1687.5	15	1125
8	1800.0	16	1200
8½	1912.5	17	1275
9	2025.0	18	1350
9½	2137.5	19	1425
10	2250.0	20	1500
20	4500.0	24	1575
40	9000.0	28	2100
50	11250.0	32	2400
60	13500.0	36	2700
75	16875.0	40	3000

TIMBER

TIMBER

Weights of Timber

Material	kg/m³	lb/cu ft
General	806 (avg)	50 (avg)
Douglas fir	479	30
Yellow pine, spruce	479	30
Pitch pine	673	42
Larch, elm	561	35
Oak (English)	724 to 959	45 to 60
Teak	643 to 877	40 to 55
Jarrah	959	60
Greenheart	1040 to 1204	65 to 75
Quebracho	1285	80
Material	kg/m² per mm thickness	lb/sq ft per inch thickness
Wooden boarding and blocks		
Softwood	0.48	2.5
Hardwood	0.76	4
Hardboard	1.06	5.5
Chipboard	0.76	4
Plywood	0.62	3.25
Blockboard	0.48	2.5
Fibreboard	0.29	1.5
Wood-wool	0.58	3
Plasterboard	0.96	5
Weather boarding	0.35	1.8

TIMBER

Conversion Tables (for timber only)

Inches	Millimetres	Feet	Metres
1	25	1	0.300
2	50	2	0.600
3	75	3	0.900
4	100	4	1.200
5	125	5	1.500
6	150	6	1.800
7	175	7	2.100
8	200	8	2.400
9	225	9	2.700
10	250	10	3.000
11	275	11	3.300
12	300	12	3.600
13	325	13	3.900
14	350	14	4.200
15	375	15	4.500
16	400	16	4.800
17	425	17	5.100
18	450	18	5.400
19	475	19	5.700
20	500	20	6.000
21	525	21	6.300
22	550	22	6.600
23	575	23	6.900
24	600	24	7.200

Planed Softwood
The finished end section size of planed timber is usually 3/16" less than the original size from which it is produced. This however varies slightly depending upon availability of material and origin of the species used.

Standards (timber) to cubic metres and cubic metres to standards (timber)

Cubic metres	Cubic metres standards	Standards
4.672	1	0.214
9.344	2	0.428
14.017	3	0.642
18.689	4	0.856
23.361	5	1.070
28.033	6	1.284
32.706	7	1.498
37.378	8	1.712
42.050	9	1.926
46.722	10	2.140
93.445	20	4.281
140.167	30	6.421
186.890	40	8.561
233.612	50	10.702
280.335	60	12.842
327.057	70	14.982
373.779	80	17.122

Tables and Memoranda

TIMBER

1 cu metre = 35.3148 cu ft = 0.21403 std

1 cu ft = 0.028317 cu metres

1 std = 4.67227 cu metres

Basic sizes of sawn softwood available (cross-sectional areas)

Thickness (mm)	Width (mm)								
	75	100	125	150	175	200	225	250	300
16	X	X	X	X					
19	X	X	X	X					
22	X	X	X	X					
25	X	X	X	X	X	X	X	X	X
32	X	X	X	X	X	X	X	X	X
36	X	X	X	X					
38	X	X	X	X	X	X	X		
44	X	X	X	X	X	X	X	X	X
47*	X	X	X	X	X	X	X	X	X
50	X	X	X	X	X	X	X	X	X
63	X	X	X	X	X	X	X		
75	X	X	X	X	X	X	X	X	
100		X		X		X		X	X
150				X		X			X
200						X			
250								X	
300									X

* This range of widths for 47 mm thickness will usually be found to be available in construction quality only

Note: The smaller sizes below 100 mm thick and 250 mm width are normally but not exclusively of European origin. Sizes beyond this are usually of North and South American origin

Basic lengths of sawn softwood available (metres)

1.80	2.10	3.00	4.20	5.10	6.00	7.20
	2.40	3.30	4.50	5.40	6.30	
	2.70	3.60	4.80	5.70	6.60	
		3.90			6.90	

Note: Lengths of 6.00 m and over will generally only be available from North American species and may have to be recut from larger sizes

TIMBER

Reductions from basic size to finished size by planning of two opposed faces

	Purpose		Reductions from basic sizes for timber			
		15–35 mm	36–100 mm	101–150 mm	over 150 mm
a)	Constructional timber	3 mm	3 mm	5 mm	6 mm
b)	Matching interlocking boards	4 mm	4 mm	6 mm	6 mm
c)	Wood trim not specified in BS 584	5 mm	7 mm	7 mm	9 mm
d)	Joinery and cabinet work	7 mm	9 mm	11 mm	13 mm

Note: The reduction of width or depth is overall the extreme size and is exclusive of any reduction of the face by the machining of a tongue or lap joints

Maximum Spans for Various Roof Trusses

Maximum permissible spans for rafters for Fink trussed rafters

Basic size	Actual size	Pitch (degrees)								
(mm)	(mm)	15 (m)	17.5 (m)	20 (m)	22.5 (m)	25 (m)	27.5 (m)	30 (m)	32.5 (m)	35 (m)
38 × 75	35 × 72	6.03	6.16	6.29	6.41	6.51	6.60	6.70	6.80	6.90
38 × 100	35 × 97	7.48	7.67	7.83	7.97	8.10	8.22	8.34	8.47	8.61
38 × 125	35 × 120	8.80	9.00	9.20	9.37	9.54	9.68	9.82	9.98	10.16
44 × 75	41 × 72	6.45	6.59	6.71	6.83	6.93	7.03	7.14	7.24	7.35
44 × 100	41 × 97	8.05	8.23	8.40	8.55	8.68	8.81	8.93	9.09	9.22
44 × 125	41 × 120	9.38	9.60	9.81	9.99	10.15	10.31	10.45	10.64	10.81
50 × 75	47 × 72	6.87	7.01	7.13	7.25	7.35	7.45	7.53	7.67	7.78
50 × 100	47 × 97	8.62	8.80	8.97	9.12	9.25	9.38	9.50	9.66	9.80
50 × 125	47 × 120	10.01	10.24	10.44	10.62	10.77	10.94	11.00	11.00	11.00

TIMBER

Sizes of Internal and External Doorsets

Description	Internal size (mm)	Permissible deviation	External size (mm)	Permissible deviation
Coordinating dimension: height of door leaf height sets	2100		2100	
Coordinating dimension: height of ceiling height set	2300 2350 2400 2700 3000		2300 2350 2400 2700 3000	
Coordinating dimension: width of all doorsets S = Single leaf set D = Double leaf set	600 S 700 S 800 S&D 900 S&D 1000 S&D 1200 D 1500 D 1800 D 2100 D		900 S 1000 S 1200 D 1800 D 2100 D	
Work size: height of door leaf height set	2090	± 2.0	2095	± 2.0
Work size: height of ceiling height set	2285 2335 2385 2685 2985	± 2.0	2295 2345 2395 2695 2995	± 2.0
Work size: width of all doorsets S = Single leaf set D = Double leaf set	590 S 690 S 790 S&D 890 S&D 990 S&D 1190 D 1490 D 1790 D 2090 D	± 2.0	895 S 995 S 1195 D 1495 D 1795 D 2095 D	± 2.0
Width of door leaf in single leaf sets F = Flush leaf P = Panel leaf	526 F 626 F 726 F&P 826 F&P 926 F&P	± 1.5	806 F&P 906 F&P	± 1.5
Width of door leaf in double leaf sets F = Flush leaf P = Panel leaf	362 F 412 F 426 F 562 F&P 712 F&P 826 F&P 1012 F&P	± 1.5	552 F&P 702 F&P 852 F&P 1002 F&P	± 1.5
Door leaf height for all doorsets	2040	± 1.5	1994	± 1.5

ROOFING

ROOFING

Total Roof Loadings for Various Types of Tiles/Slates

	Roof load (slope) kg/m²		
	Slate/Tile	Roofing underlay and battens²	Total dead load kg/m
Asbestos cement slate (600 × 300)	21.50	3.14	24.64
Clay tile interlocking	67.00	5.50	72.50
plain	43.50	2.87	46.37
Concrete tile interlocking	47.20	2.69	49.89
plain	78.20	5.50	83.70
Natural slate (18" × 10")	35.40	3.40	38.80
	Roof load (plan) kg/m²		
Asbestos cement slate (600 × 300)	28.45	76.50	104.95
Clay tile interlocking	53.54	76.50	130.04
plain	83.71	76.50	60.21
Concrete tile interlocking	57.60	76.50	134.10
plain	96.64	76.50	173.14

ROOFING

Tiling Data

Product		Lap (mm)	Gauge of battens	No. slates per m²	Battens (m/m²)	Weight as laid (kg/m²)
CEMENT SLATES						
Eternit slates	600 × 300 mm	100	250	13.4	4.00	19.50
(Duracem)		90	255	13.1	3.92	19.20
		80	260	12.9	3.85	19.00
		70	265	12.7	3.77	18.60
	600 × 350 mm	100	250	11.5	4.00	19.50
		90	255	11.2	3.92	19.20
	500 × 250 mm	100	200	20.0	5.00	20.00
		90	205	19.5	4.88	19.50
		80	210	19.1	4.76	19.00
		70	215	18.6	4.65	18.60
	400 × 200 mm	90	155	32.3	6.45	20.80
		80	160	31.3	6.25	20.20
		70	165	30.3	6.06	19.60
CONCRETE TILES/SLATES						
Redland Roofing						
Stonewold slate	430 × 380 mm	75	355	8.2	2.82	51.20
Double Roman tile	418 × 330 mm	75	355	8.2	2.91	45.50
Grovebury pantile	418 × 332 mm	75	343	9.7	2.91	47.90
Norfolk pantile	381 × 227 mm	75	306	16.3	3.26	44.01
		100	281	17.8	3.56	48.06
Renown interlocking tile	418 × 330 mm	75	343	9.7	2.91	46.40
'49' tile	381 × 227 mm	75	306	16.3	3.26	44.80
		100	281	17.8	3.56	48.95
Plain, vertical tiling	265 × 165 mm	35	115	52.7	8.70	62.20
Marley Roofing						
Bold roll tile	420 × 330 mm	75	344	9.7	2.90	47.00
		100	–	10.5	3.20	51.00
Modern roof tile	420 × 330 mm	75	338	10.2	3.00	54.00
		100	–	11.0	3.20	58.00
Ludlow major	420 × 330 mm	75	338	10.2	3.00	45.00
		100	–	11.0	3.20	49.00
Ludlow plus	387 × 229 mm	75	305	16.1	3.30	47.00
		100	–	17.5	3.60	51.00
Mendip tile	420 × 330 mm	75	338	10.2	3.00	47.00
		100	–	11.0	3.20	51.00
Wessex	413 × 330 mm	75	338	10.2	3.00	54.00
		100	–	11.0	3.20	58.00
Plain tile	267 × 165 mm	65	100	60.0	10.00	76.00
		75	95	64.0	10.50	81.00
		85	90	68.0	11.30	86.00
Plain vertical tiles (feature)	267 × 165 mm	35	110	53.0	8.70	67.00
		34	115	56.0	9.10	71.00

ROOFING

Slate Nails, Quantity per Kilogram

Length	Type			
	Plain wire	**Galvanized wire**	**Copper nail**	**Zinc nail**
28.5 mm	325	305	325	415
34.4 mm	286	256	254	292
50.8 mm	242	224	194	200

Metal Sheet Coverings

Thicknesses and weights of sheet metal coverings

Lead to BS 1178

BS Code No	3	4	5	6	7	8		
Colour code	Green	Blue	Red	Black	White	Orange		
Thickness (mm)	1.25	1.80	2.24	2.50	3.15	3.55		
Density kg/m^2	14.18	20.41	25.40	30.05	35.72	40.26		

Copper to BS 2870

Thickness (mm)		0.60	0.70					
Bay width								
Roll (mm)		500	650					
Seam (mm)		525	600					
Standard width to form bay	600	750						
Normal length of sheet	1.80	1.80						

Zinc to BS 849

Zinc Gauge (Nr)	9	10	11	12	13	14	15	16
Thickness (mm)	0.43	0.48	0.56	0.64	0.71	0.79	0.91	1.04
Density (kg/m^2)	3.1	3.2	3.8	4.3	4.8	5.3	6.2	7.0

Aluminium to BS 4868

Thickness (mm)	0.5	0.6	0.7	0.8	0.9	1.0	1.2	
Density (kg/m^2)	12.8	15.4	17.9	20.5	23.0	25.6	30.7	

ROOFING

Type of felt	Nominal mass per unit area (kg/10 m)	Nominal mass per unit area of fibre base (g/m²)	Nominal length of roll (m)
Class 1			
1B fine granule	14	220	10 or 20
surfaced bitumen	18	330	10 or 20
	25	470	10
1E mineral surfaced bitumen	38	470	10
1F reinforced bitumen	15	160 (fibre) 110 (hessian)	15
1F reinforced bitumen, aluminium faced	13	160 (fibre) 110 (hessian)	15
Class 2			
2B fine granule surfaced bitumen asbestos	18	500	10 or 20
2E mineral surfaced bitumen asbestos	38	600	10
Class 3			
3B fine granule surfaced bitumen glass fibre	18	60	20
3E mineral surfaced bitumen glass fibre	28	60	10
3E venting base layer bitumen glass fibre	32	60*	10
3H venting base layer bitumen glass fibre	17	60*	20

* Excluding effect of perforations

GLAZING

GLAZING

Nominal thickness (mm)	Tolerance on thickness (mm)	Approximate weight (kg/m²)	Normal maximum size (mm)
Float and polished plate glass			
3	+ 0.2	7.50	2140 × 1220
4	+ 0.2	10.00	2760 × 1220
5	+ 0.2	12.50	3180 × 2100
6	+ 0.2	15.00	4600 × 3180
10	+ 0.3	25.00)	6000 × 3300
12	+ 0.3	30.00)	
15	+ 0.5	37.50	3050 × 3000
19	+ 1.0	47.50)	3000 × 2900
25	+ 1.0	63.50)	
Clear sheet glass			
2 *	+ 0.2	5.00	1920 × 1220
3	+ 0.3	7.50	2130 × 1320
4	+ 0.3	10.00	2760 × 1220
5 *	+ 0.3	12.50)	2130 × 2400
6 *	+ 0.3	15.00)	
Cast glass			
3	+ 0.4		
	− 0.2	6.00)	2140 × 1280
4	+ 0.5	7.50)	
5	+ 0.5	9.50	2140 × 1320
6	+ 0.5	11.50)	3700 × 1280
10	+ 0.8	21.50)	
Wired glass			
(Cast wired glass)			
6	+ 0.3	−)	
	− 0.7	)	3700 × 1840
7	+ 0.7	−)	
(Polished wire glass)			
6	+ 1.0	−	330 × 1830

* The 5 mm and 6 mm thickness are known as *thick drawn sheet*. Although 2 mm sheet glass is available it is not recommended for general glazing purposes

METAL

METAL

Weights of Metals

Material	kg/m³	lb/cu ft
Metals, steel construction, etc.		
Iron		
– cast	7207	450
– wrought	7687	480
– ore – general	2407	150
– (crushed) Swedish	3682	230
Steel	7854	490
Copper		
– cast	8731	545
– wrought	8945	558
Brass	8497	530
Bronze	8945	558
Aluminium	2774	173
Lead	11322	707
Zinc (rolled)	7140	446

	g/mm² per metre	lb/sq ft per foot
Steel bars	7.85	3.4

Structural steelwork	Net weight of member @ 7854 kg/m³	
riveted	+ 10% for cleats, rivets, bolts, etc.	
welded	+ 1.25% to 2.5% for welds, etc.	
Rolled sections		
beams	+ 2.5%	
stanchions	+ 5% (extra for caps and bases)	
Plate		
web girders	+ 10% for rivets or welds, stiffeners, etc.	

	kg/m	lb/ft
Steel stairs: industrial type		
1 m or 3 ft wide	84	56
Steel tubes		
50 mm or 2 in bore	5 to 6	3 to 4
Gas piping		
20 mm or ¾ in	2	1¼

METAL

Universal Beams BS 4: Part 1: 2005

Designation	Mass	Depth of section	Width of section	Thickness		Surface area
	(kg/m)	(mm)	(mm)	Web (mm)	Flange (mm)	(m²/m)
1016 × 305 × 487	487.0	1036.1	308.5	30.0	54.1	3.20
1016 × 305 × 438	438.0	1025.9	305.4	26.9	49.0	3.17
1016 × 305 × 393	393.0	1016.0	303.0	24.4	43.9	3.15
1016 × 305 × 349	349.0	1008.1	302.0	21.1	40.0	3.13
1016 × 305 × 314	314.0	1000.0	300.0	19.1	35.9	3.11
1016 × 305 × 272	272.0	990.1	300.0	16.5	31.0	3.10
1016 × 305 × 249	249.0	980.2	300.0	16.5	26.0	3.08
1016 × 305 × 222	222.0	970.3	300.0	16.0	21.1	3.06
914 × 419 × 388	388.0	921.0	420.5	21.4	36.6	3.44
914 × 419 × 343	343.3	911.8	418.5	19.4	32.0	3.42
914 × 305 × 289	289.1	926.6	307.7	19.5	32.0	3.01
914 × 305 × 253	253.4	918.4	305.5	17.3	27.9	2.99
914 × 305 × 224	224.2	910.4	304.1	15.9	23.9	2.97
914 × 305 × 201	200.9	903.0	303.3	15.1	20.2	2.96
838 × 292 × 226	226.5	850.9	293.8	16.1	26.8	2.81
838 × 292 × 194	193.8	840.7	292.4	14.7	21.7	2.79
838 × 292 × 176	175.9	834.9	291.7	14.0	18.8	2.78
762 × 267 × 197	196.8	769.8	268.0	15.6	25.4	2.55
762 × 267 × 173	173.0	762.2	266.7	14.3	21.6	2.53
762 × 267 × 147	146.9	754.0	265.2	12.8	17.5	2.51
762 × 267 × 134	133.9	750.0	264.4	12.0	15.5	2.51
686 × 254 × 170	170.2	692.9	255.8	14.5	23.7	2.35
686 × 254 × 152	152.4	687.5	254.5	13.2	21.0	2.34
686 × 254 × 140	140.1	383.5	253.7	12.4	19.0	2.33
686 × 254 × 125	125.2	677.9	253.0	11.7	16.2	2.32
610 × 305 × 238	238.1	635.8	311.4	18.4	31.4	2.45
610 × 305 × 179	179.0	620.2	307.1	14.1	23.6	2.41
610 × 305 × 149	149.1	612.4	304.8	11.8	19.7	2.39
610 × 229 × 140	139.9	617.2	230.2	13.1	22.1	2.11
610 × 229 × 125	125.1	612.2	229.0	11.9	19.6	2.09
610 × 229 × 113	113.0	607.6	228.2	11.1	17.3	2.08
610 × 229 × 101	101.2	602.6	227.6	10.5	14.8	2.07
533 × 210 × 122	122.0	544.5	211.9	12.7	21.3	1.89
533 × 210 × 109	109.0	539.5	210.8	11.6	18.8	1.88
533 × 210 × 101	101.0	536.7	210.0	10.8	17.4	1.87
533 × 210 × 92	92.1	533.1	209.3	10.1	15.6	1.86
533 × 210 × 82	82.2	528.3	208.8	9.6	13.2	1.85
457 × 191 × 98	98.3	467.2	192.8	11.4	19.6	1.67
457 × 191 × 89	89.3	463.4	191.9	10.5	17.7	1.66
457 × 191 × 82	82.0	460.0	191.3	9.9	16.0	1.65
457 × 191 × 74	74.3	457.0	190.4	9.0	14.5	1.64
457 × 191 × 67	67.1	453.4	189.9	8.5	12.7	1.63
457 × 152 × 82	82.1	465.8	155.3	10.5	18.9	1.51
457 × 152 × 74	74.2	462.0	154.4	9.6	17.0	1.50
457 × 152 × 67	67.2	458.0	153.8	9.0	15.0	1.50
457 × 152 × 60	59.8	454.6	152.9	8.1	13.3	1.50
457 × 152 × 52	52.3	449.8	152.4	7.6	10.9	1.48
406 × 178 × 74	74.2	412.8	179.5	9.5	16.0	1.51
406 × 178 × 67	67.1	409.4	178.8	8.8	14.3	1.50
406 × 178 × 60	60.1	406.4	177.9	7.9	12.8	1.49

METAL

Designation	Mass (kg/m)	Depth of section (mm)	Width of section (mm)	Thickness		Surface area (m²/m)
				Web (mm)	Flange (mm)	
406 × 178 × 50	54.1	402.6	177.7	7.7	10.9	1.48
406 × 140 × 46	46.0	403.2	142.2	6.8	11.2	1.34
406 × 140 × 39	39.0	398.0	141.8	6.4	8.6	1.33
356 × 171 × 67	67.1	363.4	173.2	9.1	15.7	1.38
356 × 171 × 57	57.0	358.0	172.2	8.1	13.0	1.37
356 × 171 × 51	51.0	355.0	171.5	7.4	11.5	1.36
356 × 171 × 45	45.0	351.4	171.1	7.0	9.7	1.36
356 × 127 × 39	39.1	353.4	126.0	6.6	10.7	1.18
356 × 127 × 33	33.1	349.0	125.4	6.0	8.5	1.17
305 × 165 × 54	54.0	310.4	166.9	7.9	13.7	1.26
305 × 165 × 46	46.1	306.6	165.7	6.7	11.8	1.25
305 × 165 × 40	40.3	303.4	165.0	6.0	10.2	1.24
305 × 127 × 48	48.1	311.0	125.3	9.0	14.0	1.09
305 × 127 × 42	41.9	307.2	124.3	8.0	12.1	1.08
305 × 127 × 37	37.0	304.4	123.3	7.1	10.7	1.07
305 × 102 × 33	32.8	312.7	102.4	6.6	10.8	1.01
305 × 102 × 28	28.2	308.7	101.8	6.0	8.8	1.00
305 × 102 × 25	24.8	305.1	101.6	5.8	7.0	0.992
254 × 146 × 43	43.0	259.6	147.3	7.2	12.7	1.08
254 × 146 × 37	37.0	256.0	146.4	6.3	10.9	1.07
254 × 146 × 31	31.1	251.4	146.1	6.0	8.6	1.06
254 × 102 × 28	28.3	260.4	102.2	6.3	10.0	0.904
254 × 102 × 25	25.2	257.2	101.9	6.0	8.4	0.897
254 × 102 × 22	22.0	254.0	101.6	5.7	6.8	0.890
203 × 133 × 30	30.0	206.8	133.9	6.4	9.6	0.923
203 × 133 × 25	25.1	203.2	133.2	5.7	7.8	0.915
203 × 102 × 23	23.1	203.2	101.8	5.4	9.3	0.790
178 × 102 × 19	19.0	177.8	101.2	4.8	7.9	0.738
152 × 89 × 16	16.0	152.4	88.7	4.5	7.7	0.638
127 × 76 × 13	13.0	127.0	76.0	4.0	7.6	0.537

METAL

Universal Columns BS 4: Part 1: 2005

Designation	Mass (kg/m)	Depth of section (mm)	Width of section (mm)	Thickness Web (mm)	Flange (mm)	Surface area (m²/m)
356 × 406 × 634	633.9	474.7	424.0	47.6	77.0	2.52
356 × 406 × 551	551.0	455.6	418.5	42.1	67.5	2.47
356 × 406 × 467	467.0	436.6	412.2	35.8	58.0	2.42
356 × 406 × 393	393.0	419.0	407.0	30.6	49.2	2.38
356 × 406 × 340	339.9	406.4	403.0	26.6	42.9	2.35
356 × 406 × 287	287.1	393.6	399.0	22.6	36.5	2.31
356 × 406 × 235	235.1	381.0	384.8	18.4	30.2	2.28
356 × 368 × 202	201.9	374.6	374.7	16.5	27.0	2.19
356 × 368 × 177	177.0	368.2	372.6	14.4	23.8	2.17
356 × 368 × 153	152.9	362.0	370.5	12.3	20.7	2.16
356 × 368 × 129	129.0	355.6	368.6	10.4	17.5	2.14
305 × 305 × 283	282.9	365.3	322.2	26.8	44.1	1.94
305 × 305 × 240	240.0	352.5	318.4	23.0	37.7	1.91
305 × 305 × 198	198.1	339.9	314.5	19.1	31.4	1.87
305 × 305 × 158	158.1	327.1	311.2	15.8	25.0	1.84
305 × 305 × 137	136.9	320.5	309.2	13.8	21.7	1.82
305 × 305 × 118	117.9	314.5	307.4	12.0	18.7	1.81
305 × 305 × 97	96.9	307.9	305.3	9.9	15.4	1.79
254 × 254 × 167	167.1	289.1	265.2	19.2	31.7	1.58
254 × 254 × 132	132.0	276.3	261.3	15.3	25.3	1.55
254 × 254 × 107	107.1	266.7	258.8	12.8	20.5	1.52
254 × 254 × 89	88.9	260.3	256.3	10.3	17.3	1.50
254 × 254 × 73	73.1	254.1	254.6	8.6	14.2	1.49
203 × 203 × 86	86.1	222.2	209.1	12.7	20.5	1.24
203 × 203 × 71	71.0	215.8	206.4	10.0	17.3	1.22
203 × 203 × 60	60.0	209.6	205.8	9.4	14.2	1.21
203 × 203 × 52	52.0	206.2	204.3	7.9	12.5	1.20
203 × 203 × 46	46.1	203.2	203.6	7.2	11.0	1.19
152 × 152 × 37	37.0	161.8	154.4	8.0	11.5	0.912
152 × 152 × 30	30.0	157.6	152.9	6.5	9.4	0.901
152 × 152 × 23	23.0	152.4	152.2	5.8	6.8	0.889

METAL

Joists BS 4: Part 1: 2005 (retained for reference, Corus have ceased manufacture in UK)

Designation	Mass (kg/m)	Depth of section (mm)	Width of section (mm)	Thickness		Surface area (m²/m)
				Web (mm)	Flange (mm)	
254 × 203 × 82	82.0	254.0	203.2	10.2	19.9	1.210
203 × 152 × 52	52.3	203.2	152.4	8.9	16.5	0.932
152 × 127 × 37	37.3	152.4	127.0	10.4	13.2	0.737
127 × 114 × 29	29.3	127.0	114.3	10.2	11.5	0.646
127 × 114 × 27	26.9	127.0	114.3	7.4	11.4	0.650
102 × 102 × 23	23.0	101.6	101.6	9.5	10.3	0.549
102 × 44 × 7	7.5	101.6	44.5	4.3	6.1	0.350
89 × 89 × 19	19.5	88.9	88.9	9.5	9.9	0.476
76 × 76 × 13	12.8	76.2	76.2	5.1	8.4	0.411

Parallel Flange Channels

Designation	Mass (kg/m)	Depth of section (mm)	Width of section (mm)	Thickness		Surface area (m²/m)
				Web (mm)	Flange (mm)	
430 × 100 × 64	64.4	430	100	11.0	19.0	1.23
380 × 100 × 54	54.0	380	100	9.5	17.5	1.13
300 × 100 × 46	45.5	300	100	9.0	16.5	0.969
300 × 90 × 41	41.4	300	90	9.0	15.5	0.932
260 × 90 × 35	34.8	260	90	8.0	14.0	0.854
260 × 75 × 28	27.6	260	75	7.0	12.0	0.79
230 × 90 × 32	32.2	230	90	7.5	14.0	0.795
230 × 75 × 26	25.7	230	75	6.5	12.5	0.737
200 × 90 × 30	29.7	200	90	7.0	14.0	0.736
200 × 75 × 23	23.4	200	75	6.0	12.5	0.678
180 × 90 × 26	26.1	180	90	6.5	12.5	0.697
180 × 75 × 20	20.3	180	75	6.0	10.5	0.638
150 × 90 × 24	23.9	150	90	6.5	12.0	0.637
150 × 75 × 18	17.9	150	75	5.5	10.0	0.579
125 × 65 × 15	14.8	125	65	5.5	9.5	0.489
100 × 50 × 10	10.2	100	50	5.0	8.5	0.382

METAL

Equal Angles BS EN 10056-1

Designation	Mass (kg/m)	Surface area (m²/m)
200 × 200 × 24	71.1	0.790
200 × 200 × 20	59.9	0.790
200 × 200 × 18	54.2	0.790
200 × 200 × 16	48.5	0.790
150 × 150 × 18	40.1	0.59
150 × 150 × 15	33.8	0.59
150 × 150 × 12	27.3	0.59
150 × 150 × 10	23.0	0.59
120 × 120 × 15	26.6	0.47
120 × 120 × 12	21.6	0.47
120 × 120 × 10	18.2	0.47
120 × 120 × 8	14.7	0.47
100 × 100 × 15	21.9	0.39
100 × 100 × 12	17.8	0.39
100 × 100 × 10	15.0	0.39
100 × 100 × 8	12.2	0.39
90 × 90 × 12	15.9	0.35
90 × 90 × 10	13.4	0.35
90 × 90 × 8	10.9	0.35
90 × 90 × 7	9.61	0.35
90 × 90 × 6	8.30	0.35

Unequal Angles BS EN 10056-1

Designation	Mass (kg/m)	Surface area (m²/m)
200 × 150 × 18	47.1	0.69
200 × 150 × 15	39.6	0.69
200 × 150 × 12	32.0	0.69
200 × 100 × 15	33.7	0.59
200 × 100 × 12	27.3	0.59
200 × 100 × 10	23.0	0.59
150 × 90 × 15	26.6	0.47
150 × 90 × 12	21.6	0.47
150 × 90 × 10	18.2	0.47
150 × 75 × 15	24.8	0.44
150 × 75 × 12	20.2	0.44
150 × 75 × 10	17.0	0.44
125 × 75 × 12	17.8	0.40
125 × 75 × 10	15.0	0.40
125 × 75 × 8	12.2	0.40
100 × 75 × 12	15.4	0.34
100 × 75 × 10	13.0	0.34
100 × 75 × 8	10.6	0.34
100 × 65 × 10	12.3	0.32
100 × 65 × 8	9.94	0.32
100 × 65 × 7	8.77	0.32

Tables and Memoranda

METAL

Structural Tees Split from Universal Beams BS 4: Part 1: 2005

Designation	Mass (kg/m)	Surface area (m²/m)
305 × 305 × 90	89.5	1.22
305 × 305 × 75	74.6	1.22
254 × 343 × 63	62.6	1.19
229 × 305 × 70	69.9	1.07
229 × 305 × 63	62.5	1.07
229 × 305 × 57	56.5	1.07
229 × 305 × 51	50.6	1.07
210 × 267 × 61	61.0	0.95
210 × 267 × 55	54.5	0.95
210 × 267 × 51	50.5	0.95
210 × 267 × 46	46.1	0.95
210 × 267 × 41	41.1	0.95
191 × 229 × 49	49.2	0.84
191 × 229 × 45	44.6	0.84
191 × 229 × 41	41.0	0.84
191 × 229 × 37	37.1	0.84
191 × 229 × 34	33.6	0.84
152 × 229 × 41	41.0	0.76
152 × 229 × 37	37.1	0.76
152 × 229 × 34	33.6	0.76
152 × 229 × 30	29.9	0.76
152 × 229 × 26	26.2	0.76

Universal Bearing Piles BS 4: Part 1: 2005

Designation	Mass (kg/m)	Depth of Section (mm)	Width of Section (mm)	Thickness Web (mm)	Thickness Flange (mm)
356 × 368 × 174	173.9	361.4	378.5	20.3	20.4
356 × 368 × 152	152.0	356.4	376.0	17.8	17.9
356 × 368 × 133	133.0	352.0	373.8	15.6	15.7
356 × 368 × 109	108.9	346.4	371.0	12.8	12.9
305 × 305 × 223	222.9	337.9	325.7	30.3	30.4
305 × 305 × 186	186.0	328.3	320.9	25.5	25.6
305 × 305 × 149	149.1	318.5	316.0	20.6	20.7
305 × 305 × 126	126.1	312.3	312.9	17.5	17.6
305 × 305 × 110	110.0	307.9	310.7	15.3	15.4
305 × 305 × 95	94.9	303.7	308.7	13.3	13.3
305 × 305 × 88	88.0	301.7	307.8	12.4	12.3
305 × 305 × 79	78.9	299.3	306.4	11.0	11.1
254 × 254 × 85	85.1	254.3	260.4	14.4	14.3
254 × 254 × 71	71.0	249.7	258.0	12.0	12.0
254 × 254 × 63	63.0	247.1	256.6	10.6	10.7
203 × 203 × 54	53.9	204.0	207.7	11.3	11.4
203 × 203 × 45	44.9	200.2	205.9	9.5	9.5

METAL

Hot Formed Square Hollow Sections EN 10210 S275J2H & S355J2H

Size (mm)	Wall thickness (mm)	Mass (kg/m)	Superficial area (m²/m)
40 × 40	2.5	2.89	0.154
	3.0	3.41	0.152
	3.2	3.61	0.152
	3.6	4.01	0.151
	4.0	4.39	0.150
	5.0	5.28	0.147
50 × 50	2.5	3.68	0.194
	3.0	4.35	0.192
	3.2	4.62	0.192
	3.6	5.14	0.191
	4.0	5.64	0.190
	5.0	6.85	0.187
	6.0	7.99	0.185
	6.3	8.31	0.184
60 × 60	3.0	5.29	0.232
	3.2	5.62	0.232
	3.6	6.27	0.231
	4.0	6.90	0.230
	5.0	8.42	0.227
	6.0	9.87	0.225
	6.3	10.30	0.224
	8.0	12.50	0.219
70 × 70	3.0	6.24	0.272
	3.2	6.63	0.272
	3.6	7.40	0.271
	4.0	8.15	0.270
	5.0	9.99	0.267
	6.0	11.80	0.265
	6.3	12.30	0.264
	8.0	15.00	0.259
80 × 80	3.2	7.63	0.312
	3.6	8.53	0.311
	4.0	9.41	0.310
	5.0	11.60	0.307
	6.0	13.60	0.305
	6.3	14.20	0.304
	8.0	17.50	0.299
90 × 90	3.6	9.66	0.351
	4.0	10.70	0.350
	5.0	13.10	0.347
	6.0	15.50	0.345
	6.3	16.20	0.344
	8.0	20.10	0.339
100 × 100	3.6	10.80	0.391
	4.0	11.90	0.390
	5.0	14.70	0.387
	6.0	17.40	0.385
	6.3	18.20	0.384
	8.0	22.60	0.379
	10.0	27.40	0.374

METAL

Size (mm)	Wall thickness (mm)	Mass (kg/m)	Superficial area (m²/m)
120 × 120	4.0	14.40	0.470
	5.0	17.80	0.467
	6.0	21.20	0.465
	6.3	22.20	0.464
	8.0	27.60	0.459
	10.0	33.70	0.454
	12.0	39.50	0.449
	12.5	40.90	0.448
140 × 140	5.0	21.00	0.547
	6.0	24.90	0.545
	6.3	26.10	0.544
	8.0	32.60	0.539
	10.0	40.00	0.534
	12.0	47.00	0.529
	12.5	48.70	0.528
150 × 150	5.0	22.60	0.587
	6.0	26.80	0.585
	6.3	28.10	0.584
	8.0	35.10	0.579
	10.0	43.10	0.574
	12.0	50.80	0.569
	12.5	52.70	0.568
Hot formed from seamless hollow	16.0	65.2	0.559
160 × 160	5.0	24.10	0.627
	6.0	28.70	0.625
	6.3	30.10	0.624
	8.0	37.60	0.619
	10.0	46.30	0.614
	12.0	54.60	0.609
	12.5	56.60	0.608
	16.0	70.20	0.599
180 × 180	5.0	27.30	0.707
	6.0	32.50	0.705
	6.3	34.00	0.704
	8.0	42.70	0.699
	10.0	52.50	0.694
	12.0	62.10	0.689
	12.5	64.40	0.688
	16.0	80.20	0.679
200 × 200	5.0	30.40	0.787
	6.0	36.20	0.785
	6.3	38.00	0.784
	8.0	47.70	0.779
	10.0	58.80	0.774
	12.0	69.60	0.769
	12.5	72.30	0.768
	16.0	90.30	0.759
250 × 250	5.0	38.30	0.987
	6.0	45.70	0.985
	6.3	47.90	0.984
	8.0	60.30	0.979
	10.0	74.50	0.974
	12.0	88.50	0.969
	12.5	91.90	0.968
	16.0	115.00	0.959

METAL

Size (mm)	Wall thickness (mm)	Mass (kg/m)	Superficial area (m²/m)
300 × 300	6.0	55.10	1.18
	6.3	57.80	1.18
	8.0	72.80	1.18
	10.0	90.20	1.17
	12.0	107.00	1.17
	12.5	112.00	1.17
	16.0	141.00	1.16
350 × 350	8.0	85.40	1.38
	10.0	106.00	1.37
	12.0	126.00	1.37
	12.5	131.00	1.37
	16.0	166.00	1.36
400 × 400	8.0	97.90	1.58
	10.0	122.00	1.57
	12.0	145.00	1.57
	12.5	151.00	1.57
	16.0	191.00	1.56
(Grade S355J2H only)	20.00*	235.00	1.55

Note: * SAW process

METAL

Hot Formed Square Hollow Sections JUMBO RHS: JIS G3136

Size (mm)	Wall thickness (mm)	Mass (kg/m)	Superficial area (m²/m)
350 × 350	19.0	190.00	1.33
	22.0	217.00	1.32
	25.0	242.00	1.31
400 × 400	22.0	251.00	1.52
	25.0	282.00	1.51
450 × 450	12.0	162.00	1.76
	16.0	213.00	1.75
	19.0	250.00	1.73
	22.0	286.00	1.72
	25.0	321.00	1.71
	28.0 *	355.00	1.70
	32.0 *	399.00	1.69
500 × 500	12.0	181.00	1.96
	16.0	238.00	1.95
	19.0	280.00	1.93
	22.0	320.00	1.92
	25.0	360.00	1.91
	28.0 *	399.00	1.90
	32.0 *	450.00	1.89
	36.0 *	498.00	1.88
550 × 550	16.0	263.00	2.15
	19.0	309.00	2.13
	22.0	355.00	2.12
	25.0	399.00	2.11
	28.0 *	443.00	2.10
	32.0 *	500.00	2.09
	36.0 *	555.00	2.08
	40.0 *	608.00	2.06
600 × 600	25.0 *	439.00	2.31
	28.0 *	487.00	2.30
	32.0 *	550.00	2.29
	36.0 *	611.00	2.28
	40.0 *	671.00	2.26
700 × 700	25.0 *	517.00	2.71
	28.0 *	575.00	2.70
	32.0 *	651.00	2.69
	36.0 *	724.00	2.68
	40.0 *	797.00	2.68

Note: * SAW process

METAL

Hot Formed Rectangular Hollow Sections: EN10210 S275J2h & S355J2H

Size (mm)	Wall thickness (mm)	Mass (kg/m)	Superficial area (m²/m)
50 × 30	2.5	2.89	0.154
	3.0	3.41	0.152
	3.2	3.61	0.152
	3.6	4.01	0.151
	4.0	4.39	0.150
	5.0	5.28	0.147
60 × 40	2.5	3.68	0.194
	3.0	4.35	0.192
	3.2	4.62	0.192
	3.6	5.14	0.191
	4.0	5.64	0.190
	5.0	6.85	0.187
	6.0	7.99	0.185
	6.3	8.31	0.184
80 × 40	3.0	5.29	0.232
	3.2	5.62	0.232
	3.6	6.27	0.231
	4.0	6.90	0.230
	5.0	8.42	0.227
	6.0	9.87	0.225
	6.3	10.30	0.224
	8.0	12.50	0.219
76.2 × 50.8	3.0	5.62	0.246
	3.2	5.97	0.246
	3.6	6.66	0.245
	4.0	7.34	0.244
	5.0	8.97	0.241
	6.0	10.50	0.239
	6.3	11.00	0.238
	8.0	13.40	0.233
90 × 50	3.0	6.24	0.272
	3.2	6.63	0.272
	3.6	7.40	0.271
	4.0	8.15	0.270
	5.0	9.99	0.267
	6.0	11.80	0.265
	6.3	12.30	0.264
	8.0	15.00	0.259
100 × 50	3.0	6.71	0.292
	3.2	7.13	0.292
	3.6	7.96	0.291
	4.0	8.78	0.290
	5.0	10.80	0.287
	6.0	12.70	0.285
	6.3	13.30	0.284
	8.0	16.30	0.279

METAL

Size (mm)	Wall thickness (mm)	Mass (kg/m)	Superficial area (m²/m)
100 × 60	3.0	7.18	0.312
	3.2	7.63	0.312
	3.6	8.53	0.311
	4.0	9.41	0.310
	5.0	11.60	0.307
	6.0	13.60	0.305
	6.3	14.20	0.304
	8.0	17.50	0.299
120 × 60	3.6	9.70	0.351
	4.0	10.70	0.350
	5.0	13.10	0.347
	6.0	15.50	0.345
	6.3	16.20	0.344
	8.0	20.10	0.339
120 × 80	3.6	10.80	0.391
	4.0	11.90	0.390
	5.0	14.70	0.387
	6.0	17.40	0.385
	6.3	18.20	0.384
	8.0	22.60	0.379
	10.0	27.40	0.374
150 × 100	4.0	15.10	0.490
	5.0	18.60	0.487
	6.0	22.10	0.485
	6.3	23.10	0.484
	8.0	28.90	0.479
	10.0	35.30	0.474
	12.0	41.40	0.469
	12.5	42.80	0.468
160 × 80	4.0	14.40	0.470
	5.0	17.80	0.467
	6.0	21.20	0.465
	6.3	22.20	0.464
	8.0	27.60	0.459
	10.0	33.70	0.454
	12.0	39.50	0.449
	12.5	40.90	0.448
200 × 100	5.0	22.60	0.587
	6.0	26.80	0.585
	6.3	28.10	0.584
	8.0	35.10	0.579
	10.0	43.10	0.574
	12.0	50.80	0.569
	12.5	52.70	0.568
	16.0	65.20	0.559
250 × 150	5.0	30.40	0.787
	6.0	36.20	0.785
	6.3	38.00	0.784
	8.0	47.70	0.779
	10.0	58.80	0.774
	12.0	69.60	0.769
	12.5	72.30	0.768
	16.0	90.30	0.759

METAL

Size (mm)	Wall thickness (mm)	Mass (kg/m)	Superficial area (m²/m)
300 × 200	5.0	38.30	0.987
	6.0	45.70	0.985
	6.3	47.90	0.984
	8.0	60.30	0.979
	10.0	74.50	0.974
	12.0	88.50	0.969
	12.5	91.90	0.968
	16.0	115.00	0.959
400 × 200	6.0	55.10	1.18
	6.3	57.80	1.18
	8.0	72.80	1.18
	10.0	90.20	1.17
	12.0	107.00	1.17
	12.5	112.00	1.17
	16.0	141.00	1.16
450 × 250	8.0	85.40	1.38
	10.0	106.00	1.37
	12.0	126.00	1.37
	12.5	131.00	1.37
	16.0	166.00	1.36
500 × 300	8.0	98.00	1.58
	10.0	122.00	1.57
	12.0	145.00	1.57
	12.5	151.00	1.57
	16.0	191.00	1.56
	20.0	235.00	1.55

Tables and Memoranda

METAL

Hot Formed Circular Hollow Sections EN 10210 S275J2H & S355J2H

Outside diameter (mm)	Wall thickness (mm)	Mass (kg/m)	Superficial area (m²/m)
21.3	3.2	1.43	0.067
26.9	3.2	1.87	0.085
33.7	3.0	2.27	0.106
	3.2	2.41	0.106
	3.6	2.67	0.106
	4.0	2.93	0.106
42.4	3.0	2.91	0.133
	3.2	3.09	0.133
	3.6	3.44	0.133
	4.0	3.79	0.133
48.3	2.5	2.82	0.152
	3.0	3.35	0.152
	3.2	3.56	0.152
	3.6	3.97	0.152
	4.0	4.37	0.152
	5.0	5.34	0.152
60.3	2.5	3.56	0.189
	3.0	4.24	0.189
	3.2	4.51	0.189
	3.6	5.03	0.189
	4.0	5.55	0.189
	5.0	6.82	0.189
76.1	2.5	4.54	0.239
	3.0	5.41	0.239
	3.2	5.75	0.239
	3.6	6.44	0.239
	4.0	7.11	0.239
	5.0	8.77	0.239
	6.0	10.40	0.239
	6.3	10.80	0.239
88.9	2.5	5.33	0.279
	3.0	6.36	0.279
	3.2	6.76	0.27
	3.6	7.57	0.279
	4.0	8.38	0.279
	5.0	10.30	0.279
	6.0	12.30	0.279
	6.3	12.80	0.279
114.3	3.0	8.23	0.359
	3.2	8.77	0.359
	3.6	9.83	0.359
	4.0	10.09	0.359
	5.0	13.50	0.359
	6.0	16.00	0.359
	6.3	16.80	0.359

METAL

Outside diameter (mm)	Wall thickness (mm)	Mass (kg/m)	Superficial area (m²/m)
139.7	3.2	10.80	0.439
	3.6	12.10	0.439
	4.0	13.40	0.439
	5.0	16.60	0.439
	6.0	19.80	0.439
	6.3	20.70	0.439
	8.0	26.00	0.439
	10.0	32.00	0.439
168.3	3.2	13.00	0.529
	3.6	14.60	0.529
	4.0	16.20	0.529
	5.0	20.10	0.529
	6.0	24.00	0.529
	6.3	25.20	0.529
	8.0	31.60	0.529
	10.0	39.00	0.529
	12.0	46.30	0.529
	12.5	48.00	0.529
193.7	5.0	23.30	0.609
	6.0	27.80	0.609
	6.3	29.10	0.609
	8.0	36.60	0.609
	10.0	45.30	0.609
	12.0	53.80	0.609
	12.5	55.90	0.609
219.1	5.0	26.40	0.688
	6.0	31.50	0.688
	6.3	33.10	0.688
	8.0	41.60	0.688
	10.0	51.60	0.688
	12.0	61.30	0.688
	12.5	63.70	0.688
	16.0	80.10	0.688
244.5	5.0	29.50	0.768
	6.0	35.30	0.768
	6.3	37.00	0.768
	8.0	46.70	0.768
	10.0	57.80	0.768
	12.0	68.80	0.768
	12.5	71.50	0.768
	16.0	90.20	0.768
273.0	5.0	33.00	0.858
	6.0	39.50	0.858
	6.3	41.40	0.858
	8.0	52.30	0.858
	10.0	64.90	0.858
	12.0	77.20	0.858
	12.5	80.30	0.858
	16.0	101.00	0.858

METAL

Outside diameter (mm)	Wall thickness (mm)	Mass (kg/m)	Superficial area (m²/m)
323.9	5.0	39.30	1.02
	6.0	47.00	1.02
	6.3	49.30	1.02
	8.0	62.30	1.02
	10.0	77.40	1.02
	12.0	92.30	1.02
	12.5	96.00	1.02
	16.0	121.00	1.02
355.6	6.3	54.30	1.12
	8.0	68.60	1.12
	10.0	85.30	1.12
	12.0	102.00	1.12
	12.5	106.00	1.12
	16.0	134.00	1.12
406.4	6.3	62.20	1.28
	8.0	79.60	1.28
	10.0	97.80	1.28
	12.0	117.00	1.28
	12.5	121.00	1.28
	16.0	154.00	1.28
457.0	6.3	70.00	1.44
	8.0	88.60	1.44
	10.0	110.00	1.44
	12.0	132.00	1.44
	12.5	137.00	1.44
	16.0	174.00	1.44
508.0	6.3	77.90	1.60
	8.0	98.60	1.60
	10.0	123.00	1.60
	12.0	147.00	1.60
	12.5	153.00	1.60
	16.0	194.00	1.60

METAL

Spacing of Holes in Angles

Nominal leg length (mm)	Spacing of holes						Maximum diameter of bolt or rivet		
	A	B	C	D	E	F	A	B and C	D, E and F
200		75	75	55	55	55		30	20
150		55	55					20	
125		45	60					20	
120									
100	55						24		
90	50						24		
80	45						20		
75	45						20		
70	40						20		
65	35						20		
60	35						16		
50	28						12		
45	25								
40	23								
30	20								
25	15								

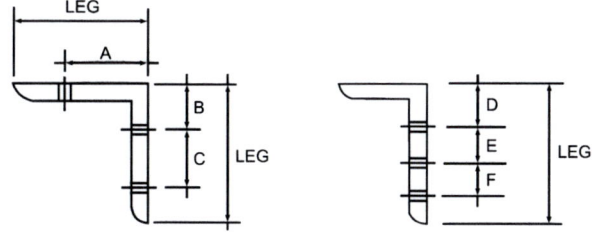

KERBS, PAVING, ETC.

KERBS/EDGINGS/CHANNELS

Precast Concrete Kerbs to BS 7263

Straight kerb units: length from 450 to 915 mm

150 mm high × 125 mm thick		
bullnosed	type BN	
half battered	type HB3	
255 mm high × 125 mm thick		
45° splayed	type SP	
half battered	type HB2	
305 mm high × 150 mm thick		
half battered	type HB1	
Quadrant kerb units		
150 mm high × 305 and 455 mm radius to match	type BN	type QBN
150 mm high × 305 and 455 mm radius to match	type HB2, HB3	type QHB
150 mm high × 305 and 455 mm radius to match	type SP	type QSP
255 mm high × 305 and 455 mm radius to match	type BN	type QBN
255 mm high × 305 and 455 mm radius to match	type HB2, HB3	type QHB
225 mm high × 305 and 455 mm radius to match	type SP	type QSP
Angle kerb units		
305 × 305 × 225 mm high × 125 mm thick		
bullnosed external angle	type XA	
splayed external angle to match type SP	type XA	
bullnosed internal angle	type IA	
splayed internal angle to match type SP	type IA	
Channels		
255 mm wide × 125 mm high flat	type CS1	
150 mm wide × 125 mm high flat type	CS2	
255 mm wide × 125 mm high dished	type CD	

KERBS, PAVING, ETC.

Transition kerb units			
from kerb type SP to HB	left handed	type TL	
	right handed	type TR	
from kerb type BN to HB	left handed	type DL1	
	right handed	type DR1	
from kerb type BN to SP	left handed	type DL2	
	right handed	type DR2	

Number of kerbs required per quarter circle (780 mm kerb lengths)

Radius (m)	Number in quarter circle
12	24
10	20
8	16
6	12
5	10
4	8
3	6
2	4
1	2

Precast Concrete Edgings

Round top type ER	Flat top type EF	Bullnosed top type EBN
150 × 50 mm	150 × 50 mm	150 × 50 mm
200 × 50 mm	200 × 50 mm	200 × 50 mm
250 × 50 mm	250 × 50 mm	250 × 50 mm

KERBS, PAVING, ETC.

BASES

Cement Bound Material for Bases and Subbases

CBM1:	very carefully graded aggregate from 37.5–75 mm, with a 7-day strength of 4.5 N/mm^2
CBM2:	same range of aggregate as CBM1 but with more tolerance in each size of aggregate with a 7-day strength of 7.0 N/mm^2
CBM3:	crushed natural aggregate or blast furnace slag, graded from 37.5 mm–150 mm for 40 mm aggregate, and from 20–75 mm for 20 mm aggregate, with a 7-day strength of 10 N/mm^2
CBM4:	crushed natural aggregate or blast furnace slag, graded from 37.5 mm–150 mm for 40 mm aggregate, and from 20–75 mm for 20 mm aggregate, with a 7-day strength of 15 N/mm^2

INTERLOCKING BRICK/BLOCK ROADS/PAVINGS

Sizes of Precast Concrete Paving Blocks

Type R blocks
200 × 100 × 60 mm
200 × 100 × 65 mm
200 × 100 × 80 mm
200 × 100 × 100 mm

Type S
Any shape within a 295 mm space

Sizes of clay brick pavers
200 × 100 × 50 mm
200 × 100 × 65 mm
210 × 105 × 50 mm
210 × 105 × 65 mm
215 × 102.5 × 50 mm
215 × 102.5 × 65 mm

Type PA: 3 kN
Footpaths and pedestrian areas, private driveways, car parks, light vehicle traffic and over-run

Type PB: 7 kN
Residential roads, lorry parks, factory yards, docks, petrol station forecourts, hardstandings, bus stations

KERBS, PAVING, ETC.

PAVING AND SURFACING

Weights and Sizes of Paving and Surfacing

Description of item	Size	Quantity per tonne
Paving 50 mm thick	900 × 600 mm	15
Paving 50 mm thick	750 × 600 mm	18
Paving 50 mm thick	600 × 600 mm	23
Paving 50 mm thick	450 × 600 mm	30
Paving 38 mm thick	600 × 600 mm	30
Path edging	914 × 50 × 150 mm	60
Kerb (including radius and tapers)	125 × 254 × 914 mm	15
Kerb (including radius and tapers)	125 × 150 × 914 mm	25
Square channel	125 × 254 × 914 mm	15
Dished channel	125 × 254 × 914 mm	15
Quadrants	300 × 300 × 254 mm	19
Quadrants	450 × 450 × 254 mm	12
Quadrants	300 × 300 × 150 mm	30
Internal angles	300 × 300 × 254 mm	30
Fluted pavement channel	255 × 75 × 914 mm	25
Corner stones	300 × 300 mm	80
Corner stones	360 × 360 mm	60
Cable covers	914 × 175 mm	55
Gulley kerbs	220 × 220 × 150 mm	60
Gulley kerbs	220 × 200 × 75 mm	120

KERBS, PAVING, ETC.

Weights and Sizes of Paving and Surfacing

Material	kg/m³	lb/cu yd
Tarmacadam	2306	3891
Macadam (waterbound)	2563	4325
Vermiculite (aggregate)	64–80	108–135
Terracotta	2114	3568
Cork – compressed	388	24
	kg/m²	lb/sq ft
Clay floor tiles, 12.7 mm	27.3	5.6
Pavement lights	122	25
Damp-proof course	5	1
	kg/m² per mm thickness	lb/sq ft per inch thickness
Paving slabs (stone)	2.3	12
Granite setts	2.88	15
Asphalt	2.30	12
Rubber flooring	1.68	9
Polyvinyl chloride	1.94 (avg)	10 (avg)

Coverage (m²) Per Cubic Metre of Materials Used as Subbases or Capping Layers

Consolidated thickness laid in (mm)	Square metre coverage		
	Gravel	Sand	Hardcore
50	15.80	16.50	–
75	10.50	11.00	–
100	7.92	8.20	7.42
125	6.34	6.60	5.90
150	5.28	5.50	4.95
175	–	–	4.23
200	–	–	3.71
225	–	–	3.30
300	–	–	2.47

KERBS, PAVING, ETC.

Approximate Rate of Spreads

Average thickness of course (mm)	Description	Approximate rate of spread			
		Open Textured		Dense, Medium & Fine Textured	
		(kg/m²)	(m²/t)	(kg/m²)	(m²/t)
35	14 mm open textured or dense wearing course	60–75	13–17	70–85	12–14
40	20 mm open textured or dense base course	70–85	12–14	80–100	10–12
45	20 mm open textured or dense base course	80–100	10–12	95–100	9–10
50	20 mm open textured or dense, or 28 mm dense base course	85–110	9–12	110–120	8–9
60	28 mm dense base course, 40 mm open textured or dense base course or 40 mm single course as base course	–	8–10	130–150	7–8
65	28 mm dense base course, 40 mm open textured or dense base course or 40 mm single course	100–135	7–10	140–160	6–7
75	40 mm single course, 40 mm open textured or dense base course, 40 mm dense roadbase	120–150	7–8	165–185	5–6
100	40 mm dense base course or roadbase	–	–	220–240	4–4.5

KERBS, PAVING, ETC.

Surface Dressing Roads: Coverage (m²) per Tonne of Material

Size in mm	Sand	Granite chips	Gravel	Limestone chips
Sand	168	–	–	–
3	–	148	152	165
6	–	130	133	144
9	–	111	114	123
13	–	85	87	95
19	–	68	71	78

Sizes of Flags

Reference	Nominal size (mm)	Thickness (mm)
A	600 × 450	50 and 63
B	600 × 600	50 and 63
C	600 × 750	50 and 63
D	600 × 900	50 and 63
E	450 × 450	50 and 70 chamfered top surface
F	400 × 400	50 and 65 chamfered top surface
G	300 × 300	50 and 60 chamfered top surface

Sizes of Natural Stone Setts

Width (mm)		Length (mm)		Depth (mm)
100	×	100	×	100
75	×	150 to 250	×	125
75	×	150 to 250	×	150
100	×	150 to 250	×	100
100	×	150 to 250	×	150

SEEDING/TURFING AND PLANTING

SEEDING/TURFING AND PLANTING

Topsoil Quality

Topsoil grade	Properties
Premium	Natural topsoil, high fertility, loamy texture, good soil structure, suitable for intensive cultivation.
General purpose	Natural or manufactured topsoil of lesser quality than Premium, suitable for agriculture or amenity landscape, may need fertilizer or soil structure improvement.
Economy	Selected subsoil, natural mineral deposit such as river silt or greensand. The grade comprises two subgrades; 'Low clay' and 'High clay' which is more liable to compaction in handling. This grade is suitable for low-production agricultural land and amenity woodland or conservation planting areas.

Forms of Trees

Standards:	Shall be clear with substantially straight stems. Grafted and budded trees shall have no more than a slight bend at the union. Standards shall be designated as Half, Extra light, Light, Standard, Selected standard, Heavy, and Extra heavy.
Sizes of Standards	
Heavy standard	12–14 cm girth × 3.50 to 5.00 m high
Extra heavy standard	14–16 cm girth × 4.25 to 5.00 m high
Extra heavy standard	16–18 cm girth × 4.25 to 6.00 m high
Extra heavy standard	18–20 cm girth × 5.00 to 6.00 m high
Semi-mature trees:	Between 6.0 m and 12.0 m tall with a girth of 20 to 75 cm at 1.0 m above ground.
Feathered trees:	Shall have a defined upright central leader, with stem furnished with evenly spread and balanced lateral shoots down to or near the ground.
Whips:	Shall be without significant feather growth as determined by visual inspection.
Multi-stemmed trees:	Shall have two or more main stems at, near, above or below ground.

Seedlings grown from seed and not transplanted shall be specified when ordered for sale as:

1+0	one year old seedling
2+0	two year old seedling
1+1	one year seed bed, one year transplanted = two year old seedling
1+2	one year seed bed, two years transplanted = three year old seedling
2+1	two year seed bed, one year transplanted = three year old seedling
1u1	two years seed bed, undercut after one year = two year old seedling
2u2	four years seed bed, undercut after two years = four year old seedling

SEEDING/TURFING AND PLANTING

Cuttings

The age of cuttings (plants grown from shoots, stems, or roots of the mother plant) shall be specified when ordered for sale. The height of transplants and undercut seedlings/cuttings (which have been transplanted or undercut at least once) shall be stated in centimetres. The number of growing seasons before and after transplanting or undercutting shall be stated.

0 + 1	one year cutting
0 + 2	two year cutting
0 + 1 + 1	one year cutting bed, one year transplanted = two year old seedling
0 + 1 + 2	one year cutting bed, two years transplanted = three year old seedling

Grass Cutting Capacities in m² per hour

Speed mph	Width of cut in metres												
	0.5	0.7	1.0	1.2	1.5	1.7	2.0	2.0	2.1	2.5	2.8	3.0	3.4
1.0	724	1127	1529	1931	2334	2736	3138	3219	3380	4023	4506	4828	5472
1.5	1086	1690	2293	2897	3500	4104	4707	4828	5069	6035	6759	7242	8208
2.0	1448	2253	3058	3862	4667	5472	6276	6437	6759	8047	9012	9656	10944
2.5	1811	2816	3822	4828	5834	6840	7846	8047	8449	10058	11265	12070	13679
3.0	2173	3380	4587	5794	7001	8208	9415	9656	10139	12070	13518	14484	16415
3.5	2535	3943	5351	6759	8167	9576	10984	11265	11829	14082	15772	16898	19151
4.0	2897	4506	6115	7725	9334	10944	12553	12875	13518	16093	18025	19312	21887
4.5	3259	5069	6880	8690	10501	12311	14122	14484	15208	18105	20278	21726	24623
5.0	3621	5633	7644	9656	11668	13679	15691	16093	16898	20117	22531	24140	27359
5.5	3983	6196	8409	10622	12834	15047	17260	17703	18588	22128	24784	26554	30095
6.0	4345	6759	9173	11587	14001	16415	18829	19312	20278	24140	27037	28968	32831
6.5	4707	7322	9938	12553	15168	17783	20398	20921	21967	26152	29290	31382	35566
7.0	5069	7886	10702	13518	16335	19151	21967	22531	23657	28163	31543	33796	38302

Number of Plants per m²: For Plants Planted on an Evenly Spaced Grid

Planting distances

mm	0.10	0.15	0.20	0.25	0.35	0.40	0.45	0.50	0.60	0.75	0.90	1.00	1.20	1.50
0.10	100.00	66.67	50.00	40.00	28.57	25.00	22.22	20.00	16.67	13.33	11.11	10.00	8.33	6.67
0.15	66.67	44.44	33.33	26.67	19.05	16.67	14.81	13.33	11.11	8.89	7.41	6.67	5.56	4.44
0.20	50.00	33.33	25.00	20.00	14.29	12.50	11.11	10.00	8.33	6.67	5.56	5.00	4.17	3.33
0.25	40.00	26.67	20.00	16.00	11.43	10.00	8.89	8.00	6.67	5.33	4.44	4.00	3.33	2.67
0.35	28.57	19.05	14.29	11.43	8.16	7.14	6.35	5.71	4.76	3.81	3.17	2.86	2.38	1.90
0.40	25.00	16.67	12.50	10.00	7.14	6.25	5.56	5.00	4.17	3.33	2.78	2.50	2.08	1.67
0.45	22.22	14.81	11.11	8.89	6.35	5.56	4.94	4.44	3.70	2.96	2.47	2.22	1.85	1.48
0.50	20.00	13.33	10.00	8.00	5.71	5.00	4.44	4.00	3.33	2.67	2.22	2.00	1.67	1.33
0.60	16.67	11.11	8.33	6.67	4.76	4.17	3.70	3.33	2.78	2.22	1.85	1.67	1.39	1.11
0.75	13.33	8.89	6.67	5.33	3.81	3.33	2.96	2.67	2.22	1.78	1.48	1.33	1.11	0.89
0.90	11.11	7.41	5.56	4.44	3.17	2.78	2.47	2.22	1.85	1.48	1.23	1.11	0.93	0.74
1.00	10.00	6.67	5.00	4.00	2.86	2.50	2.22	2.00	1.67	1.33	1.11	1.00	0.83	0.67
1.20	8.33	5.56	4.17	3.33	2.38	2.08	1.85	1.67	1.39	1.11	0.93	0.83	0.69	0.56
1.50	6.67	4.44	3.33	2.67	1.90	1.67	1.48	1.33	1.11	0.89	0.74	0.67	0.56	0.44

SEEDING/TURFING AND PLANTING

Grass Clippings Wet: Based on 3.5 m³/tonne

Annual kg/100 m²	Average 20 cuts kg/100 m²	m²/tonne	m²/m³
32.0	1.6	61162.1	214067.3

Nr of cuts	22	20	18	16	12	4
kg/cut	1.45	1.60	1.78	2.00	2.67	8.00
Area capacity of 3 tonne vehicle per load						
m²	206250	187500	168750	150000	112500	37500
Load m³	100 m² units/m³ of vehicle space					
1	196.4	178.6	160.7	142.9	107.1	35.7
2	392.9	357.1	321.4	285.7	214.3	71.4
3	589.3	535.7	482.1	428.6	321.4	107.1
4	785.7	714.3	642.9	571.4	428.6	142.9
5	982.1	892.9	803.6	714.3	535.7	178.6

Transportation of Trees

To unload large trees a machine with the necessary lifting strength is required. The weight of the trees must therefore be known in advance. The following table gives a rough overview. The additional columns with root ball dimensions and the number of plants per trailer provide additional information for example about preparing planting holes and calculating unloading times.

Girth in cm	Rootball diameter in cm	Ball height in cm	Weight in kg	Numbers of trees per trailer
16–18	50–60	40	150	100–120
18–20	60–70	40–50	200	80–100
20–25	60–70	40–50	270	50–70
25–30	80	50–60	350	50
30–35	90–100	60–70	500	12–18
35–40	100–110	60–70	650	10–15
40–45	110–120	60–70	850	8–12
45–50	110–120	60–70	1100	5–7
50–60	130–140	60–70	1600	1–3
60–70	150–160	60–70	2500	1
70–80	180–200	70	4000	1
80–90	200–220	70–80	5500	1
90–100	230–250	80–90	7500	1
100–120	250–270	80–90	9500	1

Data supplied by Lorenz von Ehren GmbH
The information in the table is approximate; deviations depend on soil type, genus and weather

FENCING AND GATES

FENCING AND GATES

Types of Preservative

Creosote (tar oil) can be 'factory' applied	by pressure to BS 144: pts 1&2 by immersion to BS 144: pt 1 by hot and cold open tank to BS 144: pts 1&2
Copper/chromium/arsenic (CCA)	by full cell process to BS 4072 pts 1&2
Organic solvent (OS)	by double vacuum (vacvac) to BS 5707 pts 1&3 by immersion to BS 5057 pts 1&3
Pentachlorophenol (PCP)	by heavy oil double vacuum to BS 5705 pts 2&3
Boron diffusion process (treated with disodium octaborate to BWPA Manual 1986)	

Note: Boron is used on green timber at source and the timber is supplied dry

Cleft Chestnut Pale Fences

Pales	Pale spacing	Wire lines	
900 mm	75 mm	2	temporary protection
1050 mm	75 or 100 mm	2	light protective fences
1200 mm	75 mm	3	perimeter fences
1350 mm	75 mm	3	perimeter fences
1500 mm	50 mm	3	narrow perimeter fences
1800 mm	50 mm	3	light security fences

Close-Boarded Fences

Close-boarded fences 1.05 to 1.8 m high
Type BCR (recessed) or BCM (morticed) with concrete posts 140 × 115 mm tapered and Type BW with timber posts

Palisade Fences

Wooden palisade fences
Type WPC with concrete posts 140 × 115 mm tapered and Type WPW with timber posts

For both types of fence:
Height of fence 1050 mm: two rails
Height of fence 1200 mm: two rails
Height of fence 1500 mm: three rails
Height of fence 1650 mm: three rails
Height of fence 1800 mm: three rails

FENCING AND GATES

Post and Rail Fences

Wooden post and rail fences
Type MPR 11/3 morticed rails and Type SPR 11/3 nailed rails
Height to top of rail 1100 mm
Rails: three rails 87 mm, 38 mm

Type MPR 11/4 morticed rails and Type SPR 11/4 nailed rails
Height to top of rail 1100 mm
Rails: four rails 87 mm, 38 mm

Type MPR 13/4 morticed rails and Type SPR 13/4 nailed rails
Height to top of rail 1300 mm
Rail spacing 250 mm, 250 mm, and 225 mm from top
Rails: four rails 87 mm, 38 mm

Steel Posts

Rolled steel angle iron posts for chain link fencing

Posts	Fence height	Strut	Straining post
1500 × 40 × 40 × 5 mm	900 mm	1500 × 40 × 40 × 5 mm	1500 × 50 × 50 × 6 mm
1800 × 40 × 40 × 5 mm	1200 mm	1800 × 40 × 40 × 5 mm	1800 × 50 × 50 × 6 mm
2000 × 45 × 45 × 5 mm	1400 mm	2000 × 45 × 45 × 5 mm	2000 × 60 × 60 × 6 mm
2600 × 45 × 45 × 5 mm	1800 mm	2600 × 45 × 45 × 5 mm	2600 × 60 × 60 × 6 mm
3000 × 50 × 50 × 6 mm	1800 mm	2600 × 45 × 45 × 5 mm	3000 × 60 × 60 × 6 mm
with arms			

Concrete Posts

Concrete posts for chain link fencing

Posts and straining posts	Fence height	Strut
1570 mm 100 × 100 mm	900 mm	1500 mm × 75 × 75 mm
1870 mm 125 × 125 mm	1200 mm	1830 mm × 100 × 75 mm
2070 mm 125 × 125 mm	1400 mm	1980 mm × 100 × 75 mm
2620 mm 125 × 125 mm	1800 mm	2590 mm × 100 × 85 mm
3040 mm 125 × 125 mm	1800 mm	2590 mm × 100 × 85 mm (with arms)

FENCING AND GATES

Rolled Steel Angle Posts

Rolled steel angle posts for rectangular wire mesh (field) fencing

Posts	Fence height	Strut	Straining post
1200 × 40 × 40 × 5 mm	600 mm	1200 × 75 × 75 mm	1350 × 100 × 100 mm
1400 × 40 × 40 × 5 mm	800 mm	1400 × 75 × 75 mm	1550 × 100 × 100 mm
1500 × 40 × 40 × 5 mm	900 mm	1500 × 75 × 75 mm	1650 × 100 × 100 mm
1600 × 40 × 40 × 5 mm	1000 mm	1600 × 75 × 75 mm	1750 × 100 × 100 mm
1750 × 40 × 40 × 5 mm	1150 mm	1750 × 75 × 100 mm	1900 × 125 × 125 mm

Concrete Posts

Concrete posts for rectangular wire mesh (field) fencing

Posts	Fence height	Strut	Straining post
1270 × 100 × 100 mm	600 mm	1200 × 75 × 75 mm	1420 × 100 × 100 mm
1470 × 100 × 100 mm	800 mm	1350 × 75 × 75 mm	1620 × 100 × 100 mm
1570 × 100 × 100 mm	900 mm	1500 × 75 × 75 mm	1720 × 100 × 100 mm
1670 × 100 × 100 mm	600 mm	1650 × 75 × 75 mm	1820 × 100 × 100 mm
1820 × 125 × 125 mm	1150 mm	1830 × 75 × 100 mm	1970 × 125 × 125 mm

Cleft Chestnut Pale Fences

Timber Posts

Timber posts for wire mesh and hexagonal wire netting fences

Round timber for general fences

Posts	Fence height	Strut	Straining post
1300 × 65 mm dia.	600 mm	1200 × 80 mm dia.	1450 × 100 mm dia.
1500 × 65 mm dia.	800 mm	1400 × 80 mm dia.	1650 × 100 mm dia.
1600 × 65 mm dia.	900 mm	1500 × 80 mm dia.	1750 × 100 mm dia.
1700 × 65 mm dia.	1050 mm	1600 × 80 mm dia.	1850 × 100 mm dia.
1800 × 65 mm dia.	1150 mm	1750 × 80 mm dia.	2000 × 120 mm dia.

Squared timber for general fences

Posts	Fence height	Strut	Straining post
1300 × 75 × 75 mm	600 mm	1200 × 75 × 75 mm	1450 × 100 × 100 mm
1500 × 75 × 75 mm	800 mm	1400 × 75 × 75 mm	1650 × 100 × 100 mm
1600 × 75 × 75 mm	900 mm	1500 × 75 × 75 mm	1750 × 100 × 100 mm
1700 × 75 × 75 mm	1050 mm	1600 × 75 × 75 mm	1850 × 100 × 100 mm
1800 × 75 × 75 mm	1150 mm	1750 × 75 × 75 mm	2000 × 125 × 100 mm

FENCING AND GATES

Steel Fences to BS 1722: Part 9: 1992

	Fence height	Top/bottom rails and flat posts	Vertical bars
Light	1000 mm	40 × 10 mm 450 mm in ground	12 mm dia. at 115 mm cs
	1200 mm	40 × 10 mm 550 mm in ground	12 mm dia. at 115 mm cs
	1400 mm	40 × 10 mm 550 mm in ground	12 mm dia. at 115 mm cs
Light	1000 mm	40 × 10 mm 450 mm in ground	16 mm dia. at 120 mm cs
	1200 mm	40 × 10 mm 550 mm in ground	16 mm dia. at 120 mm cs
	1400 mm	40 × 10 mm 550 mm in ground	16 mm dia. at 120 mm cs
Medium	1200 mm	50 × 10 mm 550 mm in ground	20 mm dia. at 125 mm cs
	1400 mm	50 × 10 mm 550 mm in ground	20 mm dia. at 125 mm cs
	1600 mm	50 × 10 mm 600 mm in ground	22 mm dia. at 145 mm cs
	1800 mm	50 × 10 mm 600 mm in ground	22 mm dia. at 145 mm cs
Heavy	1600 mm	50 × 10 mm 600 mm in ground	22 mm dia. at 145 mm cs
	1800 mm	50 × 10 mm 600 mm in ground	22 mm dia. at 145 mm cs
	2000 mm	50 × 10 mm 600 mm in ground	22 mm dia. at 145 mm cs
	2200 mm	50 × 10 mm 600 mm in ground	22 mm dia. at 145 mm cs

Notes: Mild steel fences: round or square verticals; flat standards and horizontals. Tops of vertical bars may be bow-top, blunt, or pointed. Round or square bar railings

Timber Field Gates to BS 3470: 1975

Gates made to this standard are designed to open one way only
All timber gates are 1100 mm high
Width over stiles 2400, 2700, 3000, 3300, 3600 and 4200 mm
Gates over 4200 mm should be made in two leaves

Steel Field Gates to BS 3470: 1975

All steel gates are 1100 mm high
Heavy duty: width over stiles 2400, 3000, 3600 and 4500 mm
Light duty: width over stiles 2400, 3000 and 3600 mm

FENCING AND GATES

Domestic Front Entrance Gates to BS 4092: Part 1: 1966

Metal gates:	Single gates are 900 mm high minimum, 900 mm, 1000 mm and 1100 mm wide

Domestic Front Entrance Gates to BS 4092: Part 2: 1966

Wooden gates:	All rails shall be tenoned into the stiles Single gates are 840 mm high minimum, 801 mm and 1020 mm wide Double gates are 840 mm high minimum, 2130, 2340 and 2640 mm wide

Timber Bridle Gates to BS 5709:1979 (Horse or Hunting Gates)

Gates open one way only Minimum width between posts Minimum height	 1525 mm 1100 mm

Timber Kissing Gates to BS 5709:1979

Minimum width	700 mm
Minimum height	1000 mm
Minimum distance between shutting posts	600 mm
Minimum clearance at mid-point	600 mm

Metal Kissing Gates to BS 5709:1979

Sizes are the same as those for timber kissing gates Maximum gaps between rails 120 mm

Categories of Pedestrian Guard Rail to BS 3049:1976

Class A for normal use Class B where vandalism is expected Class C where crowd pressure is likely

DRAINAGE

DRAINAGE

Width Required for Trenches for Various Diameters of Pipes

Pipe diameter (mm)	Trench n.e. 1.50 m deep	Trench over 1.50 m deep
n.e. 100 mm	450 mm	600 mm
100–150 mm	500 mm	650 mm
150–225 mm	600 mm	750 mm
225–300 mm	650 mm	800 mm
300–400 mm	750 mm	900 mm
400–450 mm	900 mm	1050 mm
450–600 mm	1100 mm	1300 mm

Weights and Dimensions – Vitrified Clay Pipes

Product	Nominal diameter (mm)	Effective length (mm)	BS 65 limits of tolerance min (mm)	max (mm)	Crushing strength (kN/m)	Weight (kg/pipe)	(kg/m)
Supersleve	100	1600	96	105	35.00	14.71	9.19
	150	1750	146	158	35.00	29.24	16.71
Hepsleve	225	1850	221	236	28.00	84.03	45.42
	300	2500	295	313	34.00	193.05	77.22
	150	1500	146	158	22.00	37.04	24.69
Hepseal	225	1750	221	236	28.00	85.47	48.84
	300	2500	295	313	34.00	204.08	81.63
	400	2500	394	414	44.00	357.14	142.86
	450	2500	444	464	44.00	454.55	181.63
	500	2500	494	514	48.00	555.56	222.22
	600	2500	591	615	57.00	796.23	307.69
	700	3000	689	719	67.00	1111.11	370.45
	800	3000	788	822	72.00	1351.35	450.45
Hepline	100	1600	95	107	22.00	14.71	9.19
	150	1750	145	160	22.00	29.24	16.71
	225	1850	219	239	28.00	84.03	45.42
	300	1850	292	317	34.00	142.86	77.22
Hepduct (conduit)	90	1500	–	–	28.00	12.05	8.03
	100	1600	–	–	28.00	14.71	9.19
	125	1750	–	–	28.00	20.73	11.84
	150	1750	–	–	28.00	29.24	16.71
	225	1850	–	–	28.00	84.03	45.42
	300	1850	–	–	34.00	142.86	77.22

Tables and Memoranda

DRAINAGE

Weights and Dimensions – Vitrified Clay Pipes

Nominal internal diameter (mm)	Nominal wall thickness (mm)	Approximate weight (kg/m)
150	25	45
225	29	71
300	32	122
375	35	162
450	38	191
600	48	317
750	54	454
900	60	616
1200	76	912
1500	89	1458
1800	102	1884
2100	127	2619

Wall thickness, weights and pipe lengths vary, depending on type of pipe required

The particulars shown above represent a selection of available diameters and are applicable to strength class 1 pipes with flexible rubber ring joints

Tubes with Ogee joints are also available

DRAINAGE

Weights and Dimensions – PVC-u Pipes

	Nominal size	Mean outside diameter (mm)		Wall thickness	Weight
		min	max	(mm)	(kg/m)
Standard pipes	82.4	82.4	82.7	3.2	1.2
	110.0	110.0	110.4	3.2	1.6
	160.0	160.0	160.6	4.1	3.0
	200.0	200.0	200.6	4.9	4.6
	250.0	250.0	250.7	6.1	7.2
Perforated pipes heavy grade	As above	As above	As above	As above	As above
thin wall	82.4	82.4	82.7	1.7	–
	110.0	110.0	110.4	2.2	–
	160.0	160.0	160.6	3.2	–

Width of Trenches Required for Various Diameters of Pipes

Pipe diameter (mm)	Trench n.e. 1.5 m deep (mm)	Trench over 1.5 m deep (mm)
n.e. 100	450	600
100–150	500	650
150–225	600	750
225–300	650	800
300–400	750	900
400–450	900	1050
450–600	1100	1300

DRAINAGE

DRAINAGE BELOW GROUND AND LAND DRAINAGE

Flow of Water Which Can Be Carried by Various Sizes of Pipe

Clay or concrete pipes

	Gradient of pipeline							
	1:10	1:20	1:30	1:40	1:50	1:60	1:80	1:100
Pipe size	Flow in litres per second							
DN 100 15.0	8.5	6.8	5.8	5.2	4.7	4.0	3.5	
DN 150 28.0	19.0	16.0	14.0	12.0	11.0	9.1	8.0	
DN 225 140.0	95.0	76.0	66.0	58.0	53.0	46.0	40.0	

Plastic pipes

	Gradient of pipeline							
	1:10	1:20	1:30	1:40	1:50	1:60	1:80	1:100
Pipe size	Flow in litres per second							
82.4 mm i/dia.	12.0	8.5	6.8	5.8	5.2	4.7	4.0	3.5
110 mm i/dia.	28.0	19.0	16.0	14.0	12.0	11.0	9.1	8.0
160 mm i/dia.	76.0	53.0	43.0	37.0	33.0	29.0	25.0	22.0
200 mm i/dia.	140.0	95.0	76.0	66.0	58.0	53.0	46.0	40.0

Vitrified (Perforated) Clay Pipes and Fittings to BS En 295-5 1994

Length not specified		
75 mm bore	250 mm bore	600 mm bore
100	300	700
125	350	800
150	400	1000
200	450	1200
225	500	

Precast Concrete Pipes: Prestressed Non-pressure Pipes and Fittings: Flexible Joints to BS 5911: Pt. 103: 1994

Rationalized metric nominal sizes: 450, 500	
Length:	500–1000 by 100 increments
	1000–2200 by 200 increments
	2200–2800 by 300 increments
Angles: length:	450–600 angles 45, 22.5,11.25°
	600 or more angles 22.5, 11.25°

DRAINAGE

Precast Concrete Pipes: Unreinforced and Circular Manholes and Soakaways to BS 5911: Pt. 200: 1994

Nominal sizes:	
Shafts:	675, 900 mm
Chambers:	900, 1050, 1200, 1350, 1500, 1800, 2100, 2400, 2700, 3000 mm
Large chambers:	To have either tapered reducing rings or a flat reducing slab in order to accept the standard cover
Ring depths:	1. 300–1200 mm by 300 mm increments except for bottom slab and rings below cover slab, these are by 150 mm increments
	2. 250–1000 mm by 250 mm increments except for bottom slab and rings below cover slab, these are by 125 mm increments
Access hole:	750 × 750 mm for DN 1050 chamber
	1200 × 675 mm for DN 1350 chamber

Calculation of Soakaway Depth

The following formula determines the depth of concrete ring soakaway that would be required for draining given amounts of water.

$$h = \frac{4ar}{3\pi D^2}$$

h = depth of the chamber below the invert pipe
a = the area to be drained
r = the hourly rate of rainfall (50 mm per hour)
π = pi
D = internal diameter of the soakaway

This table shows the depth of chambers in each ring size which would be required to contain the volume of water specified. These allow a recommended storage capacity of $\frac{1}{3}$ (one third of the hourly rainfall figure).

Table Showing Required Depth of Concrete Ring Chambers in Metres

Area m^2	50	100	150	200	300	400	500
Ring size							
0.9	1.31	2.62	3.93	5.24	7.86	10.48	13.10
1.1	0.96	1.92	2.89	3.85	5.77	7.70	9.62
1.2	0.74	1.47	2.21	2.95	4.42	5.89	7.37
1.4	0.58	1.16	1.75	2.33	3.49	4.66	5.82
1.5	0.47	0.94	1.41	1.89	2.83	3.77	4.72
1.8	0.33	0.65	0.98	1.31	1.96	2.62	3.27
2.1	0.24	0.48	0.72	0.96	1.44	1.92	2.41
2.4	0.18	0.37	0.55	0.74	1.11	1.47	1.84
2.7	0.15	0.29	0.44	0.58	0.87	1.16	1.46
3.0	0.12	0.24	0.35	0.47	0.71	0.94	1.18

DRAINAGE

Precast Concrete Inspection Chambers and Gullies to BS 5911: Part 230: 1994

Nominal sizes:	375 diameter, 750, 900 mm deep
	450 diameter, 750, 900, 1050, 1200 mm deep
Depths:	from the top for trapped or untrapped units:
	centre of outlet 300 mm
	invert (bottom) of the outlet pipe 400 mm
Depth of water seal for trapped gullies:	
	85 mm, rodding eye int. dia. 100 mm
Cover slab:	65 mm min

Bedding Flexible Pipes: PVC-u Or Ductile Iron

Type 1 =	100 mm fill below pipe, 300 mm above pipe: single size material
Type 2 =	100 mm fill below pipe, 300 mm above pipe: single size or graded material
Type 3 =	100 mm fill below pipe, 75 mm above pipe with concrete protective slab over
Type 4 =	100 mm fill below pipe, fill laid level with top of pipe
Type 5 =	200 mm fill below pipe, fill laid level with top of pipe
Concrete =	25 mm sand blinding to bottom of trench, pipe supported on chocks, 100 mm concrete under the pipe, 150 mm concrete over the pipe

DRAINAGE

Bedding Rigid Pipes: Clay or Concrete
(for vitrified clay pipes the manufacturer should be consulted)

Class D:	Pipe laid on natural ground with cut-outs for joints, soil screened to remove stones over 40 mm and returned over pipe to 150 m min depth. Suitable for firm ground with trenches trimmed by hand.
Class N:	Pipe laid on 50 mm granular material of graded aggregate to Table 4 of BS 882, or 10 mm aggregate to Table 6 of BS 882, or as dug light soil (not clay) screened to remove stones over 10 mm. Suitable for machine dug trenches.
Class B:	As Class N, but with granular bedding extending half way up the pipe diameter.
Class F:	Pipe laid on 100 mm granular fill to BS 882 below pipe, minimum 150 mm granular fill above pipe: single size material. Suitable for machine dug trenches.
Class A:	Concrete 100 mm thick under the pipe extending half way up the pipe, backfilled with the appropriate class of fill. Used where there is only a very shallow fall to the drain. Class A bedding allows the pipes to be laid to an exact gradient.
Concrete surround:	25 mm sand blinding to bottom of trench, pipe supported on chocks, 100 mm concrete under the pipe, 150 mm concrete over the pipe. It is preferable to bed pipes under slabs or wall in granular material.

PIPED SUPPLY SYSTEMS

Identification of Service Tubes From Utility to Dwellings

Utility	Colour	Size	Depth
British Telecom	grey	54 mm od	450 mm
Electricity	black	38 mm od	450 mm
Gas	yellow	42 mm od rigid 60 mm od convoluted	450 mm
Water	may be blue	(normally untubed)	750 mm

ELECTRICAL SUPPLY/POWER/LIGHTING SYSTEMS

ELECTRICAL SUPPLY/POWER/LIGHTING SYSTEMS

Electrical Insulation Class En 60.598 BS 4533

Class 1:	luminaires comply with class 1 (I) earthed electrical requirements
Class 2:	luminaires comply with class 2 (II) double insulated electrical requirements
Class 3:	luminaires comply with class 3 (III) electrical requirements

Protection to Light Fittings

BS EN 60529:1992 Classification for degrees of protection provided by enclosures.
(IP Code – International or Ingress Protection)

1st characteristic: against ingress of solid foreign objects

The figure	2	indicates that fingers cannot enter
	3	that a 2.5 mm diameter probe cannot enter
	4	that a 1.0 mm diameter probe cannot enter
	5	the fitting is dust proof (no dust around live parts)
	6	the fitting is dust tight (no dust entry)

2nd characteristic: ingress of water with harmful effects

The figure	0	indicates unprotected
	1	vertically dripping water cannot enter
	2	water dripping 15° (tilt) cannot enter
	3	spraying water cannot enter
	4	splashing water cannot enter
	5	jetting water cannot enter
	6	powerful jetting water cannot enter
	7	proof against temporary immersion
	8	proof against continuous immersion
Optional additional codes:		A–D protects against access to hazardous parts
	H	High voltage apparatus
	M	fitting was in motion during water test
	S	fitting was static during water test
	W	protects against weather
Marking code arrangement:		(example) IPX5S = IP (International or Ingress Protection)
		X (denotes omission of first characteristic)
		5 = jetting
		S = static during water test

RAIL TRACKS

	kg/m of track	lb/ft of track
Standard gauge Bull head rails, chairs, transverse timber (softwood) sleepers etc.	245	165
Main lines Flat-bottom rails, transverse prestressed concrete sleepers, etc.	418	280
Add for electric third rail	51	35
Add for crushed stone ballast	2600	1750

	kg/m²	lb/sq ft
Overall average weight – rails connections, sleepers, ballast, etc.	733	150

	kg/m of track	lb/ft of track
Bridge rails, longitudinal timber sleepers, etc.	112	75

RAIL TRACKS

Heavy Rails

British Standard Section No.	Rail height (mm)	Foot width (mm)	Head width (mm)	Min web thickness (mm)	Section weight (kg/m)
Flat Bottom Rails					
60 A	114.30	109.54	57.15	11.11	30.62
70 A	123.82	111.12	60.32	12.30	34.81
75 A	128.59	114.30	61.91	12.70	37.45
80 A	133.35	117.47	63.50	13.10	39.76
90 A	142.88	127.00	66.67	13.89	45.10
95 A	147.64	130.17	69.85	14.68	47.31
100 A	152.40	133.35	69.85	15.08	50.18
110 A	158.75	139.70	69.85	15.87	54.52
113 A	158.75	139.70	69.85	20.00	56.22
50 'O'	100.01	100.01	52.39	10.32	24.82
80 'O'	127.00	127.00	63.50	13.89	39.74
60R	114.30	109.54	57.15	11.11	29.85
75R	128.59	122.24	61.91	13.10	37.09
80R	133.35	127.00	63.50	13.49	39.72
90R	142.88	136.53	66.67	13.89	44.58
95R	147.64	141.29	68.26	14.29	47.21
100R	152.40	146.05	69.85	14.29	49.60
95N	147.64	139.70	69.85	13.89	47.27
Bull Head Rails					
95R BH	145.26	69.85	69.85	19.05	47.07

Light Rails

British Standard Section No.	Rail height (mm)	Foot width (mm)	Head width (mm)	Min web thickness (mm)	Section weight (kg/m)
Flat Bottom Rails					
20M	65.09	55.56	30.96	6.75	9.88
30M	75.41	69.85	38.10	9.13	14.79
35M	80.96	76.20	42.86	9.13	17.39
35R	85.73	82.55	44.45	8.33	17.40
40	88.11	80.57	45.64	12.3	19.89
Bridge Rails					
13	48.00	92	36.00	18.0	13.31
16	54.00	108	44.50	16.0	16.06
20	55.50	127	50.00	20.5	19.86
28	67.00	152	50.00	31.0	28.62
35	76.00	160	58.00	34.5	35.38
50	76.00	165	58.50	–	50.18
Crane Rails					
A65	75.00	175.00	65.00	38.0	43.10
A75	85.00	200.00	75.00	45.0	56.20
A100	95.00	200.00	100.00	60.0	74.30
A120	105.00	220.00	120.00	72.0	100.00
175CR	152.40	152.40	107.95	38.1	86.92

RAIL TRACKS

Fish Plates

British Standard Section No.	Overall plate length		Hole diameter	Finished weight per pair	
	4 Hole (mm)	6 Hole (mm)	(mm)	4 Hole (kg/pair)	6 Hole (kg/pair)
For British Standard Heavy Rails: Flat Bottom Rails					
60 A	406.40	609.60	20.64	9.87	14.76
70 A	406.40	609.60	22.22	11.15	16.65
75 A	406.40	–	23.81	11.82	17.73
80 A	406.40	609.60	23.81	13.15	19.72
90 A	457.20	685.80	25.40	17.49	26.23
100 A	508.00	–	pear	25.02	–
110 A (shallow)	507.00	–	27.00	30.11	54.64
113 A (heavy)	507.00	–	27.00	30.11	54.64
50 'O' (shallow)	406.40	–	–	6.68	10.14
80 'O' (shallow)	495.30	–	23.81	14.72	22.69
60R (shallow)	406.40	609.60	20.64	8.76	13.13
60R (angled)	406.40	609.60	20.64	11.27	16.90
75R (shallow)	406.40	–	23.81	10.94	16.42
75R (angled)	406.40	–	23.81	13.67	–
80R (shallow)	406.40	609.60	23.81	11.93	17.89
80R (angled)	406.40	609.60	23.81	14.90	22.33
For British Standard Heavy Rails: Bull Head Rails					
95R BH (shallow)	–	457.20	27.00	14.59	14.61
For British Standard Light Rails: Flat Bottom Rails					
30M	355.6	–	–	–	2.72
35M	355.6	–	–	–	2.83
40	355.6	–	–	3.76	–

FRACTIONS, DECIMALS AND MILLIMETRE EQUIVALENTS

FRACTIONS, DECIMALS AND MILLIMETRE EQUIVALENTS

Fractions	Decimals	(mm)	Fractions	Decimals	(mm)
1/64	0.015625	0.396875	33/64	0.515625	13.096875
1/32	0.03125	0.79375	17/32	0.53125	13.49375
3/64	0.046875	1.190625	35/64	0.546875	13.890625
1/16	0.0625	1.5875	9/16	0.5625	14.2875
5/64	0.078125	1.984375	37/64	0.578125	14.684375
3/32	0.09375	2.38125	19/32	0.59375	15.08125
7/64	0.109375	2.778125	39/64	0.609375	15.478125
1/8	0.125	3.175	5/8	0.625	15.875
9/64	0.140625	3.571875	41/64	0.640625	16.271875
5/32	0.15625	3.96875	21/32	0.65625	16.66875
11/64	0.171875	4.365625	43/64	0.671875	17.065625
3/16	0.1875	4.7625	11/16	0.6875	17.4625
13/64	0.203125	5.159375	45/64	0.703125	17.859375
7/32	0.21875	5.55625	23/32	0.71875	18.25625
15/64	0.234375	5.953125	47/64	0.734375	18.653125
1/4	0.25	6.35	3/4	0.75	19.05
17/64	0.265625	6.746875	49/64	0.765625	19.446875
9/32	0.28125	7.14375	25/32	0.78125	19.84375
19/64	0.296875	7.540625	51/64	0.796875	20.240625
5/16	0.3125	7.9375	13/16	0.8125	20.6375
21/64	0.328125	8.334375	53/64	0.828125	21.034375
11/32	0.34375	8.73125	27/32	0.84375	21.43125
23/64	0.359375	9.128125	55/64	0.859375	21.828125
3/8	0.375	9.525	7/8	0.875	22.225
25/64	0.390625	9.921875	57/64	0.890625	22.621875
13/32	0.40625	10.31875	29/32	0.90625	23.01875
27/64	0.421875	10.71563	59/64	0.921875	23.415625
7/16	0.4375	11.1125	15/16	0.9375	23.8125
29/64	0.453125	11.50938	61/64	0.953125	24.209375
15/32	0.46875	11.90625	31/32	0.96875	24.60625
31/64	0.484375	12.30313	63/64	0.984375	25.003125
1/2	0.5	12.7	1.0	1	25.4

IMPERIAL STANDARD WIRE GAUGE (SWG)

IMPERIAL STANDARD WIRE GAUGE (SWG)

SWG	Diameter		SWG	Diameter	
No.	(inches)	(mm)	No.	(inches)	(mm)
7/0	0.5	12.7	23	0.024	0.61
6/0	0.464	11.79	24	0.022	0.559
5/0	0.432	10.97	25	0.02	0.508
4/0	0.4	10.16	26	0.018	0.457
3/0	0.372	9.45	27	0.0164	0.417
2/0	0.348	8.84	28	0.0148	0.376
1/0	0.324	8.23	29	0.0136	0.345
1	0.3	7.62	30	0.0124	0.315
2	0.276	7.01	31	0.0116	0.295
3	0.252	6.4	32	0.0108	0.274
4	0.232	5.89	33	0.01	0.254
5	0.212	5.38	34	0.009	0.234
6	0.192	4.88	35	0.008	0.213
7	0.176	4.47	36	0.008	0.193
8	0.16	4.06	37	0.007	0.173
9	0.144	3.66	38	0.006	0.152
10	0.128	3.25	39	0.005	0.132
11	0.116	2.95	40	0.005	0.122
12	0.104	2.64	41	0.004	0.112
13	0.092	2.34	42	0.004	0.102
14	0.08	2.03	43	0.004	0.091
15	0.072	1.83	44	0.003	0.081
16	0.064	1.63	45	0.003	0.071
17	0.056	1.42	46	0.002	0.061
18	0.048	1.22	47	0.002	0.051
19	0.04	1.016	48	0.002	0.041
20	0.036	0.914	49	0.001	0.031
21	0.032	0.813	50	0.001	0.025
22	0.028	0.711			

PIPES, WATER, STORAGE, INSULATION

WATER PRESSURE DUE TO HEIGHT

Imperial

Head (feet)	Pressure (lb/in²)		Head (feet)	Pressure (lb/in²)
1	0.43		70	30.35
5	2.17		75	32.51
10	4.34		80	34.68
15	6.5		85	36.85
20	8.67		90	39.02
25	10.84		95	41.18
30	13.01		100	43.35
35	15.17		105	45.52
40	17.34		110	47.69
45	19.51		120	52.02
50	21.68		130	56.36
55	23.84		140	60.69
60	26.01		150	65.03
65	28.18			

Metric

Head (m)	Pressure (bar)		Head (m)	Pressure (bar)
0.5	0.049		18.0	1.766
1.0	0.098		19.0	1.864
1.5	0.147		20.0	1.962
2.0	0.196		21.0	2.06
3.0	0.294		22.0	2.158
4.0	0.392		23.0	2.256
5.0	0.491		24.0	2.354
6.0	0.589		25.0	2.453
7.0	0.687		26.0	2.551
8.0	0.785		27.0	2.649
9.0	0.883		28.0	2.747
10.0	0.981		29.0	2.845
11.0	1.079		30.0	2.943
12.0	1.177		32.5	3.188
13.0	1.275		35.0	3.434
14.0	1.373		37.5	3.679
15.0	1.472		40.0	3.924
16.0	1.57		42.5	4.169
17.0	1.668		45.0	4.415

1 bar	=	14.5038 lbf/in²
1 lbf/in²	=	0.06895 bar
1 metre	=	3.2808 ft or 39.3701 in
1 foot	=	0.3048 metres
1 in wg	=	2.5 mbar (249.1 N/m²)

PIPES, WATER, STORAGE, INSULATION

Dimensions and Weights of Copper Pipes to BSEN 1057, BSEN 12499, BSEN 14251

Outside diameter (mm)	Internal diameter (mm)	Weight per metre (kg)	Internal diameter (mm)	Weight per metre (kg)	Internal diameter (mm)	Weight per metre (kg)
	Formerly Table X		Formerly Table Y		Formerly Table Z	
6	4.80	0.0911	4.40	0.1170	5.00	0.0774
8	6.80	0.1246	6.40	0.1617	7.00	0.1054
10	8.80	0.1580	8.40	0.2064	9.00	0.1334
12	10.80	0.1914	10.40	0.2511	11.00	0.1612
15	13.60	0.2796	13.00	0.3923	14.00	0.2031
18	16.40	0.3852	16.00	0.4760	16.80	0.2918
22	20.22	0.5308	19.62	0.6974	20.82	0.3589
28	26.22	0.6814	25.62	0.8985	26.82	0.4594
35	32.63	1.1334	32.03	1.4085	33.63	0.6701
42	39.63	1.3675	39.03	1.6996	40.43	0.9216
54	51.63	1.7691	50.03	2.9052	52.23	1.3343
76.1	73.22	3.1287	72.22	4.1437	73.82	2.5131
108	105.12	4.4666	103.12	7.3745	105.72	3.5834
133	130.38	5.5151	–	–	130.38	5.5151
159	155.38	8.7795	–	–	156.38	6.6056

Dimensions of Stainless Steel Pipes to BS 4127

Outside diameter (mm)	Maximum outside diameter (mm)	Minimum outside diameter (mm)	Wall thickness (mm)	Working pressure (bar)
6	6.045	5.940	0.6	330
8	8.045	7.940	0.6	260
10	10.045	9.940	0.6	210
12	12.045	11.940	0.6	170
15	15.045	14.940	0.6	140
18	18.045	17.940	0.7	135
22	22.055	21.950	0.7	110
28	28.055	27.950	0.8	121
35	35.070	34.965	1.0	100
42	42.070	41.965	1.1	91
54	54.090	53.940	1.2	77

PIPES, WATER, STORAGE, INSULATION

Dimensions of Steel Pipes to BS 1387

Nominal size	Approx. outside diameter	Outside diameter				Thickness		
		Light		Medium & heavy		Light	Medium	Heavy
		Max	Min	Max	Min			
(mm)	(mm)	(mm)	(mm)	(mm)	(mm)	(mm)	(mm)	(mm)
6	10.20	10.10	9.70	10.40	9.80	1.80	2.00	2.65
8	13.50	13.60	13.20	13.90	13.30	1.80	2.35	2.90
10	17.20	17.10	16.70	17.40	16.80	1.80	2.35	2.90
15	21.30	21.40	21.00	21.70	21.10	2.00	2.65	3.25
20	26.90	26.90	26.40	27.20	26.60	2.35	2.65	3.25
25	33.70	33.80	33.20	34.20	33.40	2.65	3.25	4.05
32	42.40	42.50	41.90	42.90	42.10	2.65	3.25	4.05
40	48.30	48.40	47.80	48.80	48.00	2.90	3.25	4.05
50	60.30	60.20	59.60	60.80	59.80	2.90	3.65	4.50
65	76.10	76.00	75.20	76.60	75.40	3.25	3.65	4.50
80	88.90	88.70	87.90	89.50	88.10	3.25	4.05	4.85
100	114.30	113.90	113.00	114.90	113.30	3.65	4.50	5.40
125	139.70	–	–	140.60	138.70	–	4.85	5.40
150	165.1*	–	–	166.10	164.10	–	4.85	5.40

* 165.1 mm (6.5 in) outside diameter is not generally recommended except where screwing to BS 21 is necessary
All dimensions are in accordance with ISO R65 except approximate outside diameters which are in accordance with ISO R64
Light quality is equivalent to ISO R65 Light Series II

Approximate Metres Per Tonne of Tubes to BS 1387

Nom. Size	BLACK						GALVANIZED					
	Plain/screwed ends			Screwed & socketed			Plain/screwed ends			Screwed & socketed		
	L	M	H	L	M	H	L	M	H	L	M	H
(mm)	(m)	(m)	(m)	(m)	(m)	(m)	(m)	(m)	(m)	(m)	(m)	(m)
6	2765	2461	2030	2743	2443	2018	2604	2333	1948	2584	2317	1937
8	1936	1538	1300	1920	1527	1292	1826	1467	1254	1811	1458	1247
10	1483	1173	979	1471	1165	974	1400	1120	944	1386	1113	939
15	1050	817	688	1040	811	684	996	785	665	987	779	661
20	712	634	529	704	628	525	679	609	512	673	603	508
25	498	410	336	494	407	334	478	396	327	474	394	325
32	388	319	260	384	316	259	373	308	254	369	305	252
40	307	277	226	303	273	223	296	268	220	292	264	217
50	244	196	162	239	194	160	235	191	158	231	188	157
65	172	153	127	169	151	125	167	149	124	163	146	122
80	147	118	99	143	116	98	142	115	97	139	113	96
100	101	82	69	98	81	68	98	81	68	95	79	67
125	–	62	56	–	60	55	–	60	55	–	59	54
150	–	52	47	–	50	46	–	51	46	–	49	45

The figures for 'plain or screwed ends' apply also to tubes to BS 1775 of equivalent size and thickness
Key:
L – Light
M – Medium
H – Heavy

PIPES, WATER, STORAGE, INSULATION

Flange Dimension Chart to BS 4504 & BS 10

Normal Pressure Rating (PN 6) 6 Bar

Nom. Size	Flange Outside Dia.	Table 6/2 Forged Welding Neck	Table 6/3 Plate Slip on	Table 6/4 Forged Bossed Screwed	Table 6/5 Forged Bossed Slip on	Table 6/8 Plate Blank	Raised Face Dia.	T'ness	Nr. Bolt Hole	Size of Bolt
15	80	12	12	12	12	12	40	2	4	M10 × 40
20	90	14	14	14	14	14	50	2	4	M10 × 45
25	100	14	14	14	14	14	60	2	4	M10 × 45
32	120	14	16	14	14	14	70	2	4	M12 × 45
40	130	14	16	14	14	14	80	3	4	M12 × 45
50	140	14	16	14	14	14	90	3	4	M12 × 45
65	160	14	16	14	14	14	110	3	4	M12 × 45
80	190	16	18	16	16	16	128	3	4	M16 × 55
100	210	16	18	16	16	16	148	3	4	M16 × 55
125	240	18	20	18	18	18	178	3	8	M16 × 60
150	265	18	20	18	18	18	202	3	8	M16 × 60
200	320	20	22	–	20	20	258	3	8	M16 × 60
250	375	22	24	–	22	22	312	3	12	M16 × 65
300	440	22	24	–	22	22	365	4	12	M20 × 70

Normal Pressure Rating (PN 16) 16 Bar

Nom. Size	Flange Outside Dia.	Table 6/2 Forged Welding Neck	Table 6/3 Plate Slip on	Table 6/4 Forged Bossed Screwed	Table 6/5 Forged Bossed Slip on	Table 6/8 Plate Blank	Raised Face Dia.	T'ness	Nr. Bolt Hole	Size of Bolt
15	95	14	14	14	14	14	45	2	4	M12 × 45
20	105	16	16	16	16	16	58	2	4	M12 × 50
25	115	16	16	16	16	16	68	2	4	M12 × 50
32	140	16	16	16	16	16	78	2	4	M16 × 55
40	150	16	16	16	16	16	88	3	4	M16 × 55
50	165	18	18	18	18	18	102	3	4	M16 × 60
65	185	18	18	18	18	18	122	3	4	M16 × 60
80	200	20	20	20	20	20	138	3	8	M16 × 60
100	220	20	20	20	20	20	158	3	8	M16 × 65
125	250	22	22	22	22	22	188	3	8	M16 × 70
150	285	22	22	22	22	22	212	3	8	M20 × 70
200	340	24	24	–	24	24	268	3	12	M20 × 75
250	405	26	26	–	26	26	320	3	12	M24 × 90
300	460	28	28	–	28	28	378	4	12	M24 × 90

PIPES, WATER, STORAGE, INSULATION

Minimum Distances Between Supports/Fixings

Material	BS Nominal pipe size		Pipes – Vertical	Pipes – Horizontal on to low gradients
	(inch)	(mm)	Support distance in metres	Support distance in metres
Copper	0.50	15.00	1.90	1.30
	0.75	22.00	2.50	1.90
	1.00	28.00	2.50	1.90
	1.25	35.00	2.80	2.50
	1.50	42.00	2.80	2.50
	2.00	54.00	3.90	2.50
	2.50	67.00	3.90	2.80
	3.00	76.10	3.90	2.80
	4.00	108.00	3.90	2.80
	5.00	133.00	3.90	2.80
	6.00	159.00	3.90	2.80
muPVC	1.25	32.00	1.20	0.50
	1.50	40.00	1.20	0.50
	2.00	50.00	1.20	0.60
Polypropylene	1.25	32.00	1.20	0.50
	1.50	40.00	1.20	0.50
uPVC	–	82.40	1.20	0.50
	–	110.00	1.80	0.90
	–	160.00	1.80	1.20
Steel	0.50	15.00	2.40	1.80
	0.75	20.00	3.00	2.40
	1.00	25.00	3.00	2.40
	1.25	32.00	3.00	2.40
	1.50	40.00	3.70	2.40
	2.00	50.00	3.70	2.40
	2.50	65.00	4.60	3.00
	3.00	80.40	4.60	3.00
	4.00	100.00	4.60	3.00
	5.00	125.00	5.50	3.70
	6.00	150.00	5.50	4.50
	8.00	200.00	8.50	6.00
	10.00	250.00	9.00	6.50
	12.00	300.00	10.00	7.00
	16.00	400.00	10.00	8.25

PIPES, WATER, STORAGE, INSULATION

Litres of Water Storage Required Per Person Per Building Type

Type of building	Storage (litres)
Houses and flats (up to 4 bedrooms)	120/bedroom
Houses and flats (more than 4 bedrooms)	100/bedroom
Hostels	90/bed
Hotels	200/bed
Nurses homes and medical quarters	120/bed
Offices with canteen	45/person
Offices without canteen	40/person
Restaurants	7/meal
Boarding schools	90/person
Day schools – Primary	15/person
Day schools – Secondary	20/person

Recommended Air Conditioning Design Loads

Building type	Design loading
Computer rooms	500 W/m² of floor area
Restaurants	150 W/m² of floor area
Banks (main area)	100 W/m² of floor area
Supermarkets	25 W/m² of floor area
Large office block (exterior zone)	100 W/m² of floor area
Large office block (interior zone)	80 W/m² of floor area
Small office block (interior zone)	80 W/m² of floor area

PIPES, WATER, STORAGE, INSULATION

Capacity and Dimensions of Galvanized Mild Steel Cisterns – BS 417

Capacity (litres)	BS type (SCM)	Dimensions		
		Length (mm)	Width (mm)	Depth (mm)
18	45	457	305	305
36	70	610	305	371
54	90	610	406	371
68	110	610	432	432
86	135	610	457	482
114	180	686	508	508
159	230	736	559	559
191	270	762	584	610
227	320	914	610	584
264	360	914	660	610
327	450/1	1220	610	610
336	450/2	965	686	686
423	570	965	762	787
491	680	1090	864	736
709	910	1070	889	889

Capacity of Cold Water Polypropylene Storage Cisterns – BS 4213

Capacity (litres)	BS type (PC)	Maximum height (mm)
18	4	310
36	8	380
68	15	430
91	20	510
114	25	530
182	40	610
227	50	660
273	60	660
318	70	660
455	100	760

PIPES, WATER, STORAGE, INSULATION

Minimum Insulation Thickness to Protect Against Freezing for Domestic Cold Water Systems (8 Hour Evaluation Period)

Pipe size (mm)	Insulation thickness (mm)					
	Condition 1			Condition 2		
	λ = 0.020	λ = 0.030	λ = 0.040	λ = 0.020	λ = 0.030	λ = 0.040
Copper pipes						
15	11	20	34	12	23	41
22	6	9	13	6	10	15
28	4	6	9	4	7	10
35	3	5	7	4	5	7
42	3	4	5	8	4	6
54	2	3	4	2	3	4
76	2	2	3	2	2	3
Steel pipes						
15	9	15	24	10	18	29
20	6	9	13	6	10	15
25	4	7	9	5	7	10
32	3	5	6	3	5	7
40	3	4	5	3	4	6
50	2	3	4	2	3	4
65	2	2	3	2	3	3

Condition 1: water temperature 7°C; ambient temperature –6°C; evaluation period 8 h; permitted ice formation 50%; normal installation, i.e. inside the building and inside the envelope of the structural insulation
Condition 2: water temperature 2°C; ambient temperature –6°C; evaluation period 8 h; permitted ice formation 50%; extreme installation, i.e. inside the building but outside the envelope of the structural insulation
λ = thermal conductivity [W/(mK)]

Insulation Thickness for Chilled And Cold Water Supplies to Prevent Condensation

On a Low Emissivity Outer Surface (0.05, i.e. Bright Reinforced Aluminium Foil) with an Ambient Temperature of +25°C and a Relative Humidity of 80%

Steel pipe size (mm)	t = +10			t = +5			t = 0		
	Insulation thickness (mm)			Insulation thickness (mm)			Insulation thickness (mm)		
	λ = 0.030	λ = 0.040	λ = 0.050	λ = 0.030	λ = 0.040	λ = 0.050	λ = 0.030	λ = 0.040	λ = 0.050
15	16	20	25	22	28	34	28	36	43
25	18	24	29	25	32	39	32	41	50
50	22	28	34	30	39	47	38	49	60
100	26	34	41	36	47	57	46	60	73
150	29	38	46	40	52	64	51	67	82
250	33	43	53	46	60	74	59	77	94
Flat surfaces	39	52	65	56	75	93	73	97	122

t = temperature of contents (°C)
λ = thermal conductivity at mean temperature of insulation [W/(mK)]

PIPES, WATER, STORAGE, INSULATION

Insulation Thickness for Non-domestic Heating Installations to Control Heat Loss

Steel pipe size (mm)	t = 75 Insulation thickness (mm)			t = 100 Insulation thickness (mm)			t = 150 Insulation thickness (mm)		
	λ = 0.030	λ = 0.040	λ = 0.050	λ = 0.030	λ = 0.040	λ = 0.050	λ = 0.030	λ = 0.040	λ = 0.050
10	18	32	55	20	36	62	23	44	77
15	19	34	56	21	38	64	26	47	80
20	21	36	57	23	40	65	28	50	83
25	23	38	58	26	43	68	31	53	85
32	24	39	59	28	45	69	33	55	87
40	25	40	60	29	47	70	35	57	88
50	27	42	61	31	49	72	37	59	90
65	29	43	62	33	51	74	40	63	92
80	30	44	62	35	52	75	42	65	94
100	31	46	63	37	54	76	45	68	96
150	33	48	64	40	57	77	50	73	100
200	35	49	65	42	59	79	53	76	103
250	36	50	66	43	61	80	55	78	105

t = hot face temperature (°C)

λ = thermal conductivity at mean temperature of insulation [W/(mK)]

Index

Design for Outdoor Recreation

2nd Edition

S. Bell

Design for Outdoor Recreation takes a detailed look at all aspects of design of facilities needed by visitors to outdoor recreation destinations. The book is a comprehensive manual for planners, designers and managers of recreation taking them through the processes of design and enabling them to find the most appropriate balance between visitor needs and the capacity of the landscape. A range of different aspects are covered including car parking, information signing, hiking, waterside activities, wildlife watching and camping.

This second edition incorporates new examples from overseas, including Australia, New Zealand, Japan and Eastern Europe as well as focusing on more current issues such as accessibility and the changing demands for recreational use.

July 2008: 276x219: 240pp
Pb: 978-0-415-44172-8: £60.00

To Order: Tel: +44 (0) 1235 400524 Fax: +44 (0) 1235 400525
or Post: Taylor and Francis Customer Services,
Bookpoint Ltd, Unit T1, 200 Milton Park, Abingdon, Oxon, OX14 4TA UK
Email: book.orders@tandf.co.uk

For a complete listing of all our titles visit:
www.tandf.co.uk

Landscape Architect's Pocket Book

Second Edition

Siobhan Vernon, Rachel Tennant & Nicola Garmory

An indispensable tool for all landscape architects, this time-saving guide answers the most frequently asked questions in one pocket-sized volume. It is a concise, easy-to-read reference that gives instant access to a wide range of information needed on a daily basis, both out on site and in the office.

Covering all the major topics, including hard landscaping, soft landscaping as well as planning and legislation, the pocket book also includes a handy glossary of important terms, useful calculations and helpful contacts. Not only an essential tool for everyday queries on British standards and procedures, this is a first point of reference for those seeking more extensive, supplementary sources of information, including websites and further publications.

This new edition incorporates updates and revisions from key planning and environmental legislation, guidelines and national standards.

April 2013 186x123: 340pp
Pb: 978-0-415-63084-9: £19.99

To Order: Tel: +44 (0) 1235 400524 Fax: +44 (0) 1235 400525
or Post: Taylor and Francis Customer Services,
Bookpoint Ltd, Unit T1, 200 Milton Park, Abingdon, Oxon, OX14 4TA UK
Email: book.orders@tandf.co.uk

For a complete listing of all our titles visit:
www.tandf.co.uk

Taylor & Francis
Taylor & Francis Group

Guidelines for Landscape and Visual Impact Assessment

Third Edition

Landscape Institute & I.E.M.A.

Landscape and Visual Impact Assessment (LVIA) can be key to planning decisions by identifying the effects of new developments on views and on the landscape itself.

This fully revised edition of the industry standard work on LVIA presents an authoritative statement of the principles of assessment. Offering detailed advice on the process of assessing the landscape and visual effects of developments and their significance, it also includes a new expanded chapter on cumulative effects and updated guidance on presentation.

Written by professionals for professionals, the third edition of this widely respected text provides an essential tool for landscape practitioners, developers, legal advisors and decision-makers.

March 2010: 276x219: 170pp
Hb: 978-0-415-68004-2: £49.99

To Order: Tel: +44 (0) 1235 400524 Fax: +44 (0) 1235 400525
or Post: Taylor and Francis Customer Services,
Bookpoint Ltd, Unit T1, 200 Milton Park, Abingdon, Oxon, OX14 4TA UK
Email: book.orders@tandf.co.uk

For a complete listing of all our titles visit:
www.tandf.co.uk

Taylor & Francis
Taylor & Francis Group

Spon's Asia-Pacific Construction Costs Handbook

Fourth Edition

Davis Langdon

Spon's Asia Pacific Construction Costs Handbook includes construction cost data for twenty countries. This new edition has been extended to include Pakistan and Cambodia. Australia, UK and America are also included, to facilitate comparison with construction costs elsewhere.

Information is presented for each country in the same way, as follows:

- key data on the main economic and construction indicators.
- an outline of the national construction industry, covering structure, tendering and contract procedures, materials cost data, regulations and standards
- labour and materials cost data
- measured rates for a range of standard construction work items
- approximate estimating costs per unit area for a range of building types
- price index data and exchange rate movements against £ sterling, $US and Japanese Yen.

The book also includes a Comparative Data section to facilitate country-to-country comparisons. Figures from the national sections are grouped in tables according to national indicators, construction output, input costs and costs per square metre for factories, offices, warehouses, hospitals, schools, theatres, sports halls, hotels and housing.

This unique handbook will be an essential reference for all construction professionals involved in work outside their own country and for all developers or multinational companies assessing comparative development costs.

April 2010: 234x156: 480pp
Hb: 978-0-415-46565-6: £140.00

To Order: Tel: +44 (0) 1235 400524 Fax: +44 (0) 1235 400525
or Post: Taylor and Francis Customer Services,
Bookpoint Ltd, Unit T1, 200 Milton Park, Abingdon, Oxon, OX14 4TA UK
Email: book.orders@tandf.co.uk

For a complete listing of all our titles visit:
www.tandf.co.uk

Ebook Single-User Licence Agreement

We welcome you as a user of this Spon Price Book ebook and hope that you find it a useful and valuable tool. Please read this document carefully. **This is a legal agreement** between you (hereinafter referred to as the "Licensee") and Taylor and Francis Books Ltd. (the "Publisher"), which defines the terms under which you may use the Product. **By accessing and retrieving the access code on the label inside the front cover of this book you agree to these terms and conditions outlined herein. If you do not agree to these terms you must return the Product to your supplier intact, with the seal on the label unbroken and with the access code not accessed.**

1.   **Definition of the Product**
      The product which is the subject of this Agreement, (the "Product") consists of online and offline access to the VitalSource ebook edition of *Spon's External Works & Landscape Price Book 2014*.

2.   **Commencement and licence**
      2.1   This Agreement commences upon the breaking open of the document containing the access code by the Licensee (the "Commencement Date").
      2.2   This is a licence agreement (the "Agreement") for the use of the Product by the Licensee, and not an agreement for sale.
      2.3   The Publisher licenses the Licensee on a non-exclusive and non-transferable basis to use the Product on condition that the Licensee complies with this Agreement. The Licensee acknowledges that it is only permitted to use the Product in accordance with this Agreement.

3.   **Multiple use**
      For more than one user or for a wide area network or consortium, use of the Product is only permissible with the purchase from the Publisher of additional access codes..

4.   **Installation and Use**
      4.1   The Licensee may provide access to the Product for individual study in the following manner: The Licensee may install the Product on a secure local area network on a single site for use by one user.
      4.2   The Licensee shall be responsible for installing the Product and for the effectiveness of such installation.
      4.3   Text from the Product may be incorporated in a coursepack. Such use is only permissible with the express permission of the Publisher in writing and requires the payment of the appropriate fee as specified by the Publisher and signature of a separate licence agreement.
      4.4   The Product is a free addition to the book and the Publisher is under no obligation to provide any technical support.

5.   **Permitted Activities**
      5.1   The Licensee shall be entitled to use the Product for its own internal purposes;
      5.2   The Licensee acknowledges that its rights to use the Product are strictly set out in this Agreement, and all other uses (whether expressly mentioned in Clause 6 below or not) are prohibited.

6.   **Prohibited Activities**
      The following are prohibited without the express permission of the Publisher:
      6.1   The commercial exploitation of any part of the Product.
      6.2   The rental, loan, (free or for money or money's worth) or hire purchase of this product, save with the express consent of the Publisher.
      6.3   Any activity which raises the reasonable prospect of impeding the Publisher's ability or opportunities to market the Product.
      6.4   Any networking, physical or electronic distribution or dissemination of the product save as expressly permitted by this Agreement.
      6.5   Any reverse engineering, decompilation, disassembly or other alteration of the Product save in accordance with applicable national laws.
      6.6   The right to create any derivative product or service from the Product save as expressly provided for in this Agreement.
      6.7   Any alteration, amendment, modification or deletion from the Product, whether for the purposes of error correction or otherwise.

7. General Responsibilities of the License

7.1 The Licensee will take all reasonable steps to ensure that the Product is used in accordance with the terms and conditions of this Agreement.

7.2 The Licensee acknowledges that damages may not be a sufficient remedy for the Publisher in the event of breach of this Agreement by the Licensee, and that an injunction may be appropriate.

7.3 The Licensee undertakes to keep the Product safe and to use its best endeavours to ensure that the product does not fall into the hands of third parties, whether as a result of theft or otherwise.

7.4 Where information of a confidential nature relating to the product of the business affairs of the Publisher comes into the possession of the Licensee pursuant to this Agreement (or otherwise), the Licensee agrees to use such information solely for the purposes of this Agreement, and under no circumstances to disclose any element of the information to any third party save strictly as permitted under this Agreement. For the avoidance of doubt, the Licensee's obligations under this sub-clause 7.4 shall survive the termination of this Agreement.

8. Warrant and Liability

8.1 The Publisher warrants that it has the authority to enter into this agreement and that it has secured all rights and permissions necessary to enable the Licensee to use the Product in accordance with this Agreement.

8.2 The Publisher warrants that the Product as supplied on the Commencement Date shall be free of defects in materials and workmanship, and undertakes to replace any defective Product within 28 days of notice of such defect being received provided such notice is received within 30 days of such supply. As an alternative to replacement, the Publisher agrees fully to refund the Licensee in such circumstances, if the Licensee so requests, provided that the Licensee returns this copy of *Spon's External Works & Landscape Price Book 2014* to the Publisher. The provisions of this sub-clause 8.2 do not apply where the defect results from an accident or from misuse of the product by the Licensee.

8.3 Sub-clause 8.2 sets out the sole and exclusive remedy of the Licensee in relation to defects in the Product.

8.4 The Publisher and the Licensee acknowledge that the Publisher supplies the Product on an "as is" basis. The Publisher gives no warranties:

8.4.1 that the Product satisfies the individual requirements of the Licensee; or

8.4.2 that the Product is otherwise fit for the Licensee's purpose; or

8.4.3 that the Product is compatible with the Licensee's hardware equipment and software operating environment.

8.5 The Publisher hereby disclaims all warranties and conditions, express or implied, which are not stated above.

8.6 Nothing in this Clause 8 limits the Publisher's liability to the Licensee in the event of death or personal injury resulting from the Publisher's negligence.

8.7 The Publisher hereby excludes liability for loss of revenue, reputation, business, profits, or for indirect or consequential losses, irrespective of whether the Publisher was advised by the Licensee of the potential of such losses.

8.8 The Licensee acknowledges the merit of independently verifying the price book data prior to taking any decisions of material significance (commercial or otherwise) based on such data. It is agreed that the Publisher shall not be liable for any losses which result from the Licensee placing reliance on the data under any circumstances.

8.9 Subject to sub-clause 8.6 above, the Publisher's liability under this Agreement shall be limited to the purchase price.

9. Intellectual Property Rights

9.1 Nothing in this Agreement affects the ownership of copyright or other intellectual property rights in the Product.

9.2 The Licensee agrees to display the Publishers' copyright notice in the manner described in the Product.

9.3 The Licensee hereby agrees to abide by copyright and similar notice requirements required by the Publisher, details of which are as follows:

"© 2014 Taylor & Francis. All rights reserved. All materials in *Spon's External Works & Landscape Price Book 2014* are copyright protected. All rights reserved. No such materials may be used, displayed, modified, adapted, distributed, transmitted, transferred, published or otherwise reproduced in any form or by any means now or hereafter developed other than strictly in accordance with the terms of the licence agreement enclosed with *Spon's External Works & Landscape Price Book 2014*. However, text and images may be printed and copied for research and private study within the preset program limitations. Please note the copyright notice above, and that any text or images printed or copied must credit the source."

9.4 This Product contains material proprietary to and copyedited by the Publisher and others. Except for the licence granted herein, all rights, title and interest in the Product, in all languages, formats and media throughout the world, including copyrights therein, are and remain the property of the Publisher or other copyright holders identified in the Product.

10. Non-assignment

This Agreement and the licence contained within it may not be assigned to any other person or entity without the written consent of the Publisher.

11. Termination and Consequences of Termination.

11.1 The Publisher shall have the right to terminate this Agreement if:

11.1.1 the Licensee is in material breach of this Agreement and fails to remedy such breach (where capable of remedy) within 14 days of a written notice from the Publisher requiring it to do so; or

11.1.2 the Licensee becomes insolvent, becomes subject to receivership, liquidation or similar external administration; or

11.1.3 the Licensee ceases to operate in business.

11.2 The Licensee shall have the right to terminate this Agreement for any reason upon two month's written notice. The Licensee shall not be entitled to any refund for payments made under this Agreement prior to termination under this sub-clause 11.2.

11.3 Termination by either of the parties is without prejudice to any other rights or remedies under the general law to which they may be entitled, or which survive such termination (including rights of the Publisher under sub-clause 7.4 above).

11.4 Upon termination of this Agreement, or expiry of its terms, the Licensee must destroy all copies and any back up copies of the product or part thereof.

12. General

12.1 *Compliance with export provisions*

The Publisher hereby agrees to comply fully with all relevant export laws and regulations of the United Kingdom to ensure that the Product is not exported, directly or indirectly, in violation of English law.

12.2 *Force majeure*

The parties accept no responsibility for breaches of this Agreement occurring as a result of circumstances beyond their control.

12.3 *No waiver*

Any failure or delay by either party to exercise or enforce any right conferred by this Agreement shall not be deemed to be a waiver of such right.

12.4 *Entire agreement*

This Agreement represents the entire agreement between the Publisher and the Licensee concerning the Product. The terms of this Agreement supersede all prior purchase orders, written terms and conditions, written or verbal representations, advertising or statements relating in any way to the Product.

12.5 *Severability*

If any provision of this Agreement is found to be invalid or unenforceable by a court of law of competent jurisdiction, such a finding shall not affect the other provisions of this Agreement and all provisions of this Agreement unaffected by such a finding shall remain in full force and effect.

12.6 *Variations*

This agreement may only be varied in writing by means of variation signed in writing by both parties.

12.7 *Notices*

All notices to be delivered to: Spon's Price Books, Taylor & Francis Books Ltd., 3 Park Square, Milton Park, Abingdon, Oxfordshire, OX14 4RN, UK.

12.8 *Governing law*

This Agreement is governed by English law and the parties hereby agree that any dispute arising under this Agreement shall be subject to the jurisdiction of the English courts.

If you have any queries about the terms of this licence, please contact:

Spon's Price Books
Taylor & Francis Books Ltd.
3 Park Square, Milton Park, Abingdon, Oxfordshire, OX14 4RN

Multiple-user use of the Spon Press Software and eBook

For more than one user or for a wide area network or consortium, use of *Spon's Online* is only permissible with the purchase from the Publisher of a multiple-user licence and adherence to the terms and conditions of that licence.

To buy a licence and set up access to *Spon's Online* for a number of users of in your organisation please contact:

Spon's Price Books
eBooks & Online Sales
Taylor & Francis Books Ltd.
3 Park Square, Milton Park, Oxfordshire, OX14 4RN
Tel: +44 (0) 207 017 6000
www.pricebooks.co.uk

Number of users	Licence cost
2	£235
5	£540
10	£1050
other	Please contact Spon for details